BIOMASS AS A SUSTAINABLE ENERGY SOURCE FOR THE FUTURE

BIOMASS AS A SUSTAINABLE ENERGY SOURCE FOR THE FUTURE

Fundamentals of Conversion Processes

Edited By

WIEBREN DE JONG

J. RUUD VAN OMMEN

WILEY

Published by John Wiley & Sons, Inc., Hoboken, New Jersey. All rights reserved

Published by John Wiley & Sons, Inc., Hoboken, New Jersey
Published simultaneously in Canada

For general information on our other products and services or for technical support, please contact
our Customer Care Department within the United States at (800) 762-2974, outside the United States
at (317) 572-3993 or fax (317) 572-4002.

Wiley also publishes its books in a variety of electronic formats. Some content that appears in print
may not be available in electronic formats. For more information about Wiley products, visit our
web site at www.wiley.com.

Library of Congress Cataloging-in-Publication Data:

Biomass as a sustainable energy source for the future : fundamentals of conversion processes /
edited by, Wiebren de Jong and J. Ruud van Ommen.
  pages cm
Includes bibliographical references and index.
ISBN 978-1-118-30491-4 (cloth)
1. Biomass energy. I. Jong, Wiebren de, 1968– II. Ommen, J. Ruud van, 1973–
TP339.B5474 2014
662'.88–dc23

                                                                        2014015277

Printed in the United States of America

10 9 8 7 6 5 4 3 2 1

CONTENTS

PREFACE

This book deals with bioenergy as a versatile, renewable source. Ever since the dawn of mankind, people have been using wood and other biogenic sources for heating, cooking, and lighting. Trade of biomass came up in historic times (think about the silk route for example). Even industrial iron making via metal reduction was based on biomass utilization (carbonization). However, that very application also led to substantial deforestation, which was clear in the United Kingdom, and demonstrated that using biomass does not guarantee a sustainable energy supply. Therefore, the industrial revolution introduced the large-scale application of fossil fuel, starting with the use of coal.

The steam engine became the workhorse of the nineteenth century. Coal also became the basis of the chemical industry at that time. Oil was initially used for lamps, but later it appeared to be the choice of raw material for petrol and diesel in Otto and Diesel engines, respectively. Wood and other sources came back into the picture during the interbellum period and the Second World War when oil was scarce, in particular on the European continent. At that time, cars, trucks, and ships made use of the gas extracted from fixed bed wood gasification installations. Also, chemicals supply and materials were increasingly supported by wood-based processes. After WWII, the cheap oil era was entered and such routes were largely abandoned. After the oil crises of the 1970s, biomass came back into the picture as an energy source, reinforced by environmental concerns about the use of fossil energy sources due to their associated CO_2 emissions stimulating the greenhouse effect. At present, biomass is seriously back as part of a sustainable energy mix, in combination with materials and chemicals supply, and a wide world of biorefinery options has opened up.

The field of biomass to energy supply is multidisciplinary and offers a wealth of integration of knowledge to young engineers starting their careers. The technologies strongly lean on chemical engineering skills, but also on physics, mechanical engineering, and agricultural sciences among others. Not only technology issues determine the success of biomass for our energy supply, there are many hurdles to be taken into the nontechnical domain such as logistics (trade and handling), infrastructure, and politics (subsidies, rural development, employment generation, etc.) to name a few.

This book is divided into four parts, covering broad areas of the field of biomass conversion technology chains. Part I starts with the socioeconomic and environmental context and biomass basics. It gives insight into the boundary conditions and the playing field bioenergy supply has. Moreover, it provides a deeper look into what biomass really is. Part II covers the chemical engineering basics to provide the engineer with tools to solve problems in the domain, design new biomass-based processes, and evaluate conversion subprocesses. The tools range from setting up balances, evaluating the mass and heat transport phenomena, thermodynamics and kinetics, to reactor and process design. Part III deals with the study of different biomass conversion processes, ranging from nonreactive pretreatment via combustion processes, gasification, hydrothermal processing, pyrolysis, and torrefaction, to biochemical conversion processes and biorefinery integration of such technologies. Finally, Part IV treats the end use of primary biomass conversion products, for example, power production via fuel cells, transportation fuel production (e.g., via the Fischer–Tropsch process), and platform chemicals production via organic chemistry to substitute the conventional petrochemical pathways offered today.

We were inspired to write this book by the course "Energy from Biomass" that we have been teaching for a number of years in the M.Sc. program of Sustainable Energy Technology at Delft University, a program that is part of the 3TU cooperation between the technical universities of Delft, Eindhoven, and Twente. After teaching the course for some years using the lecture notes prepared by our Eindhoven colleagues Rob Bastiaans, Jeroen van Oijen, and Mark Prins, we thought it would be worthy to further improve the course material. Since the students in Sustainable Energy Technology have a very diverse background, we have devoted Part II of the book to giving the reader enough background in chemical engineering for reading the more specialized chapters. This means that this book is useful for everyone with a B.Sc. in any engineering discipline. Apart from students at the M.Sc. level, professionals in the biomass field may also find this book as a knowledgeable source, for example, for designing and evaluating novel biorefinery systems and conversion components.

Delft, December 2013   WIEBREN DE JONG AND J. RUUD VAN OMMEN

He will be like a tree planted by the water that sends out its roots by the stream. It does not fear when heat comes; its leaves are always green. It has no worries in a year of drought and never fails to bear fruit. – Bible, Jeremiah 17:8.

Biomass is forever – Prof. David Hall[†]

ACKNOWLEDGMENTS

This book would not have been published without the contributions from many people. First of all, we would like to acknowledge all the students that followed our course "Energy from Biomass" over the years: they inspired us to compose this book. We are very glad that many of our colleagues agreed to contribute chapters to this book. It was great to work with this team of co-authors, all bringing in their specific expertise to cover the broad field of energy from biomass. A big thanks to all of you! For some chapters, the additional input from others is specifically acknowledged. Likun Ma is kindly acknowledged for his contribution to the examples in Chapter 4; Ryan Bogaars for his suggestions concerning Chapter 10; Xiangmei Meng and Onursal Yakaboylu for contributing some of the examples of Chapter 10; Richard Eijsberg for the first generation ethanol process figures and data in Chapter 13; Tim Geraedts and Elze Oude Lansink for the project in Chapter 15; Fred van Rantwijk for valuable input and discussions on Chapter 18; and Adrea Fabre for her advices regarding the writing. We are also grateful for the willingness of many colleagues to review chapters in order to find mistakes and make suggestions for further improvements. Our reviewers were, in alphabetic order: Rob Bakker, Sune Bengtsson, Pouyan Boukany, Anthony Bridgwater, Harry Croezen, Lilian de Martín, Jorge Gascon, Hans Geerlings, Johan Grievink, Sef Heijnen, Kas Hemmes, Paulien Herder, Truls Liliedahl, Gabrie Meesters, Bart Merci, Kyriakos Panopoulos, Wolter Prins, Sina Sartipi, Fabrizio Scala, Tilman Schildhauer, Andrzej Stankiewicz, Georgios Stefanidis, Bob Ursem, Henk van den Berg, Theo van der Meer, Jules van Lier, Marit van Lieshout, and Stanislav Vassilev. A special word of thanks should go to Annelies van Diepen. When the chapters were complete, she made a great effort to harmonize them, for example, in figures, symbols, lay-out, and wording. She also has caught numerous mistakes that were still present in earlier versions. We would like to thank

Jan Leen Kloosterman (Director of Education Sustainable Energy Technology, SET) for the financial support from the SET program for editorial assistance. We would also like to thank the people at Wiley for the smooth cooperation during the preparation of the manuscript. Finally, we would like to thank the ones close to us—Klarine (WdJ) and Ceciel, Fenne and Chris (JRvO)—for their understanding and support during all the evenings and weekends that the writing and editing took.

LIST OF CONTRIBUTORS

P.V. Aravind, Dr.ir. Department of Process and Energy, Energy Technology Section, Faculty of Mechanical, Maritime and Materials Engineering, Delft University of Technology, Delft, the Netherlands

Isabel W.C.E. Arends, Prof.dr. Department of Biotechnology, Biocatalysis Group, Faculty of Applied Sciences, Delft University of Technology, Delft, the Netherlands

Rob J.M. Bastiaans, Dr.ir. Department of Mechanical Engineering, Combustion Technology Section, Eindhoven University of Technology, Eindhoven, the Netherlands

Maria C. Cuellar, Dr. Department of Biotechnology, BioProcess Engineering Group, Faculty of Applied Sciences, Delft University of Technology, Delft, the Netherlands

Martina Fantini, Dr.ir. Department of Process and Energy, Energy Technology Section, Faculty of Mechanical, Maritime and Materials Engineering, Delft University of Technology, Delft, the Netherlands

Johan Grievink, Prof. ir. Department of Chemical Engineering, Product & Process Engineering Group, Faculty of Applied Sciences, Delft University of Technology, Delft, the Netherlands

Arno H.H. Janssen, Ir. ECN, Biomass & Energy Efficiency, Petten, the Netherlands

Wiebren de Jong, Dr.ir. Department of Process and Energy, Energy Technology Section, Faculty of Mechanical, Maritime and Materials Engineering, Delft University of Technology, Delft, the Netherlands

Yash Joshi, Ir. Department of Process and Energy, Energy Technology Section, Faculty of Mechanical, Maritime and Materials Engineering, Delft University of Technology, Delft, the Netherlands

Sascha R.A. Kersten, Prof.dr.ir. Sustainable Process Technology Group, Faculty of Science and Technology, University of Twente, Enschede, the Netherlands

Jaap H.A. Kiel, Prof.dr.ir. ECN, Biomass & Energy Efficiency, Petten, and Department of Process and Energy, Energy Technology Section, Faculty of Mechanical, Maritime and Materials Engineering, Delft University of Technology, Delft, the Netherlands

Robbert Kleerebezem, Dr.ir. Department of Biotechnology, Environmental Biotechnology Group, Faculty of Applied Sciences, Delft University of Technology, Delft, the Netherlands

Ming Liu, Dr.ir. Department of Process and Energy, Energy Technology Section, Faculty of Mechanical, Maritime and Materials Engineering, Delft University of Technology, Delft, the Netherlands

Lilian de Martín, Dr. Department of Chemical Engineering, Product & Process Engineering Group, Faculty of Applied Sciences, Delft University of Technology, Delft, the Netherlands

Jeroen A. van Oijen, Dr.ir. Department of Mechanical Engineering, Combustion Technology Section, Eindhoven University of Technology, Eindhoven, the Netherlands

J. Ruud van Ommen, Dr.ir. Department of Chemical Engineering, Product & Process Engineering Group, Faculty of Applied Sciences, Delft University of Technology, Delft, the Netherlands

Stijn R.G. Oudenhoven, Ir. Sustainable Process Technology Group, Faculty of Science and Technology, University of Twente, Enschede, the Netherlands

Dirk J.E.M. Roekaerts, Prof.dr. Department of Process and Energy, Fluid Mechanics Section, Faculty of Mechanical, Maritime and Materials Engineering, Delft University of Technology, Delft, the Netherlands

Adrie J.J. Straathof, Dr.ir. Department of Biotechnology, BioProcess Engineering Group, Faculty of Applied Sciences, Delft University of Technology, Delft, the Netherlands

Pieter L.J. Swinkels, Ir. Faculty of Applied Sciences, Delft Product & Process Design Institute, Delft University of Technology, Delft, the Netherlands

PART I

SOCIAL CONTEXT AND STRUCTURAL BASIS OF BIOMASS AS A RENEWABLE ENERGY SOURCES

1

INTRODUCTION: SOCIOECONOMIC ASPECTS OF BIOMASS CONVERSION

WIEBREN DE JONG[1] AND J. RUUD VAN OMMEN[2]

[1]*Department of Process and Energy, Energy Technology Section, Faculty of Mechanical, Maritime and Materials Engineering, Delft University of Technology, Delft, the Netherlands*
[2]*Department of Chemical Engineering, Product & Process Engineering Group, Faculty of Applied Sciences, Delft University of Technology, Delft, the Netherlands*

ACRONYMS

CDM	clean development mechanism
CFCs	chlorofluorocarbons
dLUC	direct land use change
GDP	gross domestic product
GHG	greenhouse gas
iLUC	indirect land use change
JI	joint implementation
LCA	life cycle assessment
LUC	land use change
R/P ratio	reserves-to-production ratio [y]
TOE	tonnes of oil equivalent(s) (= 41.87 GJ)
UNFCCC	United Nations Framework Convention on Climate Change

Biomass as a Sustainable Energy Source for the Future: Fundamentals of Conversion Processes,
First Edition. Edited by Wiebren de Jong and J. Ruud van Ommen.
© 2015 American Institute of Chemical Engineers, Inc. Published 2015 by John Wiley & Sons, Inc.

1.1   ENERGY SUPPLY: ECONOMIC AND ENVIRONMENTAL CONSIDERATIONS

1.1.1   Introduction: The Importance of Energy Supply

In the past two centuries, since the Industrial Revolution in the 1700s that was initiated by the invention of the steam turbine, the world has undergone a drastic change due to the steeply increased contribution of fossil fuels (coal, oil, and natural gas) to modern societies' energy supply (McKay, 2009). Though the Chinese society already used coal for energy supply in approximately 1000 BC and the Romans prior to AD 400 (World-Coal-Institute, 2005), the first written references indicating its use are from about the thirteenth century and beyond (Hubbert, 1949). These hydrocarbon fuels so far have been considered essential, as they are comparatively cheap and convenient energy carriers used for heating, cooking, lighting, and mechanical as well as electric power production and have been widely used as transportation fuels and feedstocks for the manufacture of bulk and fine chemicals as well as other materials with a wide range of applications. Rapid global population growth, expansion of economies, and higher standards have caused an enormous increase in worldwide energy consumption, which was partly made possible by the supply of cheap fossil fuels.

1.1.2   Development of Global Energy Demand

Figure 1.1 shows a scenario toward the year 2030 presented by the oil company BP concerning population growth in relation to developments in total primary energy utilization and gross domestic product (GDP). The figure shows that global energy

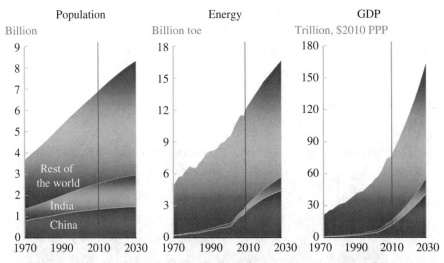

FIGURE 1.1   Prospected global growth rates in population, energy demand, and GDP; 1 toe = 4.1868.10^4 MJ. (Source: Adapted from BP, see tinyurl.com/7hlmqxn.)

demand will rise substantially from the current level with an increasing share from China and India. This rise of the primary energy demand is projected to be larger than the population growth, and this will cause a stress on the limited global resources. The projected GDP even increases stronger, so it is expected that average living standards increase, which will result in additional strain on the available resources.

1.1.3 Sustainability of Energy Supply

One of the major questions in the world, arising from the general picture sketched in Section 1.1.2, is how mankind can ensure a global sustainable development for the (near) future. In this context, sustainability of our energy supply is of paramount importance. The key issues are discussed in the following text, both from a point of view of global socioeconomics and ecological sustainability.

1.1.3.1 Socioeconomic Sustainability As one of the most important economic drivers to secure and improve the living standards of people in the world, *energy supply security* is of crucial value for current and future generations. Fossil fuels run out sooner or later as can be seen in Figure 1.2; they are not renewable on an acceptable time scale.

This figure depicts the so-called R/P ratios for different sources. The R/P ratio is the ratio of the current proven reserves to production level. The unit is years and it is a measure of the expected time a certain fuel source is expected to be available.

On a global scale, it appears that oil and natural gas reserves will be available—given the figures of 2012—for an expected approximately 55 years, and coal substantially longer (>100 years). Of course, new contributions to the reserves may be

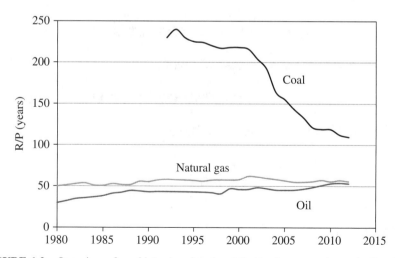

FIGURE 1.2 Overview of world (top) and regional (bottom) reserves-to-production (R/P) ratios for oil, natural gas, and coal, respectively (end 2012 status). Figures are based on data from BP (2013).

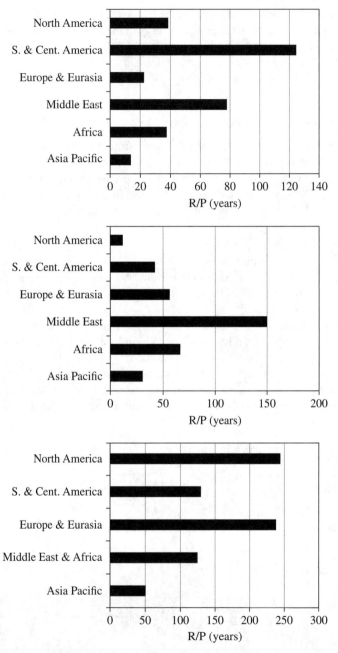

FIGURE 1.2   (*Continued*)

discovered in the (near) future, but that does not change the inherently limited supply nature of the fossil fuel sources. Regionally, there are also significant differences, which is important in the context of energy policy developments on the different continents.

For the price developments of the fossil fuels, not only their forecasted availability is of importance but also the market development in a landscape highly determined by politics. Already well before the last resources of a fuel will have been depleted the market will be severely stressed. For the economies in the world, *fuel cost development* is therefore also a primary point of concern. From past developments, particularly regarding oil, it has been shown that substantial fuel price fluctuations (volatility) occur, which has an impact on the global economy (e.g., food prices) that is difficult to predict. Supply and demand will determine the price evolution for each fuel source, but the development of the market structure is also essential: there is a large difference between a free market and an oligopoly or monopoly situation. In this respect, diversification of fuel sources with associated differentiation in suppliers is advantageous as it makes societies less prone to price manipulation by, e.g., cartel formation and sudden disruptions of supply (Johansson et al., 1993).

Self-sufficiency concerning energy supply is often mentioned as target of countries for (longer-term) sustainable economic development. However, not all countries have access to resources within their territories that are sufficient for such a target; other countries, on the other hand, have a structural surplus. Relief of *trade barriers* can help mitigate this structural discrepancy. Also, in the context of economic sustainability, a good trade balance should be maintained in relation to the energy supply within nations.

Regarding *social sustainability* in the context of energy supply, *reduction of poverty* should be mentioned first; a good supply structure of energy carriers is one of the basic requirements for such a development, next to access to clean drinking water and good soil for agricultural activity. Associated herewith, expectedly substantial *health improvement* should result from a good energy supply infrastructure. *Job creation* and maintenance is another aspect of social sustainability, and certain energy supply forms can contribute significantly to this. Also, maintaining (or improving) societies' *social cohesion* is an aspect that can be impacted by the energy supply structure.

1.1.3.2 Ecological Sustainability The energy supply structure should not compromise the sound development of our environment both from a local and global perspective. One of the major issues in this respect is global warming, which is for the main part attributed to the release of greenhouse gases (GHG) from fossil fuel combustion. Other issues are related to local emissions of acid rain precursors and particulate matter (PM).

Climate Change, the Greenhouse Effect, and Greenhouse Gas Emission Reduction The greenhouse effect occurs naturally to a large extent. Without this effect, the Earth's average global temperature would reach only a low $-18°C$, rather than the current approximate $+15°C$. Water vapor is the largest contributor to this effect, with a complex role for clouds, but also CO_2 in the atmosphere plays a significant role. More than a century ago, Arrhenius (1896) already identified this role in the Earth's temperature control. Ice core studies reveal that on millennial time scales, changes in CO_2 content recorded are highly correlated with changes in temperature, although some temperature changes have occurred without a significant CO_2 concentration change, but the opposite does not appear to have happened (Falkowski et al., 2000). Less pronounced roles are played by CH_4, N_2O (nitrous oxide), and several types of chlorofluorocarbons (CFCs) and SF_6. It is the CO_2, CH_4, N_2O, and CFC concentrations in the atmosphere upon which man's industrial

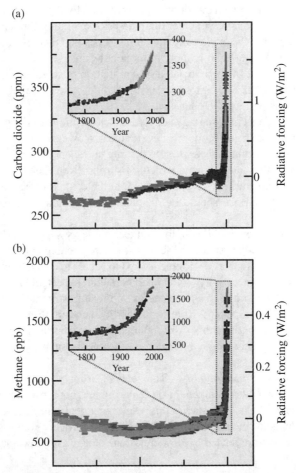

FIGURE 1.3   Atmospheric concentrations of CO_2, CH_4, and N_2O over the last 10,000 years (large panels) and since 1750 (inset panels). Measurements are shown from ice cores (symbols with different grey shades for different studies) and atmospheric samples (light grey lines in steep curve part, red lines in the original publication). The corresponding radiative forcings (net solar energy flux to the earth) relative to 1750 are shown on the right-hand axes of the large panels. (Source: Reproduced with permission from IPCC (2007), figure 2.3; figure SPM.1. © IPCC.)

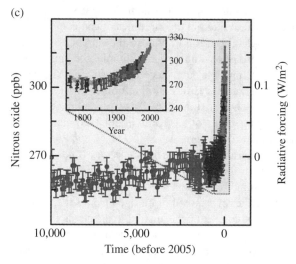

FIGURE 1.3 (*Continued*)

and household activities have a measurable impact. Scientists largely agree on the point that in the last few centuries, the activities of humans have directly or indirectly caused the concentrations of the major GHG to increase. This is exemplified by Figure 1.3. The atmospheric CO_2 concentration varies to some extent from place to place and from season to season. It has been shown that concentrations are somewhat higher in the northern hemisphere than in the southern hemisphere as most of the anthropogenic sources of CO_2 are located north of the equator. The difference in land surface covered with forests, being more concentrated north of the equator, causes larger seasonal fluctuations due to comparatively shorter growth periods than in the generally milder southern hemisphere locations that are under the influence of larger oceanic surfaces.

Oscillations of atmospheric CO_2 concentrations between about 180 and 280 ppm_v have occurred in the past approximately 480,000 years in cycles of 100,000 years, but it appears now we have abandoned this cycling behavior in a remarkably short time frame.

Studies at the NASA Goddard Institute for Space Studies in New York (United States) have shown that over the past few decades, the combined warming effect of non-CO_2 GHG should have been comparable to that of CO_2 alone. However, while each of the GHG mentioned earlier acts to warm the surface of the Earth, the long-term climatic effects of the other GHG differ from those of CO_2. Methane, e.g., has an atmospheric lifetime of only about 12 years. By comparison, newly added CO_2 will remain for a time span of tens to thousands of years. As a result, about 65% of the carbon dioxide that human activities have generated since the start of the Industrial Revolution is in the air we breathe today. A historical record of the amount of CO_2 in the atmosphere can be found in bubbles of air in arctic ice layers, dating back as far as 600,000 years. The depth of such a layer is a measure of its time of formation.

Another difference is that the principal anthropogenic sources of methane-bacterial fermentation in rice paddies and in the intestines of cattle are related to food production and, hence, are roughly proportional to the number of people on the planet. Because CH_4 has such a short atmospheric lifetime, the amount that is in the air is a good indicator of how much is being added with time. Should the global population double over the next half century, the concentration of CH_4 could also double, but it is not likely to rise by much more than that. This would add, at most, a few tenths of a degree to the mean temperature of the Earth. Future CO_2 increases could, in contrast, warm the climate by $10°C$ or more.

Nitrous oxide (N_2O) and CFCs are in some ways more like CO_2 in that once released they remain in the atmosphere for a century or more. The production of N_2O, however, is only indirectly dependent on human activities. Its principal source is a natural one, the bacterial removal of nitrogen from soils, and although the world population swells in coming years, the amount in the air should increase only slowly.

The outlook for many CFCs is even more promising. Today, the most abundant of these man-made compounds, freon-11 and freon-12, are being phased out of production altogether by international agreements because of their damaging effects on stratospheric ozone. Indeed, the concentration of one of these gases, freon-11, peaked in 1994 and is now in a slow decline that should continue for the next century or so. The freon-12 concentration has not yet leveled off, but is expected to do so within the next few years. In terms of climatic effects, the main threat from CFCs comes from other long-lived compounds that may be used to replace the ones that have been phased out and that could also act as GHG. Since these possibly harmful replacement gases are as yet present in only small amounts and since, as noted earlier, projected increases in CH_4 and N_2O are so much less severe, we shall for the rest of this discussion focus solely on the most important anthropogenic GHG: CO_2.

Some experts have estimated that the Earth's average global temperature has already increased by more than $0.5°C$ since the mid-1900s due to this human-enhanced greenhouse effect; also impacts on sea level (rising) and snow coverage (tending to decrease) have been investigated, the results of which are summarized in Figure 1.4.

Like most other planets and planetoids in the universe, the Earth contains a great deal of carbon, which is slowly and continually transported from the mantle to the crust and back again, in the course of volcanic eruption and subduction phenomena. The portion that finds itself near the surface is continually exchanged and recycled among plants, animals, soil, air, and oceans. In some of these temporary stocks, carbon is more securely held, while in others it more readily combines with oxygen in the air to form CO_2. In order to predict how atmospheric CO_2 levels and climate may change in the future, it is important to understand where carbon is stored and what its dynamic cycling behavior looks like. The carbon reservoirs that are most relevant to global warming are listed in Table 1.1, with the total amount of carbon that they contained in 2000.

The atmosphere contains approximately 720 Gt C in the form of CO_2; current measured atmospheric CO_2 concentrations are nearly 400 ppm_v. The rate of change in this carbon stock not only depends on human activities but also on biogeochemical and climatological processes and their interactions with the global carbon cycle.

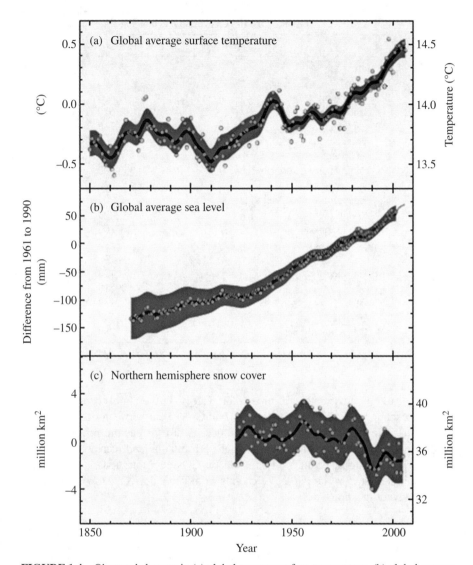

FIGURE 1.4   Observed changes in (a) global average surface temperature, (b) global average sea level from tide gauge (light grey circles; blue circles in the original publication) and satellite (light grey line through the data points and extending from early nineties to beyond the year 2000; red line in the original publication) data, and (c) northern hemisphere snow cover for March and April. All differences are relative to corresponding averages for the period 1961–1990. Smoothed curves represent decadal averaged values, while circles show yearly values. The shaded areas are the uncertainty intervals estimated from a comprehensive analysis of known uncertainties (a) and (b) and from the time series (c). (Source: Reproduced with permission from IPCC (2007), figure 1.1; figure SPM.3. © IPCC.)

TABLE 1.1   Global carbon reservoir overview

Pools	Quantity (Gt^a)
Atmosphere	720
Oceans	38,400
Total inorganic	37,400
Surface layer	670
Deep layer	36,730
Total organic	1,000
Lithosphere	
Sedimentary carbonates	>60,000,000
Kerogens	15,000,000
Terrestrial biosphere (total)	~2,000
Living biomass	600–1,000
Dead biomass	1,200
Aquatic biosphere	1–2
Fossil fuels	4,130
Coal	3,510
Oil	230
Gas	140
Other (mainly peat)	250

Reproduced with permission from Falkowski et al. (2000). © AAAS.
^{a} 1 Gt is 10^9 metric tons $= 10^{12}$ kg.

Oceans store approximately 50 times more CO_2 than the atmosphere. There is a dynamic exchange by diffusion of about 90 Gt C per year between oceans and the atmosphere, showing the importance of the oceans in CO_2 capture processes. This capture by dissolution leads to ionic bicarbonate formation and increased acidity of the water. Buffering of changes in atmospheric CO_2 concentrations is limited and is very much related to the release of cations from comparatively slow rock weathering. Relevant reactions are:

$$CO_2 + H_2O \leftrightarrow HCO_3^- + H^+ \qquad (RX.1.1)$$

$$HCO_3^- + H_2O \leftrightarrow CO_3^{2-} + H^+ \qquad (RX.1.2)$$

$$CaSiO_3 + CO_2 \rightarrow CaCO_3 + SiO_2 \qquad (RX.1.3)$$

In the deep ocean (below ~300 m), the concentration of carbonates is higher than at the surface level, which is a consequence of two processes, called the oceanic "solubility pump" and "biological pump" mechanisms. The first is related to the better solubility of CO_2 in cold saline waters, which are in particular present at the Earth's higher latitudes. These take up CO_2, sink to lower levels, and redistribute the solution laterally. However, a CO_2-induced global temperature rise leads to a more stable height profile of water (a phenomenon called stratification), by which transport of

CO_2 from higher levels to the deep sea is hampered. Increased saturation levels and stratification will weaken absorption capabilities of the oceans.

The "biological pump" is related to fixation of CO_2 by phytoplankton in photosynthesis, which is of the order of 11–16 Gt C per year. This leads to a partial conversion to $CaCO_3$ by plankton species sinking down to the lower oceanic levels. This process, though sequestering some of the original atmospheric CO_2 in solid form, also leads to partial CO_2 release again.

Question: Why would this be the case?

The biological pump mechanism thus counteracts, decreasing action of the solubility pump. If anthropogenic CO_2 release is to be effectively counteracted, then this mechanism somehow should be boosted. However, understanding of the mechanism and its implications is lacking, and one cannot rely on this process to be effectively controllable in a foreseeable time frame.

The carbon stored in the lithosphere, which is the crust and mantle part of our Earth, is present in both organic and inorganic forms. Inorganic carbon deposits in the lithosphere constitute natural carbonate rock materials (e.g., limestone, dolomite), also present in coal inclusions and oil shales. Organically bound carbon in the lithosphere includes litter, humic soil substances, and other organic compounds. CO_2 is released by different natural geophysical and geochemical processes, such as via volcanic eruptions. On the other hand, carbonates in sediments and sedimentary rocks are also removed from the crust by subduction (due to differences in density of continental and ocean tectonic plates) to lower lithosphere levels and partially molten beneath tectonic boundaries.

Terrestrial vegetation, another carbon reservoir, contains about 600–1000 Gt C, stored mostly as cellulose in the stems and branches of trees. Carbon fluxes related to terrestrial respiration and decay of CO_2 comprise a value of approximately 61 Gt C·$year^{-1}$ (Falkowski et al., 2000). Photosynthesis is the chemical process by which chlorophyll-containing plants and some bacteria can capture CO_2 and organically convert it with water under the influence of irradiated solar energy. This process is vital for life on Earth by balancing the amounts of CO_2 and O_2 in the atmosphere (Raven et al., 2005). This chemical reaction forming a carbohydrate polymer can be described by the following simple equation:

$$6nCO_2 + (5n+1)H_2O + \text{solar energy} \rightarrow C_{6n}(H_2O)_{5n+1} + 6nO_2 \qquad \text{(RX.1.4)}$$

A number of processes return CO_2 partially back to the atmosphere, namely:

- Autotrophic plant respiration, due to consumption by the plant of part of the carbohydrate polymer to sustain its metabolism
- Heterotrophic respiration, whereby soil microbes oxidize plant-derived organic matter
- Fires

The main terrestrial carbon storage capacity, being about three times the current atmospheric carbon quantity, is predominantly situated in forests, with resistance against plant matter decay being very important. As one of the major organic constituents of plant biomass, lignin plays a key role in preventing microbiological decay (see also Chapter 2). Terrestrial net primary production (NPP) is an important indicator of natural carbon sequestration and is defined as the difference between gross primary biomass production and respiration. The terrestrial carbon sequestration degree is not yet reaching the limitation posed by current CO_2 concentrations (Schimel, 1995). Thus, it forms a potential sink for anthropogenic carbon emissions, although, depending on plant type, the saturation level is reached sooner or later. Furthermore, nutrient availability (fixed nitrogen and phosphor) may hamper growth already before the saturation level has been reached.

In the context of the United Nations Framework Convention on Climate Change (UNFCCC) on December 11, 1997, the Kyoto Protocol was adopted, aimed at the "stabilization of greenhouse gas concentrations in the atmosphere at a level that would prevent dangerous anthropogenic interference with the climate system" (see tinyurl. com/35t7t7). By September 2011, 191 countries had signed and ratified the Protocol; this has been extended recently at Doha to the year 2020. The reference year for the emissions counted is 1990. The European Union as a whole targets for a −8% difference compared to that year. Countries have a certain degree of flexibility in how they make and measure their GHG emission reductions. In particular, an international "emissions trading" regime will be established, allowing industrialized countries to buy and sell emission credits among themselves. This mechanism is targeted toward accounting for external costs, which would otherwise be overlooked by the market, into the price of energy sources. Countries will also be able to acquire "emission reduction units" by financing certain kinds of projects in other developed countries through a mechanism known as Joint Implementation (JI). In addition, a "Clean Development Mechanism" (CDM) for promoting sustainable development will enable industrialized countries to finance emission reduction projects in developing countries and receive credit for doing so.

Operational guidelines will pursue emission cuts in a wide range of economic sectors. The Protocol encourages governments to cooperate with one another, improve energy efficiency, reform the energy and transportation sectors, promote renewable forms of energy, phase out inappropriate fiscal measures and market imperfections, limit methane emissions from waste management and energy systems, and protect forests and other carbon "sinks." The measurement of changes in net emissions (calculated as emissions minus CO_2 removals) from forests is methodologically complex and still needs to be clarified. The Protocol will advance the implementation of existing commitments by all countries. Under the Convention, both developed and developing countries agree to take measures to limit emissions and promote adaptation to future climate change impacts; submit information on their national climate change programs and inventories; promote technology transfer; cooperate on scientific and technical research; and promote public awareness, education, and training. The Protocol also reiterates the need to provide "new and additional" financial resources to meet the "agreed full costs" incurred by developing countries in carrying out these commitments.

Other Emissions from Energy Conversion Processes Leading to Air Pollution   The chemical conversion of fuels to generate (mechanical) power and heat, generally involving a combustion step, also results in emissions other than CO_2. For example, *acid rain precursor emissions*, such as NO_x (NO and NO_2) and SO_x (SO_2 and SO_3), are of concern. These gaseous compounds lead to the formation of nitric acid and sulfuric acid in aqueous droplets in the atmosphere. These acids can precipitate both wet (e.g., via droplets) or dry (adhered to solids that deposit on Earth), causing water and soil acidification. In particular, forests and lakes are impacted by this effect. Also, infrastructural works, such as buildings, are prone to decay due to chemical attack by acid rain on metals (corrosion), paints, and stone materials. Though recognized already in the nineteenth century, further research followed upon increased awareness and action was initiated in the 1970s (see, e.g., Likens and Bormann, 1974). Substantial abatement efforts and stringent policies to reduce NO_x and SO_x emissions have already been carried out worldwide since the 1970s; this resulted in reduced emissions in large parts of the world (Wright and Schindler, 1995). Still, there is concern for acidic environmental pollution in emerging economy areas such as China and India, where fossil fuel burning and motorized traffic have shown remarkable growths (Larssen et al., 2006). Emission abatement of acid rain precursor gases is dealt with in Chapter 9.

Another type of emission that is to be seriously considered, from both human, animal, and plant health and climate change perspectives, is *PM* (particulate matter). PM is released into the air in energy conversion processes that comprise combustion, in particular of solid and liquid fuels but also of gaseous fuels under nonoptimized combustion conditions. Especially, the respirable and very fine particles dispersed in the air, also termed aerosols, are of concern. Nature also generates them via natural processes such as fires, dust bowls, and storms (e.g., winds over salty seas) as well as volcanic activity. Aerosols have several effects with regard to atmospheric radiation (tinyurl.com/48sqk6o). Two effects are briefly described here. The so-called direct effect of PM comprises the effective scattering of radiation, causing negative radiative forcing (e.g., by fine salt aerosols), and the absorption of light, contributing to global warming (carbon black and sooty material). The "indirect effect" of PM relates to cloud formation. Clouds consist of droplets that form with initiation (partly) based on preexisting aerosols, so-called cloud condensation nuclei. When these particles increase in number concentration, increased scattering of radiation in the shortwave domain of frequency occurs. This effect is also attributed as cloud albedo effect. Another effect of increased number concentration of aerosols and thus droplets in clouds is the suppression of precipitation formation and augmentation of the lifetime of clouds. This is also referred to as the Albrecht effect.

Apart from these environmental effects, PM also has a direct negative impact on human health: particulates can lead to respiratory diseases, including lung cancer, and cardiovascular problems.

Several countries and regions in the world have set restrictions on PM emissions. These restrictions distinguish between PM10 and PM2.5, being PM with sizes up to 10 and 2.5 μm, respectively. Chapter 9 deals with technologies for the reduction of PM in the energy conversion industry where combustion plays a key role.

1.2   WAYS TO MITIGATE THREATS TO A SUSTAINABLE ENERGY SUPPLY

As discussed in the previous section, fossil fuel utilization puts pressure on our global ecosystem by contributing to global warming and harmful emissions. Moreover, the reserves of fossil fuels are finite. There are three ways to deal with this global challenge, and they all require drastic innovation development in the respective technologies:

- The energy efficiency of conversion systems should be drastically improved, going hand in hand with reduced use (savings); related to this is also a moderate population growth.
- Renewable energy sources should be used in order to supply nonfossil-based energy.
- Clean use of fossil fuels also will be needed for several decades to come, including carbon capture and storage (CCS). Nuclear energy usage also largely reduces CO_2 emission, but the generation of nuclear energy is associated with the creation of radioactive waste (fission) and potential issues with unwanted extended military (nonproliferation) or even terrorist usage.

The principle of these three energy utilization strategies is also called the "Trias Energetica," as illustrated by Figure 1.5.

This book concentrates on the pathway of sustainable development via utilization of renewable energy sources and focuses particularly on one type: *energy from biomass* or *bioenergy*. Biomass consists of material that has an organic origin. It includes

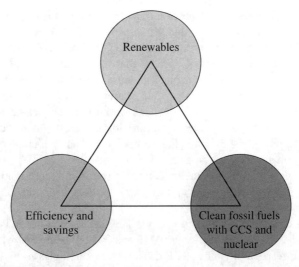

FIGURE 1.5   Concept of the "Trias Energetica," based on Lysen (1996).

matter derived from plants and animals and their waste or residual material and, in the broader sense, all conversion products such as paper or cellulose, organic residuals from the food industry, and organic waste from households, trade, and industry. Distinction from fossil fuels starts with peat, which is defined not to belong anymore to biomass (Kaltschmitt and Hartmann, 2001; Spliethoff, 2009). The definition of biomass for energy given in the European Directive 2009/28/EC is the following: "the biodegradable fraction of products, waste and residues from biological origin from agriculture (including vegetal and animal substances), forestry and related industries including fisheries and aquaculture, as well as the biodegradable fraction of industrial and municipal waste."

The major sources of renewable energy are:

- Solar energy (solar thermal and photovoltaic (PV) technologies)
- Wind energy
- Hydropower
- Tidal energy
- "Blue energy" (based on osmosis driven by the salt concentration difference of two water streams)
- Geothermal energy
- Bioenergy

The use of biomass for the generation of energy has many positive aspects, but there are also critical issues associated with its use (see, e.g., Giuntoli, 2010). The pros and cons are listed below.

Pros

- Biomass is abundantly available stored solar energy, which is thus indirectly used.
- Biomass is available as a comparatively constant supply source as it can be stored under certain biomass-specific conditions, so it may act as a "natural battery." This is important in view of enabling energy supply security. Solar energy and wind energy, on the contrary, are available as fluctuating sources for which energy storage is still an area of development.
- Bioenergy is the only renewable energy source that can be coprocessed with fossil fuels in existing energy conversion systems (such as oil refineries or coal gasification plants) so as to ensure a gradual energy transition to a renewable energy source.
- As biomass is formed (indirectly) on a relatively short time scale via photosynthesis from CO_2 and water, which are released again in energy conversion systems, in theory, one can speak of a "carbon-neutral" fuel.
- Waste streams and biogenic by-products can be valorized into valuable power, heat, and chemicals.

- Biomass is already grown for food, animal feed, and natural fiber applications as well as in forestry with its derived products; it is comparatively easily accessible, and man has experience in dealing with this source. The impact of extended growth on the environment usually is much less an issue than that of, e.g., hydro-power installations, which affect the ecology and living surroundings of people and animals.

- Biomass growing, harvesting, storage and transportation, trading, and processing to end use for energy conversion purposes can enhance rural economic develop-ment (Nag, 2009), via the creation of additional jobs, more than in the fossil fuel processing sector. This leads to extra income for rural regions of both developed and developing countries and eventually offers a way to counteract the constant depopulation of such areas.

- For obtaining energy from biomass, nonscarce materials are used. In contrast, several other major sustainable energy sources rather make use of rare materials to construct the required energy conversion devices (e.g., gallium and indium in solar PV cells and niobium and neodymium in wind turbines).

Cons

- Conversion of solar energy into biomass is generally low (of the order of 1%, depending on the species); this means that relatively large surfaces are needed to harvest sufficient material for application in energy conversion.

- Biomass is characterized by a low energy density (see Chapter 2 for more details) compared to fossil fuels, which challenges logistics. As a consequence of this and the previous point, biomass has a limited effective availability, and therefore, it can only partly (but still substantially) contribute to the world's energy demand.

- Although biomass itself is renewable, during the whole life cycle, fossil fuels are commonly used to produce fertilizers and pesticides, grow and collect the plants, transport the harvested material, and, finally, upgrade it to an actual fuel. There-fore, biofuels still have a fossil carbon footprint to some extent.

- Widely different disciplines are involved in the successful implementation of the whole integrated chain from seeding to final conversion and use, ranging from policy development and logistics to chemical, process technology, and agricul-tural sciences; in this light, implementation of effective policies is complex as bioenergy policies might conflict with other existing environmental or economic policies. For other energy sources, often, the only key is the energy conversion technology, whereas for biomass the whole system is crucial (Sims, 2002).

- There is concern with respect to biomass usage for energy supply as it may com-pete with food production (Diouf, 2008), depending on the types of biomass and technologies used.

- Issues with deforestation, associated with serious loss of biodiversity and carbon stock, may result if energy plantations are realized at the expense of, e.g., tropical rain forests.

- Competition for scarcening water sources (Falkenmark, 1998) may become a serious issue.

- A drawback of using biomass as a direct source for electricity production would be that useful work (exergy) is destroyed. It might be more efficient from an energy system's point of view to use biomass as a primary source for (petro) chemical synthesis.

Figure 1.6 presents the forecasted energy demand per energy resource and the role of different strategies to mitigate the CO_2 emissions according to recent IEA scenarios (OECD/IEA, 2012). The 6DS scenario is characterized by the absence of efforts to reduce GHG emissions, leading to a projected 6°C increase of the global mean temperature. The 4DS scenario includes pledges that countries have made to reduce such emissions as well as to increase efforts to improve energy efficiency. It is consistent with the World Energy Outlook's New Policies Scenario through 2035 (IEA, 2011) and would lead to an expected global mean temperature increase of 4°C. The 2DS scenario, projecting a long-term global temperature rise of 2°C, is targeted by IEA in their Energy Technology Perspectives 2012 (OECD/IEA, 2012) and comprises a cut of energy-associated CO_2 emissions by more than 50% in 2050, as compared

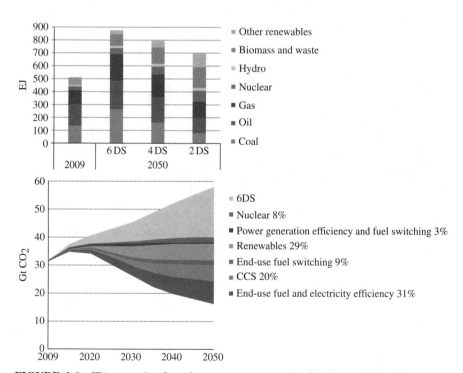

FIGURE 1.6 IEA scenarios for primary energy use and mitigation of CO_2. World total primary energy supply per energy source in 2009 and in three scenarios for 2050, with DS being °C global temperature increase perspective (top). Key technologies for reducing CO_2 emissions (bottom); note that percentages in the legend represent cumulative contributions to emission reductions relative to the 4DS scenario. (Source: Reproduced with permission from OECD/IEA (2012). © IEA.)

to 2009. This requires a substantial restructuring of the energy sector. Figure 1.6 shows the different contributions of mitigation actions in such a way that if all were to be realized, then the 2DS scenario would be followed with a probability of 80%. As can be seen, biomass and waste are expected to play a significant role both in future energy supply and in reducing the GHG emission problem.

1.3   WHAT IS SUSTAINABLE SUPPLY OF BIOMASS?

A globally raised awareness confrontation with the limitations of the oil availability in the early 1970s as a result of the first oil crisis and ecological concerns (see, e.g., Meadows et al., 1972) gave rise to a first renewed interest in biomass for energy supply and also virtually all other forms of renewable energy. After the oil crisis, however, the prices of fossil fuels decreased again due to their higher availability, and this impeded the further development of bioenergy technology. In the 1980s and 1990s, the concern grew that global warming and the resulting climate change were enhanced by CO_2 emissions. This led to the Kyoto Protocol (UNFCCC, 1997) aimed at a reduction of the emission of GHG described in Section 1.1.3 and again stimulated research in the area of renewables and in particular biomass as one of the key carbon mitigating sources.

In order to enable the large-scale introduction of sustainably produced bioenergy and biomass-derived products, a number of technical and nontechnical issues have to be addressed and solved, such as the configuration of the production technology chain, storage and transportation options, integration into the existing energy system, and social acceptance. The transition from the present fossil fuel-based energy system to a sustainable bio-based energy system is expected to be fragmented and to involve a diverse mixture of fossil and renewable energy sources.

A "development which meets the needs of the present without compromising the ability for future generations to meet their own needs" has been a globally referred definition of sustainability. It was first characterized as such by the UN Brundtland Commission in 1987 (Brundtland, 1987). Figure 1.7 shows an overview of the economic, environmental, and social aspects that play a role in the sustainability of bioenergy supply systems. In Sections 1.3.1 and 1.3.2, the socioeconomic impact and ecological implications of biomass for energy supply are addressed.

1.3.1   Sustainable Biomass in Terms of Socioeconomic Considerations

Economic development relies on a secure energy supply. Related to this is the price development of alternative fuels, in this case biofuels. Costs associated with introducing bioenergy technologies are investments to be made in capital for process equipment and infrastructure as well as the needed human capital. Furthermore, of importance for a sustainable development in terms of economic parameters are the allowed financing schemes in countries promoting bioenergy technologies. There is a wide international debate on certification, dealing with the quality of biomass products and related practices; this determines availability and thus pricing. Finally,

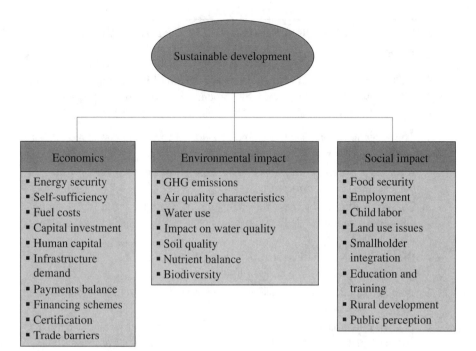

FIGURE 1.7　Aspects of sustainable development related to the implementation of bioenergy supply in society.

countries still use tariffs for importing and exporting biomass products, which causes trade barriers that impact the economics of bioenergy supply and also hampers countries with large potential for export to make a business out of biomass.

Social impact is another important aspect of sustainability. Food (including drinking water) is of first importance to life, and biomass growing schemes must not compromise this. Labor opportunities offered in the field of biomass growing and use for energy supply can substantially contribute to employment, which is of social benefit to mankind. Care should be taken, however, that this is not at the cost of child development, and biomass growth and processing should not be based on child labor. In order to stimulate agricultural practices, smallholders (farmers) should be supported in their activities to cultivate biomass and develop the land. Herein, education and training regarding the bioenergy whole chain approach should be pursued to sustain the practices. This further drives rural development and community building, thereby partly mitigating extreme urbanization with its associated problems. Finally, for any activity to be sustainable in the sense of social development, the bioenergy-related activities should be such that the people's perception remains positive and supportive.

Important for a sustainable utilization and expansion of biofuels and bioelectricity is a stable, long-term sociopolitical framework to increase investor confidence and to stimulate agricultural enterprises. Simultaneously, countries should remove subsidies for fossil fuel utilization. Moreover, internationally agreed sustainability criteria for

---

BOX 1.1: EXAMPLE OF SUSTAINABILITY CRITERIA IN NATIONAL POLICIES

In the Netherlands, a committee under the supervision of former minister Cramer has worked out criteria for sustainable bioenergy development, the so-called Cramer criteria (Cramer et al., 2007). They relate to nine principles:

1. The greenhouse gas balance of the production chain and application of the biomass must be positive.
2. Biomass production must not be at the expense of important carbon sinks in the vegetation and in the soil.
3. The production of biomass for energy must not endanger the food supply and local biomass applications (energy supply, medicines, building materials).
4. Biomass production must not affect protected or vulnerable biodiversity and will, where possible, have to strengthen biodiversity.
5. In the production and processing of biomass the soil and the soil quality are retained or improved.
6. In the production and processing of biomass ground and surface water must not be depleted and the water quality must be maintained or improved.
7. In the production and processing of biomass the air quality must be maintained or improved.
8. The production of biomass must contribute towards local prosperity.
9. The production of biomass must contribute towards the social well-being of the employees and the local population.

---

the sourcing, logistics, and use of biomass should be further worked out without creating unwanted trade barriers such as import and export tariffs (Tanaka, 2011). In this respect, also similar criteria should be applied to existing fossil fuel sourcing and distribution practices to create a similar level playing field.

1.3.2 Sustainable Biomass in Terms of Ecological Considerations

Regarding the *environmental impact*, the GHG footprint has already been discussed in Section 1.1.3 as an important point of attention when introducing certain biomass-to-energy supply schemes.

Not only GHG but also other emissions to the air determine the acceptability in terms of sustainable bioenergy development. These concern, e.g., acid rain precursors, NO_x and SO_x. Nitrogen and sulfur are bound in biomass, and oxidation processes lead to the emission of species that are further converted to acidic species in the atmosphere, as has been discussed in Section "Other Emissions from Energy Conversion Processes Leading to Air Pollution." Also, trace elements in biomass, such as Cu,

Zn, and Hg (present in sewage sludge in relatively high amounts), lead to potentially harmful emissions when thermally converted. These are often related to the coemission of fine PM, which should be cleaned. Moreover, when fertilization is applied to lands in excessive amounts, *eutrophication* of the environment may take place, which means that nitrates and/or phosphates are released into the water. This will lead to excessive growth of phytoplankton, algae, and certain water plants, resulting in a serious oxygen depletion so that water life (e.g., fish population) is endangered tremendously. A so-called life cycle assessment (LCA) can be used to quantify the aforementioned impacts on introducing a bioenergy supply chain in a certain region. We refer the reader to a number of books for more in-depth information on LCA (Guinée et al., 2002; Wrisberg and Udo de Haes, 2002).

Another important point is concern for the utilization of and quality impact on water resources of the introduction of bioenergy; this is crucial in many countries where water supply is a serious bottleneck for any development. Impact on soil characteristics should also be considered. This may turn out to be positive, though, like growth of perennial grasses that can improve the soil structure so as to counteract fatal phenomena of erosion by wind or flooding by rainfall; also carbon stock characteristics may be improved in this way. As this is not always the case in changing land use, it is a point of attention. Moreover, nutrient supply and biomass harvesting should be kept in balance to be sure that agricultural practice can be sustained. Finally, the conservation of biodiversity should be respected so as not to run the risk of failure through pests and diseases to which single crops might be subjected; a diverse environment should be ensured also for mankind, flora, and fauna to flourish.

Growing biomass for energy supply is not inherently beneficent or even GHG emission neutral. Of course, in order to be so, there should be a replacement of fossil fuels, preferably the ones with high CO_2-equivalent emissions per GJ. The climate change mitigation also depends on the geographical location. Land use strategies form an important factor in the dynamics of carbon sequestration and release. On the one hand, stimulation of higher productivity on all forms of land helps to reduce pressure on land use change (LUC). On the other hand, the world faces population growth, which leads to increasing need of agricultural lands and fertilization. In this respect, it is important to address the aspect of LUC, which can be either direct or indirect. LUC may have impact on GHG emission in several ways (Berndes et al., 2011):

- Burning of biomass in the field for the purpose of land clearing.
- Land management practices may be such that carbon stocks in soils and vegetation change.
- Land use intensity may change, causing increased fertilizer intensity with associated N_2O emissions.
- LUC may be associated with alterations regarding carbon sequestration rates as a result of different CO_2 assimilation behavior of the new plantation.

In addition to impacts on GHG emissions, LUC may influence the Earth's reflection of sunlight; depending on the changes made, this can either increase or decrease the effect.

1.3.2.1   Direct Land Use Change   Direct land use change (dLUC) is the case when other biomass is grown on land that was previously forest or other high carbon stock land. This may lead to a (usually negative) change in the soil carbon content. The impact of such a change is still comparatively straightforward when considering the next form of LUC.

1.3.2.2   Indirect Land Use Change   Indirect land use change (iLUC) occurs when farmers start to grow certain biomass on a land with the purpose of being taken up in a bioenergy supply chain, but with an associated displacement of, e.g., food production that then is started on land elsewhere where forest or other high carbon stock land is converted to arable land. Considerable debate exists among scientists about the extent of the impact of iLUC on GHG emissions.

When biomass is grown for energy supply on purpose, it should not compete with food, feed, and fiber production; therefore, cultivation on lower-quality lands with comparatively low carbon stock is advisable. One can distinguish the following such land types (Kampman et al., 2010):

- Marginal land: this is a land that is currently not cultivated as cropland. Although it is technically feasible to produce crops, its yields are too low and costs too high for competitive agricultural practice.
- Degraded land: this is a land that has been cultivated in the past but has become marginal due to degradation of its soil, erosion, or other causes as a consequence of inappropriate management or external factors such as climate change.
- Abandoned land: this is a degraded land with low productivity and a land with high productivity that is currently not in use.

Figure 1.8 illustrates the difference between dLUC and iLUC.

1.3.2.3   Upfront Carbon Debt Creation   Linked to the issue of LUC is the point of establishing carbon debt for different scenarios of biomass growth for energy supply. The assumption—often widely taken for granted—that the use of bioenergy always results in zero GHG emissions, i.e., that use of biomass is always carbon neutral regardless of the time horizon considered, is incorrect. Land clearing usually leads to CO_2-equivalent emissions that create a carbon loss from the beginning, the so-called *upfront carbon debt* creation. It is clear that all sources of woody bioenergy (replacing fossil energy) from sustainably managed forests will result in emission reductions in the long term, but diverse bioenergy sources have various impacts on the short-medium term. Therefore, some sources of wood for the generation of bioenergy do not contribute to reducing GHG emissions within the time frame of climate mitigation policies, whereas other sources may have this potential (Zanchi et al., 2012).

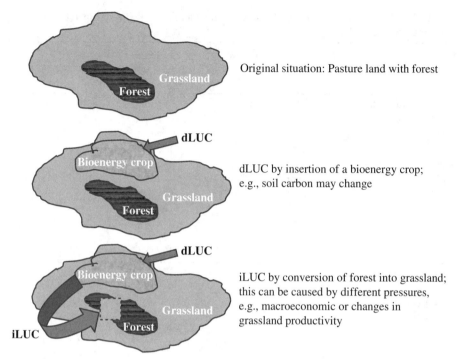

Original situation: Pasture land with forest

dLUC by insertion of a bioenergy crop;
e.g., soil carbon may change

iLUC by conversion of forest into grassland;
this can be caused by different pressures,
e.g., macroeconomic or changes in
grassland productivity

FIGURE 1.8   Illustration of the differences in LUC forms, based on Berndes et al. (2011).

1.4   RESOURCES AND SUSTAINABLE POTENTIAL OF BIOMASS

Prior to the discovery and large-scale utilization of inexpensive fossil fuels during the Industrial Revolution, mankind was strongly depending on plant biomass to meet its energy demands. In prehistoric times, already, biomass was not only the key source for the supply of food, animal feed, and materials for clothing but also for heating, cooking, and lighting. Later in preindustrial times, biomass served quite diverse purposes, such as early charcoal production for iron making and heat generation for the processing of metals.

Biomass is available in abundance and more evenly distributed over the world than fossil fuels. Energy originating from the sun is stored in biological species, and this comprises the oldest stored energy source known to mankind. Its current use is still mainly traditional, and often unhealthy and laborious practices have traditionally been established for the combustion of biomass to generate heat for cooking. This practice makes up for around 22% of the energy used in developing countries (Bauen et al., 2009) and for almost 50% of the primary energy used in Africa (Karekezi et al., 2004). In modern societies, the use of biomass as an energy source was gradually abandoned after the discovery of huge amounts of cheap fossil fuels that were easier to process and use. Furthermore, the "modern age" fuels

TABLE 1.2   The role of biomass in some global energy supply scenarios for years between 2025 and 2050

Scenario	Target year	Energy demand (EJ)	Renewables contribution (EJ)	Biomass contribution (EJ)	Notes
BP Energy Outlook	2030	~712	~85	<50%	a
EXXON 2012 The Outlook for Energy	2025	637	85	53	b
	2040	696	102	53	b
IEA "New Policies"	2030	679	96	74	c
IEA "Current Policies"	2030	719	84	68	c
IEA "450 Scenario"	2030	610	120	86	c
Shell "Blueprints"	2030	692	138	59	d
	2050	769	232	57	d
Shell "Scramble"	2030	734	174	92	d
	2050	880	326	131	d

[a] tinyurl.com/7hlmqxn
[b] tinyurl.com/6reyxlb.
[c] IEA (2011).
[d] tinyurl.com/5aamp8

allowed applications that were not so easy to realize with biomass: transportation based on liquid fuels from crude oil and cooking on fossil gases. The future for biomass as an energy source appeared to be deemed to fade away, although at the very origin of modern engines, biomass-derived fuels were proposed and used, e.g., peanut oil in Rudolf Diesel's diesel engine and ethanol in internal spark ignition Otto cycle-based engines (Henry Ford's trials). Nowadays, biomass is again becoming recognized as a green source for a multitude of energy conversion processes and as a leading practical option for the near and midterm future according to several studies, although also critical comments concerning costs and biomass supply limitations have been made regarding this route.

In different future global primary energy demand scenarios, some of which are shown in Table 1.2, biomass plays a significant role. A substantial part is expected to be covered by liquid transportation fuels, such as ethanol and biodiesel, increasingly from nonedible biomass parts.

A variety of biomass sources can be converted to enable energy supply. These can be divided into six categories, based on their properties (Khan et al., 2009):

1. Forestry products, waste, and residues, which can be subdivided into:
   - Stem-wood logs and wood chips
   - Primary forestry residues:
     - Logging residues
     - Stumps

- Secondary forestry residues:
  - Wood processing by-products and residues
  - Bark, cutter chips, and sawdust
- Tertiary residues:
  - Demolition wood (e.g., from furniture)

2. Agricultural residues and wastes (also called herbaceous species), which are subdivided into three categories:
   - Primary (or direct harvest-related) residues, e.g., straw and vineyard residues
   - Secondary residues, which are generated after processing harvested material:
     - Bagasse (residue from sugar production from sugarcane)
     - Molasses and vinasse
     - Nutshells
     - Press cakes or pulp (from, e.g., olive and other vegetable oil processing)
     - Rice husks
   - Tertiary residues: manures from (domesticated) animals, such as chickens, cows, and pigs—dung and litter

3. Industrial and municipal organic wastes, e.g.:
   - Biodegradable part of municipal solid waste
   - Biogenic part of refuse-derived fuel

4. Derivatives, e.g.:
   - Residues from the food processing industry
   - Waste from the pulp and paper industry ("black liquor")

5. Aquatic species, namely:
   - Microalgae
   - Macroalgae (seaweeds)

6. Energy crops, which are grown with the aim to supply energy carriers, e.g.:
   - Sugar-producing crops: sugar beet, sugarcane, and sweet sorghum
   - Crops rich in starch: barley, cassava, corn, potato, and rye
   - Vegetable oil-containing crops: *Jatropha*, palm oil, rapeseed, soy, and sunflower
   - Fast-growing reed and grass plants, such as hemp, kenaf, and miscanthus (these are sometimes called "energy plants")
   - Short rotation wood (e.g., eucalyptus, poplar, willow)

Slade et al. (2011) have published an overview of many studies regarding the global availability of biomass in the near and further future. A summary of the findings is schematically illustrated in Figure 1.9. This figure shows that biomass can substantially contribute to primary energy supply, but the extent of the relative share depends on factors such as population growth, diet development and its associated meat-producing farming practice development, sustainable agriculture and forestry

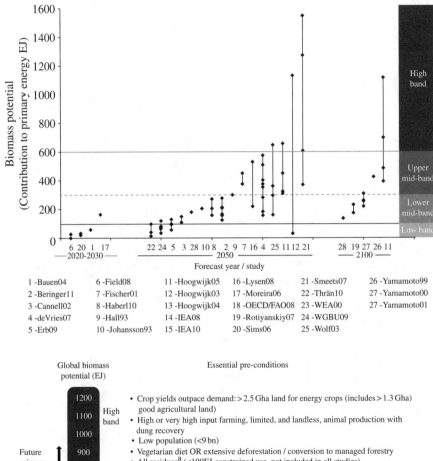

FIGURE 1.9 Biomass potential as primary energy source. (Source: Reproduced with permission from Slade et al. (2011). © Imperial College; the numbers below the top figure are references that can be found in the report.)

practice improvements, and residue usage. The study, though, does not point toward substantially opening up large nonland energy supply schemes based on the cultivation of (macro)algae, which is currently seriously considered.

1.5   A BRIEF INTRODUCTION TO MULTIPRODUCT BIOMASS CONVERSION TECHNIQUES

Biomass is a complex, heterogeneous, and versatile fuel source (see Chapter 2), so not surprisingly, the technology chain from source (Part I of this book) to multiple possible end uses (Part IV of this book) is quite complicated as can been seen in Figure 1.10. This integrated schematic also reflects the topics related to conversion technologies dealt with in this book (Part III with fundamentals of chemical engineering and process design dealt with in Part II).

The research, development, deployment, and implementation in society of bioenergy solutions thus comprise a broad, complicated, and challenging working field and are therefore fascinating to work on.

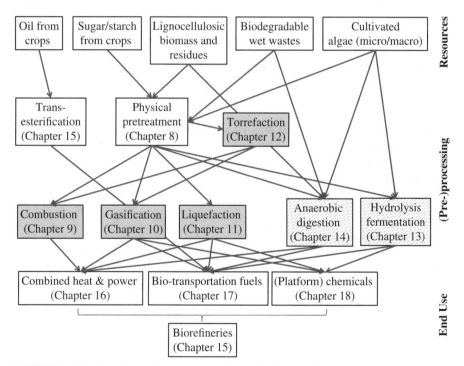

FIGURE 1.10   Overview of source to end use of biomass for energy supply; the dark gray boxes represent thermochemical conversion technologies, and the light gray boxes (bio)chemical conversion technologies.

CHAPTER SUMMARY AND STUDY GUIDE

This chapter describes the global energy situation in terms of expected population development, associated demand for energy supply, and implications of energy supply schemes on the environmental development. A major issue in this respect is climate change, which has been recognized to be partly induced by man's activities involving utilization of fossil fuels. The GHG effect is described and ways to mitigate its extent. Ways to establish sustainable development include energy savings, utilization of renewable energy sources and clean fossil, and nuclear energy usage. Regarding the second aspect, bioenergy as part of a sustainable energy mix is dealt with in this book. At present, biomass is again valued as an important contributor to secure and clean energy supply and is expected to form a substantial part of the (near) future energy mix. Types of biomass sources and their potential are discussed. In addition, aspects of economy, social context, and environmental impact of biomass for energy supply are illustrated and discussed.

KEY CONCEPTS

Primary energy demand, different scenarios
Population development
Relevant emissions from energy conversion processes
Global warming
Greenhouse gas effect
GHG emission mitigation strategies
Sustainability (economics, social aspects, and environmental impact) criteria
Trias Energetica
Biomass availability
Biomass source types
Conversion processes

SHORT-ANSWER QUESTIONS

1.1 How can modern biomass-to-energy conversion technologies mitigate poverty as compared to traditional biomass processing (heating/cooking)?

1.2 Which factors limit the RP ratio as a factor to predict how long a fossil energy will be available?

1.3 Explain in your own words the greenhouse effect and the human contribution to it.

1.4 Why is CO_2 the most important anthropogenic source of all greenhouse gases?

1.5 Explain why naturally occurring fires to some extent contribute to carbon sequestration.

1.6 What are the main sources of methane emissions? What can be done to reduce these?

1.7 What are the main sources of N_2O emissions? What can be done to reduce these?

1.8 In which ways can mankind adapt to climate change impacts, and which roles can biomass play in such adaptations?

1.9 Are there also natural sources that lead to acidification of water?

1.10 Show, by means of a chemical reaction equation, how marble is attacked chemically by acid rain caused by SO_x emissions.

1.11 Particulate matter (PM) emitted by combustion processes is of concern to both health and climate. Explain why.

1.12 Explain the effects of direct land use change (dLUC) and indirect land use change (iLUC) on the emission of GHG compounds. Give an example of a GHG emission increase scenario for both forms of LUC.

1.13 Would the use of biomass residues that otherwise would be considered as waste lead to LUC or changes in GHG emissions?

1.14 What are the differences between two policy strategies to mitigate CO_2 emissions, Clean Development Mechanism (CDM) and Joint Implementation (JI)?

1.15 Which mechanism (CDM or JI) do you think would be more effective in GHG emission reduction?

1.16 In literature, biomass is often referred to as first-, second-, and third-generation biomass. What is meant by these terms? Can you identify some biomass types for each of these generations?

1.17 What is the phenomenon of eutrophication; is there a risk of this when growing biomass for energy supply?

1.18 Why is maintaining biodiversity so important when planning crop implementation for bioenergy production?

1.19 There are multiple ways to classify biomass into different categories; compare some of these and mention pros and cons of such approaches.

1.20 How would you categorize cotton used for energy conversion? Is it wise to use this plant species?

1.21 Miscanthus is a fast-growing energy crop species; it is characterized as a "C4 plant"; what is meant by this term and what distinguishes it from "C3 plants?"

1.22 Figure 1.9 shows the biomass potential for primary energy supply on a global scale. Why do the reported magnitudes among the references differ (even if they are given by the same authors)?

1.23 What should be done in order to decrease biomass feedstock costs?

1.24   Would it be possible to realize a net removal of CO_2 from the atmosphere using biomass conversion processes?

PROBLEMS

1.1   The current primary energy use is approximately 12 GTOE (Gigatonnes of oil equivalent); what is this expressed in EJ?

1.2   From Figure 1.1, estimate the energy use per capita in the world in 2030 relative to that in 2010. How does this compare to the values of China and India, respectively?

1.3   Consider that man emits 7 $Gt \cdot year^{-1}$ carbon into the atmosphere in the form of CO_2. Assume the atmosphere is 20 km high. What will be the yearly increase of the CO_2 mass fraction in the atmosphere if no additional sorption in the seas and land can be accommodated?

1.4   If energy crops are going to be used to supply 100 EJ of global primary energy input and given that an energy crop has a yield of 15 $t \cdot ha^{-1}$ with an energy content of 15 $MJ \cdot kg^{-1}$ (as received basis), how many hectares (ha) are needed as cropland? Comment on the result of your calculation. What might be complications in such a scenario?

1.5   In 1977, Marchetti (1977) introduced a type of analysis of fuel source market penetration for the global energy market, characterized by the following equation (with F being the fraction of the market penetrated):

$$\frac{1}{F} \cdot \frac{dF}{dt} = \alpha \cdot (1 - F)$$

a. What is F as a function of t?
b. Comment on the predictions made by this researcher concerning the current global energy source mix.

1.6   Regarding resource availability, Hubbert's "peak oil" development is a well-known model (Brandt, 2007). Look up in the literature what it comprises and provide the expression. Is this model always valid for natural resources?

PROJECTS

P1.1   Investigate the literature about the usage of biomass for energy supply to manufacturing processes before the Industrial Revolution. Concentrate, e.g., on the Chinese, Greek, and Roman Empire or on Europe in medieval times.

P1.2   **a.** Make an overview for your country of available biomass types (quantity and quality in terms of heating value) and their prospected prices. Make a graphic representation of your findings.

      b. Compare your findings with the potential of solar and wind energy. What can you conclude? What could be reasons to use biomass for energy supply in the context of the situation in your country?

P1.3   Compare the world's current need for food supply and energy supply on an energy content basis.

P1.4   Based on a literature survey, make a critical comparison between different energy demand and supply scenarios for the future; take 2050 as the reference time scale.

P1.5   Review the literature on primary energy consumption expectations, now distinguishing the world regions.

P1.6   If you have access to analytical equipment in your lab, with, e.g., an infrared spectrophotometer (online, gas specific) or an FTIR spectrophotometer, you can determine the CO_2 concentration in the air and monitor the cycle in a year's time. What is the background of this yearly variation?

INTERNET REFERENCES

tinyurl.com/7hlmqxn
http://www.bp.com/liveassets/bp_internet/globalbp/STAGING/global_assets/downloads/O/
   2012_2030_energy_outlook_booklet.pdf

tinyurl.com/35t7t7
http://unfccc.int/essential_background/convention/background/items/1353.php

tinyurl.com/48sqk6o
http://en.wikipedia.org/wiki/Particulate_matter

tinyurl.com/6reyxlb
http://www.exxonmobil.com/Corporate/Files/news_pub_eo2012.pdf

tinyurl.com/5aamp8
http://www.shell.com/global/future-energy/scenarios.html

REFERENCES

Arrhenius S. On the influence of carbonic acid in the air upon the temperature of the ground. Philos Mag J Sci 1896;41:237–276.

Bauen A, Berndes G, Junginger M, Londo M, Vuille F. Bioenergy—a sustainable and reliable energy source: a review of status and prospects. Rotorua: International Energy Agency; 2009.

Berndes G, Bird N, Cowie A. Bioenergy, land use change and climate change mitigation—background technical report. Rotorua: IEA-Bioenergy; 2011.

BP. June 2013. BP statistical review of world energy. Available at www.bp.com/statisticalreview. Accessed May 16, 2014.

Brandt AR. Testing hubbert. Energy Policy 2007;35:3074–3088.

Brundtland GH. *Report of the World Commission on Environment and Development: "Our Common Future."* New York: United Nations; 1987.

Cramer J, Wissema E, de Bruijne M, Lammers E, Dijk D, Jager H. Testing framework for sustainable biomass (final report). Utrecht: Project Group Sustainable Production of Biomass; 2007.

Diouf J. *The State of Food and Agriculture: Biofuels: Prospects, Risks and Opportunities.* Rome: FAO; 2008. p 128.

Falkenmark M. Meeting water requirements of an expanding world. Philos Trans R Soc Lond 1998;352:929–936.

Falkowski P, Scholes RJ, Boyle E, Canadell J, Canfield D, Elser J, Gruber N, Hibbard K, Högberg P, Linder S, Mackenzie FT, Moore B 3rd, Pedersen T, Rosenthal Y, Seitzinger S, Smetacek V, Steffen W. The global carbon cycle: a test of our knowledge of earth as a system. Science 2000;290:291–296.

Giuntoli J. Characterization of 2nd generation biomass under thermal conversion and the fate of nitrogen [PhD thesis]. Delft: Delft University of Technology; 2010.

Guinée JB, Gorrée M, Heijungs R, Huppes G, Kleijn R, de Koning A, van Oers L, Wegener Sleeswijk A, Suh S, Udo de Haes HA, de Bruijn H, van Duin R, Huijbregts MAJ. *Handbook on Life Cycle Assessment. Operational Guide to the ISO Standards.* Dordrecht: Kluwer Academic Publishers; 2002.

Hubbert MK. Energy from fossil fuels. Science 1949;109(2):103–109.

Van der Hoeven M. World energy outlook. Paris: International Energy Agency; 2011. p 659.

Intergovernmental Panel on Climate Change (IPCC). Climate change 2007: synthesis report. Contribution of Working Groups I, II and III to the Fourth Assessment Report of the IPCC. Core Writing Team, Pachauri RK, Reisinger A, editors. Geneva: IPCC; 2007. p 104.

Johansson TB, Kelly H, Reddy AKN, Williams RH. *Renewable Energy.* Washington, DC: Island Press; 1993.

Kaltschmitt M, Hartmann H. *Energie aus Biomasse—Grundlagen, Techniken und Verfahren.* Berlin: Springer-Verlag; 2001.

Kampman B, Bergsma G, Schepers B, Croezen H, Fritsche UR, Henneberg K, Huenecke K, Molenaar JW, Kessler JJ, Slingerland S, van der Linde C. *BUBE: Better Use of Biomass for Energy.* Delft/Darmstadt: CE Delft/Öko-Institut; 2010.

Karekezi S, Lata K, Teixeira S. Traditional biomass energy: improving its use and moving to modern energy use. International Conference for Renewable Energies; Bonn, Germany; 2004.

Khan AA, De Jong W, Jansens PJ, Spliethoff H. Biomass combustion in fluidized bed boilers: potential problems and remedies. Fuel Process Technol 2009;90:21–50.

Larssen T, Lydersen E, Tang DG, He Y, Gao J, Liu H, Duan L, Seip HM, Vogt RD, Mulder J, Shao M, Wang Y, Shang H, Zhang X, Solberg S, Aas W, Okland T, Eilertsen O, Angell V, Liu Q, Zhao D, Xiang R, Xiao J, Luo J. Acid rain in China. Environ Sci Technol 2006;40 (2):418–425.

Likens GE, Bormann FH. Acid rain: a serious regional environmental problem. Science 1974;184:1176–1179.

Lysen EH. The Trias Energetica: solar energy strategies for developing countries. In: Goetzberger A, Luther J, editors. Eurosun Conference 1996; Freiburg: DGS Sonnenenergie Verlag-GmbH; 1996.

Marchetti C. Primary energy substitution models: on the interaction between energy and society. Technol Forecast Soc Change 1977;10(4):345–356.

McKay DJC. *Sustainable Energy—Without the Hot Air*. Cambridge: UIT Cambridge Ltd; 2009.

Meadows DH, Meadows DL, Randers J, Behrens III WW. The limits to growth. A Report to the Club of Rome. New York: Universe Books; 1972.

Nag A. *Biosystems Engineering*. New York: McGraw-Hill; 2009.

OECD/IEA. *Energy Technology Perspectives 2012—Pathways to a Clean Energy System*. Paris: International Energy Agency; 2012.

Raven PH, Johnson GB, Losos JB, Singer SR. *Biology*. 7th ed. Boston: McGraw-Hill; 2005.

Schimel DS. Terrestrial ecosystems and the carbon cycle. Glob Change Biol 1995;1(1):77–91.

Sims REH. *The Brilliance of Bioenergy: In Business and in Practice*. London: James & James; 2002.

Slade R, Saunders R, Gross R, Bauen A. *Energy from Biomass: The Size of the Global Resource—An Assessment of the Evidence that Biomass Can Make a Major Contribution to Future Global Energy Supply*. London: Imperial College Centre for Energy Policy and Technology and UK Energy Research Centre; 2011.

Spliethoff H. *Power Generation from Solid Fuels*. Berlin-Heidelberg: Springer Verlag; 2009.

Tanaka N. *Technology Roadmap—Biofuels for Transport*. Paris: International Energy Agency; 2011.

United Nations Framework Convention on Climate Change (UNFCCC). Kyoto Protocol by the 3rd Conference of the Parties; Kyoto, Japan. New York: United Nations; 1997. p 60.

World Coal Institute. The coal resource: a comprehensive overview of coal. London: World Coal Institute; 2005. p 44.

Wright RF, Schindler DW. Interaction of acid rain and global changes: effects on terrestrial and aquatic ecosystems. Water Air Soil Pollut 1995;85:89–99.

Wrisberg N, Udo de Haes HA, Triebswetter U, Eder P, Clift R, editors. *Analytical Tools for Environmental Design and Management in a Systems Perspective: The Combined Use of Analytical Tools*. Dordrecht: Kluwer Academic Publishers; 2002.

Zanchi G, Pena N, Bird N. Is woody bioenergy carbon neutral? A comparative assessment of emissions from consumption of woody bioenergy and fossil fuel. GCB Bioenergy 2012;4:761–772.

2

BIOMASS COMPOSITION, PROPERTIES, AND CHARACTERIZATION

WIEBREN DE JONG

Department of Process and Energy, Energy Technology Section, Faculty of Mechanical, Maritime and Materials Engineering, Delft University of Technology, Delft, the Netherlands

ACRONYMS

ar	as received basis
ADP	adenosine diphosphate
ATP	adenosine triphosphate
BDL	below the lower detection limit
DDGS	dried distiller's grains and solubles
daf	dry and ash-free basis
db	dry basis
DNA	deoxyribonucleic acid
FAME	fatty acid methyl ester
FC	fixed carbon
GCV	gross calorific value
HHV	higher heating value
LHV	lower heating value
MBM	meat and bone meal

Biomass as a Sustainable Energy Source for the Future: Fundamentals of Conversion Processes,
First Edition. Edited by Wiebren de Jong and J. Ruud van Ommen.
© 2015 American Institute of Chemical Engineers, Inc. Published 2015 by John Wiley & Sons, Inc.

NCV   net calorific value
ND   not determined
PCDD/F   polychlorinated dibenzo-p-dioxin/furan
RNA   ribonucleic acid
TGA   thermogravimetric analysis (or analyzer)
VM   volatile matter
XRD   X-ray diffraction
XRF   X-ray fluorescence

SYMBOLS

a_v specific surface area [$m^2 \cdot m^{-3}$]
HHV   higher heating value [$J \cdot kg^{-1}$]
LHV   lower heating value [$J \cdot kg^{-1}$]
MW   molecular weight [$kg \cdot kmol^{-1}$]
Y   mass fraction [–]

2.1 PHYSICOCHEMICAL PROPERTIES

The entire processing track of biomass as renewable energy source (its supply chain, pretreatment, conversion techniques, and emissions) is influenced by the type of biomass and its physical characteristics and chemical composition. In the past few decennia, comprehensive studies of the physical characteristics and chemical composition of biomass fuels have been carried out, and these are still being continued to obtain more insight into the reactive behavior of these fuels under different processing conditions.

Table 2.1 presents an overview of the most important physical properties of solid biomass (and derived solid products) that are relevant for its behavior in the chain from harvesting to conversion processes. There is a great variation among the property values for different biomasses and their derived products. Some properties are directly related to the biomass type (e.g., particle density and porosity), but fuel preparation changes such characteristics as particle size and shape. A more in-depth discussion on pretreatment of biomass is presented in Chapter 8.

The organic part of plants mainly consists of carbon, hydrogen, and oxygen, with a minor portion of nitrogen, sulfur, and phosphorus, which are vital for the plant's metabolism and physiology (Jenkins et al., 1998). Table 2.2 shows an overview of the elements composing biomass and their major effects in the bioenergy conversion chain.

Figure 2.1 shows a so-called "Van Krevelen" diagram that presents the atomic H/C ratio on the vertical axis and the O/C ratio on the horizontal axis. The diagram maps different coal types, with the anthracites near the origin and young biomass (constituents) in the upper right corner. Biomass is characterized by a relatively large content

TABLE 2.1   Physical properties of solid biomass and their possible effects in processing

Physical property	Effect
Bulk density	Logistics (storage, transportation, handling)
Electrical conductivity	Microwave processing, particle cleaning via electrostatic precipitation (ash), or fine particulate matter repulsion by plants, e.g., *sea buckthorn* (tinyurl.com/luyld2g)
Hygroscopy	Logistics (storage, transportation, handling)
Particle density[a]	Conversion processes (e.g., segregation)
Particle porosity[a]	Formation of fines in processing, intraparticle heat and mass transfer impacted and so conversion
Particle shape (distribution)	Storage behavior (dimension/shape of a heap, bridging in bunkers, self-ignition), transportation (conveying) characteristics, mass and heat transfer behavior in conversion processes
Particle size (distribution)	Storage behavior (dimension/shape of a heap, bridging in bunkers, self-ignition), transportation (conveying) characteristics, mass and heat transfer behavior in conversion processes
Thermal conductivity	Physicochemical processing (heat transfer)

[a] These properties are partly linked to each other.

TABLE 2.2   Elements in solid biofuels and their possible main effects in energy conversion processes

Element	Effect
C, carbon	Heating value, possible emission of CO
H, hydrogen	Heating value
O, oxygen	Heating value (negatively impacting)
N, nitrogen	Emission of NO, NO_2 (together termed NO_x), and N_2O
Cl, chlorine	Emission of HCl and polychlorinated dibenzo-p-dioxin/furan (PCDD/F), causing corrosion and catalyst poisoning
S, sulfur	Emission of SO_2, SO_3 (both named SO_x), causing corrosion and catalyst poisoning
F, fluor	Emission of HF, causing corrosion
K, potassium	Corrosion, ash melting, ash utilization, aerosol formation
Na, sodium	Corrosion, ash melting, ash utilization, aerosol formation
Mg, magnesium	Ash melting, ash utilization, deposits formation
Ca, calcium	Ash melting, ash utilization, deposits formation
P, phosphorus	Ash utilization, deposits formation
Trace elements	Emissions, ash utilization, aerosol formation

of oxygen, which decreases with aging of the fuel. Coal and biomass have similar ancestors, and fossil fuels in general could be marked as very old biomass. The process of coalification starts when plant matter dies and soil covers it; then long-term (simultaneous) effects of heat, pressure, and microbial action cause the material to be deprived from oxygen and hydrogen so that it is gradually enriched in carbon.

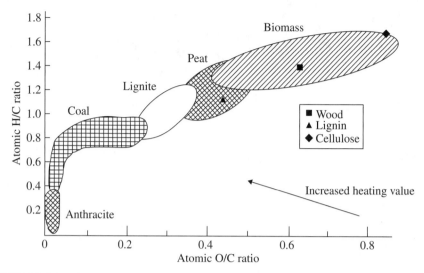

FIGURE 2.1 Van Krevelen diagram characterizing the coalification process. (Source: Reproduced with permission from Prins et al. (2007). © Elsevier.)

This process first forms peat, subsequently lignites and subbituminous coals, then bituminous coals, and finally anthracites, the most carbon-rich end stage. During this coalification process, the heating value of the organic material gradually increases.

Table 2.3 presents an overview of fuels from young biomass to highly coalified anthracite based on their proximate analysis, which determines the percentages of moisture, volatile matter (VM), fixed carbon (FC), and ash, and ultimate analysis (or elemental analysis), which determines the main elemental composition (C, H, O, N). These analysis methods are discussed in Section 2.5. The data show that biogenic sources are already heterogeneous with respect to the main elements in the fuel. Some biomasses, such as sewage sludge and meat and bone meal (MBM), have lower oxygen contents on a dry fuel basis mainly due to their high ash contents.

As the table shows, there is a wide variety in the chemical composition among biomass species. The natural biomass composition is impacted by the following main aspects (see also Vassilev et al., 2010 and references therein):

1. Biomass type, species, or specific plant part; growth characteristics; uptake of certain compounds from the environment (air/water/soil); and their transport to and deposition in dedicated parts of the plant
2. External conditions of biomass growth with particular roles of:
   a. Sunlight
   b. Climate, including seasonal fluctuations
   c. Soil type
   d. Water availability
   e. pH

TABLE 2.3 Typical proximate and elemental analyses of various biomass fuels, biomass residues, and coals

Fuel	Proximate Analysis (wt% as received (ar))				Ultimate Analysis (wt% dry and ash-free (daf))					
	Moisture	VM	FC	Ash	C	H	O	N	S	Cl
Wood pellets, clean (Khan et al., 2009)	4.9	80.4	14.5	0.2	45.6	6.6	47.8	BDL	BDL	BDL
Wood pellets, demolition (Khan et al., 2009)	9.1	69.6	19.7	1.7	51.2	7.1	40.6	1.0	BDL	0.1
Wheat straw (Arvelakis and Koukios, 2002)	8.5	69.5	15.0	7.0	47.3	5.5	45.3	0.9	0.5	0.5
Sunflower pellets (Khan et al., 2009)	11.2	65.2	19.5	4.1	52.1	6.1	41.0	0.6	0.1	0.1
Olive cake pellets (Khan et al., 2009)	11.9	64.2	15.7	8.2	52.7	6.3	38.9	1.6	0.1	0.4
Pepper plant residue (Khan et al., 2009)	6.5	60.5	19.5	13.5	42.3	5.0	48.9	3.1	0.6	0.1
Greenhouse residue[a] (Khan et al., 2009)	2.5	61.0	5.5	31.0	70.8	11.1	16.4	1.5	BDL	0.2
Sewage sludge, dried (Nilsson et al., 2012)	8.7	47.2	4.7	39.4	54.3	7.7	27.4	8.4	2.2	ND
MBM (Tortosa Masia et al., 2007)	2.5	61.7	12.4	23.4	56.6	8.0	20.6	12.0	1.7	1.1
Microalgae (tinyurl.com/km7dh8h)	5.2	77.5	14.9	2.4	54.1	7.4	29.6	8.2	0.5	0.2
Macroalgae (seaweed) (Zhou et al., 2010)	8.0	42.4	19.5	30.1	41.1	7.5	46.2	5.2	6.3[c]	21.7[c]
Peat (young surface), dry (Kurkela et al., 1995)	12.5[b]	63.1	22	2.4	52.6	5.8	40.6	0.9	0.1	0.0
Brown coal, dried (De Jong et al., 2003)	15.6	44.1	36.0	4.3	56.3	5.0	37.6	0.6	0.4	0.1
Bituminous coal (Kurkela et al., 1995)	5.5	30.1	56.6	7.8	82.3	5.1	10.3	1.4	0.8	0.1
Anthracite (tinyurl.com/km7dh8h)	1.9	7.6	87.7	2.8	91.6	3.5	2.4	1.6	0.8	0.1

BDL, below the lower detection limit; ND, not determined.

[a] Including plastics.

[b] Average value.

[c] Wt% in dry ash (S and Cl contents can be high (Mageswaran et al., 1985), report values 0.7 < S < 3.9 wt% dry basis (db) and 1.3 < Cl < 7.9 wt% db).

    f. Nutrient availability and fertilization regime

    g. Pesticide dosing regime

    h. Location (including distance from polluting sites as cities, highways, etc.)

3. Age, harvesting season, harvesting collection technology, pickup of extraneous material (e.g., dust, dirt, soil), transport, handling, and storage

4. Blending strategies

In the following sections, more details concerning the molecular structure and minor composing elements are dealt with. Biomass in general is composed of mainly organic matter but in conjunction with a smaller fraction of inorganic compounds containing a variety of intimately associated phases or minerals with different origins. These have formed by natural processes, both authigenic (formed in biomass) and detrital (formed outside biomass but fixed in/on biomass), as well as by anthropogenic (formed in or outside biomass and fixed in/on biomass) processes. In this respect, one can discriminate between presyngenesis, syngenesis, epigenesis, and postepigenesis (see Figure 2.2). The phase composition can be summarized as follows (Vassilev et al., 2010):

1. Organic matter

    a. Solid, noncrystalline—structural constituents, e.g., (hemi)cellulose, lignin, and extractives

    b. Solid, crystalline—organic (combined with inorganic) minerals such as Ca–Mg–K–Na oxalates

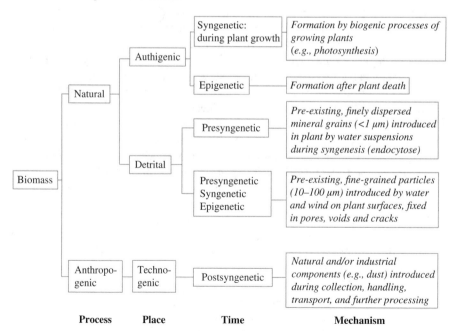

FIGURE 2.2 Origin of phases in biomass, classified based on process, place, time, and mechanism of formation (based on the table presented by Vassilev et al., 2010).

2. Inorganic matter
   a. Solid, crystalline—minerals, e.g., consisting of phosphates, carbonates, silicates, chlorides, sulfates, oxyhydroxides, and nitrates
   b. Solid, semicrystalline—poorly crystallized mineraloids of certain silicates, phosphates, hydroxides, etc.
   c. Solid, amorphous phases that are formed as glasses, silicates, etc.
3. Fluid matter—fluid, liquid, and gas (moisture, gas, and gas/liquid inclusions associated with both organic and inorganic matter)

2.2 MAIN STRUCTURAL ORGANIC CONSTITUENTS

The composition of biomass fuels is associated with a multitude of physical forms, but for species that belong to the plantcategory, the main structural cell wall components are nearly always cellulose, hemicellulose and lignin (Klass, 1998). Other major bio-organic polymer structures that are found in nature are starch (e.g., in maize, banana, etc.) and chitosan (e.g., shrimp shells).

2.2.1 Cellulose

Cellulose is a homopolysaccharide $(C_6H_{10}O_5)_n$ of glucose C_6 sugar (hexosan) units that constitute the main part of the cell walls of plants and synthesized in nature with approximately 10^{11} t·year^{-1} (Kamm et al., 2006); it is the world's most common organic biopolymer. It is naturally incorporated in fibrous products such as cotton and kapok, and further processed from plants, it forms the raw material for many manufactured goods, such as paper, paperboard, rayon, cellophane, and celluloid, a product that was used for making photographic and movie film material until the mid-1930s (Simon et al., 1998). Most biomass material consists of about 40–50 wt% cellulose on a dry fuel basis. The cellulose structure consists of up to 14,000 linearly coupled D-glucopyranoside units connected by β-glycosidic linkages in a 1 : 4 fashion (Mohan et al., 2006; Sjöström, 1993; Solomons, 1984). Figure 2.3 depicts this molecular structure, with the repeating unit consisting of two glucose units called "cellobiose." Thus, it is a high-molecular-weight species, with a molecular mass typically of the order of 10^6 kg·kmol^{-1}. The β-glycosidic linkages are commonly known as weak bonds that are easily broken and help initiate the degradation of the cellulose molecule.

The structure shows that −OH groups form both inter- and intramolecular hydrogen bonds, which cause different polymer chains to arrange themselves in parallel configurations to form a crystalline supermolecular structure. This resulting structure makes cellulose completely insoluble in ordinary aqueous solutions. The linear cellulose chains form bundles that tend to twist so as to make ribbonlike microfibrilic anisotropic (highly ordered) structures that are highly oriented in the cell wall structure (Balat et al., 2009). This structure ensures a relatively high stability against thermal and biochemical degradation compared to hemicellulose and starch. These degradation processes are dealt with in depth in Part III of this book.

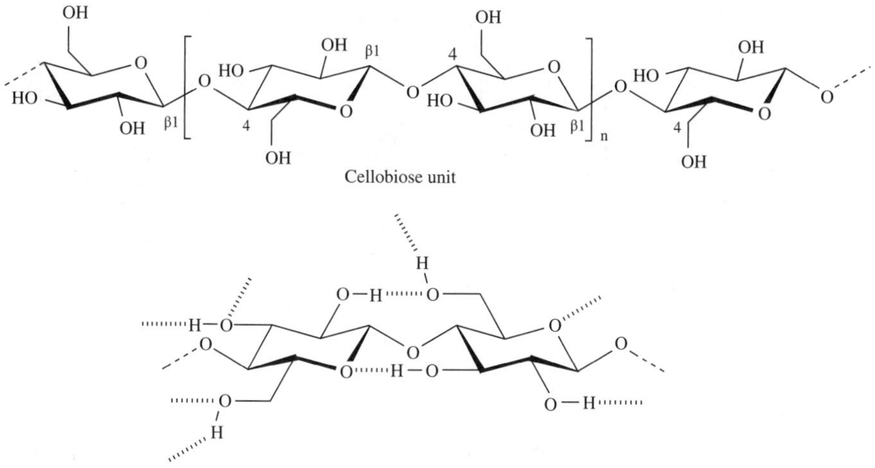

FIGURE 2.3 Cellulose—structure (top) with both inter- and intramolecular hydrogen bonding bridges (bottom).

2.2.2 Hemicellulose

Hemicelluloses (the word was derived from the Greek language hemisys = half) are heteropolysaccharides, consisting of C_5 and C_6 sugars (hexosans, pentosans) that are associated with the cellulose, and they are found in the cell wall regions of plants. It is the second most abundant biopolymer species in plant biomass, accounting for about 25–35% of dry wood, 28% of softwoods, and 35% of hardwoods (Rowell, 1984). Hemicelluloses serve as a frame cementing material in plant cell walls, holding together the cellulose micelles and fibers. This is why they are called "structural carbohydrates," together with cellulose and the less abundantly present pectin, as opposed to starch and saccharose, which serve as "storage" molecules. Hemicelluloses consist of relatively small molecules containing 50–200 monosaccharide residues and contain linkages different from those in cellulose, e.g., oligosaccharide side groups (branches) attached to the polysaccharide backbone and acetyl groups. They are classified by the kind of main monosaccharide in their structure (e.g., xylose, arabinose, mannose, and galactose). The types are D-xylans, L-arabino-D-xylans, D-mannans, D-galacto-D-mannans, D-gluco-D-mannans, and L-arabino-D-galactans, which can be grouped into xylans, glucomannans, and arabinogalactans. The glucomannans in wood have small molecular weights and are primarily linear in structure. Arabinogalactans are highly branched, water-soluble polysaccharides and are diverse regarding their physicochemical properties. The distribution of hemicellulose in plant tissue varies with different species.

Figure 2.4 shows an example of a hemicellulose structure composed of different sugars.

Hemicelluloses are easier to decompose than cellulose, both thermochemically and biochemically, due to the less stable intramolecular linkages.

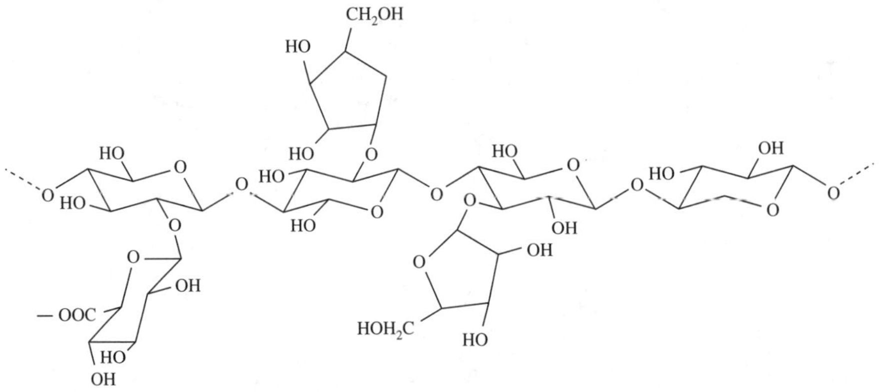

FIGURE 2.4 Hemicellulose example structure. Depicted is glucuronoarabinoxylan β-(1,4)-D-xylan substituted by glucuronic acid at the O-2 and by arabinose at the O-2 and O-3 position.

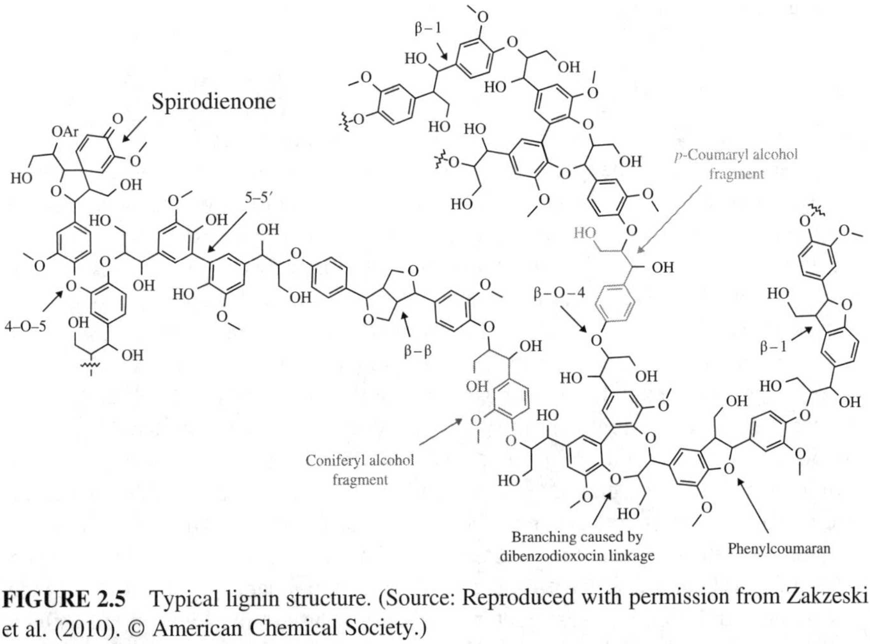

FIGURE 2.5 Typical lignin structure. (Source: Reproduced with permission from Zakzeski et al. (2010). © American Chemical Society.)

2.2.3 Lignin

Lignin (the word was derived from the Latin word lignum = wood) is an amorphous polymer consisting of a complex and heterogeneous three-dimensional network of aromatic substructures that provides coherence and toughness and, together with cellulose, forms the woody cell walls of plants and the cementing material between them. For a schematic of a representative structure, see Figure 2.5. Most biomass consists of 20–30 wt% lignin, which is deposited between the spaces of the cells, a growth-related process called lignification. The three classes of plant lignins are gymnosperm (soft-wood), angiosperm (hardwood), and grass lignins. Lignin is primarily aromatic in nature, which is obvious from the chemical structure. Its basic molecular units are linked mostly by ether bonds, but carbon-to-carbon linkages exist as well. Different functional side groups can be distinguished, such as hydroxyl, methoxyl, and carbonyl groups. Covalent bonds exist with both hemicelluloses and cellulose (Jacobsen and Wyman, 2000), rendering improved mechanical strength to the plant.

Lignin has multiple functions in plant biomass. Apart from its contribution to the mechanical strength, it gives protection to plants against attacks from, e.g., bacteria, as it is normally not digested by animal enzymes. Moreover, it stimulates the circulation of water in the vascular system of plants. Finally, it is the least degradable part of dead plants, making it the most important component of humus.

Lignin is relatively stable in thermal conversion (pyrolysis), and it has a larger higher heating value (HHV) than carbohydrates. Lignin is produced in large quantities as by-product of the pulp and paper industry, which needs cellulose fibers with a low lignin content, and it mostly serves as heat supply fuel. It can also be used as substrate for the production of high-added-value products, such as vanillin, phenol, and derived phenolic resins. Lignin also plays a role in rubber production and antioxidant generation (Pye, 2006). Chapter 15 concerning biorefineries presents some of the possibilities of lignin valorization.

The aforementioned three major biopolymers in plants combine to form so-called tracheids, which are elongated cells in vascular plants that facilitate transport of water. Figure 2.6 shows a schematic illustration of these tracheids' structure indicating the entanglement in the secondary walls of lignin, hemicelluloses, and cellulose.

To summarize the relative presence of the three main biopolymers in plant species, an example of the composition of biomass in terms of main organic constituents is given in Table 2.4.

2.2.4 Chitin and Chitosan

Chitin is a naturally occurring biopolymer, which protects insects and sea life species and for this purpose is incorporated in the form of enclosing skeletons. After cellulose, chitin is the most abundant polysaccharide in nonplant life nature and is primarily present in the exoskeletons of crustaceans (such as crabs, shrimp, lobsters, etc.) and also in various insects, worms, fungi, and mushrooms in varying amount (Arcidiacono and Kaplan, 1992). Chitin in nature is usually associated with other components to reinforce the struc-ture, such as calcium carbonate ($CaCO_3$) and proteins. Chitosan is derived from chitin by deacetylation using a sodium hydroxide solution; the structures are shown in Figure 2.7.

2.3 MINOR ORGANIC CONSTITUENTS

Together with the major support structures described in Section 2.2, biomass contains other components, such as oils, fats (lipids), proteins, starch, and sugars, as well as a spectrum of organic extractives. These compounds are widely used as (ingredients of)

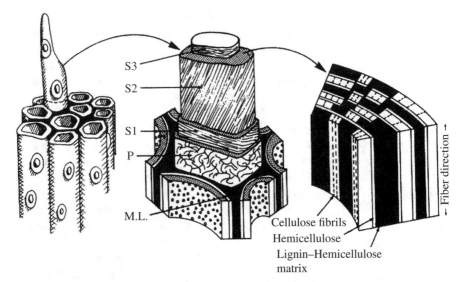

FIGURE 2.6 Schematic illustration of the molecular architecture of wood tissue, showing the relationship of contiguous cells (left), cutaway view of the cell wall layers (center), and one depiction (from Goring, 1977) of the relationship of the lignin, hemicelluloses, and cellulose in the secondary wall. Recent evidence suggests a more intimate admixture of the lignin and hemicelluloses than illustrated here (R. Atalla, personal communication.) The diameter of each cell is approximately 25 μm. S1–S3, secondary cell wall layers; P, primary wall; and M.L., middle lamella. (Source: Reproduced with permission from Kirk and Cullen (1998). © John Wiley & Sons.)

TABLE 2.4 Chemical composition of common lignocellulose feedstock (db)

Biomass type	Cellulose (wt%)	Hemicellulose-derived C_6 sugars (wt%)	Hemicellulose-derived C_5 sugars (wt%)	Lignin (wt%)
Softwood	40–48	12–15	7–10	26–31
Hardwood	30–43	2–5	17–25	20–25
Cereal straw	38–40	2–5	17–21	6–21
Maize straw	35–41	2	15–28	10–17
Rape straw	38–41	—	17–22	19–22
Recovered paper	50–70	—	6–15	15–25

Adapted from Kamm et al. (2006).

food and animal feed (e.g., olive or sunflower oil, animal fats, and sucrose from sugarcane) but also as lubricants, rubbers, pharmaceuticals (e.g., antibiotics and vitamins synthesis), dyes, and even cosmetics (e.g., *Jojoba* oil) (Biermann et al., 2006; Kripp, 2006). They are also used as feedstocks for many first-generation biofuels based on vegetable oils. When associated with the use of the edible part of the plant only for energy carrier generation, the product is called a first-generation biofuel.

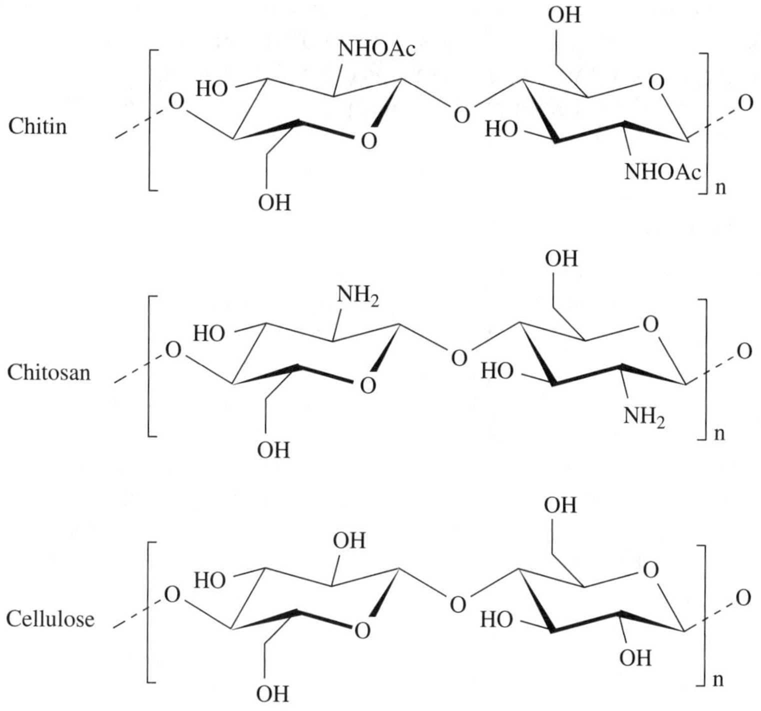

FIGURE 2.7 Molecular structures of chitin and chitosan in comparison with cellulose.

2.3.1 Oils and Fats

There are many different plants and animals that store oils and fats in their structures. These compounds are similar in structure, but their aggregation state is different; oils are liquid and fats are (partly) solid. The basic molecular structure is that of a glycerol backbone with ester bonds to three carboxylic acids (called "fatty acids"). The plant-derived oils can be used as a fuel in diesel engines, but they are viscous and cause specific problems (e.g., interaction with lubrication oils). Therefore, via the process of transesterification using methanol, resulting in fatty acid methyl esters (FAMEs), or ethanol, they are converted into biodiesel with improved properties, such as lower viscosity. Examples of oils that are used as raw materials are rapeseed, soybean, canola, sunflower, sesame, palm, and coconut oils. There is much debate about this way of utilizing vegetable oils as these can also be used for feeding humans and animals. The entire picture, though, is somewhat more complicated. *Jatropha* oil, e.g., is a specific case where the food/feed versus fuel aspect does not play a role, as the fruit of the *Jatropha curcas* plant cannot be used for human/animal consumption due to the presence of certain poisons (the contents of which vary so as to result in either a nontoxic or toxic fruit). The plant is cultivated in Central and South America, Southeast Asia, India, and Africa. It is easy to grow on widely differing and difficult soils, such as gravelly, sandy, and saline soils, that are not suitable for most other crops

(Koh and Ghazi, 2011). Moreover, the plant is useful for improving agricultural soils (root system) and acts as a drought-resistant, shade-rich hedge, thus protecting crops that are grown inside by preventing roaming cattle from entering. The hedge also prevents erosion and the species has the unique property of acting as natural "pesticide" protector.

2.3.2 Proteins

The main biomass structures, apart from chitin, in which nitrogen is incorporated, are proteins. These are biopolymers consisting of α-amino acids (20 primary species exist in nature) joined by peptide linkages. Table 2.5 shows the protein content of several relevant biomass types and biogenic residues. The table shows a broad spectrum of protein contents and related to this is the material's nitrogen content. When used as source for energy supply in thermochemical conversion processes, eventually, the high fuel-bound nitrogen content may lead to increased NO_x and/or N_2O emissions. In novel biorefinery concepts, the proteins may be retrieved from biomass sources and processed into higher-added-value products (food and feed supplements) before usage of the remaining residue for energy supply. More about this is discussed in Chapter 15.

2.3.3 Starch and Free Sugars

Apart from cellulose, another biopolymer that is composed of C_6 sugar (glucose) units is starch; the difference with cellulose is that its segments constitute α-glycosidic linkages, so that starch is a stereoisomer of cellulose. As a consequence of this structural

TABLE 2.5 Typical protein content of some biomass types

Biomass	Protein content (wt%)	References
Softwood	0.2–0.8	Leppälahti and Koljonen (1995)
Wheat straw	4[a]	Silva et al. (2011)
Corn stover	5	Dale and Kim (2006)
Olive oil cake	6	Ramachandran et al. (2007)
Needles and leaves	7–8	Leppälahti and Koljonen (1995)
Corn	10	Dale and Kim (2006)
Palm kernel cake	19	Ramachandran et al. (2007)
Alfalfa	20	Dale and Kim (2006)
Dried distiller's grains and solubles (DDGS)	25	Kim et al. (2008)
Rapeseed cake	36	Koutinas et al. (2007)
Cottonseed cake	40	Ramachandran et al. (2007)
Soybean	40	Dale and Kim (2006)
Chicken manure	47	El Boushy et al. (1985)
Soybean cake	48	Ramachandran et al. (2007)

[a] Depending on size fraction after cutting and sieving.

difference, starch is not a structure-building compound, but serves as an easily accessible storage sugar for certain plant species (e.g., corn). The bonds in starch are weaker than the cellulose intra- and intermolecular bonds, which makes starch much better digestible, e.g., by human enzymes. Two main molecules can be distinguished: the linear and helical amylase and the branched amylopectin. Starch generally consists of 10–20 wt% amylase and 80–90 wt% amylopectin (Solomons and Fryhle, 2004). Starch is usually stored in the form of tiny granules and is found in tubers (like maize, potatoes, wheat, and rice), roots, seeds (cereals), and fruits. Starch can be decomposed into sugar molecules using hydrolysis (discussed in Chapter 13). Conversion of these sugars into fuel-grade ethanol by fermentation using yeast already has been practiced on an industrial scale for some time; this is another example of a first-generation biofuel: a bioresource suitable for food or animal feed is converted into a fuel.

Some biomasses, in particular sugarcane and sugar beet, contain substantial amounts of free sugar (saccharose), which can be directly used for consumption or converted into first-generation bioethanol using fermentation. Nowadays, also other sugars—derived from hydrolysis of the polymeric carbohydrates—can be fermented to, e.g., ethanol. This topic is addressed in Chapter 13.

2.3.4   Other Organic Extractives

Apart from the aforementioned organic components, biomass contains a rich variety of minor organic molecules having very specific functions in the species' life. These comprise alkaloids, antioxidants, aromatic amines, chlorophyll, hormones, vitamins, natural dyes, and terpenes, to name a few. It is beyond the scope of this book to go into details concerning these organics. Kamm et al. (2006) in their book concerning biorefineries present a comprehensive overview of these compounds and their possible uses.

2.4   INORGANIC COMPOUNDS

The inorganic constituents of biomass are usually present in minor amounts and comprise elements that are essential for plant growth. Elements that are necessary for plants to complete their life cycle are called essential plant nutrients.

Table 2.6 gives a brief overview of such critical nutrient functions, which are required in varying amounts in plant tissue. Macronutrients (nitrogen, phosphorus, potassium, calcium, magnesium, and sulfur) are nutrients required in the largest amount in plants. Micronutrients (iron, copper, manganese, zinc, boron, molybdenum, and chlorine) are required in relatively small amounts. Additional mineral nutrient elements, which are beneficial to plants but not necessarily essential, include sodium, cobalt, vanadium, nickel, selenium, aluminum, and silicon. The nutrient elements differ with respect to the form in which they are absorbed by the plant, their functions in the plant, their mobility throughout the plant, and the plant's deficiency or toxicity symptoms characteristic for the nutrient.

TABLE 2.6   Essential plant nutrients

Element	Typical wt% relative to N in plant species' dry tissue	Plant function	Category of nutrient
Nitrogen (N)	100	Amino acids, proteins	
Potassium (K)	25	Enzyme activation, protein synthesis, osmotic pressure control, transport, cation/anion balancing	*Primary macronutrients*
Phosphorus (P)	6	Composing part of the information carriers, nucleic acids (deoxyribonucleic acid (DNA), ribonucleic acid (RNA)); energy household carrier constituent, adenosine diphosphate (ADP), adenosine triphosphate (ATP)	
Calcium (Ca)	12.5	Cell wall structural component	
Magnesium (Mg)	8	Chlorophyll constituent	*Secondary macronutrients*
Sulfur (S)	3	Amino acids, proteins	
Chlorine (Cl)	0.3	Photosynthesis reactions	
Boron (B)	0.2	Cell wall component	
Iron (Fe)	0.2	Chlorophyll synthesis	
Manganese (Mn)	0.1	Activates enzymes	*Micronutrients*
Zinc (Zn)	0.03	Activates enzymes	
Copper (Cu)	0.01	Enzyme component	
Molybdenum (Mo)	0.0001	Involved in N fixation	

Adapted from Tortosa Masia (2010).

Some of the aforementioned elements that are important for mineral- and ash-related problems in processing biomass are briefly discussed here.

Potassium is the macronutrient needed in plant life in the largest quantities after nitrogen. For optimum plant growth, its demand is of the order of 1–5 wt% dry matter, depending on the particular species, while the potassium concentration in mature plants generally does not exceed 2 wt% dry matter. Potassium has a high mobility in plants at all levels. Furthermore, it plays a key role in a large number of processes for plant growth, which include activation of enzymes, protein synthesis, photosynthesis, regulation of osmotic pressure, vascular transport, and cation/anion balancing.

The amount of *calcium* compounds incorporated in plants varies between 0.1 and >5.0 wt% (db) depending on plant growth conditions, plant species, and plant part (Marschner, 1993). This element might either be strongly bound to the plant structure or is exchangeable at the cell walls. Like potassium, calcium is a nutrient that plays an important role in many essential processes, including binding form, cell wall stabilization, secretor processes, membrane stabilization, and regulation of the osmotic pressure.

Sulfur is a macronutrient and plants need it for growth; it is increasingly being recognized as the fourth major plant nutrient after nitrogen, phosphorus, and potassium. S is present in both organic and inorganic forms (Marschner, 1993). Sulfur is contained in many organic structures that include amino acids and proteins (such as enzymes). The element is absorbed specifically by the plants' roots in the form of the sulfate ion. The sulfur content for optimal plant growth is between 0.1 and 0.5 wt% (db) and varies from plant to plant. The ratio of organic to inorganic S species may differ between plant species and may depend on local soil and growth conditions.

Chlorine in plants is available mainly as a free anion or is loosely bound to exchange sites. In addition, a number of chlorinated organic compounds have been found in plants, but the functional requirement for plant growth of most of these compounds is not known (Marschner, 1993). Similar to potassium, chlorine has a high mobility within the plant, and the average chlorine content in plants ranges from 0.2 to 2.0 wt% (db).

Silicon has a number of valuable effects in many plant species. It increases leaf erectness. In addition, a high content of silicon in leaves increases the resistance of the tissue against fungal attacks, blast infection, and insect pests. In the case of woody biomass, additional inorganic species may be present, originating from the soil ("detrital," sometimes called "adherent" materials, such as quartz sand or clay) or from other sources, such as paints or coatings added during manufacturing in the case of waste wood.

The distribution of inorganic matter within different biomass fuels varies enormously between diverse samples taken, even when, e.g., crushed and sieved. It can occur as discrete minerals, amorphous phases, and organically associated or simple salts, possibly dissolved in pore water. See Figure 2.2 for more details on the formation of the phases (Vassilev et al., 2010, 2013). In biomass, nearly all of the inorganic elements are found organically associated or present as simple inorganic salts. The inorganic material can be divided into two groups, one of which is authigenic or naturally occurring in the fuel and the other is detrital material, which has been added to the fuel through geologic or processing steps (Baxter, 1993).

The main ash-forming components embedded in the woody biomass fuels are calcium, magnesium, and potassium. A large fraction of the inherent inorganic material in lignites, but also probably the dominant fraction in many biomass fuels, is associated with oxygen-containing functional groups (carboxylic acids, hydroxyls, ethers, and ketones). These functional groups provide sites for inorganic material to become incorporated in the fuel matrix as, e.g., cations (sodium, potassium, magnesium, and calcium). The proportion of the inorganic compounds that is organically associated increases with decreasing fuel rank, due to the higher oxygen content in low-rank fuels. In the lowest-rank coals, more resembling biomass, the organically associated inorganic elements can include up to 60 wt% of the total amount of inorganics (Benson and Holm, 1985).

The second class of inorganic material in solid fuels consists of compounds that are added to the fuel from extraneous sources, mainly during harvesting. This additional material is of particulate nature, in contrast to the atomically dispersed material. It is one of the major contributors (next to, e.g., additives supplied to improve combustion processes or some sticky ash agglomerates) to fly ash particles with sizes larger than about 10 μm (Baxter, 1993).

2.5 PROXIMATE AND ULTIMATE ANALYSIS

It is common practice to characterize the composition of a given fuel by means of two standard analyses, the proximate and the ultimate analysis. These techniques are discussed in the following sections.

2.5.1 Proximate Analysis

The proximate analysis is used to determine the following constituent classes in the fuel:

- Moisture content
- VM content
- FC content
- Ash content

One of the main constituents of biomass is moisture, and this also forms a main difference with coals, and in particular high-rank (older) coals. Water is an essential molecule to sustain life as it acts as major cell liquid (cytoplasm) and plays a fundamental role in photosynthesis, as explained in Chapter 1. High moisture contents may severely impact thermal conversion, and either drying is needed (see Chapter 8) or conversion techniques using the water should be applied (see, e.g., Chapters 10, 11, 12, 13, and 14). Moisture consists of free (or surface) and inherent moisture. The moisture content is determined by placing a biomass sample on a balance in an oven filled with inert gas. The weight loss is determined after 24 h at 105°C, a temperature at which one can assume all moisture to have been evaporated, so by heating to 105°C also the inherent moisture is removed. Water that is chemically bound by hydrates is not removed in this procedure. The weight fraction of moisture measured on an "as received" basis, abbreviated as "ar," can be determined as

$$Y_{moist}^{ar} = \frac{(\text{wet weight} - \text{dry weight})}{\text{wet weight}} \qquad \text{(Eq. 2.1)}$$

Sometimes, moisture content is presented on a db:

$$Y_{moist}^{db} = \frac{(\text{wet weight} - \text{dry weight})}{\text{dry weight}} \qquad \text{(Eq. 2.2)}$$

The effect of the moisture content on the heating value is explained in Section 2.7. The amount of heat that can be recovered from the biomass drops dramatically with increasing moisture content, since the heat of vaporization of the water is usually not recovered during combustion.

To determine the VM content, a biomass sample is heated ("carbonized") in a covered crucible to 550°C (for coals to 900°C) in an inert environment. Biomass may contain appreciable amounts of volatile alkali species and also carbonates that can decompose at temperatures higher than 600°C, which limits the temperature for biomass. Under these conditions, the more instable fuel fractions are released as a complex mixture of gases and vapors. What remains is called char. The weight difference with the original sample is the VM content:

$$Y_{VM}^{db} = \frac{\text{weight loss}}{\text{dry weight}} \qquad \text{(Eq. 2.3)}$$

The ash content then is determined by controlled combustion of the char left in the previous step at a temperature of 815°C (German Standard DIN 51719) (tinyurl.com/ 26yz894). After this combustion step, the amount of residual ash is determined, and the difference with the mass at the beginning of this combustion step is called the fraction of FC:

$$Y_{ash}^{db} = \frac{\text{weight ash}}{\text{dry weight}} \qquad \text{(Eq. 2.4)}$$

$$Y_{FC}^{db} = 1 - Y_{VM}^{db} - Y_{ash}^{db} \qquad \text{(Eq. 2.5)}$$

The ash content is not exactly the same as the content of mineral matter because ash forms the residual part of the combustion process; i.e., not all inorganic elements are present in oxidic form (Spliethoff, 2009). Still, the ash content is a fairly good indicator of the ash yields that can be expected in industrial processes. As the exact biomass composition varies, in particular its moisture content, it is common use to report these analyses on a db and sometimes on a daf basis.

Using thermogravimetric analysis (TGA), one also obtains useful information regarding a solid fuel in terms of the contents of moisture, volatiles, FC, and ash of a biomass sample. The device consists of a temperature-controlled oven and balance, and it allows introduction of sweeping flow gases. Depending on the technique used for analysis of the gas phase, one is able to determine which gases are evolved upon controlled heating of the biomass sample. A schematic is presented in Figure 2.8.

A schematic overview of the TGA concept is shown in Figure 2.9. A small piece of solid fuel or ground material (biomass) is suspended on one of two balance pans (crucible) in a furnace. The temperature of the furnace is slowly increased at a known heating rate (e.g., ~20°C·min^{-1}).

The weight of the sample is measured as a function of time and therefore temperature. The derivative of the resulting weight curve corresponds to the conversion rate and usually shows several peaks corresponding to the different thermal decomposition processes associated with the main constituents of the biomass. When the temperature is raised, initially drying of the sample occurs. In fact, drying commences as soon as the furnace is purged with nitrogen (or helium or argon) to create the required inert atmosphere, since the used gas is dry.

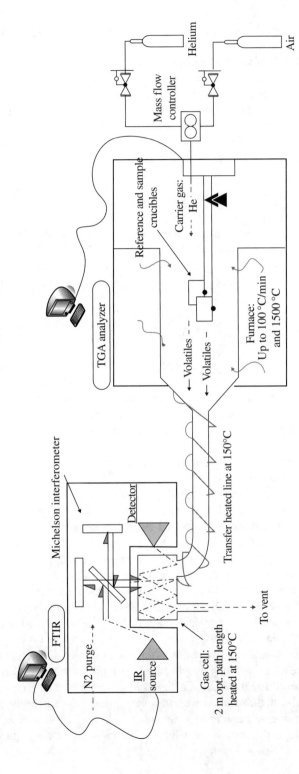

FIGURE 2.8 System of a TGA coupled to gas analysis using Fourier transform infrared (FTIR) spectroscopy. (Source: Reproduced (adapted) with permission from Giuntoli et al. (2009). © American Chemical Society.)

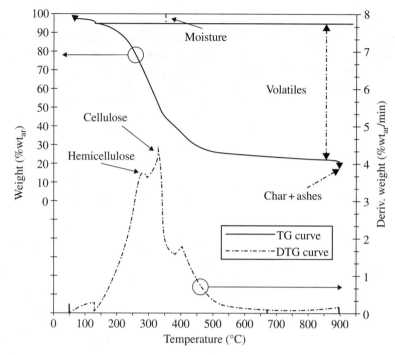

FIGURE 2.9 Example curves of mass loss (left axis) and differential mass loss observed (right axis) in a TGA (courtesy of J. Giuntoli).

As shown in Figure 2.9, at 150–200°C, devolatilization starts, and the devolatilization rate increases as the temperature is raised. There are two areas of weight loss producing a single peak with a plateau or shoulder located at the lower-temperature region. The lower-temperature region represents the decomposition of hemicellulose, and the next higher-temperature peak represents that of cellulose. Above approximately 400°C, most of the volatiles are gone and the devolatilization rate decreases rapidly. However, some devolatilization can still be observed in the temperature range of 400–600°C. This is caused by lignin decomposition, which occurs throughout the whole temperature range, although the main weight loss occurs at higher temperatures. This means that mainly lignin is responsible for the flat tailing section that can be observed for all wood species at higher temperatures.

After a TGA experiment, a fraction of the char and inorganic matter remains. If air is allowed to enter the system after devolatilization, the carbon (char) will burn, leaving ash as the final product. Each form of biomass produces slightly different quantities of char, VM, and ash. Knowledge of these quantities, and of the temperature dependencies of the reactions and associated weight losses, is useful in understanding the operation and design of biomass conversion equipment.

2.5.2 Ultimate Analysis

The ultimate (or elemental) analysis gives the chemical composition and the HHV of the fuel. This chemical analysis usually lists the carbon, hydrogen, oxygen (by difference), nitrogen, sulfur, and ash content of the dry fuel on a weight percentage basis. The biomass sample is combusted in a combustion chamber using an O_2 atmosphere with helium (He) as carrier gas. The product gases of combustion are CO_2, H_2O, NO, NO_2, SO_2, SO_3, and N_2. SO_3, NO, and NO_2 are reduced in a copper-containing section downstream the combustion compartment of the device, which ensures conversion to SO_2 and N_2, respectively. H_2O, SO_2, and CO_2 are captured in different adsorption columns. N_2 is not captured by the adsorption columns and is detected by a thermal conductivity detector (TCD). Subsequently, H_2O, SO_2, and CO_2 are released and introduced into the TCD. The weight percentage is determined integrally. The C, H, N, and S contents can then be determined from the known sample weight; O is determined from the difference between daf weights and the weights of the main other elements that have already been determined. The method has been standardized (see ASTM D3177-84 (tinyurl.com/mjqmw5a)).

The ultimate analyses of a number of biomass fuels and other solid fuels are given in Table 2.4. Note that biomass typically has very low nitrogen and sulfur contents relative to fossil fuels (especially coal). Also, the ash content of biomass is typically much less than that of coals, although some forms (e.g., sewage sludge) have a high ash content. Depending on the exact composition and quantity, this can lead to melting of the ash (known as "slagging"), which can cause severe problems in high-temperature process equipment.

Example 2.1   Calculation of the amount of air needed for the complete combustion of wood

Wood pellets with a composition given in Table 2.3 (first entry) are to be combusted with dry air with 79 vol.% of N_2 and 21 vol.% O_2 (idealized composition). Assume that the fuel is completely daf:

    a. How much air (in $kg \cdot kg^{-1}$ fuel) is needed to just completely, thus stoichiometrically, combust this fuel?
       Given: $MW_C = 12.011\ kg \cdot kmol^{-1}$, $MW_H = 1.00797\ kg \cdot kmol^{-1}$,
       $MW_N = 14.0067\ kg \cdot kmol^{-1}$, and $MW_O = 15.9994\ kg \cdot kmol^{-1}$.

    b. What is the volume percentage of CO_2 in the flue gas produced?

    c. A CO_2 analyzer needs completely dry flue gas, so the gas is cooled before entering the analyzer. What is the expected volume percentage the analyzer quantifies in the cooled flue gas?

Solution

    a. The complete combustion reaction can be written as (the factor 3.76 is the molar ratio between N_2 and O_2; air can be considered an ideal gas)

$$C_xH_yO_z + a(O_2 + 3.76N_2) \rightarrow xCO_2 + \frac{1}{2}yH_2O + 3.76a\,N_2 \quad \text{(RX. 2.1)}$$

with $a = x + 1/4y - 1/2z$
The air-to-fuel ratio ($kg \cdot kg^{-1}$) is

$$\left(\frac{air}{fuel}\right)_{stoichiometric} = \frac{4.76a}{1}\frac{MW_{air}}{MW_{fuel}}$$

Now, x, y, and z follow from calculating the number of moles of the elements i in the fuel. These are simply calculated for, e.g., 100 g of fuel as $mass_i/MW_i$:

$$x = (45.6/12.011) = 3.80, \ y = (6.6/1.00797) = 6.5, \ and \ z = (47.8/15.9994) = 2.99$$

Then, $a = 3.94$.
MW_{air} and MW_{fuel} are calculated as follows:

$$MW_{air} = 0.21 \cdot 2 \cdot 15.9994 + 0.79 \cdot 2 \cdot 15.9994 = 28.85 \ g \cdot mol^{-1}$$

$MW_{fuel} = 100 \ g \cdot mol^{-1}$ (100 g of fuel was taken as starting point of the calculation, and it concerns 1 mol of fuel considered).
Now, the required air-to-fuel ratio can be calculated as

$$\left(\frac{air}{fuel}\right)_{stoichiometric} = \frac{4.76a}{1}\frac{MW_{air}}{MW_{fuel}} = \frac{4.76 \cdot 3.99}{1}\frac{28.85}{100} = 5.41 \ kg \cdot kg^{-1}$$

b. The volume percentage of CO_2 in the wet flue gas is

$$vol.\% \ CO_2 = 100 \cdot \frac{x}{x + \frac{1}{2}y + 3.76a} = 100 \cdot \frac{3.80}{3.80 + \frac{1}{2} \cdot 6.5 + 3.76 \cdot 3.94} = 17.3\%$$

c. The volume percentage of CO_2 in the dry flue gas is

$$vol.\% \ CO_2 = 100 \cdot \frac{x}{x + 3.76a} = 100 \cdot \frac{3.80}{3.80 + 3.76 \cdot 3.94} = 20.4\%$$

Question: Does the use of a cooler lead to an entirely dry flue gas? Which process parameter determines the wetness of the gas entering the analyzer?

2.6 HEATING VALUES

The heating value (also called calorific value) of biomass is a measure of the thermal energy released upon complete combustion, and it is a key property for determining energy balances and performing flame temperature calculations concerning thermochemical conversions such as combustion. One can distinguish the HHV (also called gross calorific value (GCV)) and the lower heating value (LHV) (also named net calorific value (NCV)). The HHV assumes the water originally present as moisture but also the water chemically formed from the fuel-bound H content to be in the liquid form. In contrast, the LHV assumes that both types of water remain in the vapor phase. Thus, the difference is the latent heat of vaporization of water at the standard temperature of $25^{\circ}C$ (~2.4 MJ·kg^{-1}) of both water types released in the combustion process. The LHV can be calculated from HHV in two steps. First, one corrects for the hydrogen content in the dry (0 wt% moisture) fuel:

$$LHV^{db} = HHV - 2.4 \times 8.9 Y_H \quad \left[MJ \cdot kg^{-1} \right] \tag{Eq. 2.6}$$

Here, 2.4 MJ·kg^{-1} is the latent heat of vaporization of water and 8.9 [kg·kg^{-1}] is the stoichiometric water to H ratio, being kg water formed per kg hydrogen bound in the fuel structure. For most types of biomass, the hydrogen content is approximately 6 wt % (Y_H ~ 0.06). The second step is to correct for the wet fuel's moisture content as follows:

$$LHV^{wb} = LHV^{db} \left(1 - Y_{moisture} \right) - 2.4 Y_{moisture} \tag{Eq. 2.7}$$

Often, the heat of condensation (the reverse of vaporization) is not utilized in energy conversion processes, and therefore, usually, the LHV is used in calculations.

The HHV is experimentally determined using a so-called bomb calorimeter, a constant volume calorimeter, in which in a closed vessel a fuel portion is oxidized using pure oxygen. In this device, the heat transferred to a precisely known amount of water is measured by its temperature increase. The sample is ignited electrically. This is a standard method (e.g., DIN 51,900).

The HHV (db) can also be determined once one knows the biomass ultimate analysis, using empirical correlations; one comparatively accurate equation is given below (Gaur and Reed, 1995):

$$HHV = 34.91 Y_C + 117.83 Y_H + 10.05 Y_S - 1.51 Y_N - 10.34 Y_O - 2.11 Y_{ash} \tag{Eq. 2.8}$$

in which Y_i is the mass fraction of element i on a dry fuel basis. As can be seen from this relation, the contents of C, H, and S contribute positively to the HHV, while the contents of N, O, and ash contribute negatively. Most biomass fuels have an HHV of between 18 and 22 MJ·kg^{-1} (db); herbaceous biomass shows slightly lower LHV values than woody biomass.

Example 2.2   HHV calculation based on ultimate analysis data

Khan et al. (2008) have reported on the combustion of pepper plant residue from greenhouse cultivation in a fluidized bed. The measured HHV value (bomb calorimetry) was 16.9 MJ·kg^{-1} (db). Compare this value with Equation (2.8) given the ultimate analysis data presented in Table 2.3.

Solution

The elemental composition data in the table are on a daf basis; thus, first convert the main elemental composition to a db:

$$Y_{i,db} = (1 - Y_{ash,db}) Y_{i,daf}$$

The ash content from the proximate analysis is given on an "ar" basis, but we have to calculate it on a db:

$$Y_{ash,db} = [1/(1 - Y_{moisture,ar})] Y_{ash,ar} = [1/(1 - 0.065)]0.135 = 0.144$$

Then,

$$Y_{C,db} = (1 - 0.144)0.423 = 0.362; \quad Y_{H,db} = (1 - 0.144)0.05 = 0.043;$$
$$Y_{S,db} = (1 - 0.144)0.006 = 0.005; \quad Y_{N,db} = (1 - 0.144)0.031 = 0.027;$$
$$Y_{O,db} = (1 - 0.144)0.489 = 0.419$$

Substituting the calculated values in Equation (2.8) gives HHV = 13.1 MJ·kg^{-1} (db), a relative difference of approximately −23%.

Exercise: repeat this calculation for other wood types.
Note: This outcome indicates that care should be taken for using empirical correlations for "deviating" biomass types. In this case, the biomass contained a high amount of silica-rich (soil derived, adherent) ash.

2.7 ASH CHARACTERIZATION TECHNIQUES

Despite the comparatively low content of mineral matter in biomass, it is often the cause of diverse ash-related challenges that operators face in the (thermal) conversion of the fuel. Although in the case of firing a single fuel it is possible to predict ash behavior with relatively low effort, it becomes more difficult to predict ash behavior in the case of firing multiple fuels in proportions that vary in time due to, e.g., seasonal changes or extreme heterogeneity (Miller and Miller, 2007). Blending of fuels might lead to nonadditive behavior.

In biomass thermochemical conversion systems, ash melting (slagging) causes severe problems, such as agglomeration of the reactor contents and slagging of heat exchangers with negative implications for safe and reliable reactor operation and heat transfer performance. Therefore, the ash content and ash melting behavior are important properties of solid biomass fuels.

Ash melting behavior is determined by preparing ash samples in specific shapes, such as pyramids or cubes, and then heating these samples in a reducing or oxidizing atmosphere in an oven. The oven temperature is raised to a point not much below the expected deformation temperature. Subsequently, the oven temperature is slowly increased at a uniform heating rate of 3–7°C·min^{-1}. The shape of the samples can

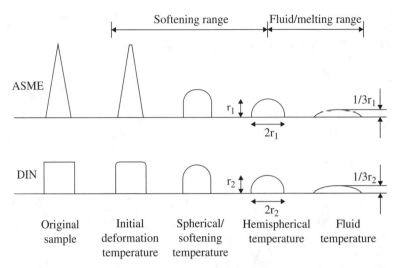

FIGURE 2.10 Ash fusion characterization according to standards of DIN and ASME. (Source: Reproduced with permission from Spliethoff (2009). © Springer.)

be observed through a visual inspection window in the furnace. The temperatures at which characteristic changes of shape occur are recorded. The different characteristic shapes are shown in Figure 2.10.

The following characteristic temperatures are defined:

- ID is the "initial deformation temperature", the temperature at which the first signs of shape changes are observed (edges and top).
- ST is the "softening temperature", the temperature at which the top part exhibits a spherically formed geometry with a height equal to the width at the shape's bottom.
- HT is the "hemispherical temperature", the temperature at which a hemisphere is formed with a geometry that is characterized by a height that is equal to half of the width.
- FT is the "fluid temperature", the final stage in which the ash is molten until an extent at which the maximum height has been reduced to one third of the height in the HT phase.

The composition of the inorganic constituents in biomass and the ash composition in terms of quantities of major elements are usually characterized using X-ray fluorescence (XRF) spectroscopy; the technique is described in, e.g., Beckhoff et al. (2006). X-ray diffraction (XRD) can be used to identify which crystal structures are present in a sample, thus enabling identification of species.

The main elemental composition of biomass (C, H, O) shows some variation, but the composition of mineral matter is subject to a much larger variety. Vassilev et al. (2010) have presented an overview of widely differing biomasses; in particular, the variation in ash constituents is illustrative, as can be seen in Figure 2.11. The figure is a triangular diagram showing on its axes the sums of acidic, high-temperature

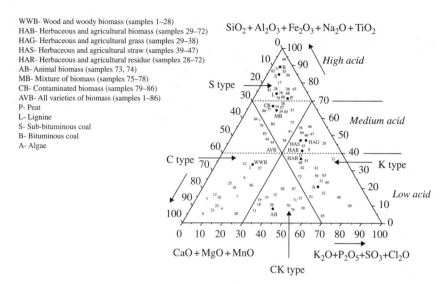

FIGURE 2.11 A way to classify inorganic matter in high-temperature ashes. (Source: Reproduced (adapted) with permission from Vassilev et al. (2010). © Elsevier.)

melting ash compounds (top part, $SiO_2 + Al_2O_3 + Fe_2O_3 + Na_2O + TiO_2$), base ash elements that form less volatiles at high temperatures ($CaO + MgO + MnO$), and base elements that tend to form more volatile species at high temperatures ($K_2O + P_2O_5 + SO_3 + Cl_2O$). Four major classes of compounds were identified. The S class mainly contains the older fuels, while woody fuels are mostly C type (base nature). Agricultural residues and algae that can form glassy agglomerates at high temperatures form the K class, and animal-derived materials (e.g., MBM) are identified as CK class.

The methods used to determine the association of inorganic compounds in fuels have developed significantly in the past few decades. Standard methods typically involve analysis after controlled combustion (also called "ashing"), whereas other (more advanced) methods, such as chemical fractionation and computer-controlled scanning electron microscopy (CCSEM), are used to quantify the abundance, size, and association of inorganic compounds in fuels (Frandsen, 2011). Chemical fractionation comprises dissolution of a sample in pure water, ammonium acetate solution, and hydrogen chloride solution, representing increasingly less volatile inorganic fractions; it can be applied to the raw fuels (Zevenhoven-Onderwater et al., 2000) or to biomass-derived ashes (with ashing performed under standardized conditions) (Tortosa Masia et al., 2007).

CHAPTER SUMMARY AND STUDY GUIDE

The approach of this chapter is to show the basic properties of biomass in its wide diversity, i.e., both physical and chemical properties. The main constituents are discussed in terms of different organic compounds, in particular the biopolymers cellulose, hemicellulose, and lignin. Moreover, the content of minor organic and inorganic species in biomass is dealt with. Characterization of biomass by means of proximate and ultimate analysis is useful for determining its global reactivity and composition data. The heating value of biomass, important for thermochemical conversion processes, is defined in terms of the HHV and LHV (GCV and NCV). Also, the fuel bases are discussed, namely, ar (or wet), db, and daf. Finally, ash melting (fusion) behavior and determination of its elemental composition are dealt with.

KEY CONCEPTS

Biomass composition
Van Krevelen diagram, coalification
Fuel bases (db, daf, ar)
Proximate analysis
Thermogravimetric analysis (TGA)
Ultimate analysis
Basic combustion calculations based on stoichiometry and elemental fuel composition
Organic constituents
Inorganic constituents (mineral matter)
Ash fusion characterization
Lower heating value (LHV), higher heating value (HHV), and their difference

SHORT-ANSWER QUESTIONS

2.1 Which effects can a high moisture content have on storage and handling, and which in combustion processes?

2.2 Why would a high porosity of the fuel lead to more fines when combusting a solid biofuel?

2.3 Discuss the effects of particle shape and size on storage and handling of biomass.

2.4 What are the main biopolymers on Earth?

2.5 What is the difference between homopolysaccharides and heteropolysaccharides? Give a few examples of both.

2.6 What are similarities and differences between starch and cellulose?

2.7 In Figure 2.3 (bottom part), indicate the intra- and intermolecular hydrogen bonds in cellulose.

2.8 Why would lignin have a higher heating value than carbohydrate constituents of biomass? In which ways does the lignin content influence biomass combustion behavior?

2.9 What distinguishes fats from oils?

2.10 Sugar extracted from sugar beets can be fermented to produce ethanol. Why is this an example of a first-generation biofuel production process?

2.11 Why is *Jatropha curcas*-derived biodiesel sometimes called a "1.5-generation" biofuel?

2.12 Which elements in biomass do you expect to be most troublesome in thermal conversion processes and in which respect?

2.13 Why is the procedure for the determination of the volatile matter content in the proximate analysis different for biomass and coals?

2.14 In TGA analysis, hardwoods (e.g., birch, beech, and acacia) show the presence of a "shouldered peak," whereas this is less visible for softwoods (e.g., spruce). What can be the reason for this difference?

2.15 In TGA analysis, nitrogen can be used as sweeping gas, but also helium might be used. What is the main difference between these gases and why is helium attractive to use?

2.16 What kind of correlation do you expect between the ignition temperature (e.g., in air) and the volatile matter (VM) content of biomass? Which other factors do you think play a role in ignition?

2.17 What would be the consequence if in a burner designed for operation on a fuel with a low VM/FC ratio (e.g., coal), this fuel were replaced by a fuel with a low VM/FC ratio (e.g., biomass)?

2.18 Explain in your own words the difference between the HHV and LHV of a fuel.

2.19 Discuss the different ash fusion methods; do they represent well the phenomena occurring in a fuel combustion reactor?

2.20 In which respects could biomass fuel blending be advantageous? Does it also have disadvantages? Explain.

2.21 Certain plant species but also (modified) chitosan can take up some elements very selectively and concentrate them. What are possible risks and opportunities of this specific uptake behavior? See Dodson et al. (2012).

PROBLEMS

2.1 For physical and chemical processing, the specific surface area of a (solid) fuel plays an important role. Now, if one wants to dry grass (blades) in a packed bed with characteristic length (l), width (w), and thickness (H), what would then be the specific surface area per m^3 of packed bed, a_v? Which dimension would be most important for determining the value of a_v and what do you assume here?

2.2 In the Van Krevelen diagram (Figure 2.1), mark the position corresponding to the composition of sunflower pellets.

2.3 Chitin is a biopolymer containing nitrogen in its structure. What is the weight percentage of N in chitin?

2.4 Table 2.3 shows the fuel composition of different biomass species; calculate the molar S to Cl ratio for olive cake pellets and sunflower residue. What possible consequences could the difference have in the combustion of these fuels?

2.5 What is the nitrogen content of pepper plant residue on a mg per MJ basis (HHV, as received fuel)?

2.6 A company exists that has developed a toilet in which the feces are burned. Consider the composition of feces to be as follows (dry basis):

C = 43.0 wt%; H = 6.3 wt%; O = 35.9 wt%; N = 2.3 wt%; ash = 12.5 wt%.

The moisture content of the "as received" material is 71 wt%.
a. What is the HHV on a dry basis (in $MJ \cdot kg^{-1}$)?
b. What is the LHV on an "as received" basis (in $MJ \cdot kg^{-1}$)?
c. How much air (in $kg \cdot kg^{-1}$ material as received) is needed to exactly completely combust the material? Assume NO to be the combustion product of N in the fuel.
d. In practice, would one use exactly this amount of air? Why (not)?

2.7 For the following fuels, calculate the emission of CO_2 in $g \cdot MJ^{-1}$ (HHV basis) for complete stoichiometric combustion:

- Wood pellets
- Peat
- Brown coal
- Bituminous coal

Which fuel gives the highest value?

2.8 A woody fuel has an HHV of 19 $MJ \cdot kg^{-1}$ (db), what is the LHV (ar) when the moisture content (ar) is 15 wt% and the ash content is 0.8 wt% (db)?

2.9 Search the literature for ash fusion temperature data related to straw; would it make sense to wash the straw with water and why (not)?

2.10 If the thermal input on an LHV basis of a straw-fired combustor is 10 MW_{th}, what is the amount of CO_2 emitted in $t \cdot year^{-1}$ in case of complete combustion of the fuel? What could be the net CO_2 savings when this thermal input replaces coal, also assuming complete combustion? Is the calculation really an exact reflection of the CO_2 savings?

2.11 An ash analysis of all the ashes discharged from a waste wood-fired thermal conversion plant revealed that this plant produces ash with a lead (Pb) content of 100 $mg \cdot kg^{-1}$. What is the Pb content of the original fuel on a dry basis (db) assuming that the fuel has been completely converted and given that the fuel contained 1 wt% ash (db)? Which assumption(s) did you have to make for this calculation and is this entirely realistic?

2.12 A thermogravimetric analyzer is a characterization device in which a sample can be heated in a controlled way (heating rate) while simultaneously measuring its weight. A typical result for an agricultural residue fuel is presented in Figure 2.9. Explain the different phenomena. What do you conclude regarding this fuel?

PROJECTS

P2.1 Design and test your own calorimeter to determine the heating value of a biomass component.

P2.2 Mill a selected biomass species and sieve it. Perform ashing in a laboratory oven at a selected temperature. What are the impacts of oven temperature and particle size classes on the determined ash contents? Explain your observations.

P2.3 Propose and perform an ash fusion test that resembles more closely what happens in a real biomass combustor.

INTERNET REFERENCES

tinyurl.com/luyld2g
http://finedustreduction.com/the-system

tinyurl.com/km7dh8h
www.ecn.nl/phyllis

tinyurl.com/26yz894
www.din.de

tinyurl.com/mjqmw5a
http://www.astm.org/Standard/index.shtml

REFERENCES

Arcidiacono S, Kaplan DL. Molecular weight distribution of chitosan isolated from Mucor rouxii under different culture and processing conditions. Biotechnol Bioeng 1992;39:281–286.

Arvelakis S, Koukios EG. Physicochemical upgrading of agroresidues as feedstocks for energy production via thermochemical conversion methods. Biomass Bioenergy 2002;22:331–348.

Balat M, Balat M, Kirtay E, Balat H. Main routes for the thermo-conversion of biomass into fuels and chemicals. Part 1: pyrolysis systems. Energy Convers Manag 2009;50:3147–3157.

Baxter LL. Ash deposition during biomass and coal combustion. A mechanistic approach. Biomass Bioenergy 1993;4(2):85–102.

Beckhoff B, Kanngiesser B, Langhoff N, Wedell R, Wolff H. *Handbook of Practical X-Ray Fluorescence Analysis*. New York: Springer; 2006.

Benson SA, Holm P. Comparison of inorganic constituents in three low rank coals. Ind Eng Chem Prod Res Dev 1985;24(1):145–149.

Biermann U, Friedt W, Lang S, Lühs W, Machmüller G, Metzger JO, Klaas MRG, Schäfer HJ, Schneider MP. New syntheses with oils and fats as renewable raw materials for the chemical industry. In: Kamm B, Gruber PR, Kamm M, editors. *Biorefineries—Industrial Processes and Products*, Volume 2. Weinheim: Wiley-VCH Verlag GmbH; 2006; p 253–289.

Dale BE, Kim S. Biomass refining global impact—the biobased economy of the 21st century. In: Kamm B, Gruber PR, Kamm M, editors. *Biorefineries—Industrial Processes and Products*, Volume 1. Weinheim: Wiley-VCH Verlag GmbH; 2006; p 41–66.

De Jong W, Ünal Ö, Andries J, Hein KRG, Spliethoff H. Biomass and fossil fuel conversion by pressurised fluidised bed gasification using hot gas ceramic filters as gas cleaning. Biomass Bioenergy 2003;25(1):59–83.

Dodson JR, Hunt AJ, Parker HL, Yang Y, Clark JH. Elemental sustainability: towards the total recovery of scarce metals. Chem Eng Process 2012;51(1):69–78.

El Boushy AR, Klaassen GJ, Ketelaars EH. Biological conversion of poultry and animal waste to a feedstuff for poultry. Worlds Poult Sci J 1985;41:133–145.

Frandsen F. Ash formation, deposition and corrosion when utilizing straw for heat and power production [Doctoral thesis]. Kongens Lyngby: Technical University of Denmark; 2011.

Gaur S, Reed TB. *An Atlas of Thermal Data for Biomass and Other Fuels*. Golden, CO: National Renewable Energy Laboratory; 1995.

Giuntoli J, Arvelakis S, Spliethoff H, de Jong W, Verkooijen AHM. Quantitative and kinetic thermogravimetric Fourier transform infrared (TG-FTIR) study of pyrolysis of agricultural residues: influence of different pretreatments. Energy Fuels 2009;23 (11):5695–5706.

Goring DA, editor. *Cellulose Chemistry and Technology*. Washington, DC: American Chemical Society; 1977.

Jacobsen SE, Wyman CE. Cellulose and hemicellulose hydrolysis models for application to current and novel pretreatment processes. Appl Biochem Biotechnol 2000;84–86:81–96.

Jenkins BM, Baxter LL, Miles Jr TR, Miles TR. Combustion properties of biomass. Fuel Process Technol 1998;54(1–3):17–46.

Kamm B, Gruber PR, Kamm M, editors. *Biorefineries—Industrial Processes and Products*. Weinheim: Wiley-VCH Verlag GmbH; 2006.

Khan AA, Aho M, de Jong W, Vainikka P, Jansens PJ, Spliethoff H. Scale-up study on combustibility and emission formation with two biomass fuels (B quality wood and pepper plant residue) under BFB conditions. Biomass Bioenergy 2008;32:1311–1321.

Khan AA, De Jong W, Jansens PJ, Spliethoff H. Biomass combustion in fluidized bed boilers: potential problems and remedies. Fuel Process Technol 2009;90:21–50.

Kim Y, Mosier NS, Hendrickson R, Ezeji T, Blaschek H, Dien B, Cotta M, Dale B, Ladisch MR. Composition of corn dry-grind ethanol by-products: DDGS, wet cake, and thin stillage. Bioresour Technol 2008;99(12):5165–5176.

Kirk TK, Cullen D. Enzymology and molecular genetics of wood degradation by wood degrading fungi. In: Young RA, Akhtar M, editors. *Environmentally Friendly Technologies for the Pulp and Paper Industry*. New York: John Wiley & Sons; 1998. p 273–307.

Klass DL. *Biomass for Renewable Energy: Fuels and Chemicals*. San Diego, CA: Academic Press; 1998.

Koh MY, Ghazi TIM. A review of biodiesel production from *Jatropha curcas* L. oil. Renew Sustain Energy Rev 2011;15(5):2240–2251.

Koutinas AA, Wang RH, Webb C. The biochemurgist—bioconversion of agricultural raw materials for chemical production. Biofuels Bioprod Biorefin 2007;1(1):24–38.

Kripp TC. Biobased consumer products for cosmetics. In: Kamm B, Gruber PR, Kamm M, editors. *Biorefineries—Industrial Processes and Products*. Weinheim: Wiley-VCH Verlag GmbH; 2006.

Kurkela E, Ståhlberg P, Luntama JL. Pressurised-fluidised bed gasification experiments with wood, peat and coal at VTT in 1991–1994. Part 2. Experiences from peat and coal gasification and hot gas filtration. Espoo: VTT; 1995. VTT Publications nr 249.

Leppälahti J, Koljonen T. Nitrogen evolution from coal, peat and wood during gasification: literature review. Fuel Process Technol 1995;43:1–45.

Mageswaran R, Balakrishnan V, Balasubramaniam S. Boron content of marine algae from the Mandaitivu and Kirinda coasts and mineral content of the nine species of algae from the Kirinda coast. J Natl Sci Council Sri Lanka 1985;13(2):131–140.

Marschner H. *Mineral Nutrition of Higher Plants*. 2nd ed. London: Academic Press; 1993.

Miller SF, Miller BG. The occurrence of inorganic elements in various biofuels and its effect on ash chemistry and behavior and use in combustion products. Fuel Process Technol 2007;88:1155–1164.

Mohan D, Pittman CU, Steele PH. Pyrolysis of wood/biomass for bio-oil: a critical review. Energy Fuels 2006;20:848–889.

Nilsson S, Gómez-Barea A, Cano DF. Gasification reactivity of char from dried sewage sludge in a fluidized bed. Fuel 2012;92:346–353.

Prins MJ, Ptasinski KJ, Janssen FJJG. From coal to biomass gasification: comparison of thermodynamic efficiency. Energy 2007;32:1248–1259.

Pye EK. Industrial lignin production and applications. In: Kamm B, Gruber PR, Kamm M, editors. *Biorefineries—Industrial Processes and Products*, Volume 2. Weinheim: Wiley-VCH Verlag GmbH; 2006.

Ramachandran S, Singh SK, Larroche C, Soccol CR, Pandey A. Oil cakes and their biotechnological applications: a review. Bioresour Technol 2007;98(10):2000–2009.

Rowell RM. *The Chemistry of Solid Wood*. Washington, DC: American Chemical Society; 1984.

Silva GGD, Guilbert S, Rouau X. Successive centrifugal grinding and sieving of wheat straw. Powder Technol 2011;208:266–270.

Simon J, Müller HP, Koch R, Müller V. Thermoplastic and biodegradable polymers of cellulose. Polym Degrad Stab 1998;59:107–115.

Sjöström E. *Wood Chemistry: Fundamentals and Applications*. Toronto/San Diego, CA: Academic Press; 1993.

Solomons TWG. *Organic Chemistry*. 3rd ed. New York: John Wiley & Sons; 1984.

Solomons TWG, Fryhle C. *Organic Chemistry*. 8th ed. New York: John Wiley & Sons; 2004.

Spliethoff H. *Power Generation from Solid Fuels*. Berlin-Heidelberg: Springer Verlag; 2009.

Tortosa Masia AA. Characterisation and prediction of deposits in biomass co-combustion [PhD thesis]. Delft: Delft University of Technology; 2010.

Tortosa Masia AA, Buhre BJP, Gupta RP, Wall TF. Characterising ash of biomass and waste. Fuel Process Technol 2007;88:1071–1081.

Vassilev SV, Baxter D, Andersen LK, Vassileva CG. An overview of the chemical composition of biomass. Fuel 2010;(89):913–933.

Vassilev SV, Baxter D, Vassileva CG. An overview of the behaviour of biomass during combustion: Part I. Phase-mineral transformations of organic and inorganic matter. Fuel 2013;112:391–449.

Zakzeski J, Bruijnincx PC, Jongerius AL, Weckhuysen BM. The catalytic valorization of lignin for the production of renewable chemicals. Chem Rev 2010;110:3552–3599.

Zevenhoven-Onderwater M, Blomquist JP, Skrifvars BJ, Backman R, Hupa M. The prediction of behaviour of ashes from five different solid fuels in fluidised bed combustion. Fuel 2000;79(11):1353–1361.

Zhou D, Zhang L, Zhang S, Fu H, Chen J. Hydrothermal liquefaction of macroalgae *Enteromorpha prolifera* to bio-oil. Energy Fuels 2010;24:4054–4061.

PART II

CHEMICAL ENGINEERING PRINCIPLES OF BIOMASS PROCESSING

3

CONSERVATION: MASS, MOMENTUM, AND ENERGY BALANCES

WIEBREN DE JONG

Department of Process and Energy, Energy Technology Section, Faculty of Mechanical, Maritime and Materials Engineering, Delft University of Technology, Delft, the Netherlands

ACRONYMS

DME   dimethyl ether
EOS   equation of state
KE   kinetic energy
PE   potential energy

SYMBOLS

c	concentration	$[mol \cdot m^{-3}]$
c_p	specific heat capacity	$[J \cdot kg^{-1} \cdot K^{-1}]$
$\overline{c_p}$	molar heat capacity	$[J \cdot mol^{-1} \cdot K^{-1}]$
E	total energy	$[J]$
e	mass-specific total energy	$[J \cdot kg^{-1}]$
f	general property	$[-]$
g	gravitational acceleration	$[m \cdot s^{-2}]$
H	enthalpy	$[J]$

Biomass as a Sustainable Energy Source for the Future: Fundamentals of Conversion Processes,
First Edition. Edited by Wiebren de Jong and J. Ruud van Ommen.

H	height (in Example 3.9)	[m]
h	mass-specific enthalpy	$[\text{J·kg}^{-1}]$
h	heat transfer coefficient	$[\text{W·m}^{-2}\text{·K}^{-1}]$
$\bar{h}$	mole-specific enthalpy	$[\text{J·mol}^{-1}]$
I	unity matrix	[–]
$\vec{j}_i$	mass flux of species i	$[\text{kg·m}^{-2}\text{·s}]$
$\vec{j}_q$	energy flux due to conduction	$[\text{W·m}^{-2}]$
k_i	reaction rate coefficient	Depends on rate expression
L	length	[m]
m	mass	[kg]
MW_i	molecular weight of species i	$[\text{kg·mol}^{-1}]$
n_i	number of moles of species i	[mol]
p	pressure	[Pa]
Q	heat supplied or extracted	[J]
$\dot{Q}$	heat flow supplied or extracted	[W]
R_i	(net) rate of production of species i	$[\text{mol·m}^{-3}\text{·s}^{-1}]$
R_u	universal gas constant	$[\text{J·mol}^{-1}\text{·K}^{-1}]$
r	radius	[m]
r	reaction rate	$[\text{mol·m}^{-3}\text{·s}^{-1}]$
s_f	source term for f	[–]
T	temperature	[K]
t	time	[s]
U	internal energy	[J]
u	mass-specific internal energy	$[\text{J·kg}^{-1}]$
u	velocity in 1D model (x-direction)	$[\text{m·s}^{-1}]$
$\bar{u}$	mole-specific internal energy	$[\text{J·mol}^{-1}]$
$\vec{V}_i$	diffusion velocity of species i	$[\text{m·s}^{-1}]$
$\vec{v}_i$	velocity of species i	$[\text{m·s}^{-1}]$
$\vec{v}$	velocity	$[\text{m·s}^{-1}]$
v	velocity in 1D model (y-direction)	$[\text{m·s}^{-1}]$
V	volume	$[\text{m}^3]$
W	work	[J]
$\dot{W}$	power	[W]
w	width	[m]
Y_i	mass fraction of species i	[–]
X_i	mole fraction of species i	[–]
z	height in a potential field	[m]
η_i	yield of component i in a reaction	[–]
η	dynamic viscosity	[Pa·s]
ξ	degree of conversion	[–]
λ	air stoichiometry coefficient (combustion)	[–]
ν_i	stoichiometric coefficient of species i	[–]

ρ	density	$[kg \cdot m^{-3}]$
σ_i	selectivity of a reaction toward species i	$[-]$
τ	residence time	$[s]$
τ	viscous stress (tensor)	$[Pa]$
υ	mass-specific volume	$[m^3 \cdot kg^{-1}]$
$\vec{\phi}_f$	flux of property f	$[-]$
φ_m	mass flow	$[kg \cdot s^{-1}]$
φ_n	mole flow	$[mol \cdot s^{-1}]$
φ_V	volume flow	$[m^3 \cdot s^{-1}]$
ζ_j	relative degree of conversion	$[-]$
$\dot{\omega}_i$	formation or destruction rate of species i	$[kg \cdot m^{-3} \cdot s^{-1}]$

Subscripts

cv	control volume
f	formation
r	radiation
ref	reference
vol	volume

3.1 GENERAL CONSERVATION EQUATION

Biomass conversion processes are subject to elementary physical conservation laws like any chemical transformation. Setting up the related balance equations forms the solid point of departure to study and understand such phenomena. These laws concern conservation of overall mass, mass of all elements involved in the conversion reactions, momentum, and energy. An extensive fundamental treatise is given in, e.g., Bird et al. (2007). The concept can even be applied to the economics of a biomass conversion process plant. The general form of the conservation law for a randomly chosen system is

Rate of accumulation = rate of supply − rate of release + rate of production (Eq. 3.1)

The general conservation equations can be further worked out depending on what level of detail is required. Balance equations can be set up with the aim to describe the detailed flow pattern and temperature distribution at every single point (x, y, z in Cartesian coordinates) in the system considered. Such balances are called "micro-scopic balances," and this way of considering balances is also termed the "differential" approach. When a (larger) finite region is considered with a balance description of the gross flow and energy exchange effects (e.g., force or torque on a body or the total energy exchange), this is called the "control volume" or "integral" method, and in this case, we deal with "macroscopic balances."

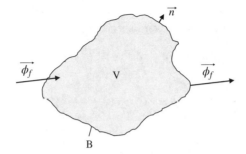

FIGURE 3.1 Domain volume V with boundary B, normal vector $\vec{n}$, and flux $\vec{\phi}_f$.

Figure 3.1 shows a randomly shaped volume V of an open-flow system with a surrounding boundary B. There is a supply flux of f (f being mass density, momentum density, or volumetric energy density) into the system and an exit flux of f out of the system.

Based on the property f, the general conservation equation can be written as

$$\frac{\partial f}{\partial t} = -\nabla \cdot \vec{\phi}_f + s_f \qquad\qquad \text{(Eq. 3.2)}$$

Here, f is a continuous scalar unit per volume, $\vec{\phi}_f$ is the flux of f (a vector), and s_f is a source term for f (a scalar quantity). The system is at steady state when $(\partial f/\partial t) = 0$.

The conservation equations derived from Equation (3.2) are described in more detail in the following sections.

3.2 CONSERVATION OF MASS

Matter is conserved in physical systems and in chemical reactors, the only exception being nuclear conversion processes, which are not dealt with here. Overall mass conservation in an open system can be given by the mass balance in macroscopic form (over a control volume):

$$\frac{dm}{dt} = \varphi_{m, in} - \varphi_{m, out} = -\Delta\varphi_m \qquad\qquad \text{(Eq. 3.3)}$$

Formulated in differential form and using Equation (3.2) as the basis, overall mass conservation can be expressed as

$$\frac{\partial \rho}{\partial t} = -\nabla \cdot (\rho \vec{v}) + 0 \Leftrightarrow \frac{\partial \rho}{\partial t} + \nabla \cdot (\rho \vec{v}) = 0 \qquad\qquad \text{(Eq. 3.4)}$$

In this equation, also called the continuity equation, with $f = \rho$, the system density and $\vec{\phi}_f = \rho \vec{v}$, the total mass flux.

A balance for a chemical species i in an arbitrarily chosen system, in terms of the generic Equation (3.1), is easily derived as

Accumulation rate of species i = rate of supply of species i – rate of release of species i
+ rate of production of species i

(Eq. 3.5)

This equation can be mathematically reformulated in macroscopic form as (Westerterp et al., 1988)

$$\frac{dm_i}{dt} = \varphi_{m,i,in} - \varphi_{m,i,out} + \langle MW_i R_i \rangle V = -\Delta \varphi_{m,i} + \langle MW_i R_i \rangle V \qquad \text{(Eq. 3.6)}$$

$MW_i R_i$ is the rate of production $(kg \cdot s^{-1})$ of species i caused by the chemical reaction with R_i, expressed as the molar conversion in time per reactor volume $(mol \cdot m^{-3} \cdot s^{-1})$ and $\langle ... \rangle$ representing the average over the space in which the reaction takes place. R_i usually depends on the concentrations of the species present and on temperature; accompanying reaction rate expressions are dealt with in Chapter 5.

The aforementioned macroscopic mass balance equation can also be rewritten so as to give the macroscopic *mole balance* for species i:

$$\frac{dn_i}{dt} = \varphi_{n,i,in} - \varphi_{n,i,out} + \langle R_i \rangle V = -\Delta \varphi_{n,i} + \langle R_i \rangle V \qquad \text{(Eq. 3.7)}$$

When chemical conversions are considered, it is important to define to which extent such reactions take place in a reactor. For this purpose, the degree of conversion (ξ) is introduced, which is based on the mass fractions of the reacting species (Y_i). It is defined as

$$\xi \equiv |Y_{i0} - Y_i| \qquad \text{(Eq. 3.8)}$$

Now, consider a reaction $|v_A|A + |v_B|B \rightarrow v_X X + v_Y Y$.

If no species are added to or removed from the reacting system, we can write

$$Y_A + Y_B + Y_X + Y_Y = \text{constant.}$$

The following relation holds with respect to the moles of the reacting species:

$$\frac{\xi_A}{|v_A|MW_A} = \frac{\xi_B}{|v_B|MW_B} = \frac{\xi_X}{v_X MW_X} = \frac{\xi_Y}{v_Y MW_Y} \qquad \text{(Eq. 3.9)}$$

For more convenience in calculations, the degree of conversion is related to the concentration of the species originally present. Thus, the relative degree of conversion of a *reactant* (ζ) is defined as

$$\zeta_i = \frac{\xi_i}{Y_{i0}} = 1 - \frac{Y_i}{Y_{i0}} \qquad \text{(Eq. 3.10)}$$

Both conversion definitions use mass fractions; however, usually, it is more convenient to use molar concentrations, which are measured with analytical instruments, or partial pressures, which are related to volume fractions in gas-phase reaction systems. For a gas-phase reaction system, one can write the molar concentration as follows (assuming that the ideal gas law holds):

$$c_i = \frac{X_i p}{R_u T} \qquad \text{(Eq. 3.11)}$$

The molar concentration and mass fraction are interrelated via

$$c_i = \frac{\rho Y_i}{MW_i} \qquad \text{(Eq. 3.12)}$$

So for reactant A, one can write

$$c_A = \rho \left\{ \left(\frac{c_{A0}}{\rho_0} \right) - \left(\frac{\xi_A}{MW_A} \right) \right\} = \frac{\rho}{\rho_0} c_{A0} (1 - \zeta_A) \qquad \text{(Eq. 3.13)}$$

Exercise: Derive Equation (3.13).

In industrial practice, two other concepts, namely, *selectivity* and *yield*, are of importance as often (undesired) by-products are formed in a reaction system. The selectivity, σ, toward reaction product X is defined as the ratio between the amount of X formed and the amount of key reactant A converted. Thus,

$$\sigma_X = \frac{\xi_X |\nu_A| MW_A}{\xi_A \nu_X MW_X} = \frac{n_{X,\text{formed}}}{n_{A,\text{reacted}}} \frac{|\nu_A|}{\nu_X} \qquad \text{(Eq. 3.14)}$$

At constant density of the reaction mixture, this is equal to

$$\sigma_X = \frac{(c_X - c_{X0}) |\nu_A|}{(c_{A0} - c_A) \nu_X} \qquad \text{(Eq. 3.15)}$$

The main aim is to realize a high yield, η, of product X from reactant A, which is the amount of X formed relative to the amount of A fed as reactant:

$$\eta_X = \sigma_X \zeta_X \qquad \text{(Eq. 3.16)}$$

Using the alternative formulation of Equation (3.2), for f, we can substitute ρ_i, the density of species i, and for $\vec{\phi}_f$, we can substitute $\rho \vec{v}_i$. Species i can be formed or consumed via a chemical reaction, so that a chemical source term exists: $s_f = \dot{\omega}_i$. This results in

$$\frac{\partial \rho_i}{\partial t} = -\nabla \cdot \left(\rho \vec{v}_i \right) + \dot{\omega}_i \Leftrightarrow \frac{\partial \rho_i}{\partial t} + \nabla \cdot \left(\rho \vec{v}_i \right) = \dot{\omega}_i \quad \text{for } i = 1, ..., N \qquad \text{(Eq. 3.17)}$$

The summation of Equation (3.7) over all species $i = 1, ..., N$ should result in total mass conservation, given by Equation (3.4). This condition is fulfilled, as the chemical source terms for all components present in the system sum up to zero:

$$\sum_{i=1}^{N} \dot{\omega}_i = 0 \qquad \text{(Eq. 3.18)}$$

Moreover, the total density of the reacting phase is the sum of the individual partial densities in the mixture, and the total mass flux is the sum of the partial mass fluxes:

$$\sum_{i=1}^{N} \rho_i = \rho \quad \text{and} \quad \sum_{i=1}^{N} \rho_i \vec{v}_i = \rho \vec{v} \qquad \text{(Eq. 3.19)}$$

With the mass fraction Y_i of component i, this can be written as

$$\sum_{i=1}^{N} Y_i = 1 \quad \text{and} \quad \sum_{i=1}^{N} Y_i \vec{v}_i = \vec{v} \qquad \text{(Eq. 3.20)}$$

This equation defines the (density-weighted) mean velocity $\vec{v}$ of the flow. The difference between the partial velocity and the mean velocity is defined as the diffusion velocity, $\vec{V}_i = \vec{v}_i - \vec{v}$, and the mass flux due to diffusion as $\vec{j}_i = \rho \vec{V}_i$. Note that the following holds:

$$\sum_{i=1}^{N} \vec{j}_i = \sum_{i=1}^{N} \left(\rho \vec{v}_i - \rho \vec{v} \right) = \rho \vec{v} - \rho \vec{v} = 0 \qquad \text{(Eq. 3.21)}$$

Using the aforementioned definitions, then Equation (3.17) can be rewritten as

$$\frac{\partial \rho Y_i}{\partial t} + \nabla \cdot \left(\rho \vec{v} Y_i \right) + \nabla \cdot \left(\vec{j}_i \right) = \dot{\omega}_i \quad \text{for } i = 1, ..., N \qquad \text{(Eq. 3.22)}$$

Example 3.1   Mass balance calculation for a well-stirred chemical reactor

Both ethanol and acetic acid can be derived from biomass, so also the product of esterification, ethyl acetate, can be produced in a biorefinery complex. The reaction equation for the esterification reaction is

$$C_2H_5OH\,(1) + CH_3COOH\,(2) \leftrightarrow CH_3COOC_2H_5\,(3) + H_2O\,(4) \qquad \text{(RX. 3.1)}$$

A well-stirred tank reactor can be used to carry out this reaction in the liquid phase. It can be assumed to operate in the steady state. Moreover, the density of the reaction mixture can be assumed to be constant. Consider a case that only the reactants are present in the feed stream in equimolar amounts. Derive an expression for the conversion of ethanol as a function of the residence time, τ, in the tank $(\tau = (V_r/\varphi_V) = (m/\varphi_m))$, with m being the mass in the tank, V_r the tank volume, and φ_m and φ_V the mass and volume flow, respectively). The reaction rate expression for the esterification is $r = k_1 c_1 c_2 - k_2 c_3 c_4$. The net rate of consumption of ethanol, R_1, is related to r via $R_1 = r.\nu_1 = r \cdot (-1) = -r$.

Solution
Start by writing the overall mass balance over the control volume of the reactor tank:

$$\frac{dm}{dt} = 0 = \varphi_{m,in} - \varphi_{m,out} \Rightarrow \varphi_{m,in} = \varphi_{m,out} = \varphi_m$$

Then an integral mass balance for species 1 (ethanol) can be set up:

$$\frac{dm_1}{dt} = 0 = \varphi_{m1,in} Y_{1,in} - \varphi_{m1,out} Y_{1,out} + MW_1 R_1 V_r = \varphi_m \left(Y_{1,in} - Y_{1,out} \right) + MW_1 R_1 V_r$$

$$\Rightarrow 0 = \varphi_m \xi_1 + MW_1 R_1 V_r = \varphi_m Y_{1,0} \zeta_1 + MW_1 R_1 V_r$$

Divide by MW_1 and express the mass fraction as concentration:

$$\Rightarrow 0 = \varphi_V c_{1,0} \zeta_1 + R_1 V_r$$

$$\Rightarrow 0 = c_{1,0} \zeta_1 + R_1 \tau_r \Leftrightarrow \zeta_1 = \left(\frac{-R_1 \tau}{c_{1,0}} \right)$$

Now, consider the expression for the conversion of ethanol, with all concentrations herein expressed as a function of the conversion of species 1, taking into account that density is constant throughout the reaction process (see Equation (3.13), in which then $\rho = \rho_0$):

$$c_1 = c_{1,0}(1 - \zeta_1)$$

For reactant 2, as $\nu_1 = \nu_2$ and as we have an equimolar reactant mixture:

$$c_2 = c_{2,0}(1 - \zeta_2) = c_{1,0}(1 - \zeta_1)$$

For products 3 and 4, as $\nu_3 = \nu_4 = -\nu_1$ and $c_{3,0} = c_{4,0} = 0$:

$$c_3 = c_{3,0} + c_{3,0} \zeta_3 = c_{1,0} \zeta_1$$

$$c_4 = c_{4,0} + c_{4,0} \zeta_4 = c_{1,0} \zeta_1$$

Thus, the expression for the conversion of ethanol simplifies to a relation based on species 1:

$$R_1 = -k_1 c_{1,0}^2 (1-\zeta_1)^2 + k_2 c_{1,0}^2 \zeta_1^2$$

Substitution of this equation in the reactor equation gives

$$0 = \zeta_1 - k_1 c_{1,0}(1-\zeta_1)^2 \tau + k_2 c_{1,0} \zeta_1^2 \tau$$

It is left to the reader to further simplify this quadratic equation, with ζ_1 explicitly expressed as a function of τ.

Example 3.2   Piston compression of syngas: Example of an unsteady-state microscopic balance

A horizontal piston compresses synthesis gas within a cylinder at a constant velocity v. The initial gas density and compartment length are ρ_0 and L_0, respectively. The gas inside the cylinder has a velocity decreasing from $u_x = v$ at the piston head to $u = 0 \, m \cdot s^{-1}$ at the position $x = L$. Let the gas density only be variable in *time*. Derive an expression for the dynamic density behavior, $\rho(t)$.

Solution
This is an unsteady-state (time-dependent) problem, for which only one dimension counts, the horizontal (x)-direction. The general continuity equation in this case simplifies to

$$\frac{\partial \rho}{\partial t} + \frac{\partial}{\partial x}(\rho u) = \frac{\partial \rho}{\partial t} + \rho \frac{\partial}{\partial x} u = 0$$

Here, $u = v(1 - x/L)$, $L = L_0 - vt$, and $\rho = \rho(t)$.
Then,

$$\frac{\partial u}{\partial x} = -\frac{v}{L}$$

Separation of variables leads to

$$\int_{\rho_0}^{\rho} \left[\frac{d\rho}{\rho}\right] = v \int_0^t \frac{dt}{(L_0 - vt)}$$

from which the following solution is obtained:

$$\ln\left(\frac{\rho}{\rho_0}\right) = -\ln\left(1 - \frac{vt}{L_0}\right) \Rightarrow \rho = \rho_0 \left(\frac{L_0}{L_0 - vt}\right)$$

3.3   CONSERVATION OF ENERGY

The conservation of energy, a scalar quantity, is given by the following general balance equation:

Accumulation rate of energy = rate of energy supply − rate of energy release
+ rate of energy production

$$\text{(Eq. 3.23)}$$

The different forms of energy considered in the balance are heat, kinetic energy (KE), and potential energy (PE) due to the presence of fields of gravity, electricity, and magnetism.

3.3.1   Energy Balance for Systems without Chemical Reactions

The first law of thermodynamics forms the basis of the principle of conservation of energy and for a closed system (a good treatise of closed and open systems can be found in Moran and Shapiro, 2010) this is simply expressed as

$$\Delta U + \Delta KE + \Delta PE = Q - W \qquad \text{(Eq. 3.24)}$$

where U is the internal energy of the system, KE ($=1/2\ mv^2$) the kinetic energy, PE the potential energy due to the action of a field (for gravitation PE = mgz, with z the height in this field), Q the heat supply (positive if added to the system, negative if removed), and W the work done by the system (positive if the system performs work, negative if work is exerted on the system by some net force action). The terms at the left-hand side of Equation (3.23) are equal to ΔE, the total energy change of the system.

For an open (flow) system with a characteristic control volume as indicated in Figure 3.2, the macroscopic energy balance can now be written as follows:

$$\frac{dE_{cv}}{dt} = \dot{Q} - \dot{W} + \varphi_{m,\,in}\left(u_{in} + \frac{v_{in}^2}{2} + gz_{in}\right) - \varphi_{m,\,out}\left(u_{out} + \frac{v_{out}^2}{2} + gz_{out}\right) \qquad \text{(Eq. 3.25)}$$

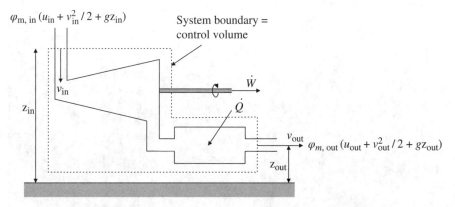

FIGURE 3.2   Energy balance illustration for an open-flow system.

In Equation (3.25), the work term is the sum of two contributions. The first is the "volume work" by the fluid, which can be expressed as

$$\dot{W}_{vol} = \varphi_{m,\,out} \cdot (p_{out} v_{out}) - \varphi_{m,\,in} \cdot (p_{in} v_{in}) \qquad \text{(Eq. 3.26)}$$

where v is the mass-specific volume (unit volume per unit mass, or $1/\rho$) and p the pressure.

The second contribution consists of all other types of work $\dot{W}_{cv}$, such as rotating shaft work, boundary displacement work, and magnetic and electric (field) work. Thus,

$$\dot{W} = \dot{W}_{cv} + \varphi_{m,\,out}(p_{out} v_{out}) - \varphi_{m,\,in}(p_{in} v_{in}) \qquad \text{(Eq. 3.27)}$$

Now, $u + pv$ ($= u + p/\rho$) has been defined as the enthalpy, h. The energy rate balance is now simplified by introduction of this quantity:

$$\frac{dE_{cv}}{dt} = \dot{Q}_{cv} - \dot{W}_{cv} + \varphi_{m,\,in}\left(h_{in} + \frac{v_{in}^2}{2} + gz_{in}\right) - \varphi_{m,\,out}\left(h_{out} + \frac{v_{out}^2}{2} + gz_{out}\right) \qquad \text{(Eq. 3.28)}$$

For the enthalpy of a mixture, the following relation holds:

$$h = \sum_{i=1}^{N} Y_i h_i \quad \text{and} \quad h_i = h_i^{ref} + \int_{T^{ref}}^{T} c_{p,i}(\tau) d\tau \qquad \text{(Eq. 3.29)}$$

The reference condition is usually defined as 1 bar and 298.15 K, and c_p for each species is often given as a polynomial function of temperature (see, e.g., Smith et al. (2005)); the website of the National Institute of Standards and Technology (NIST) provides useful data for a multitude of (inorganic) species (tinyurl.com/9s23f); where experimental data are not available, methods of estimation are employed, as described by Poling et al. (2001). The heat capacity at constant pressure for a mixture can be calculated as a mass fraction average:

$$c_p = \sum_{i=1}^{N} Y_i c_{p,i} \qquad \text{(Eq. 3.30)}$$

Regarding the density of the fluid mixture, for gases often as equation of state (EOS), the ideal gas law can be used, though this must be verified. Smith et al. (2005) give a good overview of the selection of the appropriate EOS. In cases for which the ideal gas law holds,

$$\rho = \frac{p\overline{MW}}{R_u T} \text{ with the average molecular weight } \overline{MW} = \left(\sum_{i=1}^{N} \frac{Y_i}{MW_i}\right)^{-1} \qquad \text{(Eq. 3.31)}$$

The energy balance can also be formulated in microscopic form, by returning to the approach indicated by Equation (3.2), recognizing here that the energy density $f = \rho u + \frac{1}{2}\rho v^2$, being the sum of the total internal energy density and the KE density of the flow. PE is treated as source term, like radiation; thus, $s_f = \rho \vec{v} \cdot \vec{g} + Q_r$. Heat transfer by radiation is dealt with in more details in Chapter 4. Note that chemical reactions are no source of energy for the flow, as they only convert chemical energy into thermal energy of the flow. The energy flux is now given by

$$\vec{\phi}_f = \rho \vec{v} \left(\rho u + \frac{1}{2}\rho v^2\right) + p \vec{v} - \boldsymbol{\tau} \cdot \vec{v} + \vec{j}_q \qquad \text{(Eq. 3.32)}$$

in which the first term represents the convective flux of internal energy and KE and the other three terms are work exerted by pressure and viscous forces and the energy flux due to conduction, respectively. Substitution of these terms in Equation (3.2) yields

$$\frac{\partial}{\partial t}\left(\rho u + \frac{1}{2}\rho v^2\right) + \nabla \cdot \left[\vec{v}\left(\rho u + \frac{1}{2}\rho v^2\right)\right] + \nabla \cdot \left(p \vec{v} - \boldsymbol{\tau} \cdot \vec{v}\right) + \nabla \cdot \vec{j}_q = \rho \vec{v} \cdot \vec{g} + Q_r$$

$$\text{(Eq. 3.33)}$$

Setting up and working out energy balances is useful to describe the performance of a system in terms of first law efficiencies. The concept of exergy makes use of the second law of thermodynamics and enables valuing heat streams in terms of how much work can actually be derived from them. Exergy is not conserved but is destroyed to a greater or lesser extent by irreversibilities in the system. Although an exergy analysis adds value to the analysis of the system performance with respect to energy conversion, this is considered to be out of the scope of this book. We refer the interested reader to textbooks such as Moran and Shapiro (2010), Çengel and Bolnes (2010), and Szargut (2005).

Example 3.3   Energy balance calculation (macroscopic): A continuous pyrolysis vapor condenser

A biomass pyrolysis reactor produces a pyrolysis vapor, which is a complex mixture of organic compounds and water. In this example, we consider an indirect-contact heat exchanger for complete condensation of this vapor, with water used as cooling liquid. The vapor (containing no liquid droplets) enters the condenser at its boiling point of 380 K (assumed to be constant); it leaves the heat exchanger at 45°C. The cooling water is heated up from 20 to 35°C. The heat capacity of water may be assumed to be constant at 4.18 kJ·kg^{-1}·K^{-1}.

The heat of vaporization is assumed to be a typical value for an organic acid: 400 kJ·kg^{-1}. The heat capacity at constant pressure, c_p, is assumed to be constant at 2 kJ·kg^{-1}·K^{-1}.

a. Calculate the ratio of the mass flow rates of the condensing vapor and the cooling water.

b. Calculate the rate of energy transfer from the condensing vapor to the cooling water in kJ·kg^{-1} of vapor passing through the condenser.

Solution

A schematic of the heat exchanger with system boundaries is shown in Figure 3.3. Assumptions:

1. Both control volumes indicated in Figure 3.3 are at steady state.

2. Heat transfer to the environment of the condenser is neglected, and no (shaft) work is done by the condenser system.

3. KE and PE changes of the flows can be neglected.

a. Mass balances can be set up based on the fact that the condensing vapor stream does not mix with the cooling water. Thus,

$$\varphi_{m,1} = \varphi_{m,2} \quad \text{and} \quad \varphi_{m,3} = \varphi_{m,4}$$

Now, the ratio of mass flow rates can be derived from a steady-state energy balance around the heat exchanger:

$$0 = \dot{Q}_{cv} - \dot{W}_{cv} + \varphi_{m,1}\left(h_1 + \frac{v_1^2}{2} + gz_1\right) + \varphi_{m,3}\left(h_3 + \frac{v_3^2}{2} + gz_3\right)$$

$$-\varphi_{m,2}\left(h_2 + \frac{v_2^2}{2} + gz_2\right) - \varphi_{m,4}\left(h_4 + \frac{v_4^2}{2} + gz_4\right)$$

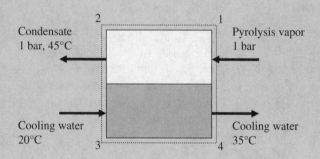

FIGURE 3.3 Schematic of the condenser; top: control volume for (a); bottom: control volume for (b).

Based on the aforementioned assumptions, the first two terms on the right-hand side of the equation are zero, and all $(v_i^2/2)$ and gz_i terms are canceled, which gives

$$\frac{\varphi_{m,3}}{\varphi_{m,1}} = \frac{\{2(380-318)+400\}}{\{4.18(308-293)\}} = 8.36$$

b. When considering the control volume of the condensing stream only, the energy rate balance at steady state is

$$0 = \dot{Q}_{cv} - \dot{W}_{cv} + \varphi_{m,1}\left(h_1 + \frac{v_1^2}{2} + gz_1\right) - \varphi_{m,2}\left(h_2 + \frac{v_2^2}{2} + gz_2\right)$$

With no work done by this subsystem and changes in KE and PE neglected, it follows that

$$\dot{Q}_{cv} = \varphi_{m,1}(h_2 - h_1) \Rightarrow \frac{\dot{Q}_{cv}}{\varphi_{m,1}} = (h_2 - h_1) = -\{400 + 2\cdot(380-318)\} kJ\cdot kg^{-1}$$
$$= 524 \ kJ\cdot kg^{-1}$$

Example 3.4   Energy balance (microscopic)

A cylindrical tube made of copper is used for heating up water to a final temperature of 30°C. The tube has an internal radius, r, of 1.5 cm. Water enters the tube at an ambient temperature of 20°C. The tube wall is kept at a constant temperature of 40°C. Consider the water density and heat capacity (c_p) to be constant with values of 1000 kg·m^{-3} and 4.18 J·g^{-1}·K^{-1}, respectively. Assume that the heat transfer from the tube wall is linearly dependent on the temperature difference and the surface area over which the heat is transferred, with the overall heat transfer coefficient (h) having a value of 3450 W·m^{-2}·K^{-1}.

With a required flow of 2232 l·h^{-1}, what is the tube length (L)?

Solution
For this problem, one needs to set up a microscopic energy balance over an infinitesimally small tube segment. The process is considered to be in steady state. The energy balance then becomes

$$0 = \varphi_V \rho c_p \overline{T}_x - \varphi_V \rho c_p \overline{T}_{x+\Delta x} + h\cdot(T_0 - \overline{T}_x)\cdot 2\pi r\cdot \Delta x$$

Now, $\overline{T}_{x+\Delta x}$ can be written as a Taylor series, $\overline{T}_{x+\Delta x} = \overline{T}_x + \dfrac{dT}{dx} \times \Delta x + \cdots$, with $\Delta x \to 0$ subsequent terms tending to zero. Then it follows that

$$0 = \varphi_V \rho c_p \left(-\frac{d\overline{T}_x}{dx}\right)\Delta x + h\cdot(T_0 - \overline{T}_x)\cdot 2\pi r\cdot \Delta x$$

Thus,

$$\varphi_V \rho c_p \frac{d\overline{T}_x}{dx} = h \cdot (T_0 - \overline{T}_x) \cdot 2\pi r \Rightarrow \frac{d\overline{T}_x}{(T_0 - \overline{T}_x)} = \frac{2\pi r h}{\varphi_V \rho c_p} \cdot dx$$

Integration from the inlet to the outlet for temperature and length of the tube gives

$$\int_{T_{in}}^{T_{out}} \frac{d\overline{T}_x}{(T_0 - \overline{T}_x)} = \int_{x=0}^{x=L} \left(\frac{2\pi r h}{\varphi_V \rho c_p} \right) \cdot dx \Rightarrow \left[-\ln(T_0 - \overline{T}_x) \right]_{T_{in}}^{T_{out}} = \frac{2\pi r h}{\varphi_V \rho c_p} (L - 0)$$

So

$$\ln\left\{ \frac{(T_0 - T_{out})}{(T_0 - T_{in})} \right\} = -\frac{2\pi r h L}{\varphi_V \rho c_p}$$

Substituting the data in this equation gives $L = 5.53$ m.

3.3.2 Energy Balances for Systems with Chemical Reactions

Here, we focus on the conservation of energy in reaction systems. This context has impact on the evaluation of the thermodynamic properties. When no chemical reaction takes place, as for the systems discussed in Section 3.3.1, the composition of a stream does not change and differences in thermodynamic properties such as internal energy (u), enthalpy (h), and entropy (s) are of importance, but the absolute values can be considered with respect to arbitrary reference states. In contrast, chemical reactions in systems cause species to be formed and destructed; therefore, changes cannot be calculated in general for all species in such systems. The thermodynamic properties now need to be univocally measured with respect to a unique standard state.

For this purpose, the choice has been made by international convention to define the enthalpy of the elements in their stable state to be zero (0) at a reference temperature, T_{ref}, of 298.15 K and reference pressure, p_{ref}, of 1 atm (absolute). For example, the stable state of some relevant, abundant species, hydrogen, nitrogen, and oxygen, is the molecular state: H_2, N_2, and O_2.

The enthalpy for a species i on a mole-specific basis can be expressed as

$$\overline{h}_i(T,p) = \overline{h}_{f,i}^0 + \left\{ \overline{h}_i(T,p) - \overline{h}_i(T_{ref}, p_{ref}) \right\} = \overline{h}_{f,i}^0 + \Delta \overline{h}_i \qquad \text{(Eq. 3.34)}$$

3.3.2.1 Open-Flow Systems For open-flow systems, the steady-state energy balance Equation (3.28) is rewritten in terms of moles and molar properties. With the assumption that changes in KE and PE are neglected, the following equation results:

$$\dot{Q}_{cv} - \dot{W}_{cv} + \varphi_{n,\text{fuel in}} \left\{ \sum_{\text{Reactants}} n_{\text{in per mole fuel}} \left(\overline{h}_f^0 + \Delta \overline{h} \right)_{\text{in}} \right.$$

$$\left. - \sum_{\text{Products}} n_{\text{out per mole fuel}} \left(\overline{h}_f^0 + \Delta \overline{h} \right)_{\text{out}} \right\} = 0 \qquad \text{(Eq. 3.35)}$$

Example 3.5   Ethanol combustion in an internal combustion engine (ICE)

Bioethanol produced by fermentation of sugars extracted from sugarcane (see Chapter 13) is combusted in a test Otto engine. The excess air amount is 5% and the combustion product gas (flue gas) leaves the engine at 800 K. The brake-specific fuel consumption (BSFC), defined as fuel consumed divided by the power produced "at the brake," is 275 $g \cdot (kWh)^{-1}$. The consumption rate of fuel is $7000 \, g \cdot h^{-1}$.
  What is the rate of heat removal (in kW)?

Data

Compound	$\bar{h}_f^o (kJ \cdot kmol^{-1})$	$\bar{h}(800 \, K) - \bar{h}_f^o (kJ \cdot kmol^{-1})$
C_2H_5OH	−229,757	—
CO_2	−393,546	22,810
H_2O (g)	−241,845	18,005
O_2	0	15,838
N_2	0	15,046

Solution
The reaction equation for stoichiometric ethanol combustion is

$$C_2H_5OH + a(O_2 + 3.76 \, N_2) \rightarrow bCO_2 + cH_2O + 3.76aN_2$$

C balance: $2 = b$
H balance: $6 = 2c$, so $c = 3$
O balance: $2a + 1 = 2b + c$, so $a = 3$
  Then

$$C_2H_5OH + 3(O_2 + 3.76 \, N_2) \rightarrow 2CO_2 + 3H_2O + 11.28N_2$$

The ethanol combustion equation for excess air situations is given by

$$C_2H_5OH + 3\lambda(O_2 + 3.76 \, N_2) \rightarrow 2CO_2 + 3H_2O + (\lambda - 1)O_2 + 11.28\lambda N_2$$

where λ is the air stoichiometry coefficient, in this case $\lambda = 1.05$.
Thus, the reaction equation is

$$C_2H_5OH + 3.15(O_2 + 3.76 \, N_2) \rightarrow 2CO_2 + 3H_2O + 0.05O_2 + 11.84N_2$$

The energy balance is

$$\dot{Q}_{cv} = \dot{W}_{cv} + \varphi_{n, \text{fuel in}} \left[\left\{ 2\left(\bar{h}_f^0 + \Delta\bar{h}\right)_{CO_2} + 3\left(\bar{h}_f^0 + \Delta\bar{h}\right)_{H_2O} + 0.05\left(\bar{h}_f^0 + \Delta\bar{h}\right)_{O_2} \right. \right.$$

$$\left. \left. + 11.84\left(\bar{h}_f^0 + \Delta\bar{h}\right)_{N_2} \right\}_{\text{prod.}} - \left\{ \left(\bar{h}_f^0 + \Delta\bar{h}\right)_{C_2H_5OH} + 3.15 \times 0 + 11.84 \times 0 \right\}_{\text{react.}} \right]$$

In this equation,

$$\dot{W}_{cv} = \frac{7000 \text{ g} \cdot \text{h}^{-1}}{275 \text{ g} \cdot (\text{kWh})^{-1}} = 25.5 \text{ kW (produced)}$$

$$\varphi_{n, \text{fuel in}} = \frac{7 \text{ kg} \cdot \text{h}^{-1}}{3600 \text{ s} \cdot \text{h}^{-1} \times 46.07 \text{ kg} \cdot \text{kmol}^{-1}} = 4.22 \times 10^{-5} \text{kmol} \cdot \text{s}^{-1}$$

Substituting the tabulated values in the term between brackets gives

$$\dot{Q}_{cv} = 25.5 \text{ kW} + 4.22 \times 10^{-5} \text{ kmol} \cdot \text{s}^{-1} \times (-1,004,298 \text{ kJ} \cdot \text{kmol}^{-1})$$
$$= -16.9 \text{ kW (to be extracted).}$$

Example 3.6   Adiabatic, stoichiometric flame temperature for DME combustion

Dimethyl ether (DME) is considered as an alternative diesel fuel for trucks. Its chemical formula is CH_3OCH_3. Consider an open-flow system in which DME is combusted at a constant pressure of 1 atm. The inlet temperature of both air and fuel is 25°C.

The $\bar{c}_p$ value of the generated flue gases is represented by the following polynomial:

$$\frac{\bar{c}_p}{R_u} = a_1 + a_2 T + a_3 T^2 + a_4 T^3 + a_5 T^4$$

$$R_u = 8.3143 \text{ kJ} \cdot (\text{kmol} \cdot \text{K})^{-1}$$

The coefficients of the polynomial are given in the table below (Kee et al., 1991).

Compound	a_1	$a_2 \cdot 10^3$	$a_3 \cdot 10^6$	$a_4 \cdot 10^{10}$	$a_5 \cdot 10^{14}$
CO_2	4.453623	3.140168	−1.2784105	2.393996	−1.6690333
H_2O	2.672145	3.056293	−0.873026	1.2009964	−0.6391618
N_2	2.926640	1.4879768	−0.568476	1.0097038	−0.6753351

The mole-specific enthalpies are given in the following table.

Compound	$\bar{h}_f^o$ (kJ·kmol^{-1})
CH_3OCH_3	−184,100
CO_2	−393,546
H_2O (g)	−241,845

Calculate the final combustion temperature when the fuel is just completely combusted with no oxygen excess under adiabatic conditions.

Solution

DME has the same overall chemical formula as ethanol (it is an isomer); thus, based on the previous example, we can easily write out the reaction equation for stoichiometric combustion:

$$CH_3OCH_3 + 3(O_2 + 3.76\,N_2) \rightarrow 2CO_2 + 3H_2O + 11.28N_2$$

Recognizing that no work is done by or on the system and that the system is adiabatic (no heat flows), the energy balance is very much simplified as

$$\sum_{\text{Reactants}} n_{\text{in per mole fuel}} \left(\bar{h}_f^0 + \Delta \bar{h} \right)_{\text{in}} = \sum_{\text{Products}} n_{\text{out per mole fuel}} \left(\bar{h}_f^0 + \Delta \bar{h} \right)_{\text{out}}$$

Thus, $H_{\text{Products}} = H_{\text{Reactants}}$.

For the left-hand side of the equation, only $\bar{h}_{f,\text{DME}}^o$ remains, leading to the equation

$$\bar{h}_{f,\text{DME}}^o = 2 \times \left(-393{,}546 + \int_{298.15}^{T_{ad}} \bar{c}_{p,CO_2}\,dT \right) + 3 \times \left(-241{,}845 + \int_{298.15}^{T_{ad}} \bar{c}_{p,H_2O}\,dT \right)$$

$$+ 11.28 \times \left(0 + \int_{298.15}^{T_{ad}} \bar{c}_{p,N_2}\,dT \right)$$

which is equal to

$$0 = -\bar{h}_{f,\text{DME}}^o + 2 \times (-393{,}546) + 3 \times (-241{,}845)$$

$$+ R_u \int_{298.15}^{T_{ad}} \left(A + BT + CT^2 + DT^3 + ET^4 \right) dT$$

with

$$A = 2a_{1,CO_2} + 3a_{1,H_2O} + 11.28a_{1,N_2}$$
$$B = 2a_{2,CO_2} + 3a_{2,H_2O} + 11.28a_{2,N_2}$$
$$C = 2a_{3,CO_2} + 3a_{3,H_2O} + 11.28a_{3,N_2}$$
$$D = 2a_{4,CO_2} + 3a_{4,H_2O} + 11.28a_{4,N_2}$$
$$E = 2a_{5,CO_2} + 3a_{5,H_2O} + 11.28a_{5,N_2}$$

Integration with partial substitution of the numbers yields

$$0 = -(-184{,}100) + 2 \times (-393{,}546) + 3 \times (-241{,}845)$$

$$+ 8.3143 \times \left[\begin{array}{l} A(T_{ad} - 298.15) + {}^1/_2 B\left(T_{ad}^2 - 298.15^2\right) + {}^1/_3 C\left(T_{ad}^3 - 298.15^3\right) \\ + {}^1/_4 D\left(T_{ad}^4 - 298.15^4\right) + {}^1/_5 E\left(T_{ad}^5 - 298.15^5\right) \end{array} \right]$$

A nonlinear equation solver (e.g., the generalized reduced gradient (GRG) nonlinear solver in Excel) can now be employed to solve for the only unknown, T_{ad}:

$$T_{ad} = 2428.4\,K = 2155.2°C$$

3.3.2.2 Closed Systems

The steady-state energy balance for closed reaction systems is as follows (ideal gas law assumed for both reactants and products):

$$0 = \sum_{Reactants} n\bar{u} - \sum_{Products} n\bar{u} + Q_{cv} - W_{cv}$$

$$\Leftrightarrow Q_{cv} - W_{cv} = \sum_{Products} n(\bar{h} - R_u T_{Products}) - \sum_{Reactants} n(\bar{h} - R_u T_{Reactants}) =$$

$$\sum_{Products} n\left(\bar{h}_f^o + \Delta\bar{h} - R_u T_{Products}\right) - \sum_{Reactants} n\left(\bar{h}_f^o + \Delta\bar{h} - R_u T_{Reactants}\right) =$$

$$\sum_{Products} n\left(\bar{h}_f^o + \Delta\bar{h}\right) - \sum_{Reactants} n\left(\bar{h}_f^o + \Delta\bar{h}\right) - R_u T_{Products} \sum_{Products} n + R_u T_{Reactants} \sum_{Reactants} n$$

(Eq. 3.36)

Example 3.7 Stoichiometric combustion of biogas in a closed vessel

A certain (idealized) biogas has the following composition: CH_4 (60 vol.%) and CO_2 (40 vol.%). Combustion is exactly complete with air and takes place in a closed vessel. The initial temperature is 25°C and the pressure then is 1 atm. Assume ideal gas conditions throughout the process. Consider the values of $\bar{c}_p$ of the products to be independent of T (evaluated at 1200 K). The mole-specific enthalpies and the heat capacities of the products are given in the following table.

Compound	$\bar{h}_f^o$ (kJ·kmol^{-1}·K^{-1})	$\bar{c}_p$ (kJ·kmol^{-1}·K^{-1})
CO_2	−393,546	56.205
H_2O	−241,845	43.874
N_2	0	33.707
CH_4	−74,850	—

Calculate the final temperature (in °C) and pressure (in atm) assuming the vessel is well insulated. Was the temperature at which the $\bar{c}_p$ of the flue gas was evaluated reasonable? What is the effect of CO_2 in the methane gas on the end pressure and temperature?

Solution
The reaction for the combustion process can be written as

$$CH_4 + 0.67 CO_2 + 2(O_2 + 3.76 N_2) \rightarrow 1.67 CO_2 + 2 H_2O + 7.52 N_2$$

The energy balance now is (no heat supply nor extraction and no work related to the system) $U_{products} - U_{reactants} = 0$:

$$0 = \left[\frac{5}{3}(\bar{h}_{CO_2} - R_u T_{ad}) + 2(\bar{h}_{H_2O} - R_u T_{ad}) + 7.52(\bar{h}_{N_2} - R_u T_{ad}) \right]$$

$$- \left[(\bar{h}_{CH_4} - R_u T_0) + 2(\bar{h}_{O_2} - R_u T_0) + 7.52(\bar{h}_{N_2} - R_u T_0) + \frac{2}{3}(\bar{h}_{CO_2} - R_u T_0) \right]$$

$$0 = \left[\frac{5}{3}\left\{ \bar{h}^o_{f,CO_2} + \bar{c}_{p,CO_2}(T_{ad} - T_0) \right\} + 2\left\{ \bar{h}^o_{f,H_2O} + \bar{c}_{p,H_2O}(T_{ad} - T_0) \right\} \right.$$

$$\left. + 7.52\left\{ \bar{h}^o_{f,N_2} + \bar{c}_{p,N_2}(T_{ad} - T_0) \right\} \right]$$

$$- \left[\bar{h}^o_{f,CH_4} + 2\bar{h}^o_{f,O_2} + 7.52\bar{h}^o_{f,N_2} + \frac{2}{3}\bar{h}^o_{f,CO_2} \right] - \left(1\frac{2}{3} + 2 + 7.52 \right) R_u(T_{ad} - T_0)$$

Now, bringing all temperature-dependent terms $(T_{ad} - T_0)$ to the left-hand side of the equation, one obtains an equation that is easy to solve with $T_{ad} - T_0$ as the only unknown. The numerical result is

$$T_{ad} - T_0 = 2346.9 \text{ K}, \quad \text{so that} \quad T_{ad} = 2645 \text{ K} = 2372°C.$$

At constant volume, the end pressure can be calculated to be

$$p_2 = T_2.p_1/T_1 = 2645\,(1/298) = 8.9 \text{ atm}.$$

There is a slight discrepancy when taking the $\bar{c}_p$ value constant at 1200 K as the average temperature, but this does not cause a very large error. In fact, assuming that no endothermic dissociation of CO_2 takes place (forming CO and ½ O_2) causes a larger discrepancy (Turns, 2000). If no CO_2 were present in this gas, then a significantly higher temperature and associated pressure would result.

3.4   CONSERVATION OF MOMENTUM

Momentum, unlike mass or energy, has a direction, and therefore, it is a vector quantity and so is momentum density, $f = \rho \vec{v}$. For each directional component of this vector, a conservation equation can be derived. However, the vector notation is used here. The external source (or force) is usually only gravity, so then $s_f = \rho \vec{g}$. As f is already a vector, the momentum flux becomes a tensor: $\vec{\varphi}_f = \rho \vec{v} \vec{v} + pI - \tau$.

Here, the first term represents the convective momentum flux, the second term is the pressure tensor with pressure p, and the third term is the viscous stress tensor. Substitution in Equation (3.2) yields

$$\frac{\partial \rho \vec{v}}{\partial t} + \nabla \cdot \left(\rho \vec{v} \vec{v} + pI - \tau \right) = \rho \vec{g} \qquad \text{(Eq. 3.37)}$$

Example 3.8   Determination of the velocity profile for liquid flow in a wide rectangular duct

Figure 3.4 shows a schematic of the situation. The liquid height is H. As a result of a pressure drop, a noncompressible liquid flows steadily (no time dependency) through the wide rectangular duct. Assume the pressure drop over the duct to be constant, dp/dx. The fluid is to be considered Newtonian, which means

$$\tau_y = -\eta \frac{dv}{dy}$$

Derive an expression for the velocity profile in the flow direction (x) as a function of the liquid depth direction (y).

Solution
In order to determine a velocity profile, one needs to set up a momentum balance. Consider the rectangular duct to be infinitely broad. Then the momentum balance over a differential volume $w \cdot dx \cdot dy$ can be written as

$$\rho v_x w dy\, v_x - \rho v_{x+dx} w dy\, v_{x+dx} - \tau_y w dx + \tau_{y+dy} w dx + p_x w dy - p_{x+dx} w dy = 0$$

The first two terms go to zero, as mass flow is constant and the medium is incompressible. For $dx \rightarrow 0$, one can write

$$\frac{d\tau_y}{dy} = + \frac{dp}{dx}$$

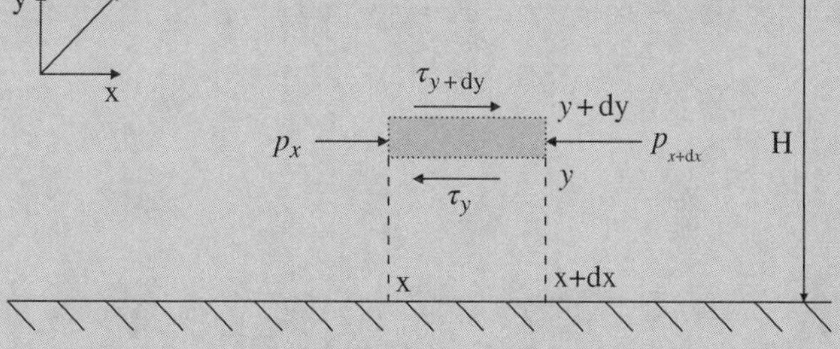

FIGURE 3.4   Schematic of the rectangular duct with liquid flow; width (w) in z-direction is infinite.

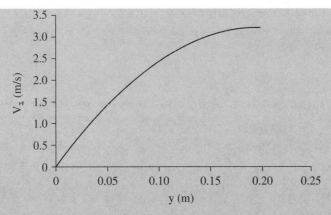

FIGURE 3.5   Velocity in x-direction as a function of the liquid height y.

Boundary conditions: $y = H$, and then $\tau_y = 0$ and thus

$$\tau_y = \frac{dp}{dx}y - \frac{dp}{dx}H$$

Now, as the liquid is Newtonian and given the boundary condition $y = 0$, $v_x = 0$ m·s⁻¹ (no slip condition at the bottom of the duct), it follows that

$$v_x = -\frac{1}{2\eta}\frac{dp}{dx}y^2 + \frac{1}{\eta}\frac{dp}{dx}Hy$$

Check whether the units of the right-hand side terms result in [m·s⁻¹]. A sketch of the velocity profile in y-direction is given in Figure 3.5.

CHAPTER SUMMARY AND STUDY GUIDE

This chapter deals with the formulation of mass, species, energy, and momentum balances. Setting up these equations for closed and open-flow systems and simplifying them as much as possible are a prerequisite for understanding any bioenergy conversion system.

KEY CONCEPTS

Systems: open and closed
Balances: mass, energy (both nonreaction and reaction systems), and momentum.
Microscopic versus macroscopic balances (differential vs. control volume approach)
Chemical conversion, selectivity, and yield

First law of thermodynamics
Internal energy
Kinetic energy
Potential energy
Work

SHORT-ANSWER QUESTIONS

3.1 When can a system be considered in "steady state"?

3.2 What is the difference between an open system and a closed system?

3.3 What is the difference between microscopic and macroscopic balances? For which situations in general do you use each of both approaches?

3.4 What is the difference between the "degree of conversion" and the "relative degree of conversion" for a chemical reaction $2A \rightarrow B$? What changes when the degree of conversion is expressed on a molar basis?

3.5 In the energy balance, the change in potential energy is usually based on the gravity field; which other fields might be relevant as well and which terms would then appear?

3.6 Indicate whether the following statements are in general *true* or *false*:

- Enthalpy is always conserved.
- Momentum is always conserved.
- The number of moles of a chemical component is always conserved in a chemical reaction.
- The mass of each element is conserved during combustion of biomass.

PROBLEMS

3.1 Using Equation (3.6), derive a material balance for a reactant in a closed system.

3.2 For a reaction mixture with a density that is linearly dependent on the conversion, with $\rho = \rho_0(1 + \varepsilon \zeta_A)$, derive a relation for the concentration of species $A(c_A)$ as a function of the relative degree of conversion ζ_A.

3.3 The cavitation number is defined as $Ca = \dfrac{p - p_{vap}}{(1/2)\rho v^2}$.

Now, for ethanol produced by sugar fermentation, a company has installed a pump for the transport of the liquid product, which is assumed to be pure. The ambient pressure is 1020 hPa and the pump is situated 1 m below a vessel from which the product is pumped through a duct of 10 cm diameter with a mass flow rate of 25 t·h^{-1}. The temperature at the pump suction side is 20°C. At this temperature, the vapor pressure is 5.7×10^3 Pa. The density of ethanol is 789 kg·m^{-3}. What is the background of cavitation? Does it occur in this situation? Which assumption(s) have you made? When will there be a possibility for this phenomenon to occur?

3.4 Present the time-dependent mass balance of a boiler drum in a steam power plant.

3.5 A steam boiler drum produces steam at a mass flow rate of 64 kg·s^{-1} at 60 bar. This stream still contains 2 wt% moisture. The feedwater from the economizer (a heat exchanger) is fed to the drum at a mass flow rate of 62 kg·s^{-1}. It contains 3 ppm$_w$ solids. Makeup water, containing 50 ppm$_w$ solids, is also fed into the drum at a mass flow rate of 2 kg·s^{-1}. Steam production is such that the solid content of the moisture leaving with the steam is 5 ppm$_w$. Blowdown, which is the release of hot liquid from the bottom of the drum, must keep the concentration of solids in the drum to 1000 ppm$_w$.

   a. Calculate the blowdown requirement in kg·s^{-1}.
   b. Calculate the heat loss associated with blowdown in kW.

3.6 A steam boiler has a horizontal 1.5 m (inside diameter) drum that is 6.2 m long. The normal water level is at the midplane. If the water level falls more than 30 cm below this point, serious dry running of some downcomers will be the result. When the steaming rate is 12 kg·s^{-1} and the pressure is 40 bar, how long an interruption in the feedwater supply could be allowed?

3.7 A massive wood block with a mass of 1 kg falls from 15 m height in a processing building with an initial temperature of 20°C; it doesn't jump up after touching the ground. What is the end temperature when the heat capacity can be considered to be constant: $c_p = 1000$ J·kg^{-1}·K^{-1}?

3.8 Power company "E" in the Netherlands operates a fluidized bed combustion-based boiler with wood residues as fuel. The plant contains a simple steam turbine. The steam conditions at the inlet of the turbine are 525°C and 100 bar. Assume that condensation of the steam takes place at a temperature of 20°C and that isentropic expansion of steam occurs in the turbine.

   a. What is the specific power (kJ·kg^{-1}) of the turbine expansion process?
   b. If the power plant generates 25 MW$_e$ and water pump work can be neglected, what is the mass flow rate of steam through the turbine?
   c. What assumptions have you made for these calculations?

3.9 Biomass chips from a hopper system are fed vertically (90° angle) onto a moving conveyer belt. The mass flow of the chips is 24 t·h^{-1}. At the end of the belt, the chips are discharged. The drive wheels have a diameter of 80 cm, and they rotate (clockwise) with a rate of 150 revolutions per minute. Friction forces as well as air drag on the particles can be neglected. Calculate the driving power of this belt.

3.10 Bio-oil flows through a duct with a radius of 2.5 cm at ambient temperature with a velocity (v_1) of $10 \, m{\cdot}s^{-1}$; the duct is followed by a permeable wall part with suction. At the end of this section (with the same radius), the velocity (v_2) has dropped to $8 \, m{\cdot}s^{-1}$. If $p_1 = 140$ kPa, estimate p_2 for the case that wall friction is negligible. What happens to p_2 in the case of significant friction?

3.11 An anaerobic digester (see Chapter 14 for this technology) produces biogas at 1.1 bar (absolute) and 25 °C. This gas, composed of 60 vol.% CH_4 and 40 vol. % CO_2, is to be compressed to 25 bar (absolute) before delivery to its end use; assume the compression to be isentropic and reversible. Calculate the temperature after compression.

3.12 Dimethyl ether (DME) is combusted in a closed cylinder with 5 mol% excess air. Assume the cylinder volume to remain constant during the (explosive) combustion process. If no heat is exchanged with the environment, what is the end temperature and pressure in the cylinder when the initial pressure is 1 atm. and the initial temperature is 25 °C?

3.13 The enthalpy of formation of a bio-oil with the empirical composition formula of $CH_{2.3}O_{0.9}$ is estimated to be $-183.7 \, kJ{\cdot}mol^{-1}$ at 20°C and 1 bar. Researchers intend to convert this gas with a plasma torch into synthesis gas via the reaction:

$$CH_{2.3}O_{0.9} + 0.3H_2O(l) \rightarrow CO + 1.25H_2 + 0.2H_2O(g)$$

What is the reaction enthalpy at 1100 K and 1 bar?

3.14 Determine the adiabatic, stoichiometric flame temperature at constant pressure for n-butanol, an alternative biofuel, given that at standard conditions (25 C, 1 atm) the enthalpy of combustion is $-2670 \, kJ{\cdot}mol^{-1}$.

PROJECTS

P3.1 Visit a local power plant that processes biomass or coal, identify the key characteristics of this plant, and set up overall mass and energy balances for the plant.

P3.2 Test the mechanical drying of biomass using a roller press (either manually or electrically driven) of Example 8.1 (Chapter 8) yourself. Measure inlet and outlet temperature accurately using, e.g., a K-type thermocouple.

INTERNET REFERENCE

tinyurl.com/9s23f
http://www.nist.gov

REFERENCES

Bird RB, Stewart WE, Lightfoot EN. *Transport Phenomena*. Revised 2nd ed. Hoboken, NJ: John Wiley & Sons, Inc.; 2007.

Çengel YA, Boles MA. *Thermodynamics—An Engineering Approach*. 7th ed. New York: McGraw-Hill; 2010.

Kee RJ, Rupley FM, Miller JA. *The Chemkin thermodynamic data base*. Sandia Report nr. SAND87-8215B. Livermore, CA: Sandia National Laboratories; 1991.

Moran MJ, Shapiro HN. *Fundamentals of Engineering Thermodynamics*. 6th ed. Hoboken, NJ: John Wiley & Sons; 2010.

Poling BE, Prausnitz JM, O'Connell JP. *The Properties of Gases and Liquids*. 5th ed. New York: McGraw-Hill; 2001.

Smith JM, Van Ness HC, Abbott MM. *Introduction to Chemical Engineering Thermodynamics*. 7th ed. New York: McGraw-Hill; 2005.

Szargut J. *Exergy Method*. Boston, MA: WIT Press; 2005.

Turns S. *An Introduction to Combustion: Concepts and Applications*. 2nd ed. Boston, MA: McGraw-Hill; 2000.

Westerterp KR, Van Swaaij WPM, Beenackers AACM. *Chemical Reactor Design and Operation*. 2nd ed. New York: John Wiley & Sons; 1988.

4

TRANSFER: BASICS OF MASS AND HEAT TRANSFER

DIRK J.E.M. ROEKAERTS

Department of Process and Energy, Fluid Mechanics Section, Faculty of Mechanical, Maritime and Materials Engineering, Delft University of Technology, Delft, the Netherlands

ACRONYMS

PDF   probability density function
RTE   radiative transfer equation

SYMBOLS

A	area across which heat/mass transfer takes place	$[m^2]$
B	solid permeability	$[m^2]$
B_M	spalding mass coefficient	$[-]$
Bi	Biot number (defined in Table 4.1)	$[-]$
c	concentration	$[mol \cdot m^{-3}]$
c_p	specific heat capacity	$[J \cdot kg^{-1} \cdot K^{-1}]$
D	diffusion coefficient	$[m^2 \cdot s^{-1}]$
D	wall thickness	$[m]$
d	diameter	$[m]$
d_p	particle diameter	$[m]$
E	radiative emissive power	$[W \cdot m^{-2}]$
$\bar{E}_a$	absorption efficiency factor	$[-]$

Biomass as a Sustainable Energy Source for the Future: Fundamentals of Conversion Processes,
First Edition. Edited by Wiebren de Jong and J. Ruud van Ommen.
© 2015 American Institute of Chemical Engineers, Inc. Published 2015 by John Wiley & Sons, Inc.

E_a	activation energy	$[\text{J}\cdot\text{mol}^{-1}]$
F	view factor or configuration factor	$[-]$
Fo	Fourier number (defined in Table 4.1)	$[-]$
G	incident radiation	$[\text{W}\cdot\text{m}^{-2}]$
g	gravitational acceleration	$[\text{m}\cdot\text{s}^{-2}]$
Gr	Grashof number (defined in Table 4.1)	$[-]$
H	irradiation	$[\text{W}\cdot\text{m}^{-2}]$
h	mass-specific enthalpy	$[\text{J}\cdot\text{kg}^{-1}]$
h	heat transfer coefficient	$[\text{W}\cdot\text{m}^{-2}\cdot\text{K}^{-1}]$
h_{fg}	heat of vaporization	$[\text{J}\cdot\text{kg}^{-1}]$
I	total radiative intensity	$[\text{W}\cdot\text{m}^{-2}\cdot\text{sr}^{-1}]$
j_D	dimensionless number for mass transfer (Chilton–Colburn)	$[-]$
j_H	dimensionless number for mass transfer (Chilton–Colburn)	$[-]$
$\vec{j}_i$	species diffusive flux of species i (vector)	$[\text{kg}\cdot\text{m}^{-2}\cdot\text{s}^{-1}]$
$\vec{j}_q$	flux of energy crossing a control volume boundary	$[\text{W}\cdot\text{m}^{-2}]$
J	radiosity	$[\text{W}\cdot\text{m}^{-2}]$
K	constant in d^2-law	$[\text{m}^2\cdot\text{s}^{-1}]$
k	mass transfer coefficient	$[\text{m}\cdot\text{s}^{-1}]$
k	reaction rate coefficient	$[\text{s}^{-1}]$
k_0	pre-exponential factor	various
L	length of domain	$[\text{m}]$
Le	Lewis number (defined in Table 4.1)	$[-]$
MW	molecular weight	$[\text{kg}\cdot\text{mol}^{-1}]$
m	mass	$[\text{kg}]$
n	real part of index of refraction	$[-]$
N	number of species in mixture	$[-]$
N_s	number of surfaces bounding enclosure	$[-]$
Nu	Nusselt number (defined in Table 4.1)	$[-]$
p	pressure or partial pressure	$[\text{Pa}]$ or $[\text{bar}]$
Pr	Prandtl number (defined in Table 4.1)	$[-]$
$\vec{q}$	radiative heat flux (vector)	$[\text{W}\cdot\text{m}^{-2}]$
Q	heat supplied or extracted	$[\text{J}]$
$\dot{Q}$	heat flow supplied or extracted	$[\text{W}]$
r	radius	$[\text{m}]$
Re	Reynolds number (defined in Table 4.1)	$[-]$
R_{SA}	ratio of the real surface area to the ideal surface area	$[-]$
R_u	universal gas constant	$[\text{J}\cdot\text{mol}^{-1}\cdot\text{K}^{-1}]$
s	source term	various
Sc	Schmidt number (defined in Table 4.1)	$[-]$
Sh	Sherwood number (defined in Table 4.1)	$[-]$
T	temperature	$[\text{K}]$
t	time	$[\text{s}]$
u	velocity in 1D model	$[\text{m}\cdot\text{s}^{-1}]$
u	mass-specific internal energy	$[\text{J}\cdot\text{kg}^{-1}]$
$\vec{v}$	mixture velocity (vector)	$[\text{m}\cdot\text{s}^{-1}]$

$\vec{V}$	diffusion velocity (vector)	$[\text{m.s}^{-1}]$
V	volume	$[\text{m}^3]$
X_i	mole fraction of species i	[–]
Y_i	mass fraction of species i	[–]

α	heat diffusion coefficient	$[\text{m}^2\cdot\text{s}^{-1}]$
α	absorptivity	[–]
β	coefficient of thermal expansion	$[\text{K}^{-1}]$
ε	porosity	[–]
ε_e	emissivity	[–]
η	dynamic viscosity	$[\text{Pa}\cdot\text{s}]$
θ_m	factor accounting for blowing effect in Equation (4.69)	[–]
θ_T	factor accounting for blowing effect in Equation (4.69)	[–]
ϑ	normalized moisture concentration	[–]
κ	absorption coefficient	$[\text{m}^{-1}]$
λ	thermal conductivity	$[\text{W}\cdot\text{m}^{-1}\cdot\text{K}^{-1}]$
λ	wavelength	$[\text{m}]$
ν	kinematic viscosity	$[\text{m}^2\cdot\text{s}^{-1}]$
ρ	density	$[\text{kg}\cdot\text{m}^{-3}]$
ρ	reflectivity	[–]
σ	scattering coefficient	$[\text{m}^{-1}]$
σ	Stefan–Boltzmann constant	$[\text{W}\cdot\text{m}^{-2}\cdot\text{K}^{-4}]$
τ	viscous stress tensor	$[\text{Pa}]$
τ	relaxation time	$[\text{s}]$
τ	tortuosity	[–]
Φ	scattering phase function	$[\text{sr}^{-1}]$
$\dot{\omega}$	species mass source term	$[\text{kg}\cdot\text{m}^{-3}\cdot\text{s}^{-1}]$

Subscripts

atm	atmospheric
b	blackbody
b	bulk
B	biomass
boil	at boiling condition
C	char
d	droplet
eff	effective
F	fuel
g	gas
G	incident mean
i	species number i
I	inert gases
j	species number j
K	Knudsen
m	mixture

n   component in direction normal to wall
p   particle
P   Planck mean
r   radiation
s   slip
s   surface
sat   saturation
T   tar
T   thermal
w   at wall, or gas–solid interface
∞   position at large distance

Superscripts

HC   Hirschfelder–Curtis approximation

4.1   INTRODUCTION

Heat and mass transfer are essential in most biomass conversion processes. In this chapter, a basic introduction to the theoretical modeling of heat and mass transfer is given. The explanation of heat and mass transfer phenomena starts from the transport equations presented in Chapter 3. The general principles are illustrated with some specific examples concerning evaporation of ethanol droplets and devolatilization of wood particles. The subject is very broad. For topics not covered here and more detailed treatment including tables of basic physical data, we refer to the textbooks listed in the bibliography.

4.2   TRANSPORT TERMS IN THE GOVERNING EQUATIONS

4.2.1   Mass Transfer

When a system contains two or more components, these components in general have a slightly different velocity. Otherwise, mixing would not occur. The velocity of a component relative to the average velocity of the mixture is called the diffusion velocity. The relative velocity in general gives rise to the transport of a component from a region of higher concentration to a region of lower concentration, called mass transfer (see Welty et al., 2001, Ch. 24).

The diffusion velocity is a factor in the diffusive mass flux $\vec{j}_i = \rho \vec{V}_i$ appearing in the transport equation for species mass fractions presented as Equation (3.22) in Chapter 3:

$$\frac{\partial \rho Y_i}{\partial t} + \vec{\nabla} \cdot \left(\rho \vec{v} \, Y_i \right) = - \vec{\nabla} \cdot \left(\vec{j}_i \right) + \dot{\omega}_i \quad \text{for } i = 1, \dots, N \qquad \text{(Eq. 4.1)}$$

The four terms in this equation, respectively, are the transient term, the advection term, the diffusion term, and the source term. The advection term describes transport of quantities carried with the fluid velocity, defined as mass-weighted average of the velocity of the species. The diffusion term describes the motion of the species relative to the mean mixture. In the literature, the advection term often is called the convection term. In the context of heat and mass transfer problems, the notion "convective" refers to combined effects of advection and diffusion, as is discussed in Section 4.4 on convective heat and mass transfer.

The diffusion velocities in general depend on the set of binary diffusion coefficients associated with each pair of components of the mixture, but are also influenced by pressure gradients, differences in body forces between components and temperature gradients; this is the so-called Soret effect (Poinsot and Veynante, 2011). Keeping only the effect of the binary diffusion coefficients, the Stefan–Maxwell equations are obtained as a set of equations determining the diffusion velocities:

$$\vec{\nabla} X_i = \sum_{j=1}^{N} \frac{X_i X_j}{D_{ij}} \left(\vec{V}_j - \vec{V}_i \right)$$
(Eq. 4.2)

Here, N is the number of species, D_{ij} is the binary mass diffusion coefficient of species i into species j, and X_i denotes the mole fraction of species i. The mole fractions can be calculated from the mass fractions and vice versa using the molecular weights (MW) of the species. The Stefan–Maxwell equations form a linear system of size N^2 to be solved in each of the three spatial directions at each point and at each instant for unsteady flow. If the mixture contains only two species ($N = 2$), the system can be solved exactly and the well-known Fick's law is obtained (Kuo, 2005):

$$\vec{V}_1 Y_1 = -D_{12} \vec{\nabla} Y_1$$
(Eq. 4.3)

Another case where Fick's law is exact is multispecies diffusion ($N > 2$) with all binary diffusion coefficients equal. In a situation with unequal binary diffusion coefficients in principle, one should solve the Stefan–Maxwell equations, but to avoid the complexity and cost for doing so, usually an approximation in the form of a generalized Fick's law is used. In this approach, for each species, a diffusion coefficient relative to the mixture $D'_{i,m}$ is considered, and one uses, e.g., $\vec{V}_i Y_i = -D'_{i,m} \vec{\nabla} Y_i$. An expression for the coefficients $D'_{i,m}$ has to be provided to complete the model. The best first-order approximation to the solution of the Stefan–Maxwell equations is the Hirschfelder and Curtis approximation (Poinsot and Veynante, 2011):

$$\vec{V}_i X_i = -D_{i,m}^{HC} \vec{\nabla} X_i$$
(Eq. 4.4)

In the absence of detailed information on the composition of the mixture as is often the case in engineering applications, an estimated effective diffusion coefficient D, often assumed equal for all species, is used.

In the species mass fraction Equation (4.1), the transient term and the advection term on the left-hand side are balanced by the diffusive term and the source term on the right-hand side. The complexity of the mass transfer problem decreases when one or more of these terms vanish. This leads to a classification of mass transfer problems. Some examples of simplification are:

Steady diffusion in the presence of reaction, described by

$$\vec{\nabla} \cdot \left(\vec{j_i} \right) = \dot{\omega}_i \ \text{ for } i = 1, \ldots, N \tag{Eq. 4.5}$$

Unsteady diffusion in the absence of convection and reaction, described by

$$\frac{\partial \rho Y_i}{\partial t} = -\vec{\nabla} \cdot \left(\vec{j_i} \right) \ \text{ for } i = 1, \ldots, N \tag{Eq. 4.6}$$

For simple flow configurations with laminar flow and simple geometry, these equations can be solved exactly. In general, however, their complexity does not allow exact solution, and it is a common approach to proceed to numerical simulation.

The species transport equations have to be solved taking into account initial conditions (for transient problems) and boundary conditions. Different boundary conditions apply at inflow boundaries, outflow boundaries, and solid walls. If species are released from a solid to a fluid or deposited on a solid from a fluid, a nonzero diffusive flux appears at the interface.

In most engineering flows, the flow is turbulent, and the velocity and composition variables show fluctuating behavior. In that case, most often, model equations are solved for mean values rather than instantaneous values. Then in addition to the laminar diffusion fluxes, also turbulent diffusion fluxes containing a turbulent diffusivity have to be considered (Poinsot and Veynante, 2011). The turbulent diffusivity is obtained from a turbulence model and added to the laminar diffusivity. In highly turbulent flow, it is sufficient to consider only the dominant turbulent diffusivity. The combined effect of advection and diffusion in a certain type of flow and geometry, especially across a boundary layer close to a wall, can often also be described using a phenomenological mass transfer law, as is in Section 4.4.

4.2.2 Energy Transfer

The transport equation for total nonchemical energy, Equation (3.33) in Chapter 3, is

$$\frac{\partial}{\partial t} \left(\rho u + \frac{1}{2} \rho v^2 \right) + \nabla \cdot \left[\vec{v} \left(\rho u + \frac{1}{2} \rho v^2 \right) \right] + \nabla \cdot \left(p \vec{v} - \tau \cdot \vec{v} \right) + \nabla \cdot \vec{j_q} = \rho \vec{v} \cdot \vec{g} + Q_r \tag{Eq. 4.7}$$

It contains the energy flux $\vec{j_q}$, describing the changes in energy contained in a small control volume by diffusion of energy across its boundary of the control volume and results from the combined effect of several physical processes. Most well known is the effect of thermal conduction, which is proportional to the temperature gradient and to the thermal conductivity λ, a material property:

$$\vec{j}_{q, cond} = -\lambda \vec{\nabla} T \tag{Eq. 4.8}$$

The rate at which heat spreads out in the medium, however, does not only depend on the thermal conductivity but also on the density and heat capacity. The diffusivity (or diffusion coefficient) of heat is given by $\alpha = (\lambda/\rho c_p)$.

In general, thermal conductivity can be direction dependent, and the heat flux can also contain contributions from other effects. In a multicomponent mixture, every component of the mixture contains a certain amount of energy per unit mass; differential diffusion of different components gives a contribution to the energy flux, called species diffusion flux, that is given by

$$\vec{j}_{q,\,species} = \rho \sum_{i=1}^{N} h_i Y_i \vec{V}_i \qquad \text{(Eq. 4.9)}$$

where h_i is the specific enthalpy of species i. This contribution to the energy flux caused by the fact that different species have a different diffusion velocity can be of importance, e.g., in gas-phase combustion processes. In addition, there can be a contribution to the energy flux proportional to species mass fraction gradients. This is called the Dufour effect and is negligible in most cases (Kuo, 2005).

As in the case of mass transfer, also heat transfer systems can be classified according to which terms in the transport equation are contributing. In the simplest case, only heat conduction is present, and steady and transient problems can be distinguished. In the presence of advection, in addition to diffusion, a convective heat transfer problem appears. Just as in the case of mass transfer, for simple flow configurations with laminar flow and simple geometry, the equations can be solved exactly, but generally, numerical simulation is required. In the case of turbulent flow, a transport equation for the mean of the fluctuating energy variable is solved, and the diffusive flux contains a contribution representing the diffusion by turbulence (Poinsot and Veynante, 2011). The combined effect of advection and diffusion is described using a phenomenological heat transfer law (Newton's law of cooling), considered in Section 4.4.

4.3 RADIATIVE HEAT TRANSFER

In the description of radiative heat transfer, a distinction has to be made between on the one hand transfer between bounding walls that emit, absorb, and reflect radiation without participation of the medium between the walls (surface transfer) and on the other hand transfer inside a medium that is interacting with the radiation by absorption, emission, and/or scattering (transfer in participating medium) (see Modest, 2003).

4.3.1 Surface Transfer

To describe radiative heat transfer between surfaces, basic quantities to be considered are the amount of radiation energy emitted per unit surface area (emissive power E) and the amount of energy per unit surface area incident on the surface (irradiation H). Both in general depend on wavelength and on direction, but here, we consider the simple case of no wavelength dependence (gray surface) and no direction dependence (diffuse surface). The quantities to be determined in surface-to-surface radiative heat transfer problems are the surface temperature and the surface radiative heat flux (defined to be positive when the net balance is that energy is leaving the surface). In any specific application, the bounding walls of a system are subdivided into parts that have constant material properties and have either constant temperature or constant heat flux. Assuming that the walls are opaque, i.e., no radiative energy is transmitted through the wall, the material properties determining the surface-to-surface transfer are emissivity ε_e and absorptivity α. In the presence of transmission (e.g., in case of a glass window), also the transmissivity would have to be considered. In the absence of transmission of radiation, the part that is not absorbed is reflected with reflectivity $\rho = 1 - \alpha$. For a gray diffuse surface, $\alpha = \varepsilon_e$, which is known as Kirchhoff's law. The emissive power of a surface is given by $E = \varepsilon_e E_b$, where E_b is the emissive power of an ideal surface absorbing all incident radiation, called black surface, which is given by

$$E_b = \sigma T^4 \qquad (Eq. 4.10)$$

with σ being the Stefan–Boltzmann constant. The part of the irradiation that is absorbed is given by αH. The radiative energy leaving the surface is called radiosity, denoted J, and is given by the sum of emitted and reflected radiation: $J = \varepsilon_e E_b + (1 - \alpha)H$. The radiative heat flux leaving the surface can be written in two different ways. In the first view, it is the balance between emitted and absorbed radiation: $q = \varepsilon_e E_b - \alpha H$. In the second view, it is the balance between the energy leaving the surface and the energy approaching the surface: $q = J - H$. In order to describe the radiative heat transfer between different parts of the solid boundaries of a certain domain, it is convenient to divide the boundary in N_s surfaces with, by assumption, uniform properties. The areas A_i of the surfaces and their emissivity $\varepsilon_{e,i} = \alpha_i (i = 1, ..., N_s)$ characterize the problem. The solution of the heat transfer problem involves both temperature T_i and heat flux $q_i (i = 1, ..., N_s)$. In general, for each surface, either temperature or heat flux is known and the other is to be determined. The radiative energy leaving surface A_j is given by $A_j J_j$, and the part of this that is intercepted by surface A_j is given by the product $F_{i-j} A_j J_j$. Here, F_{i-j} is the view factor or configuration factor defined as the fraction of the radiative energy leaving the surface number i that is intercepted by the surface number j. Here, intercepted indicates that the radiation is contributing to the irradiation H_j and not that the radiation is absorbed by the surface number j. The view factors are function of the relative position and orientation of the surfaces. The values of the configuration factors have to be determined before the surface heat transfer problem can be solved. Many view factors are available from databases, and unknown view factors can often easily be determined from known view factors using so-called view factor algebra (see Modest, 2003).

A property of great practical use is the reciprocity rule $A_i F_{i-j} = A_j F_{i-j}$. Summing all contributions to the irradiation, using the reciprocity rule, and eliminating the radiosity and the irradiation from the equations but keeping the heat flux and the emissive power in the equations, it is found that the surface heat transfer is described by the following N_s equations (see Chapter 5 in Modest, 2003):

$$\frac{q_i}{\varepsilon_{e,i}} = E_{bi} - \sum_{j=1}^{N_s} \left[E_{bj} - \left(\frac{1}{\varepsilon_{e,j}} - 1 \right) q_j \right] F_{i-j} \qquad \text{(Eq. 4.11)}$$

Hence, once view factors are known, assuming that for every surface either the temperature or the heat flux is known (and the other quantity unknown), from Equation (4.11), the unknown quantities can be found. Indeed, knowledge of the temperature T_i is equivalent to knowledge of the blackbody emissive power $E_{b,i} = \sigma T_i^4$, and Equation (4.11) gives a set of N_s linear relations between the set of values of heat fluxes q_i and the set of values of blackbody emissive power $E_{b,i}$. Then linear algebra methods provide the solution for the N_s unknown quantities. A direct formal analogy exists with problems in the domain of electricity, with heat flux corresponding to current and difference in emissive power corresponding to voltage difference, and this has been exploited in the formulation of solution methods for surface radiative heat transfer problems.

In the case that in addition to radiative heat transfer also heat transfer by conduction or convection is taking place, the equations describing all relevant processes have to be solved together.

4.3.2 Participating Medium

In the transport equation for energy Equation (4.7), a source term Q_r, representing exchange of energy between matter and radiation field, has been introduced. In this section, we derive the expression for this term as a function of the properties of the radiation field. The radiation field is an electromagnetic field that—on the macroscopic scale—can be characterized by the definition of the radiative intensity.

Radiative transfer in a participating medium is mathematically described by the radiative transfer equation (RTE), describing the rate of change of the spectral radiation intensity of a radiation beam traveling in the medium and propagating along a certain direction. It may be written as follows for an emitting–absorbing–scattering nongray medium (Modest, 2003):

$$\frac{dI_\lambda(\mathbf{r},\mathbf{s})}{ds} = -\kappa_\lambda I_\lambda(\mathbf{r},\mathbf{s}) + \kappa_\lambda I_{b\lambda}(\mathbf{r}) - \sigma_{s\lambda} I_\lambda(\mathbf{r},\mathbf{s}) + \frac{\sigma_{s\lambda}}{4\pi} \int_{4\pi} I_\lambda(\mathbf{r},\mathbf{s}^*)\, \Phi_\lambda(\mathbf{s}^*,\mathbf{s})\, d\Omega^*$$

$$\text{(Eq. 4.12)}$$

where I_λ is the spectral radiation intensity at point $\mathbf{r}$ propagating along direction $\mathbf{s}$; s is the coordinate along that direction; κ and σ_s are the absorption and scattering coefficients of the medium, respectively; and $\Phi(\mathbf{s}^*, \mathbf{s})$ is the scattering phase function. The ratio $\Phi(\mathbf{s}^*, \mathbf{s})/4\pi$ represents the probability that radiation propagating in direction $\mathbf{s}^*$ and confined within solid angle $d\Omega^*$ is scattered through the angle $\mathbf{s}^* \cdot \mathbf{s}$ into the direction $\mathbf{s}$ confined within solid angle $d\Omega$. Subscripts b and λ denote blackbody and wavelength, respectively. The term $\kappa_\lambda I_{b\lambda}$ gives the amount of energy added to the radiation field per unit distance. $I_{b\lambda}$ is the blackbody intensity, which is function of the temperature of the medium and according to Planck's law is given by

$$I_{b\lambda} = \frac{C_1}{\pi n^2 \lambda^5 \left[e^{\frac{C_2}{n\lambda T}} - 1 \right]} \qquad \text{(Eq. 4.13)}$$

Here, n is the index of refraction of the medium ($n \approx 1$ for gases), and $C_1 = 3.7419 \times 10^{-16}$ W·m^2 and $C_2 = 14,388$ μm·K are fundamental constants. The absorption coefficient κ_λ in general depends on both composition and temperature of the medium. In order to solve the RTE, also boundary conditions have to be provided. Usually, the boundaries are solid walls, and the material properties of these walls have to be specified. The RTE may be written in other forms by using the wavenumber or the frequency as the spectral variable instead of the wavelength. Whatever spectral variable is chosen, integrating the intensity over this spectral variable, the total radiation intensity is obtained. If the radiative properties of the medium are independent of the spectral variable, the medium is called gray. If also the boundary conditions are gray, the spectral dependence plays no role in the RTE, and only an equation for total intensity has to be solved.

The radiative heat flux vector, $\vec{\mathbf{q}}$ [W m^{-2}], is defined as the integral of the spectral radiative heat flux:

$$\vec{\mathbf{q}}_\lambda = \int_{4\pi} I_\lambda \, \vec{\mathbf{s}} \, d\Omega \qquad \text{(Eq. 4.14)}$$

$$\vec{\mathbf{q}} = \int_0^\infty \vec{\mathbf{q}}_\lambda \, d\lambda \qquad \text{(Eq. 4.15)}$$

Similarly, the total incident radiation, G [W·m^{-2}] (a scalar), is defined as the integral of the spectral incident radiation:

$$G_\lambda = \int_{4\pi} I_\lambda \, d\Omega \qquad \text{(Eq. 4.16)}$$

$$G = \int_0^{+\infty} G_\lambda \, d\lambda \qquad \text{(Eq. 4.17)}$$

If all the terms of Equation (4.12) are integrated over all directions, spanning a solid angle of 4π, the following equation is obtained (see Modest (2003) for details):

$$\nabla \cdot \mathbf{q}_\lambda = \kappa_\lambda \left(4\pi I_{b\lambda} - \int_{4\pi} I_\lambda \, d\Omega \right) = \kappa_\lambda (4\pi I_{b\lambda} - G_\lambda) \tag{Eq. 4.18}$$

or, after integration over the full spectrum,

$$\nabla \cdot \mathbf{q} = \int_0^\infty \kappa_\lambda (4\pi I_{b\lambda} - G_\lambda) \, d\lambda = 4\kappa_P \pi I_b - \kappa_G G \tag{Eq. 4.19}$$

where I_b is the total blackbody intensity

$$I_b = \int_0^\infty I_{b\lambda} d\lambda = \frac{1}{\pi} \sigma T^4 = \frac{1}{\pi} E_b \tag{Eq. 4.20}$$

The Planck-mean absorption coefficient κ_P and the incident-mean absorption coefficient κ_G are defined as

$$\kappa_P = \frac{1}{I_b} \int_0^\infty \kappa_\lambda I_{b\lambda} d\lambda \tag{Eq. 4.21}$$

$$\kappa_G = \frac{1}{G} \int_0^\infty \kappa_\lambda G_\lambda d\lambda \tag{Eq. 4.22}$$

Equation (4.19) expresses the conservation of total radiative energy in an elementary control volume. The divergence of the total radiative heat flux vector at an arbitrary point in space, given by Equation (4.19), represents the amount of energy per unit time and per unit volume that is gained or lost at that point as a resultant of the global radiative heat exchange. It appears as a sink term in the equation for the conservation of energy:

$$Q_r = -\vec{\nabla} \cdot \mathbf{q} \tag{Eq. 4.23}$$

Note that scattering does not contribute to Equation (4.19), since it does not cause any local change of energy, but only a redistribution of radiative energy among the directions.

The situation with a nonzero radiative source term is the case of a "participating medium." In the case of negligible absorption, emission, and scattering, the intensity does not change in the medium and travels through space until it reaches a wall.

In the simplest case of a gray medium surrounded by walls at temperature T_w and without taking into account possible reabsorption of emitted radiation, the source term takes the form

$$Q_r = -\vec{\nabla} \cdot \vec{q} = -4\kappa\sigma \left(T^4 - T_w^4 \right) \tag{Eq. 4.24}$$

4.4   CONVECTIVE HEAT AND MASS TRANSFER

In the case of heat transfer between a fluid and a solid wall, the heat flux component normal to the wall may be related to the difference between the surface temperature of the solid wall T_w and the "bulk" fluid temperature T_b:

$$j_{q,n} = h(T_w - T_b) \qquad \qquad (Eq.\ 4.25)$$

This relation is referred to as Newton's law of cooling and is a pragmatic law rather than a fundamental law. The heat flow (per unit area of the wall) is assumed to be proportional to the temperature difference. The proportionality constant h is called the heat transfer coefficient and is a phenomenological quantity. It depends on the properties of the flow near the wall and on the physical properties of the fluid. When temperature gradients are present, an appropriate mean value between values at different temperatures is to be used. For accurate results, the *1/3 rule* should be used, stating that temperature-dependent fluid properties are evaluated at an effective temperature and composition that are weighted means of the values at the surface and in the surroundings (with a weight of 1/3 for the surroundings value) (see Hubbard et al., 1975). For two or more heat transfer processes acting in parallel, heat transfer coefficients simply add. For two or more heat transfer processes connected in series, heat transfer coefficients add inversely.

Example 4.1   Heat transfer processes in series

Consider heat transferred from a fluid at bulk temperature T_{b2} to a fluid at bulk temperature T_{b1} through a flat wall with thickness D and thermal conductivity λ_w. This can be considered as three heat transfer processes in series: convective heat transfer from bulk fluid 2 to the wall, conductive heat transfer through the wall, and convective heat transfer from the other side of the wall to bulk fluid 1. For each step, the heat flux can be expressed in terms of a temperature difference, giving three expressions for the same heat flux: $j_{q,n} = h_1(T_{w1} - T_{b1}) = \dfrac{\lambda_w}{D}(T_{w2} - T_{w1}) = h_2(T_{b2} - T_{w2})$.

The first and last expressions represent the convective heat transfer at the two sides of the wall. The middle expression represents the heat transfer by conduction. By eliminating the wall temperatures from these equations, one obtains

$$j_{q,n} = h_{combined}(T_{b2} - T_{b1})$$

with the heat transfer coefficient for the combined process given by

$$\frac{1}{h_{combined}} = \frac{1}{h_1} + \frac{D}{\lambda_w} + \frac{1}{h_2}$$

Generally valid expressions for the heat transfer coefficient can be found by formulating the dependence on flow conditions and other factors using dimensionless numbers (see Table 4.1) and empirical constants. The higher the heat transfer

TABLE 4.1   Selected dimensionless numbers relevant for heat and mass transfer

Number	Definition	Interpretation
Biot number (heat)	$Bi = \dfrac{h_1 L}{\lambda_2}$	Ratio of the boundary layer thermal resistance to the internal thermal resistance of a solid; subscripts 1 and 2 denote two different media
Fourier number (heat)	$Fo = \dfrac{\alpha t}{L^2}$	Ratio of the heat conduction rate to the rate of thermal energy storage in a solid. Dimensionless time
Grashof number	$Gr = \dfrac{g\beta(T_w - T_b)L^3}{\nu^3}$	Ratio of buoyancy force to viscous force
Lewis number	$Le = \dfrac{\alpha}{D}$	Ratio of the thermal and species diffusivities
Nusselt number	$Nu = \dfrac{hL}{\lambda}$	Dimensionless temperature gradient at the surface
Prandtl number	$Pr = \dfrac{\nu}{\alpha}$	Ratio of the momentum and thermal diffusivities
Reynolds number	$Re = \dfrac{uL}{\nu}$	Ratio of inertial and viscous forces
Schmidt number	$Sc = \dfrac{\nu}{D}$	Ratio of momentum and species diffusivities
Sherwood number	$Sh = \dfrac{kL}{D}$	Dimensionless concentration gradient at the surface

coefficient, the thinner the boundary layer and the larger the temperature gradient orthogonal to the wall. The temperature gradient normalized by the temperature difference is the Nusselt number, Nu, and its value is related to the value of other dimensionless numbers characterizing the problem at hand. For example, a well-known correlation is the Ranz–Marshall correlation for flow around a solid sphere. Its form and range of validity are given by (see, e.g., Bird et al., 2007, p. 681)

$$Nu = 2 + 0.6 Re^{\frac{1}{2}} Pr^{\frac{1}{3}}, \quad 0 \leq Re Pr^{\frac{2}{3}} < 5 \times 10^4 \qquad \text{(Eq. 4.26)}$$

where Re is the Reynolds number based on the particle diameter and on the difference between the particle velocity and the velocity of the fluid ahead of the particle and Pr is the Prandtl number of the fluid.

In a similar way, in the case of mass transfer between a solid wall and a surrounding fluid, the mass flux component normal to the wall may be related to the difference between the concentration of a species at the wall surface $c_{i,w}$ and the concentration in the bulk fluid $c_{i,b}$:

$$\dot{j}_{i,n} = k\left(c_{i,w} - c_{i,b}\right) \qquad \text{(Eq. 4.27)}$$

The mass flow (per unit area of wall) is assumed proportional to the concentration difference. The proportionality constant k is called the mass transfer coefficient.

As for the heat transfer coefficient, generally valid expressions can be formulated using dimensionless numbers and empirical constants. The larger the concentration gradient orthogonal to the surface, the higher the rate of mass transfer. This is characterized by the Sherwood number, which is the normalized concentration gradient at the surface. The Ranz–Marshall correlation for mass transfer between a sphere and a surrounding fluid is given by

$$\text{Sh} = 2 + 0.6\text{Re}^{\frac{1}{2}}\text{Sc}^{\frac{1}{3}}, \quad 0 \le \text{ReSc}^{\frac{2}{3}} < 5 \times 10^4 \qquad \text{(Eq. 4.28)}$$

where Sc is the Schmidt number of the fluid.

Chilton and Colburn have introduced two additional dimensionless numbers, one for heat transfer j_H and one for mass transfer j_D (Bird et al., 2007):

$$j_H = \frac{\text{Nu}}{\text{Re Pr}^{0.33}}, \quad j_D = \frac{\text{Sh}}{\text{Re Sc}^{0.33}} \qquad \text{(Eq. 4.29)}$$

At large Reynolds number (turbulent flow), these two numbers turn out to be equal, expressing the analogy between heat transfer and mass transfer. This gives a relation for h/k:

$$\frac{h}{k} = \frac{\lambda}{D}\left(\frac{\text{Pr}}{\text{Sc}}\right)^{0.33} = \rho c_p \left(\frac{\text{Sc}}{\text{Pr}}\right)^{0.67} = \rho c_p \text{Le}^{0.67} \qquad \text{(Eq. 4.30)}$$

The transport properties in this expression are those of the flowing medium. The relation Equation (4.30) is of great practical use because in general it is easier to measure the heat transfer coefficient than the mass transfer coefficient.

4.5   TRANSFER OF HEAT AND MASS WITH PHASE CHANGE

In the case of phase changes, as is the case in melting, evaporation, pyrolysis, and combustion, the transport equations have to be set up for the separate phases (gaseous, liquid, and solid), and in addition to heat transfer between the phases, also the mass transfer has to be described. When setting up the energy balance, the latent heat involved in the phase change has to be taken into account. In the following, we discuss two cases: the evaporation of a biofuel droplet and the conversion to char of a biomass particle.

4.5.1   Evaporation of a Single-Component Fuel Droplet

Let us consider the evaporation of a spherical fuel droplet. The rate of change of the droplet diameter depends on properties of the droplet and on properties of the surroundings. The temperature and composition of the droplet and surrounding gas, and also the relative velocity, have a strong influence on the evaporation rate. The simplest situation arises when a spherical single-component droplet of diameter d_d is put in, e.g., hot air at constant temperature and composition.

The question arises whether the droplet can be assumed to be at a uniform temperature or not. If the thermal conductivity of the fuel is sufficiently large, the heat received at the droplet surface will spread so rapidly to the interior that at any time the droplet can be assumed to have uniform temperature T_d. Models using this assumption are called infinite conductivity or rapid-mixing models. In the presence of heat transfer from or to the surroundings, T_d is time dependent. In the frame of the rapid-mixing model, it is straightforward to determine the evolution in time of the droplet diameter $d_d = 2r_d$, with r_d being the radius. During the evaporation process, the droplet diameter decreases with time, and so does its surface area $A_d = 4\pi r_d^2$, volume $V_d = (4/3)\pi r_d^3$, and mass $m_d = \rho_d V_d$.

For the classical derivation of the droplet evaporation rate, we here follow Jenny et al. (2012). In the absence of gravity, buoyancy effects do not play a role, and the configuration of the evaporating droplet remains spherically symmetric. The problem becomes mathematically more tractable if it is assumed that the droplet evaporates in a very large (infinitely large) domain, where the conditions at infinity are not influenced by the presence of the droplet. To solve the problem, both the continuity equations expressing mass conservation and energy conservation have to be solved.

The flow around the droplet is caused by the vapor moving away from the droplet surface. Due to the density difference between liquid and vapor, the vapor takes up more volume and is pushed away from the droplet surface, causing the so-called Stefan flux ρu_r (ρ is the density of the gas mixture at the droplet surface and u_r the mixture velocity at the droplet surface). Diffusive mixing between vapor and the surrounding gas takes place. The surrounding gas diffuses toward the droplet surface, and the evaporated fuel diffuses into the surroundings.

The net motion of fuel vapor is the sum of convection at the mixture velocity and the diffusive velocity. The rate of change of the total droplet mass can then be expressed as

$$\frac{dm_d}{dt} = \dot{m}_d = -A_d \rho u_r = -A_d \left[\rho u_r Y + \rho D \frac{\partial Y}{\partial r} \right] \qquad \text{(Eq. 4.31)}$$

Here, Y is the fuel vapor mass fraction, and D is the diffusion coefficient of the fuel in air. If quasi-steady state is assumed, the fuel vapor conservation equation takes the form

$$\frac{\partial}{\partial r} \left(\dot{m}_d Y + 4\pi r^2 \rho D \frac{\partial Y}{\partial r} \right) = 0 \quad \text{for} \quad r \geq r_d \qquad \text{(Eq. 4.32)}$$

with boundary conditions

$$Y(r_d) = Y_s \quad \text{and} \quad Y(r \to \infty) = Y_b \qquad \text{(Eq. 4.33)}$$

Here, quasi-steady state means that the rate of regression of the surface is assumed to be so slow compared to the change in fuel vapor concentration away from the droplet that the vapor concentration profile can be found by solving the steady-state fuel vapor conservation equation, but taking into account a fixed droplet radius as boundary condition. Following a similar reasoning, provided the evaporation process is fast compared to the transport of heat and mass to the surface, the fuel mass fraction at the surface Y_s can be obtained from equilibrium thermodynamics. The equilibrium vapor mole fraction at the surface, X_s, at saturation temperature T_{sat} is obtained from the saturation pressure p_{sat} determined by the Clausius–Clapeyron equation, relating saturation temperature and saturation pressure:

$$X_s = \frac{p_{sat}}{p_{atm}} = \exp\left[\frac{h_{fg}}{R_u/MW_F}\left(\frac{1}{T_{boil}} - \frac{1}{T_{sat}}\right)\right] \qquad \text{(Eq. 4.34)}$$

where h_{fg} is the latent heat of vaporization of the fuel, R_u is the universal gas constant, and MW_F is the MW of the fuel.

The vapor mass fraction is obtained using the relation between mole fraction and mass fraction in a mixture of vapor and air:

$$Y_s = X_s \frac{MW_F}{X_s MW_F + (1-X_s)MW_{air}} \qquad \text{(Eq. 4.35)}$$

Deviations from the phase equilibrium at the surface can occur when the evaporation is very fast, occurring for small droplets and for droplets with a temperature very close to the boiling point. Bellan and Summerfield (1978) have presented expressions for evaporation rate taking into account nonequilibrium effects.

The solution of Equation (4.32) is (if ρD is assumed constant)

$$Y(r) = Y_s + (Y_b - Y_s)\frac{\exp[-(\dot{m}_d/4\pi r_d \rho D)(1-(r_d/r))]-1}{\exp[-(\dot{m}_d/4\pi r_d \rho D)]-1} \quad \text{for } r \geq r_d \qquad \text{(Eq. 4.36)}$$

From Equation (4.31) and Equation (4.36), it follows that

$$\dot{m}_d = \dot{m}_d Y_s + 4\pi r^2\left[\rho D \frac{\partial Y}{\partial r}\right]_{r_d} = \dot{m}_d Y_s + \dot{m}_d(Y_s - Y_b)\frac{1}{\exp\left[-\dfrac{\dot{m}_d}{4\pi r_d \rho D}\right]-1} \qquad \text{(Eq. 4.37)}$$

This equation can be solved for $\dot{m}_d$ with the following result:

$$\dot{m}_d = -4\pi r_d \rho D \ln(B_M + 1) \qquad \text{(Eq. 4.38)}$$

Here, B_M is the Spalding mass coefficient:

$$B_M = \frac{Y_s - Y_b}{1 - Y_s} \qquad \text{(Eq. 4.39)}$$

A key characteristic of the evaporation process at constant droplet temperature and in a steady environment is that the rate of change of the droplet diameter squared is a constant, which is known as the d^2-law:

$$\frac{d}{dt}d_d^2 = -8\frac{\rho}{\rho_d}D\ln(B_M + 1) \equiv -K \qquad \text{(Eq. 4.40)}$$

From this equation, it follows that the droplet diameter is given by $d_d^2(t) = d_d^2(0) - Kt$ and the droplet lifetime by

$$t_d = \frac{d_d^2(t=0)}{K} \qquad \text{(Eq. 4.41)}$$

In realistic process conditions, the environment is not stagnant, and the evaporating droplet can have a relative velocity compared to the surroundings, causing enhancement of heat and mass transfer and increase of the evaporation rate by forced convection effects. To take this effect into account, it is necessary to multiply the rate with the relative increase of the Nusselt number from the value 2 for stagnant conditions to the value Nu_d for flow around a sphere, leading to

$$\frac{d}{dt}d_d^2 = -8\frac{Nu_d}{2}\frac{\rho}{\rho_d}D\ln(B_M + 1) \qquad \text{(Eq. 4.42)}$$

As a first approximation, the Nusselt number Nu_d for flow around a solid sphere given by the Ranz–Marshall correlation Equation (4.26) can be used. The Prandtl number of the surrounding gas appearing in the Nusselt number correlation is given by

$$Pr = \frac{\eta c_p}{\lambda} = \frac{\nu}{\alpha} \qquad \text{(Eq. 4.43)}$$

In the preceding paragraphs, it was assumed that mass transfer from droplet to surroundings is taking place while also assuming constant physical properties at the droplet surface. In reality, the droplet temperature may change with time. This process is described by the droplet temperature evolution equation. Assuming a homogeneous droplet temperature, the droplet temperature and the surface temperature are the same $(T_d = T_s)$, and it follows from energy conservation that the droplet temperature evolves as

$$\frac{dT_d}{dt} = \frac{T_b - T_d}{\tau_{d,T}} + \frac{h_{fg}}{m_d c_{p,F}}\dot{m}_d \qquad \text{(Eq. 4.44)}$$

with $c_{p,F}$ being the specific heat capacity of the fuel and $\tau_{d,T}$ the droplet thermal relaxation time, given by

$$\tau_{d,T} = \frac{\rho_d c_{p,F}d_d^2}{6Nu_d\lambda_m} \qquad \text{(Eq. 4.45)}$$

Here, the thermal conductivity and the material properties entering the Nusselt number are evaluated as an average using the 1/3 rule. Equation (4.44) expresses that the heat transferred to the droplet surface is divided into heat used for evaporation and heat used for increasing the temperature of the droplet.

Example 4.2   Mass diffusion-controlled evaporation of a fuel droplet

The presentation of this example follows Turns (2000, p. 103) but in this case for ethanol fuel.

Given
An ethanol droplet initially at temperature 10 or 20 K below the boiling temperature is evaporating in surrounding air with a temperature of 800 K.

To be determined
Droplet lifetime, t_d:

    a. Assuming that the droplet temperature is constant in time

    b. Solving simultaneously the equation for droplet temperature and also taking into account a slip velocity, v_s, of $10 \, \mathrm{m \cdot s^{-1}}$

Discussion of the solution procedure
The balance between conductive heating and evaporative cooling determines the droplet temperature evolution. In general, the droplet temperature will change with time. In (a), as a first approximation, it is assumed that the droplet temperature remains constant during the evaporation process. In (b), this assumption is not made, but instead, an additional equation for the droplet temperature is solved.

    Physical properties of ethanol
    Boiling temperature: $T_{boil} = 351 \, \mathrm{K}$
    Latent heat: $h_{fg} = 855 \, \mathrm{kJ \cdot kg^{-1}}$
    MW: $MW_F = 46.07 \, \mathrm{g \cdot mol^{-1}}$
    Diffusion coefficient in air: $D = 1.02 \times 10^{-5} \, \mathrm{m^2 \cdot s^{-1}}$ at 273 K

    Properties of the droplet
    Diameter: $d_d = 100 \, \mu\mathrm{m}$
    Surface temperature: $T_s = T_{boil} - 10 \, \mathrm{K}$ or $T_s = T_{boil} - 20 \, \mathrm{K}$
    Density : $\rho_F = 789 \, \mathrm{kg \cdot m^{-3}}$

    Properties of the surroundings
    $p = 1 \, \mathrm{atm}$
    MW: $MW_{air} = 29 \, \mathrm{g \cdot mol^{-1}}$
    Temperature: $T_{air} = 800 \, \mathrm{K}$

Solution
    a. Assuming constant droplet temperature
       The droplet lifetime can be estimated using Equation (4.41), after calculating the evaporation constant K, from Equation (4.40).

To calculate the evaporation constant, the Spalding mass coefficient, B_M, which requires knowledge of the fuel mass fraction at the surface, has to be estimated first. The Clausius–Clapeyron equation Equation (4.34) gives the vapor mole fraction at the given droplet surface temperature, and then the vapor mass fraction follows from Equation (4.35):

$$X_s = \frac{p_{sat}}{p_{atm}} = \exp\left[\frac{h_{fg}}{R_u/MW_F}\left(\frac{1}{T_{boil}} - \frac{1}{T_{sat}}\right)\right] = \exp\left[\frac{855{,}000}{8{,}315/46.07}\left(\frac{1}{351} - \frac{1}{341}\right)\right] = 0.6732$$

$$Y_s = X_s\frac{MW_F}{X_s MW_F + (1-X_s)MW_{air}} = 0.7659$$

The Spalding mass coefficient, B_M, is then evaluated from Equation (4.39) as

$$B_M = \frac{Y_s - Y_\infty}{1 - Y_s} = 3.2718$$

Here, Y_∞ is the vapor mass fraction at infinity (i.e., far away from the droplet). To evaluate the evaporation constant, we need to estimate ρD, which we treat as $\widetilde{\rho}_{air}D(\widetilde{T} = 800\,\mathrm{K})$. Extrapolating the diffusivity from 273 to 800 K using

$$D(\widetilde{T}) = D(\widetilde{T}_0)\left(\frac{\widetilde{T}}{\widetilde{T}_0}\right)^{3/2}$$

$$= 1.02 \times 10^{-5}\left(\frac{800}{273}\right)^{3/2}\ \mathrm{m^2 s^{-1}} = 5.12 \times 10^{-5}\,\mathrm{m^2 s^{-1}}$$

and using the ideal gas law to evaluate $\widetilde{\rho}_{air}$,

$$\widetilde{\rho}_{air} = \frac{p}{\left(R_u/MW_{air}\right)T_{air}} = \frac{101{,}325}{(8{,}315/29)800}\ \mathrm{kg\cdot m^{-3}} = 0.4417\ \mathrm{kg\cdot m^{-3}}$$

Thus,

$$K = \frac{8\widetilde{\rho}_{air}D}{\rho_F}\ln(B_M + 1)$$

$$= \frac{8 \times 0.4417 \times 5.12 \times 10^{-5}}{789}\ln(3.2718 + 1) = 3.328 \times 10^{-7}\ \mathrm{m^2\cdot s^{-1}}$$

and the droplet lifetime is

$$t_d = \frac{d_d^2(0)}{K} = \frac{\left(100 \times 10^{-6}\,\mathrm{m}\right)^2}{K} = 0.030\ \mathrm{s}$$

For $T_s = T_{boil} - 20$ K, one obtains $t_d = 0.054$ s.

b. Solving the droplet temperature equation
When the droplet and air temperatures change with time, an explicit analytic solution in general is not available, but a numerical solution can be obtained. By solving the aforementioned equations until the droplet diameter reaches 0, we can obtain the droplet lifetime.

For $T_{s,i} = T_{boil} - 20 = 331$ K, $v_s = 10$ m·s^{-1}, the droplet lifetime $t_d = 0.0204$ s, and for
$T_{s,i} = T_{boil} - 10 = 341$ K, $v_s = 10$ m·s^{-1}, the droplet lifetime $t_d = 0.0209$ s.

A detailed study of this model and a sensitivity analysis of the results are presented as project at the end of this chapter. Results of such a project may look as depicted in Figures 4.1 and 4.2.

Due to the balance between heating and heat extraction for evaporation, liquid droplets usually do not reach the boiling point when they are heated. Instead, the temperature evolves toward the wet-bulb temperature. An estimation of the wet-bulb temperature can be obtained by setting the left-hand side of Equation (4.44) to zero. When droplets are situated in a very hot carrier gas, they will reach the wet-bulb

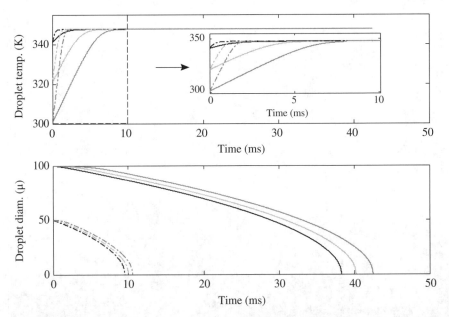

FIGURE 4.1 Droplet temperature and diameter versus time at a slip velocity between droplet and gas, v_s of 0 m·s^{-1}. Initial droplet temperatures: 301 K (dark gray), 321 K (light gray), and 341 K (black). Initial droplet diameters: 100 µm (solid lines) and 50 µm (dashed) lines (surrounding gas temperature $T_b = 800$ K).

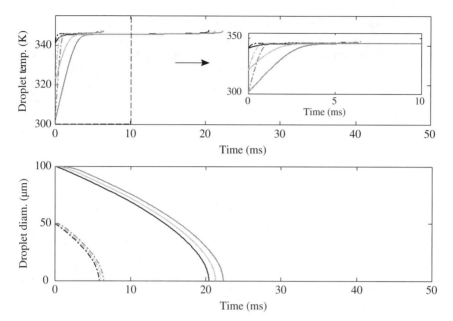

FIGURE 4.2 Droplet temperature and diameter versus time at a slip velocity between droplet and gas, v_s of 10 m·s^{-1}. Initial droplet temperatures: 301 K (dark gray), 321 K (light gray), and 341 K (black). Initial droplet diameters: 100 μm (solid lines) and 50 μm (dashed) lines (surrounding gas temperature $T_b = 800$ K).

temperature quickly, which can then be used instead of the evolution equation of the droplet temperature. When the carrier gas is hot and the boiling point of the liquid is low, the wet-bulb temperature is close to the boiling temperature of the liquid.

The following aspects should be taken into account when considering application of the simple model described in Example 4.2 to evaporation of a real biofuel:

1. The standard d^2-law together with an infinite conductivity model for the droplet temperature typically overestimates the evaporation rate, thus leading to shorter droplet lifetimes. In the framework of the infinite conductivity model, some corrections have been proposed in the literature. Several detailed studies, described in Jenny et al. (2012), show the effects of convective heat transfer on the Nusselt number and drag coefficient, and experiments suggest lower Nusselt numbers for evaporating droplets compared to the classical Ranz–Marshall correlations. A recommended correlation mentioned by Turns (2000) is

$$\mathrm{Nu}_d = 2 + 0.555 \frac{\mathrm{Re}^{\frac{1}{2}}\mathrm{Pr}^{\frac{1}{3}}}{\left[1 + 1.232\left(\mathrm{Re}\,\mathrm{Pr}^{\frac{4}{3}}\right)^{-1}\right]^{\frac{1}{2}}} \qquad \text{(Eq. 4.46)}$$

2. When the assumption of infinite conductivity cannot be made, a finite conductivity model is needed. Then the temperature and consequently also the temperature-dependent thermodynamic properties are spatially varying instead of being uniform throughout the droplet. In the simplest case, there is no convective flow pattern inside the droplet, and the temperature only depends on the radial coordinate. Using a spherical coordinate system, the conservation of energy is expressed as

$$\frac{\partial T_d}{\partial t} = \frac{\lambda_d}{\rho_d c_{p,d}} \left(\frac{\partial^2 T_d}{\partial r^2} + \frac{2}{r} \frac{\partial T_d}{\partial r} \right) \qquad \text{(Eq. 4.47)}$$

This equation is to be solved starting from an initial condition (the initial droplet temperature) and using boundary conditions at the droplet center and at the droplet surface. At the center, the flux in radial direction vanishes:

$$\left. \frac{\partial T_d}{\partial r} \right|_{r=0} = 0 \qquad \text{(Eq. 4.48)}$$

At the droplet surface, the inward heat flux just below the surface equals the sum of the inward heat flux just above the surface and the heat used for evaporation at the surface (h_{fg} being the heat of evaporation):

$$\lambda_d \left. \frac{\partial T_d}{\partial r} \right|_{inside} = \lambda_g \left. \frac{\partial T}{\partial r} \right|_{outside} - \frac{\dot{m}_d h_{fg}}{4\pi r_d^2} \qquad \text{(Eq. 4.49)}$$

3. In the case of a slip velocity between droplet and surroundings, also internal convection can be present, and then heat transfer in large liquid droplets is usually not only governed by conduction but also by internal convection. The importance of this effect increases with droplet size. To take convective effects into account, the thermal conductivity can be replaced by an effective conductivity larger than the conductivity of the liquid (see Jenny et al. (2012) for details and references).

4. Usually, the temperature of the droplets is much lower than the gas temperature, and the contribution of radiation from the droplets is negligible compared to that of the hot surroundings. The effect of radiation on heating and evaporation of the droplet can be important for relatively large droplets in hot surroundings such as spray flames. This has been investigated in detail by Sazhin et al. (2006). In the derivation of a small-scale model for the net radiative heat source of droplets, droplets are considered as semitransparent with a uniform but time-dependent temperature, and the thermal radiation from the surroundings is assumed to be that of a blackbody at temperature T_{ext}. The effect of radiative heating on droplets can be taken into account by an extra source term Q_r in the droplet temperature equation, which is of the form

$$Q_r = \frac{A_d}{V_d} \bar{E}_a \sigma T_{ext}^4 \qquad \text{(Eq. 4.50)}$$

and contains the absorption efficiency factor $\bar{E}_a = ar_d^b$ whose coefficients a and b need to be estimated from experiments.

5. Complex fuels such as diesel, kerosene, and biofuel contain many components, and the evaporation of the corresponding fuel droplets cannot accurately be described using single-component representation. Instead, models for multi-component fuels are needed. When the number of components is low, e.g., two components in methanol droplets that have absorbed water from the environment, all present species can be described by an individual mass fraction. However, if the number of components is much higher like in biofuel, it is not effective to consider transport equations for each component. Then it is better to use a statistical approach based on continuous thermodynamics. The theory of continuous thermodynamics describes the mixture in terms of a probability density function (PDF) of the composition, e.g., of the molar mass of the components. Predictions of continuous thermodynamic models differ depending on the submodel for this PDF. A good introduction and further references on the application of continuous thermodynamic methods for spray evaporation can be found in Le Clercq and Bellan (2004).

4.5.2 Devolatilization of a Biomass Particle

To further illustrate the heat and mass transfer processes taking place during biomass conversion processes, we here consider a simple model for devolatilization of a biomass (wood) particle in a hot environment without oxygen. The particle receives heat from the surroundings, causing the release of volatile matter (tar and light gases). More detailed models describing this phenomenon are considered in Chapter 11. Here, we restrict ourselves to the description of the inside of the particle and treat heat and mass exchange with the surroundings as a boundary condition, following the approach presented in Lu et al. (2010).

Five species are included in the model: biomass, char, light gas, tar, and inert gas. The first two are in the solid phase and have densities ρ_B and ρ_C. The other species are in the gas phase. The additional complexity caused by the presence of moisture and generation of water vapor is not included in this simple model. The volume of the particle is occupied by solid material and by gases. The porosity ε by definition is the fraction of the total volume occupied by gases. The fraction $(1 - \varepsilon)$ then consists of solids. The partial density of the gaseous species in the pores are denoted ρ_g. The partial density of the gaseous species in a volume containing both solid and gas then is given by $\varepsilon \rho_g$.

In the model description (see Figure 4.3), biomass is converted to light gas, tar, or char, respectively, with rate coefficients k_1, k_2, and k_3. The tar is further converted to light gas and to char, respectively, with rate coefficients k_4 and k_5.

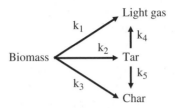

FIGURE 4.3   Global lumped component scheme for devolatilization (of wood).

Arrhenius expressions describe the temperature dependence of the kinetic rate coefficients:

$$k_i = k_{0,i} \exp\left(-\frac{E_{a,i}}{R_u T}\right) \qquad \text{(Eq. 4.51)}$$

The pre-exponential factor $k_{0,i}$ and the activation energy $E_{a,i}$ have to be determined experimentally for the type of biomass considered.

In a reference frame for the description of the biomass particle, the transport equations for species mass, momentum, and total energy can be formulated as follows. Assuming that the solid is not moving (i.e., neglecting particle swelling), the transport equations for solid species only contain a transient term and source or sink terms. Using subscripts B, C, and T to denote biomass, char, and tar, the mass conservation for biomass and char is given by

$$\frac{\partial \rho_B}{\partial t} = -(k_1 + k_2 + k_3)\rho_B \qquad \text{(Eq. 4.52)}$$

$$\frac{\partial \rho_C}{\partial t} = k_3 \rho_B + \varepsilon k_5 \rho_T \qquad \text{(Eq. 4.53)}$$

The porosity factor in front of k_5 represents the fact that the rate constants refer to the conversion per unit volume of the biomass particle, whereas ρ_T is the tar density in the pores. In a similar manner, strictly speaking, a factor $(1 - \varepsilon)$ should appear in front of k_3, but since ε usually is a small number, this factor is omitted.

The transport equations for gaseous species also contain convective and diffusive terms. Gas is leaving the particle generating an outward convection velocity. In general, the equations are three-dimensional, but for particles having symmetry, the description can be simplified. The transport model described here is one-dimensional in the sense that it is assumed that only one spatial direction is important. This coordinate is called r and, for spherical, cylindrical, and flat particles, respectively, has the meaning distance from the center of the sphere, distance from the axis of the cylinder, and distance from the midplane of the flat disk. When formulating the transport equations, the three different possible shapes of the particle are distinguished by the value of a parameter n. A spherical particle is described by $n = 2$, a cylinder particle by $n = 1$, and a flat plate particle by $n = 0$. The species mass conservation equation for gas-phase species takes the form

$$\frac{\partial}{\partial t}\left(\varepsilon \rho_g Y_i\right) + \frac{1}{r^n}\frac{\partial}{\partial r}\left(r^n \varepsilon \rho_g Y_i u\right) = \frac{1}{r^n}\frac{\partial}{\partial r}\left(r^n \varepsilon D_{eff}\rho_g \frac{\partial}{\partial r}Y_i\right) + s_i \qquad \text{(Eq. 4.54)}$$

where Y_i is the mass fraction of component i in the gas phase; u is the gas convection velocity in radial direction; D_{eff} is the effective gas diffusivity, here assumed to be independent on the species; and s_i is the source or sink term. The subscript i denotes any gas component (T for tar, G for light gas, and I for inert gas). Comparing with the general form of the species transport equation Equation (4.1), the following features can be pointed out: the effect of the presence of porosity, only one-dimensional equation in appropriate coordinate system, Fick's law for diffusion, and notation of source term changed from $\dot{\omega}_i$ to s_i.

The source terms are given by

$$\begin{aligned}
s_T &= k_2\rho_B - \varepsilon k_4\rho_T - \varepsilon k_5\rho_T \\
s_C &= k_1\rho_B + \varepsilon k_4\rho_T \\
s_I &= 0
\end{aligned} \qquad \text{(Eq. 4.55)}$$

Again, the porosity appears in order to take into account that the density of gas referred to the total volume is smaller than the intrinsic density of the gas inside the pores. The total gas-phase continuity equation is obtained by summation of the species equations and has the form

$$\frac{\partial}{\partial t}\left(\varepsilon \rho_g\right) + \frac{1}{r^n}\frac{\partial}{\partial r}\left(r^n \varepsilon \rho_g u\right) = s_g \qquad \text{(Eq. 4.56)}$$

with

$$s_g = k_1\rho_B + k_2\rho_B - \varepsilon k_5\rho_T \qquad \text{(Eq. 4.57)}$$

The flow of gas inside the particle can be described as flow in a porous medium.

In a porous medium, convection is caused solely by the pressure gradient. The area-averaged velocity through a cross-sectional plane through the medium is related to the local pressure drop by Darcy's law:

$$u = -\frac{B}{\eta}\frac{\partial p}{\partial r} \qquad \text{(Eq. 4.58)}$$

where B is the solid permeability [m^2], η is the dynamic viscosity of the gas [Pa·s], and p is the gas pressure given by the ideal gas law:

$$p = \frac{\rho_g R_u T}{MW_g} \qquad \text{(Eq. 4.59)}$$

The permeability B can be written as the mass-weighted average of the permeability of the biomass part and the char part of the porous medium:

$$B = \frac{\rho_B}{\rho_{B0}}B_B + \left(1 - \frac{\rho_B}{\rho_{B0}}\right)B_C \qquad \text{(Eq. 4.60)}$$

During conversion, the fraction of biomass decreases and the char part increases. The subscript 0 indicates the initial state, before conversion to char. The local velocity of the gas in the channels of the porous medium is related to the area-averaged velocity via

$$u_g = \frac{u}{\varepsilon} \qquad \text{(Eq. 4.61)}$$

The energy conservation equation can be written as

$$\frac{\partial}{\partial t}\left(\rho_B h_B + \rho_C h_C + \varepsilon \rho_g (Y_G h_G + Y_I h_I + Y_T h_T)\right)$$

$$+ \frac{1}{r^n}\frac{\partial}{\partial r}\left(r^n \varepsilon \rho_g u (Y_G h_G + Y_I h_I + Y_T h_T)\right) =$$

$$\frac{1}{r^n}\frac{\partial}{\partial r}\left(r^n k_{eff}\rho_g \frac{\partial T}{\partial r} + r^n \varepsilon \rho_g \left(D_{eff,G}\frac{\partial Y_G}{\partial r}h_G + D_{eff,I}\frac{\partial Y_I}{\partial r}h_I + D_{eff,T}\frac{\partial Y_T}{\partial r}h_T\right)\right)$$

$$\text{(Eq. 4.62)}$$

where h_i is the mass-specific enthalpy of component i:

$$h_i = h_i^0 + \int_{T_0}^{T} c_{p,i}dT \qquad \text{(Eq. 4.63)}$$

The effective diffusivity of the gas species in the particle depends on the pore size relative to the mean free path of the molecules. In wide pores, where the mean free path of is much larger than the pore diameter, a standard diffusivity D_m for the gas mixture applies. In narrow pores, where the mean free path is of the same order of magnitude as the diameter of the pores, the molecules collide with the wall more often than with other molecules. In the limit of negligible molecule–molecule collisions, this type of diffusion is called Knudsen diffusion. The diffusivity for Knudsen diffusion D_K can be obtained from free-molecule flow theory. The general case for any pore diameter can be covered by simply assuming additive resistances (Mills, 1999) and adding the inverses of the diffusivities for both regimes:

$$\frac{1}{D} = \frac{1}{D_m} + \frac{1}{D_K} \qquad \text{(Eq. 4.64)}$$

The diffusivity of a gaseous component in a volume with both solid and gas is given by

$$D_{eff} = \frac{\varepsilon D}{\tau} \qquad \text{(Eq. 4.65)}$$

Here, the porosity ε accounts for the reduction in cross-sectional area for diffusion posed by the solid material, and the tortuosity factor τ accounts for the increased diffusion length due to the tortuous paths of real pores and for the effects of constrictions and dead-end pores (Mills, 1999).

The thermal conductivity of a porous particle λ_{eff} depends on the conductivity of the solid and gaseous parts and the porosity ε:

$$\lambda_{eff} = \varepsilon\lambda_g + (1-\varepsilon)\left[\frac{\rho_g}{\rho_{g,0}}\lambda_B + \left(1-\frac{\rho_g}{\rho_{g,0}}\right)\lambda_C\right] \qquad \text{(Eq. 4.66)}$$

In general, also radiation can contribute to transfer of heat through the particle, and provided the temperature differences are small (less than a few 100 K), this effect can be taken into account by an additive term in the thermal conductivity (see Kaviany (1995), page 340, and Chapter 11, Appendix).

To obtain a complete description of the problem, initial conditions and boundary conditions must be known. As initial conditions, we can assume an isothermal biomass particle with only inert gas in the pores:

$$\begin{aligned}
p(t=0,r) &= p_{atm} \\
T(t=0,r) &= T_0 \\
u(t=0,r) &= 0 \\
Y_I(t=0,r) &= 1 \\
Y_T(t=0,r) &= Y_C(t=0,r) = Y_G(t=0,r) = 0
\end{aligned} \qquad \text{(Eq. 4.67)}$$

As boundary conditions, we have to impose that at the particle center of symmetry, all fluxes are zero:

$$\left.\frac{\partial p(t,r)}{\partial r}\right|_{r=0} = \left.\frac{\partial T(t,r)}{\partial r}\right|_{r=0} = \left.\frac{\partial Y_i(t,r)}{\partial r}\right|_{r=0} = 0 \qquad \text{(Eq. 4.68)}$$

$$u(t,r=0) = 0$$

As boundary conditions at the particle outer surface, we have to impose equality of the pressure and equality of the fluxes at both sides of the particle boundary. The flux at the inside is the effective diffusive flux; the flux at the outside is the sum of the convective and radiative fluxes:

$$p(t,r=r_p) = p_{atm}$$

$$\lambda_{eff}\left.\frac{\partial T(t,r)}{\partial r}\right|_{r=r_p} = \theta_T h R_{SA}(T_b - T) + R_{SA}\varepsilon_e\sigma(T_w^4 - T^4)$$

$$D_{eff}\left.\frac{\partial Y_i(t,r)}{\partial r}\right|_{r=r_p} = \theta_m k R_{SA}(Y_{i,\infty} - Y_{i,s}) \qquad \text{(Eq. 4.69)}$$

Here, h and k are heat and mass transfer coefficients valid in the absence of outflow of gas from the particle. θ_m and θ_T are factors representing the effect of outflow of gas (blowing effect). This is a similar effect as described by the difference between the Ranz–Marshall correlation Equation (4.26) and the correlation for the evaporating droplet Equation (4.31). R_{SA} is the ratio of the real surface area to the ideal surface area (sphere, cylinder, or flat plat) and hence takes into account the effect of surface roughness. T_b is the bulk gas temperature, and T_W is the temperature of the surroundings relevant for radiative heat transfer. Note that in the radiative term for simplicity, the same emissivity ε_e is used for the particle surface and the surrounding wall. For each shape, an appropriate heat transfer coefficient has to be used. In the case of devolatilization of a particle cloud, if available, a particle size distribution and effects of random particle orientation can be taken into account. In the absence of such information, the average particle dimension can be used as the characteristic length. For more details, see Lu et al. (2010).

CHAPTER SUMMARY AND STUDY GUIDE

Transfer of mass is described by the combination of general transport equations for species mass fractions and specific models for transport by diffusion (Fick's law as a particular case, Stefan–Maxwell equations are more generally applicable). Transfer of energy is described by the combination of a general transport equation for an energy variable (internal energy, total energy, enthalpy, temperature), a specific model for transport by diffusion, and a source term due to radiation. Radiative energy transport in general is described by the radiative heat transfer equation. In many cases, it is sufficient to only consider a balance between radiative energies leaving different surfaces. The complexity of heat and mass transfer processes between a fluid and a solid is described effectively using the concept of transfer coefficients. In processes involving a phase change, such as evaporation of an ethanol droplet, the energy needed for this phase change has to be taken into account. In many processes, multiphase systems appear, with solid, liquid, and/or gas present, and additional concepts such as the porosity of a biomass particle are needed.

KEY CONCEPTS

Diffusion
Convection
Radiation
Mass flux
Heat flux
Transfer coefficient
Nusselt number
Sherwood number
Evaporation
Devolatilization
Combustion
Reacting flow

SHORT-ANSWER QUESTIONS

4.1 What is Fick's law for diffusion? For which type of mixtures is Fick's law an exact law? What is meant by the generalized Fick's law?

4.2 What is the form of the mass transfer equation for a steady nonreactive balance between advection and diffusion?

4.3 Give examples of radiative heat transfer processes dominated by surface-to-surface transfer and of processes for which the effects of absorption and emission in the space between surfaces (participating medium) are essential.

4.4 Explain in what sense the set of equations Equation (4.11) provides a solution for the surface-to-surface radiative heat transfer problem. How would you solve the equations (a) in the case that the temperature of all surfaces is known and (b) in the case that for some surfaces the temperature and for other surfaces the heat flux are known?

4.5 What is the nature of Newton's law of cooling Equation (4.25)? Considering a few examples, explain how the value of the heat transfer coefficient depends on the properties of the system.

4.6 What is the d^2-law for droplet evaporation? What are the assumptions that have to be satisfied for this law to be valid?

4.7 What is the difference between the infinite conductivity model and the finite conductivity model for fuel droplet evaporation?

PROBLEMS

4.1 Equation (4.8) is valid when the heat conductivity is independent on the direction. How would a generalization look like for the case of heat conductivity being different in the three coordinate directions? Which physical properties, e.g., of biomass material, could cause such difference?

4.2 Consider a layer of biomass of thickness $2L$, with moisture removal at both sides. In a simple modeling approach, the drying process is assumed to be mass transfer limited and to proceed at constant or slowly changing temperature. Assuming that the sample does not shrink during drying, moisture removal can be expressed using Fick's law for unsteady diffusion of moisture in the direction orthogonal to the layer. Experimentally, it was found that the effective moisture diffusivity D_{eff} is temperature dependent and has the form of an Arrhenius relationship:

$$D_{eff} = D_0 \exp\left(-\frac{E_a}{R_u T}\right),$$ depending on pre-exponential factor D_0 and the activation energy E_a (Gebreegziabher et al., 2013). Let $X(x, t)$ denote the sample moisture concentration as a function of position (distance from the symmetry plane)

and time. Assume that the initial value inside the layer is $X(x, 0) = X_0$ and outside the layer is $X(x, 0) = X_e$. The normalized moisture concentration is defined as $\vartheta = \dfrac{X - X_e}{X_0 - X_e}$.

a. Write an evolution equation for $\vartheta(x, t)$ containing a transient term and a diffusion term according to Fick's law.

b. Assuming that the temperature remains constant during drying, solve the equation by the method of separation of variables to obtain the solution

$$\vartheta(x,t) = \sum_{n=1}^{\infty} \frac{(-1)^n}{(2n+1)\pi} \exp\left(-\frac{(2n+1)^2 \pi^2 D_{eff}}{4L^2}t\right) \cos\left(\left(n+\frac{1}{2}\right)\pi\frac{x}{L}\right)$$

Hint: The problem is mathematically similar to the problem of diffusion of heat from a slab considered, e.g., in Mills (1999). The factor $(D_{eff}t/L^2)$ appearing in the exponential function is the analogue of the Fourier number Fo appearing in Table 4.1 and is the dimensionless time of the process.

c. Determine the surface moisture flux at $x = L$.

d. Identify accurate approximate solutions for the case of very long drying time and very short drying time.

e. Describe a procedure to determine the pre-exponential factor and the activation energy from experiments.

4.3 The radiative transfer equation Equation (4.12) contains terms representing the effects of scattering. For the simplest case of isotropic scattering, the scattering phase function is constant $\Phi_\lambda(s^*, s) \equiv 1$. For this case, show, by integrating the RTE over all directions, that scattering does not influence the value of the total incident radiation G.

4.4 There is currently substantial interest in utilizing eukaryotic algae for the renewable production of several bioenergy carriers, including starches for alcohols, lipids for diesel fuel surrogates, and hydrogen for fuel cells. Algae can convert solar energy into fuels at high photosynthetic efficiencies and can thrive in saltwater systems. Part of an energy balance analysis of such a process is the computation of the penetration of the radiative heat flux in a pool of water.

a. Considering the surface of the sun as a black surface at a temperature of 5777 K, determine the total (i.e., integrated over all wavelengths) solar heat flux incident on the top of the Earth's atmosphere. The radius of the sun is 6.96×10^8 m, and a representative value of the sun-to-Earth distance is 1.496×10^{11} m. Compare your result to the generally accepted annual mean value of 1366 W·m^{-2} and discuss possible reasons for difference.

b. Assuming that at a certain location and time, when the sun is in the zenith, the total heat flux arriving on the Earth surface is 1100 W·m^{-2}, calculate the

TABLE 4.2 Thermal conductivity λ [W·m^{-1}·K^{-1}] of air and ethanol as function of temperature

	250 K	300 K	400 K	500 K	600 K	800 K	1000 K
Air	0.0223	0.0267	0.0331	0.0395	0.0456	0.0569	0.0672
Ethanol	0.0092	0.0147	0.0245	0.0327			

From Kaviany (2002).

balance between incoming and outgoing radiative heat flux at the surface of a pond of stagnant water when the sun is at an angle of 30° from the zenith. The total hemispherical emissivity of water is known to vary from 0.95 to 0.963 in the temperature range from 273 to 373 K (Kaviany, 2002).

c. The absorption coefficient κ_λ of clear water is known to depend on wavelength. In the visible range (0.4–0.7 µm), it varies from 0.02 to 0.6 m^{-1} (Modest 2003, p. 416). By solving the radiative transfer equation, determine the decrease of intensity of incoming sunlight in a water layer with a depth 0.1 m for the case that the sun is at an angle of 30° from the zenith.

4.5 For purposes of engineering calculations, the most appropriate reference state for the evaluation of properties determining heat and mass transfer during evaporation of a fuel droplet is a simple 1/3 rule, wherein the reference temperature and species mass fractions are, respectively, $T_r = T_s + \frac{1}{3}(T_b - T_s)$ and $Y_{i,r} = Y_{i,s} + \frac{1}{3}(Y_{i,b} - Y_{i,s})$, with subscripts b and s referring to bulk and surface (Hubbard et al. 1975). An ethanol droplet at an initial temperature $T_s = 273$ K is heated by surrounding air at temperature $T_{air} = 500$ K. What is the value of the thermal conductivity to be used according to the 1/3 rule? Physical data on the thermal conductivity of air and ethanol are given in Table 4.2.

PROJECTS

P4.1 Using an appropriate software package, solve the model equations for an evaporating ethanol droplet for a range of conditions and plot the properties of the solutions. Consider varying droplet temperature and the effect of convective heat transfer.

P4.2 Using the kinetic data and properties data available in the literature, create an implementation of the model for particle devolatilization described in Section 4.5.2. Compare your results with those presented in Lu et al. (2010), and discuss the sensitivity of the predictions to different aspects such as wood material properties, kinetic rate coefficients, and particle size and shape.

REFERENCES

Bellan J, Summerfield M. Theoretical examination of assumptions commonly used for the gas phase surrounding a burning droplet. Combust Flame 1978;33:107–122.

Bird RB, Stewart WE, Lightfoot EN. *Transport Phenomena*. Revised 2nd ed. New York: John Wiley & Sons, Inc.; 2007.

Gebreegziabher T, Oyedun AO, Hui CW. Optimum biomass drying for combustion—a modeling approach. Energy 2013;53:67–73.

Hubbard GL, Denny VE, Mills AF. Droplet evaporation: effects of transients and variable properties. Int J Heat Mass Transfer 1975;18:1003–1008.

Jenny P, Roekaerts D, Beishuizen N. Modeling of turbulent dilute spray combustion. Prog Energy Combust Sci 2012;38:846–887.

Kaviany M. *Principles of Heat Transfer in Porous Media*. 2nd ed. New York: Springer; 1995.

Kaviany M. *Principles of Heat Transfer*. New York: John Wiley & Sons; 2002.

Kuo K. *Principles of Combustion*. 2nd ed. New York: John Wiley & Sons; 2005.

Le Clercq P, Bellan J. Direct numerical simulation of a transitional temporal mixing layer laden with multicomponent-fuel evaporating drops using continuous thermodynamics. Phys Fluids 2004;16(6):1884–1907.

Lu H, Ip E, Scott J, Foster P, Vickers M, Baxter LL. Effects of particle shape and size on devolatilization of biomass particle. Fuel 2010;89:1156–1168.

Mills AF. *Basic Heat and Mass Transfer*. Upper Saddle River, NJ: Prentice-Hall; 1999.

Modest MF. *Radiative Heat Transfer*. New York: Academic Press; 2003.

Poinsot T, Veynante D. *Theoretical and Numerical Combustion*. 3rd ed; 2011. Available at http://elearning/cerfacs.fr. Accessed on May 27, 2014.

Sazhin S, Kristyadi T, Abdelghaffar WA, Heikal MR. Models for fuel droplet heating and evaporation: comparative analysis. Fuel 2006;85:1613–1630.

Turns S. *An Introduction to Combustion. Concepts and Applications*. 2nd ed. Series in Mechanical Engineering. Boston, MA: McGraw-Hill; 2000.

Welty JR, Wicks CE, Wilson RE, Rorrer G. *Fundamentals of Momentum, Heat and Mass Transfer*. 4th ed. John Wiley & Sons; 2001.

5

REACTIONS: THERMODYNAMIC ASPECTS, KINETICS, AND CATALYSIS

MARTINA FANTINI[1], WIEBREN DE JONG[1], AND J. RUUD VAN OMMEN[2]

[1]*Department of Process and Energy, Energy Technology Section, Faculty of Mechanical, Maritime and Materials Engineering, Delft University of Technology, Delft, the Netherlands*
[2]*Department of Chemical Engineering, Product & Process Engineering Group, Faculty of Applied Sciences, Delft University of Technology, Delft, the Netherlands*

SYMBOLS

a_{ij}	number of atoms of the jth element in the ith species	[–]
c	concentration	[mol·m^{-3}]
E	energy, mole specific	[J·mol^{-1}]
E_a	activation energy	[J·mol^{-1}]
f_i	fugacity of component i	[bar]
G	Gibbs free energy	[J·mol^{-1}]
H	enthalpy	[J·mol^{-1}]
K_{eq}	equilibrium constant	[reaction dependent]
K_c	equilibrium constant based on molar concentrations	[reaction dependent]
K_p	equilibrium constant based on partial pressures	[–]
K_x	equilibrium constant based on mole fractions	[reaction dependent]
k	reaction rate coefficient	[reaction dependent]
k_0	pre-exponential factor	[reaction dependent]
n	total number of chemical species in a reacting system	[–]
n_i	number of moles of species i	[mol]
p	pressure	[Pa or bar]

Biomass as a Sustainable Energy Source for the Future: Fundamentals of Conversion Processes,
First Edition. Edited by Wiebren de Jong and J. Ruud van Ommen.
© 2015 American Institute of Chemical Engineers, Inc. Published 2015 by John Wiley & Sons, Inc.

p_i	reaction order with respect to species i	[–]
R_u	universal gas constant	[=8.314 $J \cdot mol^{-1} \cdot K^{-1}$]
r	reaction rate	[$mol \cdot m^{-3} \cdot s^{-1}$]
S	entropy	[$J \cdot mol^{-1} \cdot K^{-1}$]
T	temperature	[K]
t	time	[s]
V	volume	[m^3]
x_i	number of moles of species i reacted	[mol]
y_i	mole fraction of species i	[–]
$\Delta_r H$	reaction enthalpy	[$J \cdot mol^{-1}$]
λ_j	Lagrangian multiplier	[$J \cdot mol^{-1}$]
μ_i	chemical potential of component i	[$J \cdot mol^{-1}$]
ν	stoichiometric coefficient	[–]

Subscripts

b	backward direction
env	related to the environment surrounding a system
f	forward direction
syst	system
tot	total

Superscripts

t	total

5.1   REACTION KINETICS

Combustion is an example of a chemical reaction and can thus be described by a reaction equation. For example, the combustion of methane can be written as

$$CH_4 + 2O_2 \rightarrow CO_2 + 2H_2O \qquad \text{(RX. 5.1)}$$

This *overall reaction equation* states that if a reaction takes place, this will advance according to the molar and mass balances of the reaction itself, but it says nothing about the real reaction mechanism (usually formed by many elementary reactions). From (RX. 5.1), we know that 1 mol of methane reacts with 2 mol of molecular oxygen to give 1 mol of carbon dioxide and 2 mol of water. Considering that in chemistry, species mass is computed by multiplying the atomic weight (in atomic mass units) of each element in a chemical formula by the number of atoms of that element present in the formula, then adding all of these products together gives 16 (12 + 4) g of methane and 64 (2 × 32) g of molecular oxygen form 44 (12 + 32) g of carbon dioxide and 36 (2 × 18) g of water.

It is important to highlight that in chemistry the mass conservation law is the conservation of the number of atoms and this does not always implies the conservation of the number of molecules. Molecules can be broken down or merged, but the atoms stay intact.

In RX. 5.1, nothing is said about:

- The elementary reactions that define the entire *reaction mechanism* (an elementary reaction is a reaction that proceeds through only one transition state, i.e., one mechanistic step)
- The rate of the reaction itself

In order to have a reliable and real description of the phenomena going on, we need to have a look at the elementary reactions as, e.g.,

$$CO + OH \rightarrow CO_2 + H \qquad \text{(RX. 5.2)}$$

This is an elementary reaction of the complete mechanism for the methane oxidation in the gas phase (Bowman, 1975).

An example of a complete reaction mechanism can be given for the Fischer–Tropsch synthesis. The Fischer–Tropsch process, or Fischer–Tropsch synthesis, is a collection of chemical reactions that converts a mixture of carbon monoxide and hydrogen into liquid hydrocarbons. The process, a key component of gas-to-liquid technology, produces a synthetic lubrication oil and synthetic fuel, typically from coal, natural gas, or biomass. Its complete reaction mechanism is given by Kellner and Bell (1981), where S denotes the substrate (e.g., an active site):

$$CO + S \rightleftarrows CO_S \qquad \text{(RX. 5.3)}$$

$$CO_S + S \rightleftarrows C_S + O_S \qquad \text{(RX. 5.4)}$$

$$H_2 + 2S \rightleftarrows 2H_S \qquad \text{(RX. 5.5)}$$

$$H_2 + O_S \rightleftarrows H_2O + S \qquad \text{(RX. 5.6)}$$

$$C_S + H_S \rightleftarrows CH_S + S \qquad \text{(RX. 5.7)}$$

$$CH_S + H_S \rightleftarrows CH_{2S} + S \qquad \text{(RX. 5.8)}$$

$$CH_{2S} + H_S \rightleftarrows CH_{3S} + S \qquad \text{(RX. 5.9)}$$

$$CH_{3S} + H_S \rightarrow CH_4 + 2S \qquad \text{(RX. 5.10)}$$

$$CH_{3S} + CH_{2S} \rightarrow C_2H_{5S} + S \qquad \text{(RX. 5.11)}$$

$$C_2H_{5S} + S \rightarrow C_2H_4 + H_S + S \qquad \text{(RX. 5.12)}$$

$$C_2H_{5S} + H_S \rightarrow C_2H_6 + 2S \qquad \text{(RX. 5.13)}$$

$$C_2H_{5S} + CH_{2S} \rightarrow C_3H_{7S} + S \qquad \text{(RX. 5.14)}$$

etc.

This mechanism shows the formation of the "growth monomer CH_{2S}" (reactions RX. 5.3–5.8), the formation of the "chain starter CH_S" (reaction RX. 5.9), propagation (RX. 5.11 and RX. 5.14), the termination to a paraffin (alkane, hydrocarbon without double bonds, (RX. 5.10) and (RX. 5.13)), and the termination to an olefin (alkene, hydrocarbon with at least one carbon-to-carbon double bond, (RX. 5.12)).

5.1.1   Order of Reaction and Reaction Rate

As said, in order to understand how a chemical reaction takes place, elementary reactions are needed. For a homogeneous system, a common kinetic behavior exists, which can be described by the Guldberg and Waage law (or law of mass action). This law states that the rate of an elementary reaction is proportional to the product of the concentrations of the participating molecules.

Reactions are classified based on the order of reaction, which is the number of chemical species (molecules or atoms) that determine, based on their concentration, the reaction rate of the process. In other words, the order of reaction is the index, or exponent, to which its concentration term in the rate equation is raised. It is important to highlight that these indexes can be different from the stoichiometric coefficients as the balanced reaction equation is usually different from the reaction mechanism, whose rate is determined by the slowest step.

A generic chemical reaction could be represented in the following way:

$$\sum_{i=1}^{n} \nu_i' M_i \rightarrow \sum_{i=1}^{n} \nu_i'' M_i \qquad \text{(RX. 5.15)}$$

where ν_i' are the stoichiometric coefficients of the reactants and ν_i'' are the stoichiometric coefficients of the products, n is the total number of the chemical species involved in the reaction, and M_i represents the species. For the reaction

$$2H_2 + O_2 \rightarrow 2H_2O \qquad \text{(RX. 5.16)}$$

of which the order can be determined only experimentally, we clearly have

$$n = 3(H_2, O_2, H_2O); \quad M_1 = H_2; \quad M_2 = O_2; \quad M_3 = H_2O$$
$$\nu_1' = 2; \quad \nu_2' = 1; \quad \nu_3' = 0; \quad \nu_1'' = 0; \quad \nu_2'' = 0; \quad \nu_3'' = 2$$

By convention, the stoichiometric coefficients are positive for products and negative for reactants, and we can define the reaction rate as

$$r = \frac{1}{\nu_i} \frac{d[M_i]}{dt} \qquad \text{(Eq. 5.1)}$$

The rate equation can also be written as a product of two factors, one dependent on the temperature (k_f) and the other on the concentrations of the reactants:

$$r_f = k_f \times f([M_1],[M_2],\ldots) = k_f \prod_{i=1}^{n} [M_i]^{p_i} \qquad \text{(Eq. 5.2)}$$

where k_f is called the *reaction rate coefficient* or *rate constant* (Arrhenius), p_i is the reaction order with respect to species M_i, and $\sum_{i=1}^{n} p_i$ is the total reaction order.

We also have to take into account the reverse reaction, so the reaction rate can be written as

$$r = r_f - r_b = k_f [M_1]^{p_1} [M_2]^{p_2} \cdots - k_b [M_m]^{p_m} [M_{m+1}]^{p_{m+1}} \cdots \qquad \text{(Eq. 5.3)}$$

The reaction rate is influenced by:

1. The **nature** of the reactants
2. The **concentration** of the reactants
3. The **temperature**
4. The presence of **catalysts**

The first point concerns the nature of the chemical bonds of the species involved. To understand the influence of the other factors, the way a reaction takes place should be well known. A necessary condition for a reaction to take place is the collision of the reactant molecules, having kinetic energy due to the thermal energy. Thus, the higher the concentration of the reactants, the higher the probability that a reaction takes place, because of the larger number of collisions. However, the reaction rate does not depend on the number of collisions alone, because otherwise we would observe high rates for every reaction. The collision must be a successful collision. This means that, given the right orientation of the molecules at the moment of the collision, the collision has to be strong enough to break the existing bonds and to exceed the repulsive barriers of the molecules. The minimum amount of kinetic energy needed is called the activation energy. Not every collision is successful as not every molecule has the same kinetic energy. For example, the velocities of the particles of a gas at a certain temperature continuously change due to the high number of collisions, and theoretically, they assume all the values between zero and infinite, according to the Maxwell–Boltzmann law. This law is represented in Figure 5.1a. Here, the area of each rectangle of very small size (at its limit, infinitesimal) is the number of particles with a certain kinetic energy close to a certain value E_1. The sum of all the infinitesimal rectangles on the interval zero—infinite is the total number of particles. The hatched area in Figure 5.1b represents the number of particles with a kinetic energy greater or equal to the activation energy E_a; the lower the activation energy, the higher the reaction rate as more molecules will be ready to react. If the temperature increases, the distribution curve moves to the left (the percentage of particles with higher kinetic energy

(a)   (b)

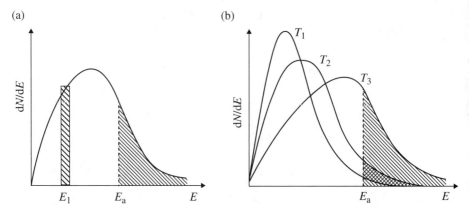

FIGURE 5.1   Maxwell–Boltzmann distribution and activation energy: (a) number of particles with an activation energy greater than E_a and (b) temperature influence (Source: Reproduced with permission from Pasquetto and Patrone, 1999, vol. 1. © Zanichelli editore S.p.A).

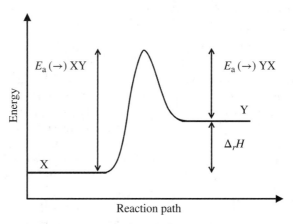

FIGURE 5.2   Relationship between the activation energy (E_a) and reaction enthalpy ($\Delta_r H$), plotted versus the reaction coordinate. The highest energy position (peak position) represents the transition state.

increases) and flattens (as the area under the distribution is constant). This shows the huge effect of the temperature on the reaction rate, which can double or even triple with a temperature increase of only 10°C.

When two molecules with a kinetic energy equal to or greater than the activation energy collide, they reach a higher energy content (peak position of Figure 5.2) called *transition state* in which the bonds of the molecules are instable.

As shown in Figure 5.2, the energy of this state is greater than both the energy of the reactants and the energy of the products. The difference between the energy of the reactants and the energy of the products is the standard enthalpy of reaction.

In the Arrhenius equation, the activation energy (E_a) is used to describe the energy required to reach the transition state:

$$k = k_0 e^{-E_a/R_u T}$$ (Eq. 5.4)

where k = reaction rate coefficient in the rate law

k_0 = pre-exponential factor (unit is reaction rate expression dependent)
E_a = activation energy [$J \cdot mol^{-1}$]
R_u = universal gas constant [$=8.314 \ J \cdot mol^{-1} \cdot K^{-1}$]
T = reaction temperature [K]

Example 5.1   Find (a) the order of reaction and (b) the reaction rate coefficient for the reaction

$$CH_3CHO(g) \rightarrow CH_4(g) + CO(g)$$

at 518°C (thus 791.15 K) using the following experimental data:

Initial concentration [$mol \cdot L^{-1}$] of acetaldehyde	Formation rate of acetaldehyde [$mol \cdot L^{-1} \cdot s^{-1}$]
0.150	1.5×10^{-7}
0.300	6.0×10^{-7}
0.600	24.0×10^{-7}

Solution
As the decomposition rate of the acetaldehyde quadruples when the initial concentration doubles, we can deduce that the reaction is a second-order reaction.
From the reaction rate law and using the experimental data, we can write

$$k = \frac{r}{[CH_3CHO]^2} = \frac{1.5 \times 10^{-7} mol \cdot L^{-1} \cdot s^{-1}}{0.150^2 mol^2 \cdot L^{-2}} = \frac{6.0 \times 10^{-7}}{0.300^2} = \frac{24 \times 10^{-7}}{0.600^2}$$

$$= 6.7 \times 10^{-6} L \cdot mol^{-1} \cdot s^{-1}$$

Example 5.2   Calculation of the activation energy of a higher-order reaction

A second-order reaction was observed. The reaction rate coefficient was found to be $8.9 \times 10^{-3} \ L \cdot mol^{-1} \cdot s^{-1}$ at 3°C and $7.1 \times 10^{-2} \ L \cdot mol^{-1} \cdot s^{-1}$ at 35°C. What is the activation energy of this reaction?

Solution

The activation energy is the amount of energy required to initiate a chemical reaction. The activation energy can be determined from the reaction rate coefficients at different temperatures from the equation

$$\ln\left(\frac{k_2}{k_1}\right) = \frac{E_a}{R_u} \cdot \left(\frac{1}{T_1} - \frac{1}{T_2}\right)$$

where

  T_1 and T_2 are absolute temperatures

  k_1 and k_2 are the reaction rate coefficients at T_1 and T_2

Step 1: Convert the temperatures from °C to K

  $T = °C + 273.15$

  $T_1 = 3 + 273.15 = 276.15$ K

  $T_2 = 35 + 273.15 = 308.15$ K

Step 2: Find E_a using the equation given previously E

$$\ln\left(\frac{7.1 \cdot 10^{-2}}{8.9 \cdot 10^{-3}}\right) = \frac{E_a}{8.3145} \cdot \left(\frac{1}{276.15} - \frac{1}{308.15}\right)$$

$$\ln(7.98) = \frac{E_a}{8.3145} \cdot 3.76 \times 10^{-4}$$

$$2.077 = E_a \cdot \left(4.52 \times 10^{-5}\right)$$

$$E_a = 4.59 \times 10^4 \text{J} \cdot \text{mol}^{-1} = 45.9 \text{kJ} \cdot \text{mol}^{-1}$$

5.1.2 Unimolecular First-Order Reactions

A reaction involving one molecular entity is called unimolecular and is given by

$$M \rightarrow \text{products} \qquad\qquad\qquad (RX.\ 5.17)$$

If this reaction is a first-order reaction, the reaction rate is proportional to the concentration of the reactant:

$$r_f = -\frac{d[M]}{dt} = -\frac{dc}{dt} = k_f c \qquad\qquad (Eq.\ 5.5)$$

With n_i^0 the number of moles of species i at t = 0, n_i the number of moles of species i at time t, and x_i the number of reacted moles at time t, we can write

$$n_i = n_i^0 - x_i \qquad\qquad\qquad (Eq.\ 5.6)$$

and for homogeneous reactions

$$r = -\frac{1}{V}\frac{dn_i}{dt} = -\frac{dc_i}{dt} = \frac{1}{V}\frac{dx_i}{dt} \qquad \text{(Eq. 5.7)}$$

Combining Equation (5.5), Equation (5.6), and Equation (5.7) and omitting subscript i:

$$r_f = \frac{1}{V}\frac{dx}{dt} = \frac{k_f}{V}\left(n^0 - x\right) \qquad \text{(Eq. 5.8)}$$

$$\int_0^x \frac{dx}{(n^0 - x)} = \int_0^t k_f dt \qquad \text{(Eq. 5.9)}$$

$$\ln\frac{n^0}{(n^0 - x)} = k_f t \qquad \text{(Eq. 5.10)}$$

$$n^0 - x = \frac{n^0}{e^{k_f t}} \Leftrightarrow x = n^0\left(1 - e^{-k_f t}\right) \qquad \text{(Eq. 5.11)}$$

from which it is clear that the quantity of the reacting species decreases exponentially.

5.1.3   Bimolecular Second-Order Reactions

Second-order reactions can be of second order in one reactant or of first order in two different reactants, i.e.,

$$2M \rightarrow \text{products} \qquad \text{(RX. 5.18)}$$

$$M_1 + M_2 \rightarrow \text{products} \qquad \text{(RX. 5.19)}$$

The reaction rate for the reaction in RX. 5.18 is given by

$$r_f = -\frac{1}{2}\frac{dc}{dt} = k_f c^2 \qquad \text{(Eq. 5.12)}$$

or

$$\frac{1}{2V}\frac{dx}{dt} = k_f \frac{(n^0 - x)^2}{V^2}$$

$$V\int_0^x \frac{dx}{(n^0 - x)^2} = 2\int_0^t k_f dt \qquad \text{(Eq. 5.13)}$$

$$\frac{Vx}{n^0(n^0 - x)} = 2k_f t$$

5.2   CHEMICAL EQUILIBRIUM

In principle, all chemical reactions are reversible: they comprise a forward and a reverse reaction. When the concentrations of the reactants and products have no further tendency to change with time, this is called chemical equilibrium: the reaction results in an equilibrium mixture of reactants and products. The condition of no variation of concentrations is necessary but not sufficient for equilibrium. We can speak about chemical equilibrium only when the forward reaction proceeds at the same rate as the reverse reaction. Since at equilibrium, the rates of the forward and reverse reactions are equal, given the general reaction equation:

$$aA + bB \rightleftarrows cC + dD \qquad \text{(RX. 5.20)}$$

the ratio of the rate coefficients is constant and known as the equilibrium constant:

$$K_{eq} = \frac{[C]^c [D]^d}{[A]^a [B]^b} \qquad \text{(Eq. 5.14)}$$

The relation in Equation (5.14) is known as the law of mass action and is valid only for concerted one-step reactions that proceed through a single transition state. It is not generally valid because rate equations do not, in general, follow the stoichiometry of the reaction as Guldberg and Waage proposed. Caused by the existence of the elementary reactions, the kinetics of the global rate law cannot be theoretically deduced but only experimentally. Therefore, also the relation between rates and equilibrium constants is not valid like shown in Equation (5.14), but has to be defined thermodynamically. Despite the failure of this derivation, the equilibrium constant for a reaction is indeed a constant, independent of the activities of the various species involved, though it does depend on temperature. Given a reaction at a certain temperature, the numerical value of the equilibrium constant depends on the units of measurement chosen. That is, depending on whether the concentrations are indicated as $mol \cdot L^{-1}$, atm, or mole fractions, K_{eq} is symbolized, respectively, by K_c, K_p, or K_x.

Adding a catalyst, a substance capable of changing the rate of a chemical reaction, will affect both the forward reaction and the reverse reaction in the same way but will not have an effect on the equilibrium constant. The catalyst will speed up both reactions, thereby increasing the rate at which equilibrium is reached.

5.2.1   Equilibrium Calculations Based on the Gibbs Free Energy

Let us investigate the thermodynamic conditions for chemical equilibrium.

To measure the spontaneity of chemical phenomena, the extensive thermodynamic property "entropy" cannot be used as the reactions usually take place in "nonisolated" systems. To overcome this difficulty, a new thermodynamic function has been defined: the *Gibbs free energy* also called *Gibbs function*, which is suitable for reactions that occur at constant pressure. The mathematical definition of this function of state is

$$G = H - TS \qquad \text{(Eq. 5.15)}$$

For spontaneous processes, the entropy will increase, i.e.,

$$\Delta S_{tot} = \Delta S_{syst} + \Delta S_{env} > 0 \qquad \text{(Eq. 5.16)}$$

Considering that

$$\Delta S_{env} = -\frac{\Delta H}{T} \qquad \text{(Eq. 5.17)}$$

substitution of Equation (5.17) in Equation (5.16)

$$\Delta S_{tot} = \Delta S_{syst} - \frac{\Delta H}{T} > 0 \qquad \text{(Eq. 5.18)}$$

and referring only to the system yields

$$T\Delta S - \Delta H > 0 \qquad \text{(Eq. 5.19)}$$

Differentiating under the condition that the temperature is constant ($dT = 0$) and considering a finite increment, we can conclude that for spontaneous processes, the free energy decreases, i.e.,

$$\Delta G_{T,P} < 0 \qquad \text{(Eq. 5.20)}$$

Considering the general reaction equation (RX. 5.20), the reaction rates of A and B (r_f) will be the highest at the beginning and will decrease in time with the formation of the products C and D; the reverse reaction rates (r_b) will increase from zero to r_f. When r_b is equal to r_f, we can say that the reaction has reached chemical equilibrium ($\Delta G_{T,P} = 0$) and the concentrations of all species are constant. This is called dynamic equilibrium: the rates of the forward and reverse reactions are not zero, but they are equal (forward and reverse reactions occur continuously at the same rate). From a thermodynamic point of view, we can say that if the sum of the standard free energies G^0 (at the standard temperature of 298.15 K and at the standard pressure of 1 bar) of the products is less than that of the reactants, ΔG^0 for the reaction is negative, and the reaction will proceed to the right spontaneously.

For a reversible process, the infinitesimal change of the free energy of the system is given by (Moran and Shapiro, 2010; Smith et al., 2005)

$$dG = Vdp - SdT + \sum_{i=1}^{j} \mu_i dn_i \qquad \text{(Eq. 5.21)}$$

which, for only mechanical work and for an isothermal process concerning an ideal gas, becomes (V related to 1 mol of gas)

$$dG = Vdp = \frac{R_u T}{p} dp \qquad \text{(Eq. 5.22)}$$

In order to estimate the finite value of the free energy variation ΔG that a mole of an ideal gas needs to isothermally pass from an initial pressure p_0 to a final pressure p, we integrate Equation (5.22)

$$\int_1^2 dG = G_2 - G_1 = \Delta G = \int_{p_0}^{p} R_u T \frac{dp}{p} = R_u T \ln \frac{p}{p_0} \qquad \text{(Eq. 5.23)}$$

Considering a standard initial state as reference, p_0 being 1 bar, Equation (5.23) can be written as

$$G = G^0 + R_u T \ln p \qquad \text{(Eq. 5.24)}$$

where G is the free energy of 1 mol of ideal gas at pressure p (in bar!) and temperature T, while G^0 is its free energy at the standard temperature and pressure (298.15 K and 1 bar).

Considering the fact that the free energy is a function of state and given Equation (5.24)), the change of the free energy for a general reaction (RX. 5.20) is expressed by the Van't Hoff equation:

$$\Delta G = \Delta G^0 + R_u T \ln \frac{p_C^c p_D^d}{p_A^a p_B^b} \qquad \text{(Eq. 5.25)}$$

The argument of the logarithm in Equation (5.25) is formally analogous to the equilibrium constant (K_p) but has the same value only when the reaction has reached equilibrium, i.e.,

$$\Delta G = \Delta G^0 + R_u T \ln K_p = 0 \quad \Rightarrow \quad \Delta G^0 = -R_u T \ln K_p \qquad \text{(Eq. 5.26)}$$

Example 5.3   Calculation of equilibrium constants, K_c and K_p

In a container with a capacity of 5 L, 1 mol of SO_2 and 1 mol of O_2 react at 1000 K. Once equilibrium has been reached, the amount of SO_3 in the container is 68 g (0.85 mol). Find K_c and K_p. $R_u = 0.082$ L.bar.K^{-1}.mol^{-1}

Solution
The reaction observed is

$$2SO_2 + O_2 \rightleftarrows 2SO_3$$

and the stoichiometric ratios are: 2:1:2 or 1:0.5:1. Thus, if x moles of SO_2 react, 1/2 x moles of O_2 reacts to form x moles of SO_3. So we can write

	$2SO_2$	O_2	$2SO_3$
Initial moles	1	1	0
Δ moles	$-x$	$-\frac{1}{2}x$	$+x$
Moles at equilibrium	$(1-x)$	$\left(1-\frac{x}{2}\right)$	x

For this problem, $x = 0.85$ mol, so the molar concentrations of the species are

$$[SO_2] = \frac{(1-x)}{V} = \frac{0.15}{5} = 0.03 \, mol \cdot L^{-1}$$

$$[O_2] = \frac{(1-(x/2))}{V} = \frac{0.575}{5} = 0.115 \, mol \cdot L^{-1}$$

$$[SO_3] = \frac{x}{V} = \frac{0.85}{5} = 0.17 \, mol \cdot L^{-1}$$

Answers: We can now calculate the equilibrium constants:

$$K_c = \frac{[SO_3]^2}{[SO_2]^2[O_2]} = \frac{0.17^2}{0.03^2 \times 0.115} = 279 \, \left[L \cdot mol^{-1}\right]$$

$$K_p = K_c(R_uT)^{\Delta n} = \frac{279}{0.082 \times 1000} = 3.4$$

Question: Why is K_p dimensionless? Hint: How is K_p derived?

Example 5.4   Calculation of the dissociation grade of a reactant

The equilibrium constant K_c of the reaction $N_2O_4 \rightleftarrows 2NO_2$ at 47°C is 0.05 $mol \cdot L^{-1}$. How many grams of NO_2 (MW = 46 $g \cdot mol^{-1}$) are present at equilibrium if we fill a container with a capacity of 1 L with 46 g of N_2O_4 (MW = 92 $g \cdot mol^{-1}$)? What is the dissociation grade of N_2O_4 at this temperature?

Solution

The stoichiometric ratio for the reaction is 1:2. If x are the moles of N_2O_4 reacted, we can write

	N_2O_4	$2NO_2$
Initial number of moles	46/92 = 0.5	0
Δ moles	$-x$	$+2x$
Moles at equilibrium	$(0.5-x)$	$2x$

For a 1 L container:

$$K_c = \frac{[NO_2]^2}{[N_2O_4]} = 0.05 \, mol \cdot L^{-1}$$

$$0.05 = \frac{(2x)^2}{(0.5-x)}$$

$$4x^2 + 5 \times 10^{-2}x - 2.5 \times 10^{-2} = 0$$

$$x_1 < 0; \quad x_2 = 0.073$$

Answers: The amount of NO_2 present is $2 \cdot 0.073 = 0.146$ mol $= 6.72$ g, and the dissociation grade is $(0.073/0.5) \cdot 100\% = 14.6\%$.

Example 5.5   Calculation of the equilibrium constant based on H^0 and S^0 data

Calculate the equilibrium constant at 25°C for the following reaction:

$$CO(g) + 2H_2(g) \rightleftarrows CH_3OH(g)$$

Data:

$CH_3OH(g): S^0 = 236.8 \, J \cdot mol^{-1} \cdot K^{-1} \quad H^0 = -202.1 \, kJ \cdot mol^{-1}$
$CO \qquad : S^0 = 197.9 \, J \cdot mol^{-1} \cdot K^{-1} \quad H^0 = -110.5 \, kJ \cdot mol^{-1}$
$H_2 \qquad : S^0 = 130.6 \, J \cdot mol^{-1} \cdot K^{-1} \quad H^0 = 0 \, kJ \cdot mol^{-1}$

Solution
Step 1: Calculate the enthalpy and entropy changes:

$$\Delta H^0 = (-202,100) - (-110,500) = -91,600 \, J \cdot mol^{-1}$$
$$\Delta S^0 = 236.8 - (261.2 + 197.9) = -222.3 \, J \cdot mol^{-1} \cdot K^{-1}$$

Step 2: Considering that the free energy is a function of state, we can use Equation (5.15) to calculate the standard free energy change:

$$\Delta G^0 = -91,600 - 298.15(-222.3) = -91,600 + 66,245 = -25,355 \, J \cdot mol^{-1}$$

Answer: The value of K_p follows from the relation Equation (5.26):

$$K_p = e^{-\frac{\Delta G^0}{R_u T}} = e^{\frac{25,355}{8.314 \times 298.15}} = 2.8 \times 10^4$$

5.2.2 Minimization of the Gibbs Free Energy

In a closed system at constant T and p, equilibrium is reached when the total Gibbs free energy attains its minimum value. Thus, if a mixture of chemical species is not in chemical equilibrium, any reaction that occurs at constant T and p must lead to a decrease in the total Gibbs free energy of the system. The total Gibbs free energy of a system is defined as

$$G^t = \sum_{i=1}^{N} n_i \mu_i \qquad \text{(Eq. 5.27)}$$

where n_i is the number of moles of species i and μ_i is the chemical potential of species i, which can be presented by

$$\mu_i = G_i^0 + R_u T \ln \left(\frac{f_i}{f_i^0} \right) \qquad \text{(Eq. 5.28)}$$

Here, f_i is the fugacity of species i. Equation (5.28) can also be presented in terms of pressure as

$$\mu_i = G_i^0 + R_u T \ln \left(\frac{\hat{\varphi}_i p_i}{p^0} \right) \qquad \text{(Eq. 5.29)}$$

where $\hat{\varphi}_i$ is the fugacity coefficient of species i (in a nonideal gas mixture). Normally, f_i and p_i take on the same value when the pressure approaches zero. At that condition, the real gas mixture also approaches an ideal gas. If all gases are considered as ideal gases at a pressure of one atmosphere and if G_i^0 is arbitrarily set equal to zero for all elements in their standard states (so, for the compounds, $G_i^0 = \Delta G_{f,i}^0$), then Equation (5.29) can be rewritten as

$$\mu_i = \Delta G_{f,i}^0 + R_u T \ln(y_i) \qquad \text{(Eq. 5.30)}$$

where y_i is the mole fraction of gas species i, which is the ratio of n_i to the total number of moles in the reaction mixture. $\Delta G_{f,i}^0$ is the standard Gibbs free energy of formation of species i. By convention, it is set equal to zero for all chemical elements. Substituting Equation (5.30) into Equation (5.27) yields

$$G^t = \sum_{i=1}^{N} n_i \Delta G_{f,i}^0 + \sum_{i=1}^{N} n_i R_u T \ln \left(\frac{n_i}{n_{tot}} \right) \qquad \text{(Eq. 5.31)}$$

Now, the problem is to find the values of n_i, which minimize the objective function G^t. A method that is usually used for the minimization of the Gibbs free energy problem is that of the Lagrange multipliers. The constraint of this problem is the elemental balance, i.e.,

$$\sum_{i=1}^{N} a_{ij} n_i = A_j, \quad j = 1, 2, 3, \ldots, k \tag{Eq. 5.32}$$

where a_{ij} is the number of atoms of element j in the species i. A_j is defined as the total number of atoms of the element j in the reaction mixture. To form the Lagrangian function (L), the Lagrange multipliers $\lambda_j = \lambda_1, \ldots, \lambda_k$, multiplied with the elemental balance constraint, and the resulting terms are added to G^t as follows:

$$L = G^t + \sum_{j=1}^{k} \lambda_j \left(\sum_{i=1}^{N} a_{ij} n_i - A_j \right) \tag{Eq. 5.33}$$

The partial derivatives of Equation (5.33) are set equal to zero in order to find the limit:

$$\left(\frac{\partial L}{\partial n_i} \right) = \left(\frac{\partial G^t}{\partial n_i} \right) + \sum_j \lambda_j a_{ij} = 0 \tag{Eq. 5.34}$$

Equation (5.34) can be written as a matrix that has i rows, and those are solved simultaneously with the constraints as defined in Equation (5.32). Because the first term on the right is the definition of the chemical potential, Equation (5.34) can be written as

$$\mu_i + \sum_j \lambda_j a_{ij} = 0 \tag{Eq. 5.35}$$

Combination of Equation (5.30) and Equation (5.35) gives

$$\Delta G_{f,i}^0 + R_u T \ln(y_i) + \sum_j \lambda_j a_{ij} = 0 \tag{Eq. 5.36}$$

Example 5.6   Calculation of the equilibrium composition based on minimization of the Gibbs free energy

Ethane and steam are fed to a steam cracker at 1000 K and a total pressure of 1 bar at a ratio of 4 mol H_2O to 1 mol ethane. Estimate the equilibrium distribution of the products (CH_4, C_2H_4, C_2H_2, CO_2, CO, O_2, H_2, H_2O, and C_2H_6). Values of $\Delta G_{f_i}^0$ at 1000 K are:

$$\Delta G_{f,CH_4}^0 = 4.61 \, \text{kcal·mol}^{-1} \Delta G_{f,CO_2}^0 = -94.61 \, \text{kcal·mol}^{-1} \Delta G_{f,CO}^0$$
$$= -47.942 \, \text{kcal·mol}^{-1}$$

$$\Delta G_{f,C_2H_4}^0 = 28.249 \, \text{kcal·mol}^{-1} \Delta G_{f,O_2}^0 = 0 \, \text{kcal·mol}^{-1} \Delta G_{f,H_2O}^0 = -46.03 \, \text{kcal·mol}^{-1}$$

$$\Delta G_{f,C_2H_2}^0 = 40.604 \, \text{kcal·mol}^{-1} \Delta G_{f,H_2}^0 = 0 \, \text{kcal·mol}^{-1} \Delta G_{f,C_2H_6}^0 = 26.13 \, \text{kcal·mol}^{-1}$$

Solution

We will construct the Gibbs function for the mixture and obtain the equilibrium composition by minimization of the function subject to the elemental mass balance constraints.

At 1 bar and 1000 K, the assumption of ideal gas is justified, and we can write Equation (5.36) for each component involved in the process:

$$CH_4 : \frac{4.61}{R_u T} + \ln\left(\frac{n_{CH_4}}{\sum_i n_i}\right) + \frac{\lambda_C}{R_u T} + \frac{4\lambda_H}{R_u T} = 0$$

$$C_2H_2 : \frac{40.604}{R_u T} + \ln\left(\frac{n_{C_2H_2}}{\sum_i n_i}\right) + \frac{2\lambda_C}{R_u T} + \frac{2\lambda_H}{R_u T} = 0$$

$$C_2H_4 : \frac{28.249}{R_u T} + \ln\left(\frac{n_{C_2H_4}}{\sum_i n_i}\right) + \frac{2\lambda_C}{R_u T} + \frac{4\lambda_H}{R_u T} = 0$$

$$C_2H_6 : \frac{26.13}{R_u T} + \ln\left(\frac{n_{C_2H_6}}{\sum_i n_i}\right) + \frac{2\lambda_C}{R_u T} + \frac{6\lambda_H}{R_u T} = 0$$

$$CO_2 : \frac{-94.61}{R_u T} + \ln\left(\frac{n_{CO_2}}{\sum_i n_i}\right) + \frac{\lambda_C}{R_u T} + \frac{2\lambda_O}{R_u T} = 0$$

$$CO : \frac{-47.942}{R_u T} + \ln\left(\frac{n_{CO}}{\sum_i n_i}\right) + \frac{\lambda_C}{R_u T} + \frac{\lambda_O}{R_u T} = 0$$

$$O_2 : \ln\left(\frac{n_{O_2}}{\sum_i n_i}\right) + \frac{2\lambda_O}{R_u T} = 0$$

$$H_2 : \ln\left(\frac{n_{H_2}}{\sum_i n_i}\right) + \frac{2\lambda_H}{R_u T} = 0$$

$$H_2O : \frac{-46.03}{R_u T} + \ln\left(\frac{n_{H_2O}}{\sum_i n_i}\right) + \frac{2\lambda_H}{R_u T} + \frac{\lambda_O}{R_u T} = 0$$

The total number of each type of atom must be the same as the number entering the reactor. This conservation of elements forms equality constraints on the equilibrium composition. The conservation equations are:

$$C : n_{CH_4} + 2n_{C_2H_4} + 2n_{C_2H_2} + n_{CO_2} + n_{CO} + 2n_{C_2H_6} = 2$$

$$H : 4n_{CH_4} + 4n_{C_2H_4} + 2n_{C_2H_2} + 2n_{H_2} + 2n_{H_2O} + 2n_{C_2H_6} = 14$$

$$O : 2n_{CO_2} + n_{CO} + 2n_{O_2} + n_{H_2O} = 4$$

$$\sum_i n_i = n_{CH_4} + n_{C_2H_4} + n_{C_2H_2} + n_{CO_2} + n_{CO} + n_{O_2} + n_{H_2} + n_{H_2O} + n_{C_2H_6}$$

The computer solution of these 13 equations produces consistent results for five species, the others are negligible. The following results are given in number of moles:

$$n_{CH_4} = 0.0664$$
$$n_{CO_2} = 0.545$$
$$n_{CO} = 1.39$$
$$n_{H_2} = 5.35$$
$$n_{H_2O} = 1.52$$

5.2.3 Considerations Regarding Chemical Equilibrium

1. The chemical equilibrium law is valid for gaseous systems, for solutions, and when the reactants have different phases (heterogeneous equilibrium), but it is not generally valid because rate equations do not, in general, follow the stoichiometry of the reaction as Guldberg and Waage proposed.

2. The concentrations in Equation (5.14) are those at equilibrium and not those at the start of the reaction.

3. In Equation (5.14), the numerator contains the concentrations of the products and the denominator the concentration of the reactants. This means that the larger K_{eq}, the higher the concentration of the products at equilibrium.

4. Given a reaction at a certain temperature, the numerical value of the equilibrium constant depends on the units of measurement chosen.

5. Given a balanced reaction, multiplying each stoichiometric coefficient with a constant value m gives a different K_{eq} value. In that case, K_{eq} of the balanced reaction with the minimum full stoichiometric coefficients becomes raised to the power m. Conventionally, K_{eq} of the balanced equation with the minimum full stoichiometric coefficients is used.

6. Independently from the quantities of the reactants, the equilibrium concentrations of reactants and products have to match with K_{eq} value at that temperature.

7. The units for K_{eq} depend upon the units used for the concentrations. Since the concentrations of reactants and products are not dimensionless, the unit of the equilibrium constant is represented by *activity*. Activity is expressed as the dimensionless ratio $[X]/c^0$ where $[X]$ signifies the molarity (concentration in $mol \cdot m^{-3}$) of the reaction and c^0 is the chosen reference state. Thus, the units are canceled and K_{eq} becomes dimensionless.

5.2.4 Factors Affecting the Chemical Equilibrium

The factors that affect the chemical equilibrium are all external factors that can somehow change the values of the equilibrium concentrations (partial pressures and molar fractions) for a certain reaction. The effect of these factors is to bring the system to a new equilibrium status with new concentration values.

As the equilibrium concentrations of reactants and products have to be such that

$$r_f = r_b \qquad\qquad (Eq. 5.37)$$

it is clear that these external factors have to influence the forward and reverse reaction rates in different ways. These factors are external change of concentrations of one of the species, temperature, and pressure.

To predict the effect of a change in conditions on a chemical equilibrium, le Chatelier's principle can be used. The principle can be summarized as follows: if a chemical system at equilibrium experiences a change in concentration, temperature, or (partial) pressure, then the equilibrium shifts to counteract the imposed change and a new equilibrium is established, i.e., more generally, any change in status quo prompts an opposing reaction in the system.

Let us analyze the effects of the three factors mentioned.

Effect of Change in Concentration:
As an example, consider the reaction

$$N_2(g) + 3H_2(g) \rightleftharpoons 2NH_3(g) \qquad\qquad (RX. 5.21)$$

which is at equilibrium at a certain temperature.

If we add any quantity of hydrogen, nitrogen, or both, in order to minimize its effect, these reactants turn into the product ammonia, and a new equilibrium, shifted to the right side of RX. 5.21, is reached. Similarly, if ammonia is removed, new ammonia is formed, which results in a new equilibrium shifted to the right side. If, on the contrary, ammonia is added, to minimize the effect of this alteration, the concentrations of the reactants increase until a new equilibrium is reached. Similarly, if hydrogen and/or nitrogen is removed, this would cause the reaction to fill the "gap" and favor the side where the species was/were reduced. This observation is supported by the collision theory. Increasing the concentration of hydrogen and/or nitrogen also increases the frequency of successful collisions of these reactants, resulting in an increase of the rate of the forward reaction and generation of ammonia.

These phenomena do not contradict the law of mass action since both the numerator and the denominator in the equation for K_{eq} change so the value of K_{eq} remains the same.

Effect of Change in Temperature: Almost all reactions are accompanied by either the release or consumption of energy. So, the effect of changing the temperature on the equilibrium can be made clear by incorporating heat as either a reactant or a product. For an exothermic reaction, i.e., a reaction releasing energy and having a negative enthalpy of reaction ($\Delta_r H$), we include heat as a product; for an endothermic reaction, which consumes energy ($\Delta_r H$ is positive), we include it as a reactant. Obviously, if the forward reaction is exothermic, the reverse reaction is endothermic and the other way around. We can determine whether increasing or decreasing the temperature favors the forward or reverse reaction by applying the same principle as with concentration changes.

For example, the forward reaction of the reversible reaction (RX. 5.21) is exothermic with $\Delta_r H = -92$ kJ·mol^{-1}; heat is a product: $N_2(g) + 3H_2(g) \rightleftharpoons 2NH_3(g) + 92$ kJ.

If we lower the temperature, the equilibrium shifts to produce more heat. Since the formation of ammonia is exothermic, this favors the production of more ammonia.

So, if we increase the temperature of a reaction at equilibrium, to minimize the effect of this change, the equilibrium shifts to the side where the heat is absorbed, to the right if the reaction is endothermic, and to the left if it is exothermic. For exothermic reactions, an increase in temperature decreases the value of the equilibrium constant, K_{eq}, whereas for endothermic reactions, an increase in temperature increases the K_{eq} value.

Effect of Change in Pressure: A change in pressure due to a change in volume of a gaseous system leads to a shift of the equilibrium. According to Avogadro's principle, which states that under the same condition of temperature and pressure, equal volumes of all gases contain the same number of molecules, we can divide gas-phase reactions in three groups:

1. Reactions for which the number of moles of the products is greater than the amount of moles of the reactants ($\Delta n > 0$). According to Avogadro's law, these kind of reactions take place with a volume increase:

$$2CH_4(g) + O_2(g) \rightleftharpoons 2CO(g) + 4H_2(g) \quad \Delta n = 6 - 3 = 3 \qquad \text{(RX. 5.22)}$$

2. Reactions by which the volume decreases ($\Delta n < 0$):

$$N_2(g) + 3H_2(g) \rightleftharpoons 2NH_3(g) \quad \Delta n = 2 - 4 = -2 \qquad \text{(RX. 5.21)}$$

3. Reactions where $\Delta n = 0$, i.e., without a volume change:

$$H_2(g) + I_2(g) \rightleftharpoons 2HI(g) \quad \Delta n = 2 - 2 = 0 \qquad \text{(RX. 5.23)}$$

An increase in system pressure due to decreasing volume causes the reaction to shift to the side with the least moles of gas. A decrease in pressure due to increasing volume causes the reaction to shift to the side with the most moles of gas. There is no effect on a reaction where the number of moles of gas is the same on each side of the reaction equation. So, increasing the pressure by decreasing the volume shifts the equilibrium to the left for reactions belonging to the first group and to the right for reactions belonging to the second group and is without any effect for reactions belonging to the third group.

5.3 CATALYSIS

The word catalysis was coined by Berzelius in 1836 to describe the acceleration of certain chemical reactions. It refers to materials (catalysts) that accelerate chemical reactions without undergoing changes themselves. This is too optimistic a definition as the properties of all real catalysts change with use. The definition is also unsatisfactory because it implies that the acceleration is brought about without direct involvement of the catalyst in the process. Actually, the fact that the reaction rate increases by the presence of a catalyst can only be explained if the catalyst combines with the reactants and the reaction takes place through elementary reactions where the transition states formed require a lower activation energy than the uncatalyzed reaction. This concept is graphically explained by Figure 5.3.

The lower the activation energy required for the molecules to react, the higher the number of molecules that can react, which results in a higher reaction rate as shown in Figure 5.4.

For a first-order reaction, the time needed to halve the reactant concentration is

$$t_{\frac{1}{2}} = \frac{\ln 2}{k} \qquad \text{(Eq. 5.38)}$$

where k is the rate coefficient in the Arrhenius expression Equation (5.4). Now, if, e.g., the activation energy of an uncatalyzed reaction at 400°C is 200,000 J·mol^{-1} and 120,000 J·mol^{-1} for its catalyzed counterpart, their rate coefficients are:

$$k_1 = k_0 e^{-\frac{200,000}{8.31 \times 673.15}} \quad \text{and} \quad k_2 = k_0 e^{-\frac{120,000}{8.31 \times 673.15}}$$

so

$$\frac{t_{1,1/2}}{t_{2,1/2}} = \frac{k_2}{k_1} = \frac{e^{-\frac{120,000}{8.31 \times 673.15}}}{e^{-\frac{200,000}{8.31 \times 673.15}}} = 1.6 \times 10^6$$

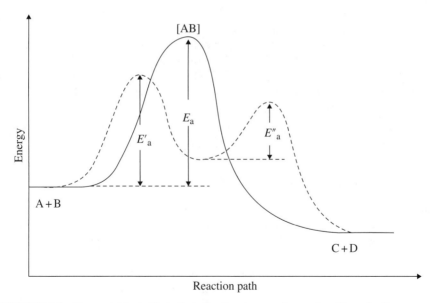

FIGURE 5.3 How a catalyst affects the activation energy of a reaction (E_a, activation energy of the noncatalyzed reaction; E'_a, E''_a, activation energies of the elementary reactions with catalyst) (Source: Reproduced with permission from Pasquetto and Patrone, 1999, vol.3. © Zanichelli editore S.p.A).

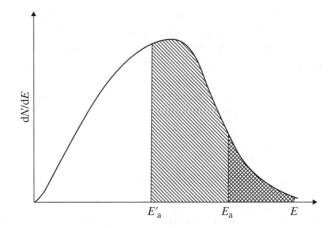

FIGURE 5.4 Activation energies for catalyzed and noncatalyzed reactions and number of molecules reacted (Source: Reproduced with permission from Pasquetto and Patrone, 1999, vol.3. © Zanichelli editore S.p.A).

This means that decreasing the activation energy with 80,000 J·mol^{-1}, the time to halve the reactant concentration decreases by more than a million times!

The fundamental properties of a catalyst are:

a. At the end of a reaction, in principle, the catalyst has to be chemically unchanged. Nothing can be said a priori about the deactivation.

b. The products of the catalyzed reaction can, at least in principle, be obtained from an uncatalyzed reaction under the same conditions. There is, therefore, no way of using catalysis to "cheat" equilibrium. In practice, however, the uncatalyzed reaction may be immeasurably slow or may yield a product distribution different from the one obtained in a catalyzed reaction, in which the catalyst may promote only some reactions.

c. Catalysts can be either homogeneous or heterogeneous, depending on whether the catalyst exists in the same phase as the reactants or not.

5.3.1 Homogeneous Catalysis

Homogeneous catalysts function in the same phase (gas or liquid) as the reactants. In this case, the reaction rate depends on the catalyst concentration. Generally, a small quantity is enough to increase the quantity of the products in a short time. The original reaction in the presence of a homogeneous catalyst takes place through a new sequence of elementary reactions where the activation energy for the formation of the unstable "catalyst–reactant" component is lower. Of particular interest is the transition metal catalysis. The transition metal ion catalyzes the original reaction by providing an alternative route between reactants and products that has a lower activation energy. It can do this because transition metals can form stable compounds in more than one oxidation state and the transition metal ions can therefore readily move between oxidation states. Life itself gives good examples of catalysis through the proteins called enzymes. Their three-dimensional shape enables them to stabilize a temporary association between substrates, the molecules that will undergo the reaction. By bringing two substrates together in the correct orientation, an enzyme lowers the activation energy required for new bonds to form.

Some examples of homogeneous catalysis are:

1. The oxidation of sulfur dioxide to sulfur trioxide in the gas phase, using NO as a catalyst.

    Without catalyst: $2SO_2(g) + O_2(g) \rightarrow 2SO_3(g)$
    With NO, initially consumed and then re-formed:

    $2NO(g) + O_2(g) \rightarrow 2NO_2(g)$
    $2SO_2(g) + 2NO_2(g) \rightarrow 2SO_3(g) + 2NO(g)$

    $$\overline{2SO_2(g) + O_2(g) \rightarrow 2SO_3(g)}$$

2. Thermal decomposition in the gas phase of acetaldehyde catalyzed by iodine I_2 vapors.

    Without catalyst: $CH_3CHO(g) \rightarrow CH_4(g) + CO(g)$
    With I_2, initially consumed and then re-formed:

    $CH_3CHO(g) + I_2(g) \rightarrow CH_3I(g) + HI(g) + CO(g)$
    $CH_3I(g) + HI(g) \rightarrow CH_4(g) + I_2(g)$

    $$\overline{CH_3CHO(g) \rightarrow CH_4(g) + CO(g)}$$

3. If the catalytic process takes place in water, the catalytic effect is often due to hydronium H_3O^+ (acid catalysis) or to hydroxyl OH^- (basic catalysis).

4. An example of how an enzyme works is the role played by carbonic anhydrase that converts dissolved carbon dioxide (CO_2) into carbonic acid, which dissociates into bicarbonate and hydrogen ions:

$$CO_2 + H_2O \rightarrow H_2CO_3 \rightarrow HCO_3^- + H^+$$

The carbonic anhydrase catalyzes this reaction through the deep cleft of its active site. Deep within the cleft are three histidines all pointed at the same place in the center of the cleft. Together, they hold a zinc ion (Zn^{++}) that will be the cutting blade of the catalytic process. Immediately adjacent to the position of the zinc atom in the cleft are a group of amino acids that recognize and bind CO_2. When the CO_2 binds to this site, it interacts with the Zn^{++} in the cleft, orienting in the plane of the cleft. Meanwhile, water bound to the zinc is rapidly converted to hydroxide ion that is now positioned to attack the CO_2. When it does so, HCO_3^- is formed and the enzyme is unchanged (Figure 5.5).

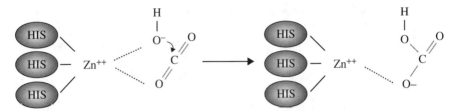

FIGURE 5.5 Carbonic anhydrase as a catalyst. It brings its two substrates into close proximity and optimizes their orientation for reaction (see also Raven et al., 2005).

5.3.2   Heterogeneous Catalysis

Heterogeneous catalysis refers to the form of catalysis where the phase of the catalyst differs from that of the reactants. The great majority of practical heterogeneous catalysts are solids (metal, oxide, or salt), and most reactants are liquids or gases. In solid-catalyzed reactions, the catalytic activity is proportional to the catalyst–reactant contact surface; this is why a pulverized or spongy catalyst is normally used. The catalyst surface has *active sites*, regions with an irregular atom distribution where the valences are not completely saturated. This is why the catalyst atoms can combine with the reactant molecules, producing transition states with lower activation energy. The atoms of the solid are in this state chemically bound to the atoms of the crystal lattice, and in this way, it is possible to win the reactant(s) chemical inertia, and the final stadium of the reaction is easily reached. A solid catalyst makes the reactants' bonds less strong by combining with the reactants' molecules, weakening the original bonds and, in this way, speeding up the products formation.

   The steps of the overall process are:

- Diffusion of the reactants to the solid catalyst surface
- Adsorption of the reactants onto this surface (physically, chemically, or chemically–physically)
- Formation of chemical bonds
- Formation of products and restoration of the catalyst
- Desorption of the products desorb from the catalyst surface
- Diffusion of the products away from the catalyst surface

Some examples of heterogeneous catalysis are:

1. Synthesis of sulfuric acid on platinum:

$$SO_2 + \frac{1}{2}O_2 \rightarrow SO_3$$

   Opening of the oxygen double bond is possible as a result of the combination with platinum (Pt^0).

2. Ammonia production based on the reaction

$$N_2 + 3H_2 \rightleftarrows 2NH_3$$

From a catalytic point of view, the key factor of this reaction is the chemical inertia of the $N \equiv N$ bond that needs 670,000 $J \cdot mol^{-1}$ to be broken (vs. the H–H bond that needs 420,000 $J \cdot mol^{-1}$). Moreover, from a stoichiometric point of view, for the uncatalyzed reaction, a collision of four molecules (which is not very probable) is needed. With the use of iron-based catalysts, an activation energy of 125,000 $J \cdot mol^{-1}$ is reached.

3. Cracking: Complex organic molecules such as kerosene or heavy hydrocarbons are broken down into simpler molecules such as light hydrocarbons by the breaking of carbon–carbon bonds in the precursors. The rate of cracking and the end products are strongly dependent on the temperature and the presence of catalysts.

The catalyzed reaction steps take place very close to the solid surface. These steps may be between gas molecules adsorbed on the catalyst surface, or the reaction may involve the topmost atomic layers of the catalyst. The influence of the solid does not effectively extend more than one atomic diameter into the gas phase, and the direct involvement of atoms below the topmost catalyst layers is not usually possible (Twigg, 1996). In Table 5.1, a list of pros and cons of homogeneous and heterogeneous catalysis is given.

5.3.3 Catalyst Deactivation

One of the major problems related to the operation of heterogeneous catalysis is *deactivation*, the loss of catalyst activity with time-on-stream. This process may be both

TABLE 5.1 Advantages and disadvantages of the use of homogeneous versus heterogeneous catalysis

Homogeneous catalysis		Heterogeneous catalysis	
Pros	Cons	Pros	Cons
• Fast transport phenomena	• Difficult separation of catalyst and products	• Catalyst and products easy to separate	• Only the surface is available
• Uniform catalyst	• Corrosion (if acid catalysts are used)	• No corrosion	• The surface can be poisoned
• High selectivity	• Low selectivity at low temperatures	• Possible at any temperature	• Lower selectivity
• Temperature easily controlled for exothermic reactions	• High costs for separation and purification of catalyst and products		• Heat removal problems

chemical and physical and occurs simultaneously with the main reaction. Deactivation can occur by a number of different mechanisms, commonly divided into four classes: poisoning, coking or fouling, sintering, and phase transformation. Other mechanisms of deactivation include masking and loss of the active elements via volatilization, erosion, and attrition. Deactivation is inevitable, but it can be slowed and some of its consequences can be avoided.

Poisoning: The loss of activity due to the strong chemisorption on the active sites of impurities present in the feed stream (Forzatti and Lietti, 1999). A poison may act simply by blocking an active site (geometric effect) or may alter the adsorptivity of other species essentially by an electronic effect. The chemical nature of the active sites may be modified, and these modified sites can no longer accelerate the reaction that the catalyst was supposed to catalyze.

Coking: For catalytic reactions involving hydrocarbons (or even carbon oxides), side reactions occur on the catalyst surface leading to the formation of carbonaceous residues (usually referred to as coke or carbon), which tend to physically cover the active surface. Sometimes, a distinction is made between coke and carbon, although the difference is somewhat arbitrary—usually carbon is considered the product of CO disproportionation:

$$2CO \rightleftharpoons C + CO_2 \qquad\qquad (Eq.\ 5.24)$$

whereas coke is referred to the material originated by decomposition (cracking) or condensation of hydrocarbons. Coke deposits may amount to 15 or even 20wt% (Forzatti and Lietti, 1999) of the catalyst, and they may deactivate the catalyst either by covering of the active sites or by pore blocking. The chemical nature of the carbonaceous deposits depends very much on how they are formed, the conditions of temperature and pressure, the age of the catalyst and the chemical nature of the feed and products formed.

Sintering: This usually refers to the loss of active surface via structural modification of the catalyst. This is generally a thermally activated process and is physical in nature. An extreme form of sintering occurring at high temperatures and leading to the transformation of one crystalline phase into a different one is solid-state transformation.

Other Deactivation Mechanisms: Other mechanisms of deactivation include masking or pore blockage, caused, e.g., by the physical deposit of substances on the outer surface of the catalyst, thus rendering the active sites inaccessible to the reactants.

Finally, loss of catalytic material due to attrition in moving or fluidized beds is a serious source of deactivation since the catalyst is continuously abraded away.

CHAPTER SUMMARY AND STUDY GUIDE

The energetical, kinetic, and chemical equilibrium aspects of chemical reactions are dealt with: the concept of enthalpy related to the heat of a chemical reaction is presented; the reaction rate, order of reaction, and elementary reactions are defined;

and chemical equilibrium is treated quite deeply for gas-phase reactions and with respect to the factors that affect the equilibrium itself. Finally, the main aspects of catalysis are introduced, in particular the difference between homogeneous and heterogeneous catalysis and catalyst deactivation.

KEY CONCEPTS

Elementary reactions
Order of reaction
Reaction rate
Activation energy
Arrhenius equation
Chemical equilibrium
Homogeneous catalysis
Heterogeneous catalysis
Catalyst deactivation

SHORT-ANSWER QUESTIONS

5.1 How can you determine whether a reaction is at chemical equilibrium?

5.2 How would you define the reaction rate? Which factors can affect it?

5.3 What is the difference between a global reaction rate expression and one based on elementary reactions?

5.4 At a certain temperature, why can a certain reaction have multiple values of the equilibrium constant?

5.5 What is the difference between K_c, K_p, and K_x? For a gas-phase reaction, what is the relation between them?

5.6 For a reaction taking place at constant p and T, why can we say that the ΔG^0 remains constant, but we cannot say the same for ΔG?

5.7 Why do catalysts not influence the chemical equilibrium?

5.8 Explain the difference between homogeneous and heterogeneous catalysis.

5.9 Does the catalyst concentration affect the reaction rate?

PROBLEMS

5.1 By integrating Equation (5.5) with $c = c^0$ at $t = 0$ and given that the reaction proceeds under isothermal conditions, determine the relation between $\ln c$ and t.

5.2   Using one of the relations Equations (5.8–5.11), find the time in which the concentration of a reactant is reduced by 50%.

5.3   For a second-order reaction, calculate the time needed to halve the initial concentration of the reactants. Which conclusion can you draw comparing this result with the one of the previous exercise?

5.4   Find the relation between k_f and V for RX. 5.19.

5.5   Find the relation between K_p and K_c for the following gas-phase reaction: $aA + bB \rightleftarrows cC + dD$

5.6   Qualitatively plot K_c versus the temperature for an endothermic and an exothermic reaction.

5.7   In a container with a volumetric capacity of 1 L, we leave 0.05 mol of PCl_5 and 5 mol of PCl_3 reacting at 760 K. At equilibrium, we have 0.043 mol of Cl_2. Find K_c for the reaction $PCl_5 \rightleftarrows PCl_3 + Cl_2$.

5.8   A container with a capacity of 1 L holds 3 mol of N_2O_4 at 343 K. At this temperature, the dissociation grade is 65%. Find K_p for the reaction $N_2O_4 \rightleftarrows 2NO_2$.

5.9   At 450°C, the equilibrium constant for the reaction $2HI \rightleftarrows H_2 + I_2$ is $K_p = 50$. Find the dissociation grade of HI.

5.10   Find the value of ΔG^0 at 865°C for the following reaction:

$$CaCO_3(s) \rightleftarrows CaO(s) + CO_2(g)$$

knowing that at that temperature the partial pressure of CO_2 is 1333 mmHg. Is this a spontaneous reaction?

PROJECTS

P5.1   *In a chemical lab, determine the decomposition rate of hydrogen peroxide with a homogeneous KI catalyst.*
The decomposition of the hydrogen peroxide with I^- takes place following the reaction equation:

$$2H_2O_2(l) \rightarrow 2H_2O(l) + O_2(g)$$

The catalyst makes the reaction follow these two steps:

$$\text{a.}\quad H_2O_2 + I^- \rightarrow IO^- + H_2O$$
$$\text{b.}\quad IO^- + H_2O_2 \rightarrow I^- + O_2 + H_2O$$

Track the decrease in weight of the beaker solution due to the formation of gaseous oxygen.

Materials needed: precision balance (0.01 g accuracy), beaker 250 mL, pipette 10 mL, H_2O_2 solution (3.3% vol), KI saturated solution, and stopwatch.
How to do: Place the beaker on the precision scale with 100 mL H_2O_2 and 2 mL of KI solution. Shake a bit and start reading, every 15 s, the decrease in weight. Repeat the experiment increasing the concentration of KI and plot the results (the O_2 in mg as a function of time for the different KI concentrations).

P5.2 *Give a Matlab formulation of the minimization of the Gibbs free energy for calculating the equilibrium composition of Example 5.6.*

1. Problem setup

```
Function main

R = 0.00198588; % kcal/mol//K
T = 1000; % K

% we store the species names in a cell array since they have different
% lengths (i.e. different number of characters in names) .
species = {'CH4' 'C2H4' 'C2H2' 'CO2' 'CO' 'O2' 'H2' 'H2O' 'C2H6'};

%
% $G_ `\ circ for each species . These are the heats of formation for each
% species
Gjo = [4.61 28.249 40.604 -94.61 -47.942 0 0 -46.03 26.13]; % kcal/mol
```

2. Gibbs free energy of a mixture

```
function G = func (nj)
    Enj = sum (nj ) ;
    G = sum(nj.*(Gjo/R/T+log(nj/Enj)));
end
```

3. Linear equality constraints for atomic mass conservation

```
Aeq = [0    0    0    2    1    2    0    1    0        % oxygen balacne
       4    4    2    0    0    0    2    2    6        % hydrogen balance
       1    2    2    1    1    0    0    0    2];       % carbon balance

% the incoming feed was 4 mol H2O and 1 mol ethane
beq = [4      % moles of oxygen atoms coming in
       14     % moles of bydrogen atoms coming in
       2];    % moles of oarbon atoms coming in
```

4. Limits on mole number
   No mole number can be negative, so we define a lower limit of zero for each mole number.

```
LB = [0 0 0 0 0 0 0 0 0]; % no mole numbers less than zero
```

5. Initial guess for the solver

```
x0 = [1e-3 1e-3 1e-3 0.993 1 1e-4 5.992 1 1e-3]; % initial guess
```

6. Setup of minimization

```
options = optimset('Algorithm', 'sqp');
[x fval] = fmincon(@func,x0,[],[],Aeq,beq,LB,[],[],options);

for i=1:numel(x)
    fprintf('%d5%10s%10.3g\n',i,species{i},x(i))
end
```

REFERENCES

Bowman CT. Non-equilibrium radical concentration in shock-initiated methane oxidation. 15th Symposium (International) on Combustion; 1975. p. 869–882.

Forzatti P, Lietti L. Catalyst deactivation. Catal Today 1999;52(2–3):165–181.

Kellner S, Bell A. Evidence for H_2/D_2 isotope effect on Fischer Tropsch synthesis over supported ruthenium catalysts. J Catal 1981;67:175–185.

Moran MJ, Shapiro HN. *Fundamentals of Engineering Thermodynamics*. 6th ed. New York: John Wiley & Sons; 2010.

Pasquetto S, Patrone L. *Chimica Fisica*. Bologna: Zanichelli; 1999.

Raven PH, Johnson GB, Singer S. *Biology*. 7th ed. Boston, MA: McGraw-Hill; 2005.

Smith JM, Van Ness HC, Abbott M. *Introduction to Chemical Engineering Thermodynamics*. 7th ed. Boston, MA: McGraw-Hill; 2005.

Twigg MW. *Catalyst Handbook*. 2nd ed. London: Manson Publishing Ltd.; 1996.

6

REACTORS: IDEALIZED CHEMICAL REACTORS

LILIAN DE MARTÍN AND J. RUUD VAN OMMEN

Department of Chemical Engineering, Product & Process Engineering Group, Faculty of Applied Sciences, Delft University of Technology, Delft, the Netherlands

SYMBOLS

a, b, c, and d	stoichiometric coefficients for reacting substances A, B, C, and D	[–]
A	area across which heat/mass transfer takes place	[m^2]
c	concentration	[mol·m^{-3}]
c_p	specific heat capacity	[J·kg^{-1}·K^{-1}]
$\bar{c}_p$	molar heat capacity	[J·mol^{-1}·K^{-1}]
d	diameter	[m]
D	axial dispersion coefficient for flowing fluid	[m^2·s^{-1}]
E_a	activation energy	[J·mol^{-1}]
E	dimensionless output to a pulse input, the exit age distribution function	[s^{-1}]
h	heat transfer coefficient	[W·m^{-2}·K^{-1}]
$H_i(T)$	enthalpy of species i at temperature T	[J·mol^{-1}]
$\Delta_r H$	enthalpy of reaction for the stoichiometry as written	[J·mol^{-1}]

Biomass as a Sustainable Energy Source for the Future: Fundamentals of Conversion Processes, First Edition. Edited by Wiebren de Jong and J. Ruud van Ommen.

k	reaction rate coefficient for reaction of order n	$[(mol \cdot m^{-3})^{1-n} \, s^{-1}]$
k_0	pre-exponential factor in Arrhenius equation for reaction of order n	$[mol \cdot m^{-3}]^{1-n} \, s^{-1}$
L	length	[m]
m	mass	[kg]
n	number of moles	[mol]
p	pressure	[Pa or bar]
$\dot{Q}$	heat flow supplied or extracted	$[J \cdot s^{-1}]$
R_u	universal gas constant (=8.314)	$[J \cdot mol^{-1} \cdot K^{-1}]$
r	reaction rate	$[mol \cdot s^{-1} \cdot m^{-3}]$
R_i	(net) rate of production of species i	$[mol \cdot s^{-1} \, m^{-3}]$
S	cross section of the reactor	$[m^2]$
t	time	[s]
T	temperature	[°C or K]
u	velocity	$[m \cdot s^{-1}]$
V	volume	$[L \text{ or } m^3]$
X	conversion	[–]
ε	expansion factor, fractional volume change on complete conversion	[–]
φ_V	volume flow	$[m^3 \cdot s^{-1}]$
φ_n	mole flow	$[mol \cdot s^{-1}]$
ξ	extent of the reaction	$[mol \text{ or } mol \cdot s^{-1}]$
τ	space time (residence time); see Equation (6.20)	[s]
ν_i	stoichiometric coefficient of species i	[–]

Subscripts

A	key species A
0	initial
i	generic species
f	fluid
f	final

6.1 PRELIMINARY CONCEPTS

In Chapter 5, it was shown how the rate law can be expressed as a function of the concentrations of the different species. In this chapter, it will be shown how to use that information to design ideal chemical reactors: reactors with a simplified flow pattern. Design in this context refers to *functional design*, i.e., the selection of the type of reactor and the estimation of its dimensions to satisfy the established mass and energy balances. The *mechanical design* of the reactor, which includes the design of the supports, construction materials, wall thickness, internals, etc., is out of the scope of this book.

A reactor is nothing else than a vessel to carry out a reaction; a pan or a pot could be considered a chemical reactor. The typical steps involved in the functional design of a chemical reactor are:

1. Defining the rate law of the reactions taking place. In this chapter, only single reactions will be considered. If the rate law depends on the concentration of more than one species, all these concentrations must be related to the concentration of a key species using stoichiometric relations.
2. Selecting a type of operation, batch, or continuous (see Table 6.1).
3. Selecting a mixing mode. In this chapter, two extreme cases will be considered, plug flow and perfect mixing.
4. Setting up the mass and energy balances for the reactor.

In the subsequent sections, we will see that the kinetic term of the molar balance can be handled either in terms of concentration or conversion. Concentration has the advantage of being a directly measurable property, which is related to the molar balances in the reactor. Conversion is a better descriptor of the evolution of the reaction, but it must be ultimately related to the concentration to get information about the molar balances in the reactor. It is defined as

$$X_i \equiv \frac{n_{i0} - n_i}{n_{i0}} \qquad \text{(Eq. 6.1)}$$

TABLE 6.1 Comparison between batch and continuous-flow reactors: (+) advantage and (−) disadvantage

Batch reactor (BR)	Continuous-flow reactor
(+) Suitable for small productions volume, such as in fine chemistry	(+) Suitable for large production volumes
(+) High flexibility. Different products can be produced in the same reactor	(−) Lower flexibility
(+) Easy startup and shutdown	(−) Difficult startup and shutdown. The startup process to steady-state operation can take weeks
(+) Capital cost relatively low	(−) Capital cost usually high. Need of a flowing system, such as pumps and pipes
(−) Downtime between batches	(+) No downtime between batches
(−) Maintaining product uniformity between batches is difficult due to unsteady-state operation	(+) Uniformity of product is easily maintained working in steady state
(−) If the BR is part of a continuous process, buffering tanks are required	(+) Steady-state operation allows easy coupling with continuous downstream operations

Please note that this definition of the mole-based conversion differs from the relative degree of conversion of a reactant Equation (3.10), which is mass based.

Consider the following reaction: $a\,A + b\,B \rightarrow c\,C + d\,D$. If the rate law depends on several species, we must relate the concentrations of the different species to each other. This can be done using the stoichiometric coefficients a, b, c, and d. The stoichiometric coefficients also give the relationship between the reaction rate r and the net production rates of the components, R_A, R_B, R_C, R_D:

$$r = \frac{(-R_A)}{a} = \frac{(-R_B)}{b} = \frac{R_C}{c} = \frac{R_D}{d} \qquad \text{(Eq. 6.2)}$$

Note that $(-R_A)$, $(-R_B)$, R_C and R_D are positive. A and B are reactants that disappear during the reaction, so the net rates of production R_A, R_B are negative. In this chapter, the analysis of the reactor will be based on the key component A.

When there are gaseous species involved in the reaction, the volume of the system might be a function of the conversion. For instance, for an ideal gas, $V = nR_uT/p$. If the pressure p and temperature T are constant and the number of moles n increases due to the reaction (in case $c + d > a + b$), the gas will expand. Since the concentration is the ratio between the number of moles and the volume that these moles occupy, a change in the volume will affect the concentration.

To account for changes in the volume of the system, we define the expansion factor ε_A as

$$\varepsilon_A = \frac{V_{X_A = 1} - V_{X_A = 0}}{V_{X_A = 0}} \qquad \text{(Eq. 6.3)}$$

Here, ε_A refers to reactant A, but it can be defined for any component. In those reactions where a change in the number of moles is associated with a change in the reaction volume, ε_A can be calculated from the stoichiometric coefficients. For instance, for the hypothetical reaction $2\,A \rightarrow 5\,B$ starting from pure A, the expansion factor would be $\varepsilon_A = (5-2)/2 = 1.5$. Starting from a mixture consisting of 50 vol.% A and 50 vol.% inerts that do not contribute to the reaction, $\varepsilon_A = (7-4)/4 = 0.75$. For most liquids, $\varepsilon = 0$ (no change in volume), while for many gases $\varepsilon \neq 0$.

The relation between the concentration and the conversion is

$$X_A = \frac{c_{A0} - c_A}{c_{A0} + \varepsilon_A c_A}; \quad dX_A = -\frac{c_{A0}(1 + \varepsilon_A)}{(c_{A0} + \varepsilon_A c_A)^2} dc_A \qquad \text{(Eq. 6.4)}$$

Together with the conversion, it is common to define the extent of the reaction ξ. For closed systems (BR), it is defined as

$$\xi = \frac{n_i - n_{i0}}{\nu_i} \qquad \text{(Eq. 6.5)}$$

where ν_i is the stoichiometric coefficient of species i and n_i is the number of moles of i. Note that ν_i is negative for reactants and positive for products. For open (or continuous flow) systems, the extent of reaction is defined as

$$\xi = \frac{\varphi_{n,i} - \varphi_{n,i0}}{\nu_i} \qquad \text{(Eq. 6.6)}$$

where $\varphi_{n,i}$ is the mole flow rate of i. Note that this means that the unit of ξ is different for closed and for open systems! The main difference between ξ and X_i is that conversion is dimensionless and describes how much of a certain reactant has reacted. It is therefore dependent on the reactant considered. In contrast, ξ describes how much the reaction has evolved and is independent of the reactant chosen for its assessment.

6.2 BATCH REACTORS (BRs)

6.2.1 Isothermal Batch Reactors

Isothermal means *constant temperature*. It is therefore a term describing those systems in which the temperature does not change. In a batch reactor (BR), during reaction, no reactants are supplied to the reactor and no products are released (Figure 6.1). The BR

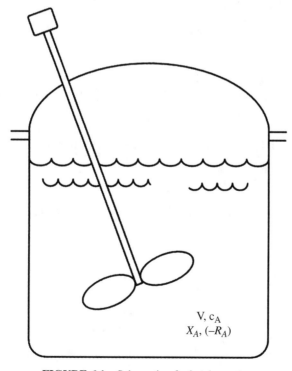

FIGURE 6.1 Schematic of a batch reactor.

can be considered as a kind of stirred pot. This type of ideal reactor works under the assumption that the reaction volume is perfectly mixed, so the composition of the mixture is exactly the same in the entire reactor. The concentration of species is *time dependent*. By definition, a BR cannot operate in steady state since the system evolves dynamically. A BR may be a constant-volume reactor (isochoric) or a variable-volume reactor if it has a piston–cylinder arrangement.

Since no reactants are supplied and no products are released, the formation of products and the disappearance of reactants will be translated as accumulation, positive for formation and negative for disappearance. For reactant A, the molar balance of Equation (3.5) then simplifies to

Rate of accumulation = rate of supply − rate of release + rate of production

(Eq. 6.7)

in which the rate of release and rate of production are zero, so

$$-\frac{dn_A}{dt} = V(-R_A) \quad \text{or} \quad n_{A0}\frac{dX_A}{dt} = V(-R_A) \tag{Eq. 6.8}$$

Since A is a reactant, the accumulation term (dn_A/dt) must be negative, indicating that the number of moles of A decreases in time.

Integration of Equation (6.8) gives the time necessary to obtain a conversion X_A of reactant A. If the reaction volume is constant, $\varepsilon_A = 0$, which means that the density is also constant because the total mass is conserved, then

$$t = -\int_{c_{A0}}^{c_A} \frac{dc_A}{(-R_A)}; \quad t = c_{A0}\int_0^{X_A} \frac{dX_A}{(-R_A)} \tag{Eq. 6.9}$$

For reactions where the volume of the reacting mixture changes with the conversion, $\varepsilon_A \neq 0$, such as single-phase reactions with significant density changes, Equation (6.9) becomes

$$t = n_{A0}\int_0^{X_A} \frac{dX_A}{(-R_A)V_0(1+\varepsilon_A X_A)}; \quad t = c_{A0}\int_0^{X_A} \frac{dX_A}{(-R_A)(1+\varepsilon_A X_A)} \tag{Eq. 6.10}$$

in which t is the necessary time to reach a conversion X_A. To solve Equations (6.9) and (6.10), the reaction term ($-R_A$) must be substituted with a kinetic rate equation, and then the equation must be integrated.

Example 6.1   Solution of the molar balance for a BR

Consider a first-order irreversible reaction with constant volume, $\varepsilon_A = 0$:

$$A \rightarrow bB + cC + \cdots - R_A = kc_A$$

Then with Equation (6.9),

$$t = -\int_{c_{A0}}^{c_A} \frac{dc_A}{kc_A} = -\frac{1}{k} \cdot \ln c_A \Big|_{c_{A0}}^{c_A}$$

resulting in

$$kt = \ln\frac{c_{A0}}{c_A} = \ln\frac{1}{1-X_A} \quad \Rightarrow \quad c_A = c_{A0}\exp(-kt) \qquad \text{(Eq. 6.11)}$$

From Equation (6.11), it is possible to obtain the variation of the concentration of A in time or the time needed to obtain a certain conversion. Note that A is related to B, C, and other products through the stoichiometric coefficients. Because only a single reaction is considered, if we know how A changes in time, we know how the concentrations of all reactants and products change in time. A solution of the molar balance for BR and different reaction types and orders can be found in Levenspiel (1998, 2002) and Schmidt (2004).

6.2.2 Non-isothermal Batch Reactors

Sometimes, it is not possible or even desirable to carry out a reaction under isothermal conditions. One reason is that if there are several reactions taking place in parallel, it is possible to vary the product distribution playing with the temperature. According to the Arrhenius equation (Atkins and de Paula, 2009), the temperature affects the reaction exponentially depending on the activation energy E_a. If there are several reactions taking place in parallel and they have different E_a values, increasing the temperature will favor the reactions with higher values of E_a.

Assuming that the volume of the BR is constant, the molar balance Equation (6.8) for the non-isothermal reaction of a reactant A reads

$$n_{A0}\frac{dX_A}{dt} = V(-R_A(X_A,T)) \qquad \text{(Eq. 6.12)}$$

In non-isothermal cases, the temperature must be accounted for in the reaction rate. Considering the first-order reaction of Example 6.1, the solution to the non-isothermal molar balance would be

$$t = -\int_{c_{A0}}^{c_A} \frac{dc_A}{k_0\exp\left(-\dfrac{E_a}{R_u T}\right)c_A} \qquad \text{(Eq. 6.13)}$$

There are two unknown terms, T and c_A, so an additional equation is necessary to solve the molar balance. That additional equation comes from the energy balance

Rate of accumulation $\left(\sum m_i c_{p,i} \dfrac{dT}{dt}\right) =$

Rate of supply $(=0)$ − rate of release $(=0)$ +

Rate of production $\left(V(-\Delta_r H)(-R_A(X_A,T))\right)$ + (heating or cooling) $(\dot{Q})$

(Eq. 6.14)

The term $\dot{Q}$ represents any addition or removal of heat from the reactor. The energy balance states that the variation of temperature in the reactor (dT/dt) depends on four key factors, namely, the rate of heat generation $V(-\Delta_r H)(-R_A(X_A, T))$, the heat added to or removed from the system $\dot{Q}$, the mass of the species m_i, and the specific heat capacity of the species involved $c_{p,i}$, which in this case is considered constant. A strongly exothermic reaction such as combustion ($\Delta_r H$ has a large negative value) will increase (dT/dt) unless the removal of heat $\dot{Q}$ counteracts the production of heat. The term $\dot{Q}$ can be manipulated to get the desired (dT/dt) in the reactor by heating the reactor or cooling it.

The solution of the energy balance leads to

$$\sum m_i c_{p,i} \frac{dT}{dt} = V(-\Delta_r H)(-R_A(X_A,T)) + \dot{Q}$$ (Eq. 6.15)

For adiabatic systems, $\dot{Q}=0$. If the reactor is heated/cooled by interchanging heat with an external fluid, e.g., through a heat exchange coil, $\dot{Q}$ will have the form $\dot{Q}=hA(T_f-T)$. T_f is the temperature of the heating/cooling fluid, T is the temperature of the reactor, h is the heat transfer coefficient, and A is the heat exchange area.

Substituting Equation (6.12) in Equation (6.15), we obtain

$$\sum m_i c_{p,i} \frac{dT}{dt} - (-\Delta_r H) n_{A0} \frac{dX_A}{dt} = \dot{Q}$$

if $=0 \rightarrow$ adiabatic (Eq. 6.16)

Thus,

$$\sum m_i c_{p,i}(T-T_0) - (-\Delta_r H) n_{A0}(X_A-X_{A0}) = \int_0^t \dot{Q} dt$$

$$= \dot{Q}t$$ (Eq. 6.17)

(if $\dot{Q}=$ constant)

if $=0 \rightarrow$ adiabatic

Although in some cases it is possible to obtain an analytical solution, most of the cases require numerical methods to solve the combined mass and energy balances.

6.3　STEADY-STATE CONTINUOUS STIRRED TANK REACTORS (CSTRs)

The continuous stirred tank reactor (CSTR) is characterized by a continuous supply of reactants and release of products (Figure 6.2). Like in the BR, the CSTR model works under the assumption that the reactor is perfectly mixed, so the composition of the mixture is exactly the same in the entire reaction volume. This type of reactor can operate at steady or nonsteady state. In the steady state, which is the case considered in this section, there are no dynamic variations in the inlet and outlet flows, and the composition inside the reactor is constant. The temperature of the reaction mixture is also constant, and the value should be such that the desired conversion and product distribution are obtained. That temperature is kept constant by adjusting the amount of heating or cooling of the system.

A material balance over the entire reactor for a reactant A is

Rate of accumulation = rate of supply − rate of release + rate of production ⇒

$$0 = \varphi_{n,A0} - \varphi_{n,Af} = \varphi_{n,A0}(1 - X_{Af}) + V(-R_A) \qquad \text{(Eq. 6.18)}$$

Note that since steady state is considered, there are no time-dependent terms. Equation (6.18) can be rewritten as

$$\frac{V}{\varphi_{n,A0}} = \frac{X_{Af}}{(-R_A)_f} \qquad \text{(Eq. 6.19)}$$

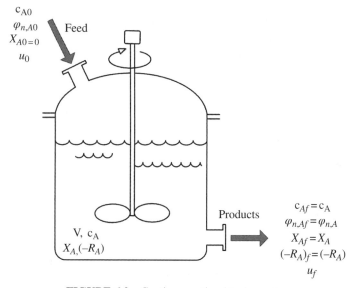

FIGURE 6.2　Continuous stirred tank reactor.

For continuous-flow reactors, in which there is a flow of reactants/products, it is common to express the conversion as a function of the space time τ, which is defined as the time needed to process a volume V of reactants up to a given conversion:

$$\tau = \frac{V}{\varphi_{V0}} = \frac{c_{A0}V}{\varphi_{n,A0}} = \frac{c_{A0}X_{Af}}{(-R_A)_f} \quad \text{for any } \varepsilon_A \qquad \text{(Eq. 6.20)}$$

If the density is constant ($\varepsilon_A = 0$), which holds for practically all liquid-phase and some gas-phase reactions, τ can be expressed as

$$\tau = \frac{V}{\varphi_{V0}} = \frac{c_{A0}X_{Af}}{(-R_A)_f} = \frac{c_{A0}-c_A}{(-R_A)_f} \quad \text{for } \varepsilon_A = 0 \qquad \text{(Eq. 6.21)}$$

The reader should be aware that most engineers refer to τ as the *residence time* (see, for instance, Chapter 3). Strictly speaking, τ is not the residence time and the reason for this is detailed in Section 6.5.

Example 6.2   Solution of the molar balance for a CSTR

Consider the first-order irreversible reaction with constant volume ($\varepsilon_A = 0$)

$$A \rightarrow bB + cC + \cdots - R_A = kc_A$$

Then

$$\tau = \frac{1}{k}\frac{X_A}{1-X_A} = \frac{c_{A0}-c_A}{kc_A} \qquad \text{(Eq. 6.22)}$$

Solutions of molar balances for CSTRs and different reaction types and orders can be found in Levenspiel (1998, 2002). Note that there are no integrals in the expressions relating τ with the conversion.

6.4   STEADY-STATE PLUG FLOW REACTORS (PFRs)

6.4.1   Isothermal PFR

In plug flow reactors (PFRs), the reactants are fed at one side of the reactor and then flow parallel to the length of the reactor, leaving the reactor at the other side. The PFR model assumes that the composition does not vary with the radius and there is no axial mixing of the species in the reactor.

Contrary to the BR and CSTR, in the PFR, the composition is not the same everywhere in the reactor but varies as a function of the axial position. At the entrance of the reactor, there is a higher concentration of reactants than at the exit of the reactor, resulting in a profile of concentrations along the flow path. If the operation takes place in steady state, this profile does not vary in time. To address

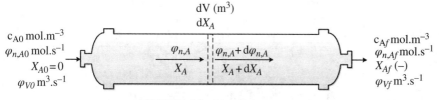

FIGURE 6.3 Plug flow reactor.

the plug flow model mathematically, it is common to divide the total volume of the reactor in an infinite number of slices, infinitely thin and with constant concentration (Figure 6.3).

The steady-state molar balance around each slice of the reactor is

Rate of accumulation = rate of supply − rate of release + rate of production ⇒

$$0 = \varphi_{n,A} - \left(\varphi_{n,A} + d\varphi_{n,A}\right) + dV(-R_A)$$ (Eq. 6.23)

After simplification, the balance reads

$$d\varphi_{n,A} = (-R_A)dV$$ (Eq. 6.24)

Using the definition of conversion,

$$\varphi_{n,A} = \varphi_{n,A0}(1 - X_A)$$ (Eq. 6.25)

Differentiation of Equation (6.25), substitution in Equation (6.24), and rearrangement lead to

$$\frac{V}{\varphi_{n,A0}} = \int_{0}^{X_{Af}} \frac{dX_A}{(-R_A)}$$ (Eq. 6.26)

Comparing Equations (6.19) and (6.26), we observe that in the CSTR there is no need of integrals (resulting in a relatively simple algebraic equation) because the term $(-R_A)$ is constant throughout the whole reaction volume, whereas in the PFR, the conversion changes with the axial position, so $(-R_A)$ also changes.

If the kinetics of the reaction is known Equation (6.26), allows calculation of the volume of the reactor necessary for a conversion X_A of a flow $\varphi_{n,A0}$ of reactant A. The time τ needed to process one reactor volume of feed can be calculated as

$$\tau = \frac{V}{\varphi_{V0}} = \frac{V c_{A0}}{\varphi_{n,A0}} = c_{A0} \int_{0}^{X_{Af}} \frac{dX_A}{(-R_A)} \quad \text{for any } \varepsilon_A$$ (Eq. 6.27)

If the density is constant,

$$\tau = \frac{V}{\varphi_{V0}} = c_{A0} \int_0^{X_{Af}} \frac{dX_A}{(-R_A)} = -\int_{c_{A0}}^{c_{Af}} \frac{dc_A}{(-R_A)} \quad \text{for} \quad \varepsilon_A = 0 \qquad \text{(Eq. 6.28)}$$

Example 6.3   Solution of the molar balances for a PFR

Consider the first-order irreversible reaction with constant volume $\varepsilon_A = 0$:

$$A \rightarrow bB + cC + \cdots - R_A = kc_A$$

Then

$$\tau = \frac{1}{k} \ln \frac{1}{1-X_A} = \frac{1}{k} \ln \frac{c_{A0}}{c_A} \qquad \text{(Eq. 6.29)}$$

Solutions of molar balances for a PFR and different reaction types and orders can be found in Levenspiel (1998, 2002).

6.4.2   Non-isothermal PFR

In a non-isothermal PFR, the temperature changes with the axial position (Figure 6.4). If the non-isothermal PFR operates in the steady state, this temperature profile does not vary in time. When the temperature varies along the reactor, an energy balance must be solved in parallel with the molar balance.

A component i that enters any slice within the reactor introduces an enthalpy to the slice H_i (T) (see Chapter 5):

$$H_i(T) = H_i(T_{ref}) + \bar{c}_{p,i}(T - T_{ref}) \Rightarrow \Delta H_i = \bar{c}_{p,i}(T - T_{ref}) \qquad \text{(Eq. 6.30)}$$

where T_{ref} is a reference temperature, $\bar{c}_{p,i}$ is the molar heat capacity of the component A, and $H_i(T_R)$ is the enthalpy at T_R (unknown). The absolute enthalpy of a system cannot be measured directly, so engineers usually work with enthalpy increments ΔH instead. ΔH is the enthalpy of a system minus the enthalpy of some reference system, which can be chosen arbitrarily but is commonly chosen in a way that simplifies the calculations.

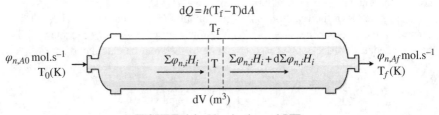

FIGURE 6.4   Non-isothermal PFR.

The enthalpy introduced by the flow to a slice of the reactor is

$$\sum \varphi_{n,i} H_i(T_{ref}) + \sum \varphi_{n,i} \bar{c}_{p,i}(T - T_{ref}) \qquad \text{(Eq. 6.31)}$$

The enthalpy that leaves the slice is

$$\sum \varphi_{n,i} H_i(T_{ref}) + \sum \varphi_{n,i} \bar{c}_{p,i}(T + dT - T_{ref}) \qquad \text{(Eq. 6.32)}$$

The enthalpy generated in the slice due to the reaction of component A is

$$dV(-\Delta_r H)(-R_A(X_A, T)) \qquad \text{(Eq. 6.33)}$$

If the reactor operates in steady state, there is no accumulation of heat in the slice. The energy balance then reads

Rate of accumulation = rate of supply − rate of release + rate of production
+ heating/cooling ⇒

$$0 = \sum \varphi_{n,i} H_i(T_{ref}) + \sum \varphi_{n,i} \bar{c}_{p,i}(T - T_{ref}) - \left(\sum \varphi_{n,i} H_i(T_{ref}) \right.$$
$$\left. + \sum \varphi_{n,i} \bar{c}_{p,i}(T + dT - T_{ref}) \right) + dV(-\Delta_r H)(-R_A(X_A, T)) + d\dot{Q} \quad \text{(Eq. 6.34)}$$

This balance neglects any shaft work introduced to the system, for instance, the work done by the turbine that pumps the flow. Simplifying the balance, we obtain

$$\sum \varphi_{n,i} \bar{c}_{p,i} dT = dV(-\Delta_r H)(-R_A(X_A, T)) + d\dot{Q} \qquad \text{(Eq. 6.35)}$$

Dividing by dV and rearranging gives

$$\frac{dT}{dV} = \frac{(-\Delta_r H)(-R_A(X_A, T)) + \dfrac{d\dot{Q}}{dV}}{\sum \varphi_{n,i} \bar{c}_{p,i}} \qquad \text{(Eq. 6.36)}$$

The heat removed from the system by the coolant, $\dot{Q} = hA(T_f - T)$, can be expressed as

$$d\dot{Q} = h(T_f - T)dA = h(4/d)(T_f - T)dV \qquad \text{(Eq. 6.37)}$$

where d is the reactor diameter, h is the heat transfer coefficient, and T_f is the temperature of the coolant. Substituting the expression for $d\dot{Q}$ in Equation (6.36), we obtain the final expression for the temperature gradient inside the reactor:

$$\frac{dT}{dV} = \frac{(-\Delta_r H)(-R_A(X_A, T)) + h(4/d)(T_f - T)}{\sum \varphi_{n,i} \bar{c}_{p,i}} \qquad \text{(Eq. 6.38)}$$

Note that in Equation (6.38), the coolant temperature is considered to be constant. This holds when the coolant is at the boiling point. If the coolant temperature changes along the length of the reactor, we must add the coolant energy balance. For more information, see Froment et al. (2010).

Example 6.4   Calculation of conversion and temperature profile in a non-isothermal PFR

A first-order exothermic reaction $A \rightarrow B$ is carried out in a PFR in the liquid phase.

Data
Enthalpy of reaction $(-\Delta_r H) = 40{,}000$ kJ·kmol^{-1}
Pre-exponential factor $k_0 = 1.2 \cdot 10^3$ s^{-1}
Activation energy $E_a = 25{,}000$ kJ·kmol^{-1}
Inlet temperature $T_0 = 300$ K
Molar heat capacity $\bar{c}_p = 132.6$ J·mol^{-1}·K^{-1}
Superficial velocity of the fluid in the tube $u = 1.5$ m·s^{-1}
Initial concentration of A $c_{A0} = 0.2$ kmol·m^{-3}
Tube diameter d = 0.05 m
Heat transfer coefficient $h = 250$ W·m^{-2}·K^{-1}
Length of the reactor L = 5 m
Universal gas constant $R_u = 8.314$ kJ·kmol^{-1}·K^{-1}

From a cost analysis, it is concluded that to make a profit the conversion of A needs to be at least 90%. The operators decide to introduce the flow at room temperature (300 K). However, B degrades at 500 K and the reaction is exothermic, so the operators fear they will have to cool the reactor.

Is it necessary to cool the reactor? If so, what coolant temperature is most suitable for the purposes of the company?

Solution
The energy balance Equation (6.38) reads

$$\frac{dT}{dV} = \frac{(-\Delta_r H)(-R_A(X_A, T)) + h(4/d)(T_f - T)}{\sum \varphi_{n,i} \bar{c}_{p,i}}$$

For simplification, the overall thermal properties of the system will be considered, so $\sum \varphi_{n,i} \bar{c}_{p,i} = \varphi_n \langle c_p \rangle$, where < > indicate average specific heat capacity. The number of moles is conserved due to the stoichiometry of the reaction so $\varphi_n \langle c_p \rangle = \varphi_{n,A0} \langle c_p \rangle$.

With V = SL, the energy balance then is

$$\frac{dT}{dL} = \frac{S}{\varphi_{n,A0}\langle c_p \rangle}(-\Delta_r H)k c_{A0}(1 - X_A) + \frac{S}{\varphi_{n,A0}\langle c_p \rangle}h(4/d)(T_f - T) \qquad \text{(Eq. 6.39)}$$

Knowing that $u = \dfrac{\varphi_{n,A0}}{S c_{A0}}$ and using Arrhenius law for the temperature dependence of the reaction rate constant,

$$\frac{dT}{dL} = \frac{(-\Delta_r H)}{u \langle c_p \rangle} k_0 \exp\left(-\frac{E_a}{R_u T}\right)(1-X_A) + \frac{4h}{d u c_{A0} \langle c_p \rangle}(T_f - T) \qquad \text{(Eq. 6.40)}$$

The molar balance is given by Equation (6.24)

$$\frac{d\varphi_{n,A}}{dV} = (-R_A)$$

Substituting the reaction rate term by the first-order kinetics, we obtain

$$\frac{d\varphi_{n,A}}{dL} = S k_0 \exp\left(-\frac{E_a}{R_u T}\right) c_{A0}(1-X_A) \qquad \text{(Eq. 6.41)}$$

Since $d\varphi_{n,A} = \varphi_V dc_A = -\varphi_V c_{A0} dX_A$ and $u = \dfrac{\varphi_V}{S}$,

$$\frac{dX_A}{dL} = \frac{1}{u} k_0 \exp\left(-\frac{E_a}{R_u T}\right)(1-X_A) \qquad \text{(Eq. 6.42)}$$

The resultant system is a system of two differential equations with two unknown variables, temperature and conversion. The solutions of the balances for different coolant temperatures are shown in Figures 6.5 and 6.6.

The conclusion is that the reactor needs to be cooled. If the temperature of the coolant is 340 K or higher, the conversion of A is greater than 90% (Figure 6.6), a value that satisfies the requirements of the company. At higher coolant temperatures, there is a temperature peak above 500 K that will cause the degradation of product B (Figure 6.5). If the temperature of the coolant is lower than 340 K, the temperature of the whole reactor will be kept below 500 K; however, at these coolant temperatures, the conversion obtained is too low. Consequently, the temperature of the coolant that best matches the cooling and conversion requirements is around 340 K.

6.5 RESIDENCE TIME AND SPACE TIME FOR FLOW REACTORS

In Section 6.3, it was explained that the parameter $\tau = V/\varphi_{V0}$ used in the balances of CSTRs and PFRs is not exactly the same as the residence time t. The next example, extracted from Levenspiel (1998), serves to illustrate the differences.

Consider a PFR for the production of popcorn (Figure 6.7) with a capacity of 1 L. In this reactor, 1 L·min^{-1} of raw corn is fed and 28 L·min^{-1} of popcorn is obtained. The volume flow of the products is larger than the flow of the reactants due to the expansion.

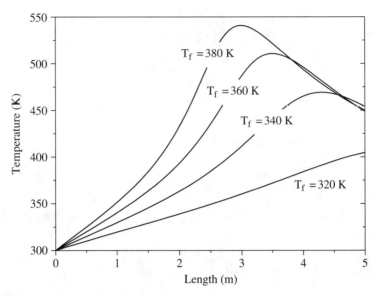

FIGURE 6.5 Variation of the temperature profile along the reactor for different coolant temperatures.

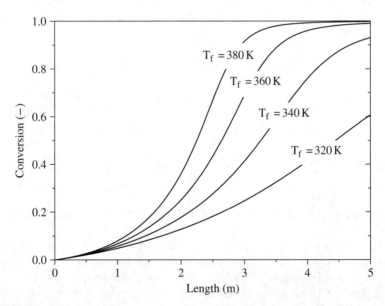

FIGURE 6.6 Variation of the conversion of reactant A along the reactor for different coolant temperatures.

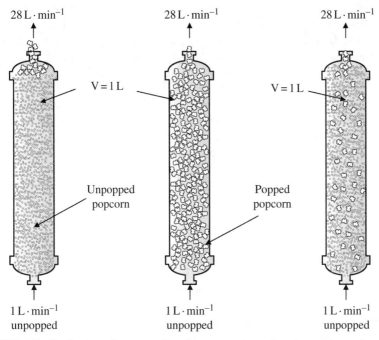

FIGURE 6.7 Production of popcorn. For the same values of τ, the residence times **t** are different in each case. Based on Levenspiel (1998).

In the three cases shown in Figure 6.7, the inlet and outlet flows are the same; the only difference is the point inside the reactor where the corn pops. In the first case, most of the corn pops close to the outlet of the reactor. In the second case, the corn pops as soon as it enters the reactor. In the third case, it pops somewhere across the reactor.

The time τ needed to convert a volume of reactor (1 L) of raw corn is the same in all three cases:

$$\tau_1 = \tau_2 = \tau_3 = \frac{V}{\varphi_{V0}} = 1 \text{ min}$$

However, in the first case, the corn moves slower than in the second case. The residence times are

$$t_1 = \frac{V}{\varphi_{V1}} = \frac{1 \text{ L}}{1 \text{ L} \cdot \text{min}^{-1}} = 1 \text{ min}$$

$$t_2 = \frac{V}{\varphi_{V2}} = \frac{1 \text{ L}}{28 \text{ L} \cdot \text{min}^{-1}} \simeq 2 \text{ s}$$

$$2 \text{ s} < t_3 < 1 \text{ min}$$

In the first case, the entire reactor is full of unpopped corn, so the residence time of the corn is determined by the volume flow rate of unpopped corn (1 L·min^{-1}). In the second case, the entire reactor is full of popped corn, so the residence time of the corn is determined by the volume flow rate of popped corn (28 L·min^{-1}). Although the production is the same in the three cases, the residence time is different.

The residence time t represents the time that molecules (or particles) spend in the reactor, whereas the space time τ is the time necessary to process a volume V of *reactants*. For constant density systems (all liquids and constant density gases), the mean residence time of flowing material in the reactor $\bar{t}$ is the same as τ. Thus, $\bar{t} = \tau = V/\varphi_{V0}$. However, as exemplified in the popcorn reactor, for systems with changing density, $\bar{t}$ depends on ε_A.

6.6   DEVIATIONS FROM PLUG FLOW AND PERFECT MIXING

So far, two flow patterns have been considered, ideal plug flow and ideally mixed flow. These flow patterns represent two extreme and opposite situations that are hardly encountered in reality. However, these two approaches are the favorites of engineers due to their simplicity. When is plug flow or perfectly mixed flow behavior a good description of the flow behavior in a real reactor? The answer to that question can be obtained either theoretically or experimentally and is determined by three factors:

1. Residence time distribution (RTD) of the material. Deviation from plug flow or perfectly mixed flow can be caused by channeling of the fluid, recycling of the fluid, or creation of stagnant regions in the vessel.
2. State of aggregation of the flowing stream. A microfluid is a fluid in which the individual molecules are free to move and intermix. Gases and ordinary, not very viscous liquids belong to this group. A macrofluid is a fluid in which the molecules are kept grouped together in aggregates or packets. Non-coalescing droplets, solid particles, and very viscous liquids belong to this group.
3. Earliness and lateness of mixing of material in the vessel. The fluid elements of a single flowing stream can mix with each other either early or late in their flow through the vessel.

Often, one or more of these factors can be ignored. Much depends on reaction time mixing time and average residence time. In some cases, it is even a good approximation to ignore all three factors and to assume a pure plug flow or perfectly mixed reactor.

6.6.1   Residence Time Distribution (RTD)

Consider the CSTR described in Section 6.3. Although it is perfectly mixed, not all the molecules that enter into the reactor spend the same time in it. There is a chance that a molecule leaves the reactor as soon as it enters, while another molecule could stay in the reactor for an indefinite amount of time. The distribution of the time that the molecules spend in the reactor is called the exit age distribution E or the RTD of the fluid.

E has the units of time^{-1}. It is common to work with a normalized E function so that the area under the E–t curve is unity:

$$\int_0^\infty E\,dt = 1 \qquad \text{(Eq. 6.43)}$$

E can be easily determined by introducing a tracer into the reactor and measuring its concentration at the exit flow of the reactor at different times. The tracer is typically a nonreactive component whose concentration can be easily determined, e.g., by measuring the (liquid's) color or the conductivity of the exit flow.

There are several common ways of introducing the tracer into the reactor, namely, as a pulse, step, periodic, or random function. Here, the pulse experiment will be detailed; for information of other methods, the reader is referred to Westerterp et al. (1988) and Rawlings and Ekerdt (2002).

Suppose we want to obtain the E curve of a reactor with volume V (m^3) by using a pulse experiment. The reactor has a continuous inlet flow of φ_{V0} m$^3\cdot$s^{-1}. Suddenly, an amount m (kg or moles) of tracer is injected at once (pulse) in the inlet flow right before it enters the reactor. Then, the concentration of tracer is measured at the outlet of the reactor at different times. The observed curve of the tracer concentration versus time has the following properties:

$$\left(\text{Area under the } c_{pulse} \text{ curve}\right) \quad \text{Area} = \int_0^\infty c\,dt \cong \sum_i c_i \Delta t_i = \frac{m}{\varphi_{V0}} \quad \left[\text{kg}\cdot\text{s}\cdot\text{m}^{-3}\right]$$

$$\text{(Eq. 6.44)}$$

$$\left(\text{Mean of the } c_{pulse} \text{ curve}\right) \quad \bar{t} = \frac{\displaystyle\int_0^\infty tc\,dt}{\displaystyle\int_0^\infty c\,dt} \cong \frac{\displaystyle\sum_i t_i c_i \Delta t_i}{\displaystyle\sum_i c_i \Delta t_i} = \frac{V}{\varphi_{V0}} \quad [\text{s}] \qquad \text{(Eq. 6.45)}$$

To obtain the E curve, simply normalize the c_{pulse} curve dividing by m/φ_{V0}:

$$E = \frac{c_{pulse}}{m/\varphi_{V0}} \qquad \text{(Eq. 6.46)}$$

Sometimes, it is useful to also normalize the time dividing by the average time $\theta = t/\bar{t}$. To keep the area under the curve unity, E must be multiplied by $\bar{t}$:

$$E_\theta = \bar{t}\cdot E = \frac{V}{\varphi_{V0}} \cdot \frac{c_{pulse}}{m/\varphi_{V0}} = \frac{V}{m} c_{pulse} \qquad \text{(Eq. 6.47)}$$

Example 6.5   Conversion in a tubular reactor using the RTD

The following liquid-phase reaction is performed in a tubular reactor:

$$A \rightarrow bB + cC + \cdots - R_A = kc_A \quad \text{with} \quad k = 0.2 \, \text{min}^{-1}$$

A pulse of tracer is introduced in the inlet of a reactor, and the tracer concentration as a function of time is measured at the reactor outlet (Table 6.2). What is the conversion of reactant A?

The area under the curve is Equation (6.44)

$$\text{Area} = \sum c_i \Delta t_i = (1 + 4 + 6 + 3 + 1) \times \Delta t = (1 + 4 + 6 + 3 + 1) \cdot 2 = 30 \, \text{g} \cdot \text{min} \cdot \text{L}^{-1}$$

The average time that the reactant spends in the reactor is Equation (6.45)

$$\bar{t} = \frac{\sum_i t_i c_i \Delta t_i}{\sum_i c_i \Delta t_i} = \frac{(2 \times 1 + 4 \times 4 + 6 \times 6 + 8 \times 3 + 1 \times 10) \times 2}{(1 + 4 + 6 + 3 + 1) \times 2} = 5.9 \, \text{min}$$

The **E** curve can be obtained from Equation (6.46) and is shown in Table 6.3.

In a plug flow tubular reactor, all the molecules spend exactly 5.9 min in the reactor. According to Equation (6.29),

$$\frac{c_A}{c_{A0}} = \exp(-kt) = \exp(-0.2 \cdot 5.9) \approx 0.31$$

The conversion of A is $X_A = 1 - 0.31 \approx 0.69$.

However, according to the **E** curve, $\approx 27\%$ of the molecules spend more than 6 min in the reactor, and $\approx 40\%$ of the molecules spend between 4 and 6 min in the reactor. To estimate the total c_A/c_{A0}, we have to analyze each fraction of molecules independently, according to their residence time (Table 6.4).

Thus, the conversion in the real reactor is slightly lower than the conversion in the ideal PFR.

TABLE 6.2   Concentration of tracer at the exit of the reactor

Time t (min)	0	2	4	6	8	10	12
Concentration of tracer (g·L^{-1})	0	1	4	6	3	1	0

TABLE 6.3   E curve obtained from the concentrations shown in Table 6.2

Time t (min)	0	2	4	6	8	10	12	
E = c/area (min^{-1})	0	0.033	0.13	0.20	0.10	0.033	0	
EΔt	0	0.067	0.27	0.40	0.20	0.067	0	=1

TABLE 6.4 Total conversion of reactant A

t (min)	E	exp(−kt)	exp(−kt)EΔt
2	0.033	0.670	0.670 × 0.033 × 2 = 0.044
4	0.133	0.449	0.120
6	0.200	0.301	0.120
8	0.100	0.202	0.040
10	0.033	0.135	0.009
			Sum $c_A/c_{A0} \approx 0.33$
			$X_A = 1 - 0.33 \approx 0.67$

6.6.2 Dispersion Model

The plug flow model detailed in Section 6.4 does not consider axial mixing of the species in the reactor. If this is the case, the shape of a pulse of tracer injected at the input will be conserved along the reactor. Then, at the exit, the tracer concentration versus time curve will have exactly the same shape as the injection of the tracer. The reader can imagine that this situation is unrealistic.

In the reactor, the tracer is spread axially, and this dispersion will be reflected in an increase of the width of the concentration versus time curve at the exit. However, sometimes, the dispersion is so small that it can be neglected, and then one may assume that the fluid flows as plug flow. The dispersion coefficient D ($m^2 \cdot s^{-1}$) represents this spreading process:

- Large D means rapid spreading of the tracer curve.
- Small D means slow spreading.
- $D = 0$ means no spreading, hence plug flow.

The parameter that measures the extent of axial dispersion is the vessel dispersion number $\left(\dfrac{D}{uL}\right)$, which is dimensionless. The number $\left(\dfrac{D}{uL}\right)$ can be determined either experimentally from tracer experiments or using correlations available in the literature.

Consider a steady-flow tubular reactor of length L through which fluid is flowing at a constant velocity u and in which material is mixing axially with a dispersion coefficient D. The molar balance in each slice of the reactor is

Rate of accumulation $(= 0) =$

Rate of supply bulk flow $\left(\varphi_{n,A}\right)$ − rate of release bulk flow $\left(\varphi_{n,A} + d\varphi_{n,A}\right)$

Rate of supply dispersion $\left(-DS\dfrac{dc_A}{dL}\Big|_z\right)$ − rate of release dispersion $\left(-DS\dfrac{dc_A}{dL}\Big|_{z+dz}\right)$

$\qquad\qquad\qquad$ + rate of production $(dV(-R_A))$

$$0 = \varphi_{n,A} - \left(\varphi_{n,A} + d\varphi_{n,A}\right) + \left(-DS\frac{dc_A}{dL}\bigg|_z\right) - \left(-DS\frac{dc_A}{dL}\bigg|_{z+dz}\right) + dV(-R_A) \Rightarrow$$

$$d\varphi_{n,A} - DS\frac{d^2c_A}{dL} + (-R_A)SdL = 0 \qquad\qquad (Eq.\ 6.48)$$

Dividing by S dL and knowing that $d\varphi_{n,A} = S\,u\,dc_A$ lead to

$$u\frac{dc_A}{dL} - D\frac{d^2c_A}{dL^2} + (-R_A) = 0 \qquad\qquad (Eq.\ 6.49)$$

To solve the molar balance, the term $(-R_A)$ must be substituted by the kinetic expression, e.g., kc_A^n, and then a numerical method must be applied to obtain the concentration as a function of the axial position in the reactor.

CHAPTER SUMMARY AND STUDY GUIDE

This chapter provides the basics of chemical reactor engineering. It describes the three most common ideal chemical reactors: the BR, the CSTR, and the PFR (Figure 6.8).

The analysis focuses on the mass and energy balances for single reactions, leaving aside complex kinetic laws and multiple reactions.

The concept of RTD is introduced as a way of quantifying the non-idealities of the flow. It is shown how to determine the RTD curve and its meaning.

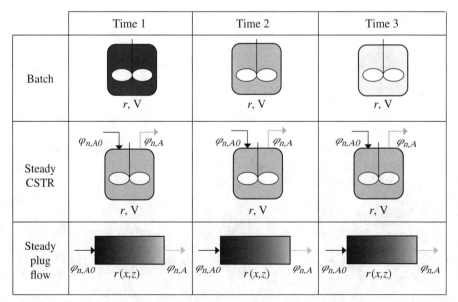

FIGURE 6.8   Scheme of the three types of ideal reactors covered in this chapter.

Although this chapter serves as a starting point to understand reaction and reactor design, the reader is referred to any of the references at the end of this chapter to deal with complex kinetics, multiple reactions, or multiphase systems.

KEY CONCEPTS

Chemical reactors
Plug flow
Perfectly mixed flow
Balances: mass, energy
Residence time distribution
Axial dispersion

SHORT-ANSWER QUESTIONS

6.1 Compare Equation (6.4) with Equation (6.1). What is the relation between conversion and concentration when $\varepsilon_A = 0$?

6.2 Compare Equations (6.22) and (6.29). Assuming a first-order reaction and the same operation conditions, which reactor needs less volume to obtain the same conversion, the CSTR or the PFR? Why?

6.3 Assume a BR and a PFR with the same residence time. Which one gives a higher conversion? Why?

6.4 Is it possible to find a steady-state batch reactor?

6.5 Is it possible to find a non-isothermal steady-state continuous stirred tank reactor and a non-isothermal steady-state plug flow reactor?

6.6 Consider an exothermic reaction carried out in a jacketed batch reactor. The cooling flow is connected to a feedback control loop programmed to create the following temperature ramp inside the reactor: $T = a \cdot t + b$. Is it necessary to simultaneously solve the energy and the molar balances in the reactor?

6.7 In Example 6.5, the conversion for the plug flow reactor is 0.69, whereas for the tubular reactor with RTD, it is 0.67. Would it be possible to find the opposite: a lower conversion in the plug flow reactor?

PROBLEMS

6.1 Repeat the calculations presented in Example 6.4 for a second-order reaction with respect to component A.

6.2 A scientist carries out a liquid-phase reaction in an isothermal ideal batch reactor. The conversion of a reactant A is 0.8 in 8 min and 0.9 in 18 min. What is the kinetic rate expression for this reaction?

6.3 For the elementary reaction A → R, the conversion of A in 30 min is 0.5 in a batch reactor. How long will it take before a conversion of 0.8 is achieved?

6.4 The elementary reaction A → R is carried out in a CSTR. What is the conversion?
Data
Total inlet mole flow $\varphi_{n0} = 40$ mol · min^{-1} with 75% A and 25% inerts
Initial concentration of A $c_{A0} = 2$ mol·L^{-1}
Volume of the reactor V = 500 L
Temperature in the reactor T = 400 K
Activation energy $E_a = 10,000$ cal·mol^{-1} and Arrhenius constant $k_0 = 3 \times 10^4$ min^{-1}

6.5 The reaction A ↔ R carried out in the liquid phase in a CSTR has a kinetic expression$(-R_A) = r = k_1 c_A - k_2 c_R$. $c_{A0} = 1$ mol · L^{-1}, $c_{R0} = 0$ mol·L^{-1}, and the equilibrium conversion is $X_{Ae} = 0.667$. The reactor is working with a conversion at the outlet $X_A = 0.333$. How should the feed rate be adjusted to obtain a conversion of $X_A = 0.5$?

6.6 A chemical plant accidentally discharges a pollutant A into a river. Fortunately, A degrades in time. Calculate how far from the plant the concentration of the pollutant has decreased by 90%. What is the best reactor model for this problem? What is the kinetic order for this reaction?
Data
Reaction rate coefficient k = 0.0008 mol$^{0.5}$·m$^{1.5}$·min^{-1}
River flow $\varphi_V = 200$ m^3·s^{-1}
Cross section of the river S = 300 m^2
Concentration of A in the plant $c_{A0} = 0.02$ mol·m^{-3}

6.7 The reaction A → bB + cC + $\cdots$ − $R_A = kc_A^n$ with k = 0.35 min^{-1}·mol$^{(1-n)}$·L$^{(n-1)}$ is carried out in a tubular reactor with $c_{A0} = 1$ mol·L^{-1} obtaining a conversion of ≈0.70.
An analysis of the residence time distribution, using the same flow and a pulse injection of tracer gives

t (min)	0	3	6	9	12	15	18	21	24	27	30	33	36
c_A (mol·L^{-1})	0	0.1	3	6	8	9	6	5	4	2	1	0.2	0

What is the order of the reaction?

PROJECT

P6.1 Download and install the free software Reactor Lab (English), available at http://www.simzlab.com/. Simulate the different reactors introduced in this chapter for different conditions and orders of reaction. You can also use this software to validate the scripts written by yourself, for instance, in Matlab.

REFERENCES

Atkins P, de Paula J. *Physical Chemistry*. 9th ed. Oxford: Oxford University Press; 2009.

Froment GF, Bischoff KB, De Wilde J. *Chemical Reactor Analysis and Design*. 3rd ed. Hoboken, NJ: John Wiley & Sons, Inc.; 2010.

Levenspiel O. *Chemical Reaction Engineering*. 3rd ed. New York: John Wiley & Sons, Inc.; 1998.

Levenspiel O. *The Chemical Reactor Omnibook*. Corvallis, OR: Oregon State University Bookstores; 2002.

Rawlings JB, Ekerdt JG. *Chemical Reactor Analysis and Design Fundamentals*. 2nd ed. Madison, WI: Nob Hill Publishing; 2002.

Schmidt LD. *The Engineering of Chemical Reactions*. 2nd ed. New York: Oxford University Press; 2004.

Westerterp KR, Van Swaaij WPM, Beenackers AACM. *Chemical Reactor Design and Operation*. 2nd ed. New York: John Wiley & Sons, Inc.; 1988.

7

PROCESSES: BASICS OF PROCESS DESIGN

JOHAN GRIEVINK[1], PIETER L.J. SWINKELS[2], AND J. RUUD VAN OMMEN[1]

[1]Department of Chemical Engineering, Product & Process Engineering Group, Faculty of Applied Sciences, Delft University of Technology, Delft, the Netherlands
[2]Faculty of Applied Sciences, Delft Product & Process Design Institute, Delft University of Technology, Delft, the Netherlands

ACRONYMS

CHP	combined heat and power
FT	Fischer–Tropsch
IP	intellectual property
SHEET	safety, health, environment, economic performance, technological performance

SYMBOLS

c_i	cost of purchase of species i	[\$ kg^{-1}]
ccf	capital charge factor	[year^{-1}]
C_i	hourly cost of purchase of species i	[\$ h^{-1}]
$C_{cap,h}$	hourly cost of capital investments	[\$ h^{-1}]
e_k	extent of reaction k for an open flow system	[kmol·s^{-1}]
EP_m	economic potential related to design level m	[\$·h^{-1}]

Biomass as a Sustainable Energy Source for the Future: Fundamentals of Conversion Processes,
First Edition. Edited by Wiebren de Jong and J. Ruud van Ommen.
© 2015 American Institute of Chemical Engineers, Inc. Published 2015 by John Wiley & Sons, Inc.

$F_{l,i}$	mass flow rate of component i in stream l	$[kg \cdot s^{-1}]$
$\Delta_r H_j^0$	enthalpy of reaction j at standard conditions	$[J \cdot mol^{-1}]$
I_{cap}	capital investment in a process plant	$[\$]$
$I_{nom}^{(unit)}$	capital investment for a unit at nominal capacity	$[\$]$
MW	average molecular weight	$[kg \cdot kmol^{-1}]$
n	total number of chemical species in a reacting system	$[-]$
n_a	average number of carbon atoms in a set of alkanes	$[-]$
p_i	sales price per amount of component i	$[\$ \, kmol^{-1}]$
P_i	hourly earnings from selling of component i	$[\$ \, h^{-1}]$
q	power of the capacity factor in an investment function	$[-]$
r	recycle factor $(0 \le r < 1)$	$[-]$
$R_{\Delta H}^0$	rate of enthalpy production by all reactions together	$[kmol \cdot s^{-1}]$
$S^{(s)}_i$	separation factor of species i $(0 < S^{(s)}_i < 1)$	$[-]$
t_{prod}	production hours per year in a process plant	$[h \cdot year^{-1}]$
$x_{l,i}$	molar fraction of species i in stream l	$[-]$
X	degree of conversion	$[-]$
α	chain growth probability	$[-]$
β	conversion factor for CO to CO_2	$[-]$
$\varepsilon_{element}$	efficiency in use of atoms of an element in a process	$[-]$
η	single-pass conversion factor for CO_2	$[-]$
$\varphi_{l,i}$	molar flow of species i in stream l	$[kmol \cdot s^{-1}]$
λ	scaling factor for efficiency of separation technology	$[-]$

Subscripts

a	averaged quantity
C	carbon related
i	index of a chemical species in a set
j	index of a chemical reaction
k	index for carbon chain length in alkane
l	index for a stream in a block diagram
WGS	water–gas shift reaction
total	related to the total flow of a stream
4+	index for a lump of alkanes, including C_4 and up

Superscripts

conv	chemical conversion unit in a process
(feed)	relating to feed
mix	mixing unit in a process
(prod)	relating to products
sep	separation unit in a process
(waste)	relating to waste

7.1   SCOPE

An introduction is given to the conceptual design of conversion processes of biomass into various forms of energy (carriers). Conceptual design is the first major stage in process design and engineering, involving mainly chemical engineering. The subsequent stages, often called basic and detailed design, deeply involve other engineering disciplines, such as mechanical, civil, control, and electrical engineering. Design by its very nature is quite different from the contents of the preceding chapters (Chapters 3–6), which present a fundamental description of physical–chemical *behavior* by means of mathematical equations at the continuum level (i.e., not distinguishing individual molecules). Understanding of this behavior is essential for design, but when designing a process, one has to deal with a broader range of activities:

- *Specification* of the *functionality* of a process, accounting for stakeholder interests
- *Synthesis* of the internal *structure* for a process, using process unit operations
- *Analysis* of the *behavior* of a process, using models (outlined in preceding chapters)
- *Evaluation* of the *performance* with respect to ecology, economy, and technology criteria, also agreed with the stakeholders
- *Selection* of the most promising design option(s)

Design involves a lot of *decision making* to obtain preferred functionality by creating suitable structures, behavior, and performance. As a result, one often finds a number of alternative solutions performing about equally well. Which solution is preferred depends on economic and social context, beyond strictly technical factors. Designing is a highly iterative process: successful designs are often based on an understanding why earlier design options failed. Being successful in design requires mastery of technical skills, good team work, well-organized work processes, and, last but not least, creativity (see Tassoul, 2009).

This chapter focuses on the introduction of technical skills as a core necessity. One could present a range of process design cases for "learning by example." While quite instructive, this is also a slow-paced learning process. In line with the principle-based approach of this part of the book, a more generic approach to process design is developed in this chapter. The expense of faster-paced learning is a higher level of abstraction than common for a first introduction to design in chemical engineering. To accommodate readers from other engineering disciplines, the approach leans more toward network and system theory.

Design requires much more than mastery of domain contents (e.g., physical, chemical, and equipment knowledge); a procedure to structure the flow of rational design decision making is also needed. This chapter presents a procedure for conceptual process design, enabling the creation of new designs and the improvement existing designs. Like most engineering design procedures, also, this one works top-down in view of mastering the complexity of a system by means of a multilevel decomposition. In this chapter, only the first few levels of design are discussed, beginning with the positioning of a process in a supply chain and ending with functional block diagrams of processes. Special attention is given to the integration of process units into the overall process design, setting a stage for the coverage of conversion units in the next part of this book. In view of the introductory nature of this chapter, the following subjects will be skipped:

a. Modeling and simulation of processes by means of process flow sheeting, e.g., ASPEN®
b. Process integration of energy, solvents, and utilities by matching suitable sources and sinks within the process
c. Equipment design
d. Safety and control engineering

Making some shortcuts in presenting this process design topic is unavoidable. General and deeper reading on chemical process design is offered by, among others, the following textbooks: Douglas (1988), Smith (2005), and Biegler et al. (1997). Textbooks focused on analysis, design, and optimization of integrated biorefineries have recently appeared (see, e.g., Stuart and El-Halwagi, 2012).

7.2   CHARACTERIZATION OF BIOMASS PROCESSING

The focus of this book is the conversion of the chemical energy stored in biomass into other forms of energy. The selection and proper modeling of such a route is the field of process design. There are a number of reasons that make the design of biomass conversion processes challenging: the wide variety of feeds, the many different processing options and the broad product spectrum.

7.2.1   Wide Variety of Feeds

As discussed in Chapter 2, there are many different types of biomass, with very different compositions, that can be used as feed for biomass conversion processes. In addition, a single type of biomass might also vary strongly in composition over time, due to seasonal influences or different places of origin. In addition to the biomass feeds, most processes will also require nonbiomass feed, e.g., air for combustion, gasification, or fermentation. In some cases, a mixture of biomass with other energy-containing feeds (fossil fuels) is used. A proper characterization of the chemical structure and composition of a biomass feed and its variability is essential for process design (see Chapter 2).

7.2.2   Wide Variety of Processing Options

There are many different ways to convert biomass, thermochemical conversion and biochemical conversion being the two most important categories, as discussed in

Chapter 1. Within each category, many different processing options are available. The selection of the most suitable process route(s) for a given task, based on predefined criteria, is one of the major goals of process design.

7.2.3 Broad Product Spectrum

Biomass-based processes can produce a number of different products: heat (often in the form of steam), electricity, fuels (energy-carrying chemicals), and other chemicals. Many processes deliver more than one product. The term cogeneration is typically used when both electricity and useful heat are produced; it is also referred to as combined heat and power (CHP). The term polygeneration is used when three or more products are delivered, e.g., heat (or cold), electricity, and synthesis gas. Synthesis gas or syngas, a mixture of CO and H_2, is used as the feedstock for many chemical processes.

7.2.4 Batch versus Continuous Processing

Another important question when designing biomass conversion processes is whether the process will be operating in batch mode or in a continuous way. In batch processing, all feed is put into a vessel, and multiple steps are carried out one after the other in order to step-by-step change the feed into the end product. This approach is used for certain small-volume, high-value products, e.g., pharmaceuticals, foods, and integrated electronic circuits. Continuous processing is more widely employed in the high-volume, bulk chemical process industry. Continuous processing works like an assembly line: a continuous flow of feed enters the process, typically consisting of multiple steps positioned one after another, all continuously operating. At the end of the process, a constant outflow of product is obtained. Continuous processing is especially more economical when handling large streams, which is typically the case in energy conversion processes. Therefore, we will focus on continuous processes in this chapter. Design of batch units and processes is covered in Douglas (1988), Smith (2005), and Biegler et al. (1997). Processing of feeds with bulk solids generates particulates that induce more fouling by deposition of solids than fluid feeds. Consequently, in the design of such processes, one must account for cleaning and regeneration operations in addition to regular production. Here, we only consider processes in steady-state operation: start-up and shutdown effects will not be taken into account.

Figure 7.1 shows a typical example of a biomass conversion process, presented as a sequence of processing units. The feedstock is wood, and the products are electricity and liquid biofuels. After pretreatment of the wood by chopping, grinding, and drying (see Chapter 8), it is converted into raw syngas by gasification with an oxygen carrier (air, pure oxygen, steam, or a mixture of air or oxygen with steam). The raw syngas is cleaned from components harmful to further processing, such as solids (carbonaceous fly ash) and gaseous components such as H_2S. The clean syngas may undergo one or more gas processing steps, after which it is converted into liquid hydrocarbon fuel

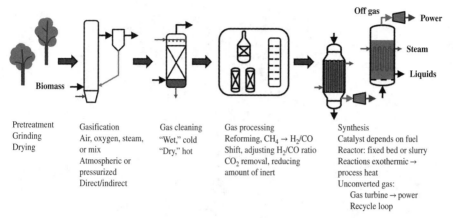

Pretreatment Gasification Gas cleaning Gas processing Synthesis
Grinding Air, oxygen, steam, "Wet," cold Reforming, $CH_4 \rightarrow H_2/CO$ Catalyst depends on fuel
Drying or mix "Dry," hot Shift, adjusting H_2/CO ratio Reactor: fixed bed or slurry
 Atmospheric or CO_2 removal, reducing Reactions exothermic $\rightarrow$
 pressurized amount of inert process heat
 Direct/indirect   Unconverted gas:
    Gas turbine $\rightarrow$ power
    Recycle loop

FIGURE 7.1 Overview of standard biomass gasification facility. The processes for "gas processing" are optional. (Source: Adapted from Olofsson et al. (2005).)

using Fischer–Tropsch (FT) synthesis. The remaining unreacted syngas is used to produce electricity. Each unit in the process has a certain *function* to fulfill, effecting changes from its inputs to outputs. All functions combined in the right order accomplish the overall change from feed to products. This process is used to illustrate the concepts discussed in this chapter.

7.3   ANALYZING THE OUTSIDE OF A PROCESS

In general, the design of a process must always start by analyzing and specifying how the process will be interacting with the outside world. Any process to convert biomass to energy is part of a physical supply chain, both upstream and downstream. Furthermore, it is also imbedded in a socioeconomic and natural environment (see Figure 7.2).

At the *upstream side* of the supply chain, one faces the following issues. Who will supply biomass to the process? What can be delivered in terms of annual amounts with seasonal effects, variability in quality, security of supply, and expected price ranges? Is the logistics of the biomass delivery feasible and what are feed storage requirements at the site of the process? What source of oxygen to use for partial combustion (air, oxygen, water)? Is cofiring with a second feedstock desirable?

At the *downstream side*, similar issues arise concerning the slate of energy products to be generated by the process. What annual amounts of energy (power, fuels, and thermal heat) can be sold to (permanent) customers and at what quality specifications? Is there a dynamic pattern in the product demand? What is the minimum uptime (=hours per year a process is capable of producing) for securing supply requirements of the customers? What are realistic past price ranges and future price scenarios? In addition to making products, any process generates some waste and

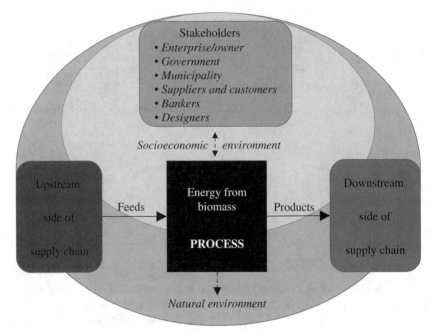

FIGURE 7.2 Connections between a process, supply chain, and socioeconomic environment.

emissions. Can other parties treat this waste for a fee? Meanwhile, environmental emissions (gas, noise, smell) must be kept under control within the limits of the permits. This touches upon a nontechnical but equally important "supply" issue, which is to obtain a societal license to build and operate a process on some local industrial complex, endorsed by legal permits by the government. Last but not least, finance is another "supply" aspect. Money must be made available, often in the form of loans, to finance the design, engineering, construction, and start-up of a process. Figure 7.3 illustrates the main connections between a process, its supply chain, and socioeconomic environment.

The enterprise and the supply chain, extending outside the enterprise, form the highest level of aggregation, at which the external specifications for the process design are set. The smallest scale in a process design deals with molecules. Figure 7.4 shows a multiscale representation plotted versus time and geometric size. A cascade arises from the enterprise and supply chain at the upper scale down to the molecular scale.

At every scale in this figure, one must create the associated structures (units, compartments, particles) by design. This requires a specification of the function of a structure at each of these scales, a choice of suitable building blocks to create the structure while connecting the blocks by streams that carry suitable physical resources (chemical species, energy, momentum) or signals (for control). This brings us to the point where the focus shifts to developing a systematic view on the inside of a process.

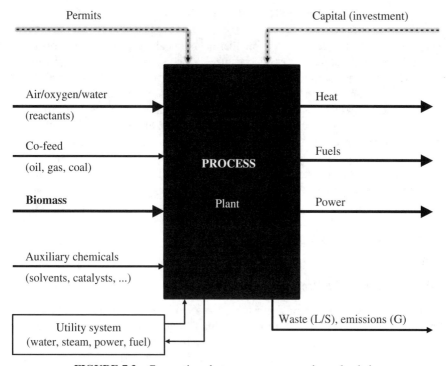

FIGURE 7.3   Connections between a process and supply chain.

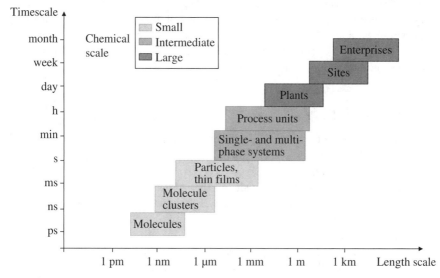

FIGURE 7.4   Multiscale view on a process plant, interior and exterior. (Source: Adapted from Grossmann and Westerberg (2000).)

7.4 ANALYZING THE INSIDE OF A PROCESS

The diagram of a process based on thermochemical gasification of biomass in Figure 7.1 shows that the process is organized as a sequence of different processing units. The process operates in a continuous production mode, and the main mass streams flow from the left to right in the diagram. Often, there is a recycle of unconverted feed as well as a coupling of heat sources and heat sinks inside the process, in order to minimize external energy requirements. Different processes have quite varying internal structures and an enormous host of processing units, each with a specific function. To get conceptually on top of this complicated situation, some abstraction is needed. It is helpful to realize that a process is an engineered system, just like an airplane, a mobile phone, or a building. Thus, one can draw from the generic terminology in the sciences of engineering to describe a process in some generic way. The inside of a process is characterized using the following key words: *function*, *structure*, *behavior*, and *performance*.

7.4.1 Functions of a Process and Process Units

Engineered systems derive their existence from being able to perform certain functions that add value to the user of the system. The *physical* function of a process is to transform the feeds into commercial products. Its *economic* function is to generate a profit for the owner by turning cheaper feeds into more expensive products. The *social* function is to provide useful products fulfilling needs in society as well as high-quality jobs for staff, which design, build, operate, and manage such processes. The overall physical function of a process is broken down into a sequence of subfunctions, each of which changes the state of the matter being processed, e.g., by grinding and drying, gasification, cleaning and conditioning of the produced gas, and chemical synthesis to produce liquid fuels. These subfunctions are conducted in processing units, also called unit operations, as they perform specific processing operations (reactions, separations, heating, cooling, compression, etc.).

7.4.2 Structure of a Process

The generic structure of an engineered system is a multilevel, multiresource *network*. The nodes in the network represent operations for the transformation or transfer of resources; the edges represent the flows of resources. Each node in the network can host a subnetwork, and some nodes in the subnetwork can be expanded into a sub-subnetwork. Often, this process of expansion can go three or four levels deep. Figure 7.5 shows a representation of a process network structure with two and three levels of expansion.

The starting point of the expansion is always an input–output diagram (level 0). This diagram can be filled in with a network with some key operations (level 1). Each key operation is often built up of more elementary operations (level 2) that involve one or more physical phases (gas–liquid–solids) in specific modes of contacting and

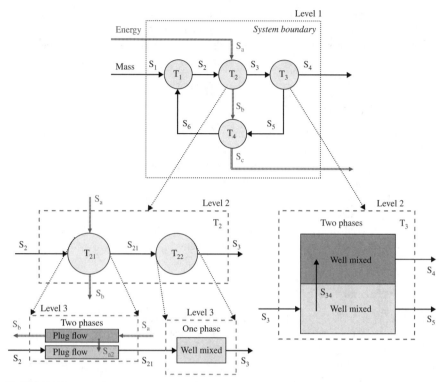

FIGURE 7.5 Multilevel network structure of a process with expanded nodes. Top: process reference network of streams and nodes. Bottom left: two levels of expansion of node T_2. Bottom right: expansion of node T_3 in a two-phase, well-mixed volume.

hydrodynamic regimes. The resources in the network flow from one node to another, coupled or separately:

- *Chemical species*, which change in identity and amount by chemical reactions.
- *Mass*, differentiated in various chemical species and in thermodynamic phases: gases, liquids (organic, aqueous), and solids (in dispersed or massive form).
- *Energy*, differentiated in various forms: chemical, thermal, and mechanical (work), electrical power, and electromagnetic field energy (radiation); potential and kinetic energy contributions are almost always small enough to be ignored in chemical processes.
- *Momentum*, considering convective flow and pressure drop.
- *Electrical charge*.
- *Information*, in the form of signals from sensors to controllers to actuators.

The nodes in the network act on one or more resources by changing their amounts. Some common examples are reactors and fuel cells (involving a change of species mass by reactions with a thermal heat effect and power generation), evaporators (turning liquid into vapor by addition of thermal energy, which is a change in chemical energy), and compressors (turning power or thermal energy (steam) into mechanical energy).

7.4.3 Physical Behavior of a Process

The process network with its nodes and edges can be modeled to analyze physical behavior in response to changes in external conditions, internal operating conditions, design parameters, and physical properties. For each of the nodes, the conservation equations for the relevant resources, the rate laws, and the thermodynamic equations can be set up (see Chapter 5). In the initial stages of conceptual design, steady-state (stationary) models suffice. Dynamic models come into play when analyzing how a design and its control system perform for various operational scenarios (e.g., start-up, shutdown, feed, and load changes) and disturbances.

7.4.4 Performance of a Process in Creating Value

A process is supposed to add value in a socioeconomic sense, without creating significant risks for the people who operate the process, the environment, the owner, and the society in general. Rather than having a single performance criterion, such as profitability, often, multiple performance measures are used. These are of an ecological, economic, and technological nature and are together abbreviated as *SHEET*. The ecological indicators involve the *safety* (*S*) of the process (immediate risk for human and animal life and property damage), the *health* (*H*) reflecting longer-term risks for human health, and the exposure of the *environment* (*E*) to emissions and waste. *Economic* performance (*E*) can be expressed in multiple ways, such as payback time or internal rate of return. The latter accounts for the changing value of money over the lifetime of a process (decades) (see Peters et al., 2003). Finally, *technological* performance (*T*) can be related to the operational capability of a design, e.g., the uptime and flexibility to respond to market changes, as well as to energy and atom efficiencies (see Chapter 18). For instance, carbon efficiency indicates which fraction of carbon atoms present in the feeds ends up in the products. Energy efficiency is the fraction of energy taken in by the process through the feeds and utilities finishing in the products. Improving one performance measure, say, increasing carbon efficiency by having more CO_2 capture and internal recycling in the process, may adversely affect another measure, such as profitability, which becomes lower due to a higher investment in the process. One may generate a so-called Pareto frontier curve for representing a trade-off between opposing performance measures. In that case, a design is represented by the values of its performance measures and plotted as a point in a 2-dimensional diagram, where the two performance measures form the orthogonal coordinate axes. A Pareto frontier curve is the set of critical points for which one cannot improve one of the performance measures without making the other worse (tinyurl.com/yocw7q). Since designs with performance points on a Pareto frontier curve cannot be further optimized, the selection of a specific design from this critical set must be guided by company policy regarding economy and ecology ("people, planet, profit").

Having analyzed the more common outside and inside aspects of a process, the scene is set for the introduction of a process design procedure. Such a procedure aims primarily at creating a well-structured flow in making rational design decisions, based on domain knowledge. Such domain knowledge of biomass processing (thermodynamics, reaction kinetics, physical transport phenomena, and equipment) can be obtained from Chapters 8–15.

7.5 A DESIGN PROCEDURE FOR BIOMASS CONVERSION PROCESSES

7.5.1 What Is the Design Problem?

The design of processes for biomass conversion is very similar to the design of other chemical processes. The key distinction is the solid-phase biomass feed, having a complex and varying composition. In any process design, one must consider the following issues and make decisions:

- Which physical, chemical, and mechanical processing operations are needed to go from feed(s) to product(s)?
- How to order and connect these processing operations? Are recycles of unconverted feed and solvents possible?
- Which solvents and catalysts are to be used as active agents in the processing?
- Should heat sources and heat sinks in the process be connected and in what way?
- In which type of equipment can each operation best be conducted?
- Which mode of operation of the equipment (batch, continuous) is most suitable?
- What are suitable operating conditions (temperature, pressure, thermodynamic phases) in the equipment?
- How much residence time is needed in the equipment per thermodynamic phase?
- How to design the equipment internally (compartmentalization, internals)?
- What should be the target uptime of the process?
- How reliable should be the equipment in its operation (as to meet this target uptime)?
- Are risks of failure (economic, ecological, technological) sufficiently small?
- Is the expected performance of the design satisfactory?

The design decision variables in many of the above design questions are of a *discrete* nature; the variable can take only a few distinct values. For example, equipment, is it type A or B or C; connectivity, should there be a solvent recycle between unit U1 and U2: Y/N? The other decision variables are *continuous* within a constrained interval: e.g., 298 < temperature (K) < 348.

This brief sketch shows that there are many design decision variables of a discrete and continuous nature, which turns design into a huge combinatorial problem.

7.5.2   What Is the Problem in Generating Designs?

Structuring of process design by means of a certain procedure is necessary, because one wants to avoid a situation in which an astronomical number of alternative designs are created, which are all logically possible and must be evaluated with respect to performance. A simple example indicates the enormity of this problem. Let there be N units in a process, and each unit has, on average, M design decision variables, where each decision variable can attain, on average, K discrete values. Let us assume that the design decision variables are fully independent; i.e., picking a value for one variable does not impair the freedom of decision for any of other decision variables. The number of logically possible designs becomes $D = K^{M \cdot N}$. If there is one choice per variable, $K = 1$, only one design is possible: $D = 1$. Things start to get out of hand when the choices (K) increase. Even for a very simple process ($K = 2$; $M = 4$; $N = 5$), a million "designs" can be logically generated, $D = 2^{20} = {\sim}10^6$, while for a more complicated process ($K = 4$; $M = 8$; $N = 20$), the outcome is $D = 4^{160} = {\sim}10^{96}$. The problem with this combinatorial approach is that almost all of the resulting alternative designs that are logically possible do not make sense from physical and economic points of view. Generating all alternatives and filtering out the very few feasible ones is a prohibitive effort. The practical failure of this brute force combinatorial approach is caused by assuming that the decision variables are independent while they are not. For instance, it does not make sense to place a product purification unit before the reactor making the product. There are many physical constraints connecting the decision variables and constraining the choices. The core challenge is how to cope with very many design decision variables with almost as many constraints between them and arrive at a small number of the better design alternatives.

7.5.3   Hierarchical Decomposition Approaches for Process Design

The time-honored practical approach is to apply a *hierarchical decomposition* approach to process design. Here, the design decision variables are partitioned in a hierarchy of many smaller clusters, as explained by Douglas (1988). The aim is to structure the design procedure in such a way that the decision variables with a potentially high impact on the economic performance are covered first and the ones with a lesser impact later. This approach implies a more sequential way of decision making, one cluster after the other. The clusters are organized according to design *levels*. The economic impact is highest at the top level, where decisions are made on the nature of products, feeds, capacity, type of processing agents (e.g., catalysts, solvents), mode of operation, and one sets targets for the degree of conversion of feeds. The impact of decision variables on economics decreases when addressing the increasingly finer details of the design at the lower levels: process unit, compartments, and equipment internals.

Along with the hierarchical approach comes top-down targeting: the design at a particular level establishes targets for a processing task, which must be achieved by the design at a subsequent level of finer detail. For example, one can target for a certain product yield of a reactor when dealing with the integration of that reactor in the entire process. When going one level down in actually designing the reactor, the product yield target becomes a fixed quantity, and the reactor variables, such as temperature and residence time, can be varied to meet this target. Then, when setting the residence time as new target, the geometric size of the reactor can be varied to meet this residence time target.

7.5.4 Design Levels for a Biomass Conversion Process

A hierarchical decomposition approach is presented here, in which the process design decisions are distributed over several levels in a top-down manner. Decisions made at an upper level restrain the choices for making decisions at the lower levels. The interactions in the opposite direction are weaker, although there is always some iterative feedback: if at a lower level only bad solutions are possible due to an apparently poor decision at an upper level, this latter decision has to be revised.

For the conceptual design of a biomass conversion process, seven levels will be considered. The first three levels are treated in this introductory chapter on design, because they offer the basis of a design with the highest impact on performance. The remaining four levels are essential to arrive at a finished conceptual process design. They are mentioned here for the sake of completeness but not further explained due to restrictions on the size of this chapter.

Level 1: Input–output type of exchanges between the process and its environment. This environment is often a "supply chain" of materials, energy, and information, from which the process extracts feedstocks and to which the products are delivered.

Level 2: Splitting the process into subprocesses, where a subprocess can buy feed from or sell product to the market, or when technologies of different license holders are used, or when subprocesses are at different locations.

Level 3: Creating a network of process units per subprocess. The focus here is on the selection of suitable process unit operations and the setting of targets for processing duties and changes in the flows of resources in a balanced way to enhance overall performance. The targets act as set points for the control system design.

Level 4: Process integration of energy, solvents, water, and common reactants (e.g., hydrogen). Such integration is done by matching internal sources and sinks. When a resource is generated in some excess in a unit, it can be transferred to units where shortages of that resource arise, provided that the quality level of the available resource is or can be made fit for the identified sink(s).

Level 5: Equipment design and engineering, enabling to achieve the established processing targets, among others, by providing enough actuators for control.

Level 6: Design of safety, information, and control systems to support plant operations.

Level 7: Sensitivity analysis and optimization of sections of the process that have a high impact on performance. This last step aims at correcting for any adverse effects arising due to hierarchical sequential decision making.

7.5.5 Application of Generic Engineering Design Steps at Each Level

Per design level, the same typical sequence of design activities emerges:

a. Setting the *scope* of the design at this level by establishing:
   - *Functions* and goals to be realized at this level
   - The boundaries with boundary conditions
   - Suitable types of "building blocks" to achieve the design goal(s)
b. Collecting and assessing *knowledge* on existing designs and building blocks by looking for patent information, empirical data, models, theory, and operational experiences and identification of gaps in knowledge to be remedied with additional experimental programs
c. *Synthesis* of alternative *structures* by selecting, ordering, and connecting suitable building blocks at an appropriate scale to meet the goals targets set in the *scope* step
d. *Analysis* of the physical *behavior* of the generated alternative structures by means of model-based simulations, showing the flows of resources in these structures
e. *Evaluation* of the *performance* of the alternative structures using as much as possible quantitative metrics for SHEET
f. *Selection* of a few of the better alternatives as candidates for further elaboration at the next lower design level
g. *Documentation* of the design decisions, rationales, models, and results in the form of drawings and numerical data for transfer to the next level of design

It is a current engineering design capability to integrate the activities in synthesis, analysis, and evaluation steps in a systematic mathematical optimization format, provided that one has enough confidence in the process models (see Biegler et al., 1997). The synthesis step gives rise to discrete decision variables, which define alternative process structures (building blocks and connectivity), while continuous variables relate to processing conditions, duties, and sizing. In the analysis step, the model equations can be rigorously set up, extended with inequalities delineating the domain in which the model equations are valid. In the evaluation step, one can add a proper design objective function and perform optimization. Thus, in mathematical terms, one obtains a mixed integer optimization problem Equation (7.1):

$$\text{Maximize}_{\underline{d}^{(l)}, \underline{c}^{(l)}} : F^{(l)}\left(\underline{x}^{(l)}, \underline{c}^{(l)}, \underline{d}^{(l)}; \underline{p}\right) \qquad \text{(design objective function)}$$

$$\text{Subject to}: \underline{g}^{(l)}\left(\underline{x}^{(l)}, \underline{c}^{(l)}, \underline{d}^{(l)}; \underline{t}^{(l-1)}, \underline{p}\right) = \underline{0} \quad \text{(process model equality equations)}$$

$$\underline{h}^{(l)}\left(\underline{x}^{(l)}, \underline{c}^{(l)}, \underline{d}^{(l)}; \underline{t}^{(l-1)}, \underline{p}\right) \geq \underline{0} \quad \text{(process model inequality equations)}$$

$\underline{c}^{(l)}$: continuous decision variables at level (l), to be optimized

$\underline{d}^{(l)}$: discrete decision variables at level (l), to be optimized

$\underline{p}$: fixed parameters in the model, prone to some uncertainty

$\underline{t}^{(l-1)}$: targets already fixed at preceeding level $(l-1)$

$\underline{x}^{(l)}$: dependent process variables at level (l) (Eq. 7.1)

It is not that a design problem simply reduces to a mathematical optimization. Often, the *key* conceptual design effort is the generation of a superstructure of alternative building blocks that can be connected by streams in multiple ways. Another challenge is the representation of this superstructure by means of discrete decision variables, along with making choices among alternative design models. Last but not least, the outcome of a design optimization must be assessed with respect to its practical significance while accounting for the underlying parametric uncertainties.

7.5.6 Conceptual Design Matrix as a Summary of the Design Procedure

The two dimensions of the design procedure (levels and engineering design steps) can be brought together in a matrix, as presented in Table 7.1. The purpose of this matrix is to serve as a mental frame for structuring the design process. It is advised to work top-down and level by level and at each level perform the activities per design step (see Section 7.5.5). The few promising design alternatives selected in steps (e) and (f) at a level [*i*] must be transferred to step (a) of level [*i* + 1] for further expansion and refinement. If at a particular level none of the generated alternatives meet the performance criteria, one may challenge the wisdom of the evaluation criteria or decide to return to preceding design levels and look harder for better options there.

The application of the design steps are illustrated for the first three levels in the following sections. The leading example to illustrate these steps is the conversion of biomass into syngas, followed by FT reactions to produce liquid fuels. The use of this example does not imply that other conversion routes for biomass are less relevant.

The passes through the "design matrix" can be visualized as a helix-like staircase (see Figure 7.6). One full cycle involves passing through the seven engineering design steps. Then one arrives at the next level of design at the initial step again.

TABLE 7.1 Conceptual design matrix for Biomass to Energy Processes

Level of design ⇩	Engineering Design steps ⇒						
	(a) **Scope** (functionality, goals, design objects)	(b) **Knowledge** (collection & assessment of objects)	(c) **Synthesis** (alternative structures & scales)	(d) **Analysis** (of physical behaviour of structures)	(e) **Evaluation** (performance indicators per structure)	(f) **Selection** (of the better performing alternatives)	(g) **Reporting** (decisions, rationales & results)
1 **Input-output structure** • Product & feed specifications • Process boundary • Boundary conditions							
2 **Division in sub-processes** • Market exchange feed/product • Other technology /sub-process • Input-output structure		Perform the associated activities for each cell				
3 **Network of units/sub-process** • Chem. conversions & recycle • Physical separations • Mechanical operations • Thermal operations							
4 **Process integration** (energy, solvents, water...)							
5 **Equipment designs** (internal structure, sizing, ..)							
6 **Safety, control & information systems**							
7 **Process flow sheet sensitivity analysis & optimization**							

See text for further explanation of design activities at the levels and in the engineering design steps.

FIGURE 7.6   Helix-like staircase: visualizing the cyclic pathway through the design levels and design engineering steps of the Conceptual Design Matrix. Digital photograph, P.L.J. Swinkels/Basílica Sagrada Família Barcelona (2013).

7.6   INTERFACE WITH SUPPLY CHAIN: INPUT–OUTPUT DIAGRAM

The first level of design establishes the function of the process in a supply chain and defines an interface between process and supply chain. Each of the design steps at this level is reviewed here and serves as an illustration of the use of the design matrix.

a. *Scope*

The main products, production capacity (average and dynamic: minimum and maximum capacity level), and suitable available feeds (and supply dynamics) must be specified, often in conjunction with a future production location. External targets and restrictions on the design of the process must be identified, such as environmental, economic, societal, and legal ones. The (conceptual) boundary of the process must be specified as well as the conditions of the streams that cross this boundary and link the process with the supply chain. Also, some main characteristics of the process itself must be specified, such as preferred processing routes and key technologies to implement such routes. Available key technologies serve as building blocks at this level.

b. *Knowledge*

Knowledge (in the form of patents, empirical data, diagrams, heuristic design rules, and models) has to be gathered about the properties of the feeds, availability of feeds, product properties and market demand, available key conversion technologies, and their performance. "Soft" knowledge, such as awareness of public perception of certain feeds and products is also vital information for creating a process that meets public acceptance. Last but not least, analysis of the performance of competitors' plants and their plans for the future is advised.

When making a design, it is necessary to know (or estimate) the chemical composition of the biomass and its variability over time (seasons). In addition to organic matter, biomass contains water, sand, and minerals. The moisture content of the biomass feed is quite important for the energy economy in a process. Furthermore, the molecular composition of the organic material is not uniform. The average elemental composition of an organic molecule is given by CH_xO_y (x hydrogen and y oxygen atoms per carbon atom). These (fractional) atom numbers x and y vary over a certain range. The feed composition can be characterized by a distribution function for the fraction of organic molecules in the feed having x hydrogen and y oxygen atoms: $0 \leq f(x, y) \leq 1$, $x_{min} \leq x \leq x_{max}$, $y_{min} \leq y \leq y_{max}$.

c. *Synthesis*

Decision items in the synthesis step are shown in a so-called input–output diagram (see Figure 7.7). The decisions on suitable feeds and products can only be made in connection with the choice for a key processing technology. Reversely, given feeds and products limit the choice of suitable technologies. The mode of production depends on the production capacity and the way the products are formed and used. From an economic perspective, a continuous mode of operation is already preferred when the mass flow capacity of a process exceeds 5 kt·year^{-1} (Douglas, 1988). A flexible continuous mode is also required for uninterrupted power supply to society, though some of the contributing processes to the power production may be of the on/off type (wind energy) or are daily varying (e.g., solar energy), necessitating internal energy storage. The uptime of a process (i.e., the time that a process functions properly and can make products) depends on the physical state of its feed. Practical experiences indicate that for well-maintained gas–liquid plants, an uptime of well over 8150 h·year^{-1} is possible (1 year $= 8760$ h), while for solid processing units and batch processes, an uptime of 6000–7000 h·year^{-1} is more realistic (Douglas, 1988).

In gasification and combustion process design, the way of oxygen supply for partial combustion of the biomass and cofeeds is an important factor. Air is free of cost, but it brings along 80% of inert nitrogen, which has to be pumped through the process. This enlarges the required volumes of piping and equipment at a very significant additional investment cost. As an alternative, one can enrich

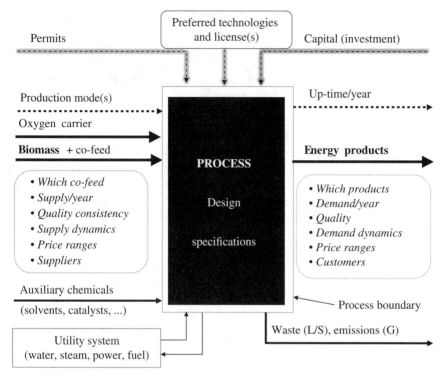

FIGURE 7.7 Input–output diagram of a process as a basis for design.

the oxygen content by putting air through a cryogenic air separation unit with partial or nearly complete removal of nitrogen. The cost of enriched air is significant but one saves much on investment cost by having a more compact process. Thirdly, oxygen from the water molecules in steam can be used for conversion, as is done for reforming reactions in the gas phase:

$$CH_xO_y + H_2O \rightleftharpoons (1-y)\,CO + y\,CO_2 + (1+x/2)\,H_2 \quad (0 \leq y < 1; 0 \leq x \leq 4)$$

$$(RX.7.1)$$

Fourthly, in dry reforming, the oxygen atoms in carbon dioxide are exploited:

$$CH_xO_y + (1-y)\,CO_2 \rightleftharpoons (2-y)\,CO + (x/2)\,H_2 \quad (0 \leq y < 1; 0 \leq x \leq 4)$$

$$(RX.7.2)$$

In all four situations, the reverse water–gas shift (WGS) reaction and the methanation reaction play an additional role, turning carbon dioxide into carbon monoxide and methane and so lowering the hydrogen-to-carbon monoxide ratio:

$$CO_2 + H_2 \rightleftharpoons CO + H_2O \qquad (RX.7.3)$$

$$CO + 3\,H_2 \rightleftharpoons CH_4 + H_2O \qquad (RX.7.4)$$

If a process is built on a site with existing utility and waste processing systems of sufficient capacity, the generation of an oxygen carrier can be kept outside the process boundary. However, the costs of using utilities and having the waste processed show up at the negative side of the process economics. For a process design on a new ("greenfield") site, the generation of utilities specific for this process must fall within the process boundary.

d. *Analysis*

One can set up mass and energy balances over the process input–output diagram, covering all feed, product, and waste streams. From these balances, one can see what fraction of a property (e.g., atoms, energy) present in the feeds ends up in the products and what fraction goes to waste. Equation 7.2 presents balance equations for a general property with an associated efficiency factor per property. These efficiency factors play a role in the evaluation of a design:

Balance equation for property p:

$$\text{IN}: \quad \sum_{l=1}^{L^{\text{feed}}} \varphi_l^{(\text{feed})} \cdot \sum_{i=1}^{N_s} \left\{ x_{l,i}^{(\text{feed})} \cdot X_{p,i} \right\} + \sum_{i=1}^{N_s} \left\{ R_i^{(\text{generation})} \cdot X_{p,i} \right\} \quad =$$

$$\text{OUT}: \quad \sum_{l=1}^{L^{\text{prod}}} \varphi_l^{(\text{prod})} \cdot \sum_{i=1}^{N_s} \left\{ x_{l,i}^{(\text{prod})} \cdot X_{p,i} \right\} + \sum_{l=1}^{L^{\text{waste}}} \varphi_l^{(\text{waste})} \cdot \sum_{i=1}^{N_s} \left\{ x_{l,i}^{(\text{waste})} \cdot X_{p,i} \right\}$$

Feed-to-product efficiency factor η for property p:

$$\eta_p = \left[\sum_{l=1}^{L^{\text{prod}}} \varphi_l^{(\text{prod})} \cdot \sum_{i=1}^{N_s} \left\{ x_{l,i}^{(\text{prod})} \cdot X_{p,i} \right\} \right] \Big/ \left[\sum_{l=1}^{L^{\text{feed}}} \varphi_l^{(\text{feed})} \cdot \sum_{i=1}^{N_s} \left\{ x_{l,i}^{(\text{feed})} \cdot X_{p,i} \right\} \right]$$

(Eq. 7.2)

$\varphi_l^{(\text{feed/prod/waste})}$: molar flow rate of a feed/product/waste stream l,

$$l = 1, \ldots, L^{(\text{feed})/(\text{prod})/(\text{waste})}$$

$x_{l,i}^{(\text{feed/prod/waste})}$: molar fraction of species i in a feed/product/waste stream l,

$$l = 1, \ldots, L^{(\text{feed})/(\text{prod})/(\text{waste})}$$

$R_i^{(\text{generation})}$: rate of generation of kmol of species i inside process

$X_{p,i}$: intensive property p of chemical species i (> 0), $i = 1, \ldots, N_S$

Species efficiency : $X_{p,i} = 1$ ($p = m$, kmol per kmol of species i)

Atom efficiency : $X_{p,i} = n_{E,i}$ ($p = E$, number of atoms of element

E per molecule of species i)

Enthalpy efficiency : $X_{p,i} = h_i^0$ ($p = h$, enthalpy per kmol of species

i at reference conditions)

These efficiency factors will play a role in the evaluation of a design.

e. *Evaluation*

In general, multiple evaluation criteria play a role (see Section 7.4.4), of which economic performance is often dominant. One can calculate the economic potential, according to Equation (7.3), based on the prices of the feed, product, and waste streams:

$$\text{EP}_1 = \left[\sum_{l=1}^{L^{\text{prod}}} \left\{ p_l^{(\text{prod})} \cdot \varphi_j^{(\text{prod})} \right\} - \sum_{l=1}^{L^{\text{feed}}} \left\{ p_l^{(\text{feed})} \cdot \varphi_j^{(\text{feed})} \right\} - \sum_{l=1}^{L^{\text{waste}}} \left\{ c_l^{(\text{waste})} \cdot \varphi_j^{(\text{waste})} \right\} \right]$$

Criterion :

$$\text{EP}_1 \geq \varepsilon_{\text{profit}} \left[\sum_{l=1}^{L^{\text{prod}}} \left\{ p_l^{(\text{prod})} \cdot \varphi_j^{(\text{prod})} \right\} \right] \tag{Eq. 7.3}$$

EP_1 : economic potential at level 1 (earning potential without capital and overhead costs)

p_l : price per unit amount of stream l, with $l = 1, \ldots, L$

c_l : cost per unit amount of waste stream l

$\varphi_l^{(\text{feed/prod})}$: flow of stream l $\left(\text{in kmol·s}^{-1} \text{ or in kg·s}^{-1}, \text{whatever is most convenient} \right)$

$\varepsilon_{\text{profit}}$: profitability margin (typical range : $0.50 - 0.75$)

This potential can be assessed against a profitability criterion. At this input–output level of design, the setting of the profitability criterion should be high enough to leave room for the costs of utilities and the capital charge due to process investments costs and for costs of overhead in operations, to be determined in later design levels. Another criterion can be atom efficiency: $\varepsilon_{\text{element}} \geq \varepsilon_{\text{element,min}}$. For instance, carbon atom efficiency is an increasingly important performance indicator, defined here as the carbon atoms in the feed that end up in products. The remainder is often lost as CO_2 in a flue gas stream.

There is a degree of freedom at this design level to minimize the losses of atom efficiency or energy efficiency by adjusting the ratios of the product streams $\varphi_l^{(\text{prod})}$ within practically feasible limits.

f. *Selecting and (g) reporting*

Often, alternative, profitable options can be identified for setting up the process, e.g., by using different feeds or another mix of products or applying a different technology. In the end, a choice is necessary on which few options to pass on to the next level of design. Continuing with all options will soon result in a prohibitive design effort, because at every level the number of design alternatives is multiplied with the number of new options. So, pruning of alternatives is necessary. It is advisable, though, to report the essence of the disregarded ones, as in case the designs of the selected options come to a dead end prematurely, one has to revisit the abandoned options at an earlier level.

7.7   DIVISION IN SUBPROCESSES

Design at the second level deals with the division of the process into certain subprocesses.

a. *Scope*
   A division into subprocesses makes sense (among others) when a subprocess has:
   - A feed or product that has commercial value and can be traded on the market. It makes the subprocess an independent economic entity and less sensitive to what happens in other subprocesses.
   - A special technology, operating under a license from another company.
   - Confidential in-house technology that the owner does not want to share with co-owners of the process.
   - A different mode of operation, e.g., batch processing, requiring intermediate storage between subprocesses.
   - A different location than the other subprocesses. For example, it is better to collect, chop, and dry biomass close to the different harvesting areas, while further processing is best done at a central location, thereby avoiding transportation of useless moisture along with the feed.

b. *Knowledge*
   An analysis of existing processing technologies for the (intermediate) products at hand is made to see if splitting is desirable and if competitors have done this and in what ways.

c. *Synthesis*
   The synthesis activities at the level of subprocess formation consist of:
   - Identifying candidate subprocesses and conversion technologies with associated intermediate feeds and products
   - Ordering the subprocesses and connecting them by streams, considering recycle structures
   - Developing an input–output diagram for each subprocess, which forms a basis for further design, similar as for the main process

Figure 7.8 shows an example of the split of a process into three subprocesses.
   Two subprocesses run in parallel, each converting a different feed into the same product (e.g., wood and natural gas to syngas), though with a different composition. The intermediate product (syngas) can be sold on the market if the joint capacity of subprocesses 1 and 2 exceeds the demand of subprocess

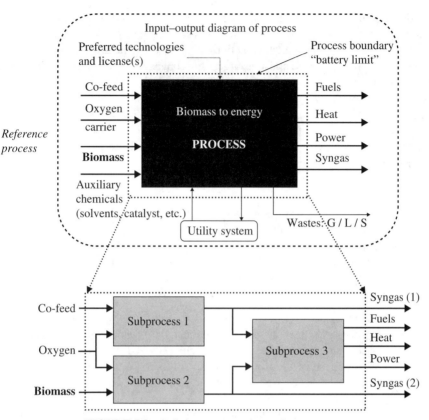

FIGURE 7.8   Splitting a process into subprocesses.

3. Reversely, the intermediate product can be imported if the demand of sub-process 3 exceeds the joint production capacity of subprocesses 1 and 2. In this example, the syngas going to subprocess 3 is converted into three products: liquid fuels, power, and heat.

d. *Analysis and (e) Evaluation*
These are done for each alternative in the same way as for the full process, ending with (f) *Selection* and (g) *Documentation* of the better options for propagation to the next level of design.

7.8   PROCESS DESIGN: FUNCTIONAL BLOCK DIAGRAM

Design at the third level generates a network of processing functions, encapsulated in process units. The focus here is on which functions are needed and how they interact in the network for an optimal performance. Targets are derived for the functional duties (e.g., degrees of conversion, selectivity, and separation). This is an *outside* view from the perspective of a unit. One also needs an *inside* view on how to structure each of the process units such that the functions and targets can be achieved or corrected if found to be infeasible. Supposing targets for conversion and selectivity have been established, the question arises as regards the following: what types of reactors should be chosen, how many, and in which configuration (parallel, sequential staging). The equipment engineering is part of level 5. Iterations between both views are likely during the design. This section deals with the creation of a network of process units (per subprocess), taking an outside view on the network. The design of the process unit (the inside view) is part of a lower level.

a. *Scope*

The goal is to design per chosen subprocess a network of process units that jointly achieve the conversion targets that were set for the subprocess at the preceding level of design. Each unit performs specific processing functions. A network in a subprocess is represented as a functional block diagram. Conversion targets are set for each of the functions, mass and energy balances are set up, and SHEET evaluations are made.

b. *Knowledge*

The processing functions operate on a material stream to affect a change. The physical state of the material can be characterized by a stream state vector. Table 7.2 presents the common stream states as well as common processing functions to change a specific state. The processing functions can be used to synthesize a network.

Targeting a specific state for a change by means of a processing function often results in another state being changed to some extent, too. This latter change may need to be undone by exerting another processing function in parallel. An example is keeping the reactor temperature constant by external

TABLE 7.2   Generic state vector of a stream with some common transformation mechanisms

State of a material stream	Processing function to affect a change
1. Thermodynamic phase(s) (G/L/S)	Evaporation/condensation, melting/solidifying, crystallization/dissolving, sublimation
2. Phase ratio	Phase change/decanting/drying
3. Temperature	Heating/cooling
4. Pressure and velocity (flow rate)	Compression/pumping/expansion
5. Chemical identity (species)	Reactions
6. Electrical charge	Reactions
7. Composition per phase	Reactions/mixing/diffusional separation
8. Particle/droplet/bubble size	Milling, grinding/dispersion/agglomeration/ coalescing
9. Mass flow	Mixing, splitting

cooling to compensate for the heat effect of the reactions taking place. The selection of different processing functions and their suitability to actually perform well are often supported by *heuristic rules* that capture practical experience and common sense. This knowledge (heuristics, data, experiences, models) must be obtained from academic textbooks, handbooks, intellectual property (IP) documents such as patents, and commercial information sources.

c. *Synthesis*

When synthesizing the processing functions, a distinction must be made between the order in which the functions are introduced in a design and the sequence by which they appear in a process. The intention is to design the functions in order of decreasing economic impact. Economic impact means the value added to a process stream by a specific operation, not the investment cost of the equipment. For example, converting 98% of the feed into product (change of chemical energy) adds more value than increasing its pressure by a few bars or raising the temperature by some 50 K. Figure 7.9 shows the approximate order of the economic impact of process operations as well as the corresponding sequence of design. This view is quite similar to the "onion diagram" for process design in Smith (2005), where the design procedure is like "reverse peeling": moving from the inside out! One begins with chemical conversions (reactors, fuel cells) to form the target chemicals (or intermediate chemicals in a sequence to arrive at the final chemical(s) for the product(s)). Then physical separations and mechanical operations on gases (compression) and solids (cutting, milling) follow. Finally, heating and cooling by thermal media are considered. With the exception of (expensive) cryogenic cooling, this finds application as a part of physical separation design.

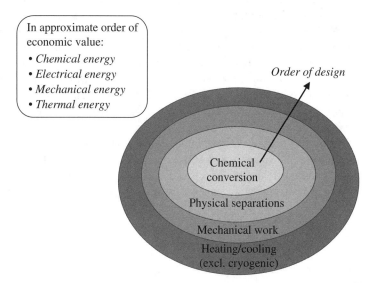

FIGURE 7.9   Order of design of processing functions. (Source: Adapted from Smith (2005), Figure 1.7.)

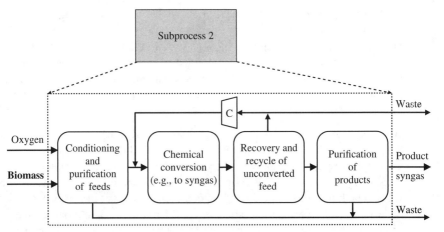

FIGURE 7.10   Expansion of subprocess 2 in Figure 7.8 with sequential order of processing functions.

The chemical conversion, being the heart of the process, can take place only if the reactant feeds are delivered under conditions at which the reactions will actually run. Thus, upstream of a reactor, the feeds must be purified and conditioned for use in the reactor. This purification involves the removal of abrasive and corrosive components, catalyst inhibitors, and solid materials that will accumulate in a reactor (like sand and ash). Downstream of a chemical conversion function, the unconverted reactants can be recovered by physical separations and recycled. *That is, the value of the recovered material must well exceed the cost of separation.* When a recycle occurs, the accumulation of the inert components in the system must be prevented by creating an exit or purge (by venting (G) or bleeding (L)). The product can be purified from side products and is brought to customer quality specs. This generic order of processing functions is shown in Figure 7.10. Here, subprocess block 2 in Figure 7.8 is expanded into a block diagram with a common order of processing activities. Mixing and splitting of streams are essential functions to direct and control the streams in a process.

Subprocess 3 is expanded in Figure 7.11, showing a sequential–parallel ordering.

In subprocess 3, syngas from biomass and from natural gas is partially mixed to match the inlet syngas composition ($H_2 : CO = 2.1$) for the FT conversion to hydrocarbon fuels. The remainder of biomass syngas is fed to a power generator, including some unconverted syngas from the FT synthesis. An alternative is not to recycle the unconverted syngas but combust it directly for power generation.

In summary, the key activities in the synthesis step are to establish:

- Which processing functions are needed per subprocess and in which order
- Targets for duties of the various functions (yield, recovery, and separation factors)
- Links between duties and costs of processing functions, allowing to balance the duties for a good overall (economic) performance of the subprocess

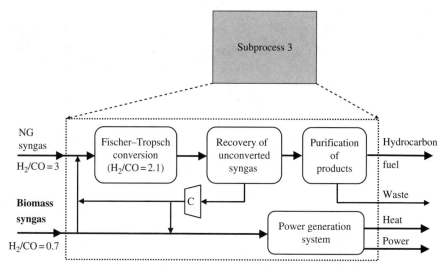

FIGURE 7.11 Expansion of subprocess 3 in Figure 7.8 with sequential and parallel order of processing functions.

- Suitable specific technologies to perform the functions
- Active agents to support processing functions, such as catalysts and solvents
- Mode of operation (continuous, semibatch, etc.) and operating conditions
- Control over levels for inert and trace components by venting and bleeding
- Alternative arrangements in the ordering of units and their connections by streams

In setting targets for the duties of the various functions, economic trade-offs occur. For instance, a decision to have a low conversion of the feed per pass through a reactor will lead to a smaller reactor and a lower investment cost. However, this decision will often give rise to much higher costs in separation for the recovery of the unconverted feed while attaining high product purity. Increasing the conversion and product yield in the reactor will increase the reactor cost but may greatly decrease separation costs and so lower the overall cost.

d. *Analysis*

Having synthesized a functional block diagram, a model can be developed to obtain the flow of resources and analyze how the flows change when design variables are altered. The process model is made up of the models of the individual blocks, connected by the streams as in the functional block diagram. The models of the functional blocks are of the input–output type. One cannot go beyond the input–output type of model because at this level of design, the inner structure of the process functional units is not known yet. It is only the target for a duty that is set per functional block, such as a desired degree of conversion of a key reactant. Thus, for each block, one can easily set up the species balances and determine the main heat effects associated with chemical reactions and phase changes. Performing rigorous energy balancing requires a lot of physical properties to determine the species enthalpies. In the latter case, one better uses a standard process flow sheet simulation program, e.g., ASPEN®, (tinyurl.com/ mfeymzk) to solve coupled mass and energy balances and determine pressure losses over process units.

e. *Evaluation*
At this level of design, the economic potential function can be refined to include estimates of investment costs for the function blocks and utility costs for balancing the heat effects (cooling and heating requirements at specified temperature levels) and for compression duties (energy). The investment costs can be expressed as capital charge estimates, scaled with throughput through the function blocks. It cannot be based yet on volume or size, as these quantities are not available at this level of design. In the next design level, it will be attempted to decrease the utility costs by heat, solvent, and water integration. Capital cost estimates can be obtained from cost estimation software for conceptual (front-end) process design, e.g., ASPEN Capital Cost Estimator®, (tinyurl. com/nhw4msc) and from books of professional engineering organizations, as issued by Dutch Association of Cost Engineers (DACE) (DACE, 2012) or IChemE (Gerrard, 2000). Parallel to an economic evaluation, one can analyze over a sequence of process units where the main losses occur in atom and energy efficiencies.

f. *Selection and (g) Documentation*
The generated alternatives are documented, and the better ones are selected and propagated to the next level of design, dealing with process integration for optimal use of common resources, such as thermal heat, solvents, and water, in the process. The process integration step is required before one can start to design the process units. It is necessary to know which resources are in excess or short supply in the processing functions and can be exchanged with other processing functions or the utility system. Such mass and heat integration in a process has often just as much impact on process economy and ecology as an improved detailed design of an individual process unit.

7.9 EXAMPLE OF ANALYSIS AND EVALUATION IN PROCESS DESIGN

A FT process is used as an example to perform an analysis and an evaluation of a design at level 3. For the analysis, one needs:

i. A physical process model, to compute flows of species and some ecological and technological performance indicators
ii. A closure of this model by means of design specifications
iii. An economic process model to compute the economic potential at level 3

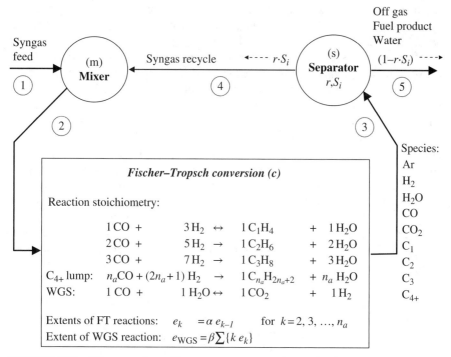

FIGURE 7.12 FT conversion with syngas recovery and recycle; the process streams are uniquely labeled 1–5.

The evaluation involves an assessment of the computed economic, ecological, and technological performance indicators.

The FT conversion unit is related to subprocess 3 and covers also the recovery and recycle of the unconverted syngas. The pertaining block diagram is shown in Figure 7.12.

The streams and the function blocks in this diagram are all uniquely labeled. Also, the FT reaction stoichiometry is shown as well as how the syngas recovery factors are defined. The compressor for the recycle of the unconverted syngas is not shown as it does not affect the mass balances, though it has definitely an energy cost. The following nine species are accounted for in the analysis model: Ar, H_2, H_2O, CO, CO_2, CH_4, C_2H_6, C_3H_8, and a "lump" C_{4+}. The use of a "lump" to collectively represent a range of species needs some explanation. In reality, the hydrocarbon species range from C_1 to ~C_{200} with a geometric distribution. The molar quantities of hydrocarbon species (C_k) with a chain length of k carbon atoms can be given as $C_{k+1} = \alpha\, C_k$, with $k > 0$. The chain growth probability, α, is of the order 0.7–0.95. The actual value of this factor can be influenced by the design of the catalyst and operating conditions in the reactor. The C_{4+} lump can be obtained as the molar sum over all species ranging from C_4 up to C_∞. The lump is represented by a single "averaged" pseudocomponent: $C_{n_a}H_{2n_a+2}$. The averaged carbon number n_a of this pseudocomponent is obtained by weighting of the relative molar amounts of these species over the chain length distribution. It is just for the sake of computational simplicity that in this example only four hydrocarbon species are considered, C_1–C_3 and the C_{4+} lump in the form of $C_{n_a}H_{2n_a+2}$ rather than having the full range of species (>100).

A remark is due regarding simplifications to be made here with respect to the FT conversion modeling. At this functional level of design (level 3), one can still avoid detailed mechanistic modeling of chemical equilibrium and rate limited reactions. Instead, some practically sound assumptions (based on prior experience with laboratory experiments and existing process plants) can be made about ranges for the degree of conversion of CO and the fraction of the converted CO ending up as CO_2. The effect of such assumptions is the replacement of nasty, nonlinear reaction equilibrium equations for the methanation and WGS reactions by much simpler (bi)linear expressions. A mechanistic approach to FT reaction kinetics is discussed in Chapter 17.

The resulting simplified model is based on molar balances for all species involved, using empirical estimates for the expected degrees of conversions. The model equations are shown in the following, aiming to make a clear distinction between the species balances, rates of change by reactions and separations, and design targets imposed by the designers.

1. *Molar Balances with Flows and Extents of Reactions in the FT Conversion Block*

   Equation 7.4 shows the molar balances for all species, including the extents of the reactions, with φ the molar flow rate. The hydrocarbon lump C_{4+}, covering the species range C_4–C_∞, is the main fuel product:

$$
\begin{aligned}
&\varphi_{2,\text{Ar}} & & & & & -\varphi_{3,\text{Ar}} &= 0\\
&\varphi_{2,\text{H}_2} & -3e_1^{(c)} - 5e_2^{(c)} - 7e_3^{(c)} - (2n_a+1)\widetilde{e}_{4+}^{(c)} & +e_{\text{WGS}}^{(c)} & & -\varphi_{3,\text{H}_2} &= 0\\
&\varphi_{2,\text{H}_2\text{O}} & + e_1^{(c)} + 2e_2^{(c)} + 3e_3^{(c)} + & n_a\widetilde{e}_{4+}^{(c)} - e_{\text{WGS}}^{(c)} & & -\varphi_{3,\text{H}_2\text{O}} &= 0\\
&\varphi_{2,\text{CO}} & - e_1^{(c)} - 2e_2^{(c)} - 3e_3^{(c)} - & n_a\widetilde{e}_{4+}^{(c)} - e_{\text{WGS}}^{(c)} & & -\varphi_{3,\text{CO}} &= 0\\
&\varphi_{2,\text{CO}_2} & & +e_{\text{WGS}}^{(c)} & & -\varphi_{3,\text{CO}_2} &= 0\\
&\varphi_{2,\text{CH}_4} & + e_1^{(c)} & & & -\varphi_{3,\text{CH}_4} &= 0\\
&\varphi_{2,\text{C}_2} & + e_2^{(c)} & & & -\varphi_{3,\text{C}_2} &= 0\\
&\varphi_{2,\text{C}_3} & + e_3^{(c)} & & & -\varphi_{3,\text{C}_3} &= 0\\
&\varphi_{2,\text{C}_{4+}} & + & \widetilde{e}_{4+}^{(c)} & & -\varphi_{3,\text{C}_{4+}} &= 0
\end{aligned}
\qquad\text{(Eq. 7.4)}
$$

The molar quantities of hydrocarbon species (C_k) with a chain length of k carbon atoms satisfy a geometric distribution: $C_{k+1} = \alpha\,C_k$, with $k > 0$ and $0 < \alpha < 1$. The composition of the hydrocarbon lump (C_{4+}) is characterized by two averaged variables:

(1) Average number of carbon atoms : $n_a = (4-3\alpha)/(1-\alpha)$

(2) Average molecular mass : $MW_{C_{4+}} = 14n_a + 2$

$\qquad\qquad\qquad\qquad\qquad\qquad\qquad\qquad\qquad$ (Eq. 7.5)

This geometric distribution also governs the relationships between the extents $e_k^{(c)}$ of the chemical reactions $C_k \rightarrow C_{k+1}$:

$$e_2^{(c)} = \alpha e_1^{(c)}$$

$$e_3^{(c)} = \alpha^2 e_2^{(c)}$$

$$\tilde{e}_{4+}^{(c)} = [\alpha^3/(1-\alpha)]e_1^{(c)}$$

$$e_{WGS}^{(c)} = \beta\left(e_1^{(c)} + 2e_2^{(c)} + 3e_3^{(c)} + n_{C,a}\tilde{e}_{4+}^{(c)}\right) \qquad \text{(Eq. 7.6)}$$

The chain growth probability parameter α and the CO-to-CO$_2$ conversion factor β are fixed inputs in this simplified model: $0.70 < \alpha < 0.95$ and $0.02 < \beta < 0.10$.

The set of equations (7.4–7.6) counts 25 variables and 15 equations. Assuming that the nine inlet flows are obtained from an upstream unit in the flow sheet, there is 1 degree of freedom remaining. It is customary for any conversion block in a flow sheet to specify the degree of conversion of one of the main reactants. In the FT conversion case, we choose the degree of conversion η of carbon monoxide (CO):

$$(1-\eta)\varphi_{2,CO} - \varphi_{3,CO} = 0 \quad \text{with } 0.25 < \eta < 0.75 \qquad \text{(Eq. 7.7)}$$

For the conversion factor η, one has to specify a target value within a preset range as an input to a design computation. The model is mathematically well posed. If model parameters α and β, the conversion target η, and the inlet flows $\{\varphi_2,*\}$ are given, the set of equations (7.4–7.7) suffices to determine the extents of the reactions $\{e\}$, the averaged composition $\{n_a, MW_a\}$ of the fuel product, and the outlet flows $\{\varphi_3,*\}$.

2. *Molar Balances and Separation Factors for Syngas Recovery and Recycle*
The effluent from the FT conversion unit is separated into a fuel product stream (C_{4+}), a water stream, a recycle syngas, and a gas purge. A purge is always needed to vent any inert components entering the process system to avoid infinite build-up. Two operations are combined in the separator block of Figure 7.12: a physical separation into liquids (some C_2 and C_3 and all C_{4+} and the bulk of the water) and gas. The gas stream is immediately split into a syngas recycle stream (fraction r) and a purge stream (fraction $1 - r$).

Equation 7.8 gives the molar balances and recycle of all species i:

Molar balances for all species i ($i = 1, \ldots, 9$): $\varphi_{3,i} - \varphi_{4,i} - \varphi_{5,i} = 0$

Recovery of species i for recycle to conversion: $rS_i^{(s)} \varphi_{3,i} - \varphi_{4,i} = 0$ (Eq. 7.8)

The separation parameters $\{0 < S_i^{(s)} < 1\}$ indicate the separation fractions of species i for recycle. The recycle factor $\{0 \le r < 1\}$ indicates the fraction actually recycled; the remainder is purged. Some separation parameters are scalable by a separation technology factor λ (see Equation (7.9)). This factor is a design variable—$0.9 < \lambda < 1.1$:

Gas: $\qquad S_{Ar}^{(s)} = S_{H_2}^{(s)} = S_{CO}^{(s)} = S_{CO_2}^{(s)} = S_{CH_4}^{(s)} = 1.0$

"Light" condensables: $\quad S_{C_2H_6}^{(s)} = 0.90\lambda; S_{C_3H_8}^{(s)} = 0.80\lambda$

"Heavy" condensables: $\quad S_{C_{4+}}^{(s)} = 0.10/\lambda; S_{H_2O}^{(s)} = 0.25/\lambda$ $\qquad$ (Eq. 7.9)

There are $4 \times 9 + 2 = 38$ variables and 27 equations in these sets (Equations (7.8) and (7.9)). The inlet flows for nine species are determined by the supply from an upstream unit. Hence, there are two design degrees of freedom (r, λ) in the model of the separator unit for syngas recovery.

3. *Molar Balances for the Mixer*
   The molar balances for the mixer are

   $$\varphi_{1,i} + \varphi_{4,i} = \varphi_{2,i} \qquad (i = 1, 2, \ldots, 9) \qquad \text{(Eq. 7.10)}$$

   and no design targets are needed; the flows entering the mixer are determined by their respective upstream units. Hence, the number of flow variables and equations match, leaving no design degree of freedom.

4. *Feed Composition Specification*
   The molar fractions of feed stream (1) are given by

   $$x_{1,i}\varphi_{1,\text{total}} - \varphi_{1,i} = 0 \qquad (i = 1, \ldots, 9)$$

   $$\sum_{i=1}^{9} \varphi_{1,i} - \varphi_{1,\text{total}} = 0 \qquad \text{(Eq. 7.11)}$$

   The feed composition is given by

   $$x_{1,Ar} = 0.02; x_{1,H_2} = 0.62; x_{1,H_2O} = 0.02; x_{1,CO} = 0.31; x_{1,CO_2} = 0.01;$$

   $$x_{1,CH_4} = 0.02; x_{1,C_2H_6} = 0.00; x_{1,C_3H_8} = 0.00; x_{1,C_{4+}} = 0.00$$

   $$\text{(Eq. 7.12)}$$

   There are only eight independent values because of normalization to unity. There are 19 variables and 18 equations in the sets of equations (7.11 and 7.12), leaving 1 degree of freedom. One may wonder if the total feed flow rate should not be specified. In process design, there are two options to specify process capacity: either the (nominal) intake rate of the main feed or the (nominal) production rate of the main product. The latter option is chosen here.

5. *Nominal Liquid Fuel Production Rate*
The mass-based production rate of the liquid fuel (C_{4+}), $P_{5,C_{4+},mass}$, is chosen as an input design parameter of this FT process. The relation between the mass ($kg \cdot s^{-1}$) and molar ($kmol \cdot s^{-1}$) liquid fuel production rate is given by

$$F_{5,C_{4+},mass} = \varphi_{5,C_{4+}} MW_{C_{4+}} \qquad \text{(Eq. 7.13)}$$

The mass fuel production rate $F_{5,C_{4+},mass}$ is a design input parameter.

6. *Nominal Values for Design Parameters and Production Rate*
Nominal values must be given as inputs to the model for the chain growth probability (α), the CO-to-CO_2 conversion factor (β), the single-pass CO conversion in the FT reactor (η), the recycle fraction (r), the separation technology factor (λ), and the C_{4+} production rate, $F_{5,C_{4+}}$ (see Table 7.3).
   The CO conversion per pass must be kept relatively low due to an operational limitation on the formation of water. Too high, a syngas conversion could result in the production of excessive amounts of water, hampering the catalytic conversion.

7. *Ecological Performance Indicators and Technological Constraints*
In a real process plant, there is one or more ecological performance indicators and often many technological constraints. Here, we will consider only one of each. In this problem, the ecological performance is represented by the carbon (atom) efficiency, indicating how much of the carbon intakes through the feeds end up in the main product. The carbon efficiency ε_C is defined as

$$\varepsilon_C = \frac{\text{carbon flow OUT}}{\text{carbon flow IN}}$$

$$\varepsilon_C \left[\varphi_{1,CO} + \varphi_{1,CO_2} + \varphi_{1,C_1} + 2\varphi_{1,C_2} + 3\varphi_{1,C_3} \right]$$
$$- \left[\varphi_{5,C_1} + 2\varphi_{5,C_2} + 3\varphi_{5,C_3} + n_a\varphi_{5,C_{4+}} \right] = 0 \qquad \text{(Eq. 7.14)}$$

The technological constraint is the molar fraction of water in the gas-phase outlet stream of the FT conversion block (stream 3). For simplicity, a simple

TABLE 7.3 Nominal input values

Design parameter	Value
Chain growth probability, $\alpha(-)$	0.90
Conversion of CO to CO_2, $\beta_{nom}(-)$	0.050
CO conversion per pass, $\eta_{nom}(-)$	0.50
Syngas recycle fraction, $r_{nom}(-)$	0.90
Separation technology factor, $\lambda_{nom}(-)$	1.0
Production rate of liquid fuel (C_{4+} lump):	
$F_{5,C_{4+}}$ ($kg \cdot s^{-1}$)	9.2
$\varphi_{5,C_{4+}}$ ($kmol \cdot s^{-1}$)	0.050

upper bound is imposed on this molar fraction. This bound puts a ceiling on syngas conversion by limiting the amount of reaction water being formed. In a mechanistic reactor model, fundamental thermodynamic equations are needed to model the gas–liquid-phase equilibrium with possible condensation of water. The simplified water constraint is given as

$$x_{3,H_2O} \sum_{i=1}^{9} \varphi_{3,i} - \varphi_{3,H_2O} = 0$$

$$x_{3,H_2O} \le x_{3,H_2O}^{(max)} \quad \text{with } x_{3,H_2O}^{(max)} = 0.22 \quad (= \text{input data}) \qquad \text{(Eq. 7.15)}$$

8. *Check on Overall Model Consistency: Square, Full-Rank Set of Equations*
   The process model contains 80 variables and 80 equations. The count of variables arises from the following: 5 streams × 9 species flows (φ) + 5 extents (e) + 2 product variables (n_a, MW_a) + 9 separation factors, S + 10 (feed fractions x + total feed flow) + 5 design factors ($\alpha, \beta, \eta, r, \lambda$) + 2 product flows (mass, molar) + 2 performance variables (x_{H_2O}, ε_C). The 80 equations are the sum of 16 equations (conversion block related) + 27 (separator block) + 9 (mixer) + 18 (feed specifications) + 1 (production rate) + 2 (performance) + 6 design specifications for $\alpha, \beta, \eta, r, \lambda$, and $F_{5,C_{4+}}$. There are no linear dependencies between the equations, so the system of equations can be solved. The model has many linear equations and a few bilinear ones. The bilinear equations reduce to linear ones if the given numerical values for the physical parameters in Equation (7.6), for the separation factors in Equation (7.9), and for the molar fractions in Equation (7.12) are substituted. One may solve the resulting linear model ($A.\underline{x} = \mathbf{b}$, where square matrix A has full rank) in parallel. A sequential approach can start with solving the CO balances and the CO conversion equation (7.7), enabling to determine all extents of reactions and thus solving the other species balances.

9. *Analysis of a Base Case Design: Species Flows in a Stream Table*
   The specifications for the base case design variables are given in Table 7.3.
   Table 7.4 specifies the reaction conditions and performance variables. The resulting flows of the main species (reactants and products) are given in a stream

TABLE 7.4   Reaction conditions and performance variables for FT process

Reaction conditions	
Average carbon number of C_{4+} lump (–)	13.0
Average molecular weight of C_{4+} lump (kg·kmol^{-1})	184.0
Extent of reaction to C_1 (kmol·s^{-1})	0.00686
Extent of reaction to C_2 (kmol·s^{-1})	0.00617
Extent of reaction to C_3 (kmol·s^{-1})	0.00556
Extent of reaction to C_{4+} lump (kmol·s^{-1})	0.05000
Extent of WGS reaction to CO_2 (kmol·s^{-1})	0.03429
Performance variables	
Water fraction in reactor outlet stream 3 (–)	0.20056
Carbon atom efficiency (–)	0.75918

TABLE 7.5   Stream table for FT process

Stream (flows: kmol·s^{-1})	CO	H$_2$	Main product (C$_{4+}$ lump)	H$_2$O	CH$_4$	CO$_2$
1. Fresh feed	0.792	1.584	0.000	0.051	0.077	0.026
2. Reactor in	1.440	2.572	0.005	0.225	0.828	0.564
3. Reactor out	0.720	1.097	0.055	0.907	0.835	0.598
4. Recycle	0.648	0.988	0.005	0.204	0.752	0.539
5. Outlet	0.072	0.110	0.050	0.703	0.084	0.060

table (Table 7.5). The stream index corresponds with the numbering of the streams in Figure 7.12.

The total flow into the conversion block (stream 2) is more than twice as large as the fresh feed intake (stream 1). The molar product (C$_{4+}$) flow is very small (0.05) relative to water (0.703). However, the C$_{4+}$ lump has a large average molecular weight, MW$_{C_{4+}}$ = 184, compared to that of water (18). Thus, on a mass basis, the outlet flows are comparable in size: 9.2 kg·s^{-1} C$_{4+}$ and 11.2 kg·s^{-1} H$_2$O.

10. *Closing Remarks on the Analysis Model*

The molar balance model discussed is an input–output one, primarily intended to get an overview of the order of magnitude of the species flows in the streams and the suitability of the target variables, such as conversion and separation factors. Such a simplification comes at a price: an inherent weakness of this input–output model is the lack of internal mechanisms. For example, the conversion and extents of reaction have not yet been mechanistically obtained from the internal rates and operating conditions in the reactor. These missing links can be introduced at a deeper level of design. At this level, first principle models are used to determine internal reactor geometry and conditions, such that the target conversion and selectivity are closely met. Another weakness is the absence of energy and momentum balances. However, the rate of change in enthalpy due to all reactions, which is a major contribution to an enthalpy balance, can easily be computed. One takes for each reaction the product of its extent and its corresponding standard reaction enthalpy and sums up over all reactions:

$$R_{\Delta H^0}^{(\text{conv})} = \sum_{j=1}^{n_{\text{reactions}}} \left[e_j^{(\text{conv})} \Delta_r H_j^0 \right] \qquad \text{(Eq. 7.16)}$$

A simplified momentum balance would cover pressure changes over the process units. Pumps and compressors are needed to overcome pressure drops in process units. The amount of work done by compressors is often another dominant term in the energy balance over the process.

11. *Evaluation of Economic, Ecological, and Technological Performances*

For the evaluation of economic performance, a simplified economic potential function is formulated: EP$_3$, the economic potential function at design level 3.

EP_3 is expressed per hour being a more practical scale for economics than per second. This EP_3 function accounts for the earnings from the products, the cost of the feed, as well as the capital costs:

$$
\begin{aligned}
EP_3 &= P_{\text{products}} - C_{\text{feed}} - C_{CO_2} - C_{\text{cap,h}} \quad (\text{in } \$ \, \text{h}^{-1}) \\
P_{\text{products}} &= P_{\text{liq.fuel}} + P_{\text{offgas}} + P_{\text{water}} \\
P_{\text{liq.fuel}} &= 3600 \, p_{C_{4+}} \varphi_{5,C_{4+}} \\
P_{\text{offgas}} &= 3600 \left[c_{\text{syngas}} \left(\varphi_{5,H_2} + \varphi_{5,CO} \right) + p_{\text{offgas}} \left(\varphi_{5,C_1} + \varphi_{5,C_2} + \varphi_{5,C_3} \right) \right] \\
P_{\text{water}} &= 3600 \, p_{H_2O} \varphi_{5,H_2O} \\
C_{\text{feed}} &= 3600 \, c_{\text{syngas}} \varphi_{1,\text{total}} \\
C_{CO_2} &= 3600 \, c_{CO_2} \left(\varphi_{5,CO_2} - \varphi_{1,CO_2} \right) \\
C_{\text{cap,h}} &= ccf \, I_{\text{cap}} / t_{\text{prod}} \quad \quad\quad\quad\quad\quad\quad\quad\quad\quad (\text{Eq. 7.17})
\end{aligned}
$$

The capital investment is built up of the sum of the investments in the process blocks. The investment for a process block scales with the state of its technology, as expressed by a relevant design parameter (η, λ), and the throughput φ with a power law. Both technology and throughput are scaled with respect to a set of nominal conditions:

$$
I_{\text{cap}} = I_{\text{actual}}^{(\text{conv})} + I_{\text{actual}}^{(\text{sep})} + I_{\text{actual}}^{(\text{mix})}
$$

with:

$$
I_{\text{actual}}^{(\text{conv})} = I_{\text{nom}}^{(\text{conv})} \left(\frac{\varphi_{\text{actual}}^{(\text{conv})}}{\varphi_{\text{nom}}^{(\text{conv})}} \right)^q \left(\frac{\eta_{\text{nom}}}{\eta} \right)^q
$$

$$
I_{\text{actual}}^{(\text{sep})} = I_{\text{nom}}^{(\text{sep})} \left(\frac{\varphi_{\text{actual}}^{(\text{sep})}}{\varphi_{\text{nom}}^{(\text{conv})}} \right)^q \left(\frac{\lambda_{\text{nom}}}{\lambda} \right)^q
$$

$$
I_{\text{actual}}^{(\text{mix})} = I_{\text{nom}}^{(\text{mix})} \left(\frac{\varphi_{\text{actual}}^{(\text{mix})}}{\varphi_{\text{nom}}^{(\text{mix})}} \right)^q \quad\quad\quad\quad (\text{Eq. 7.18})
$$

The technology parameter for the conversion block is the degree of CO conversion (η), while the separation scaling factor (λ) serves a similar goal for the separation block. The mixer block has no technology parameter assigned. The structure of these equations properly reflects trends in investment costs as a function of technology and throughput. Improvement in technology leads to smaller investment costs for the same throughput.

The numerical values of the economic and investment parameters in Table 7.6 are just estimates having a right order of magnitude. The investment cost correlations in confidential use in industry are more refined than ours. The parameter values in real design cases will vary, depending on market conditions and location.

The economic and ecological performances are presented for two different CO conversion levels (base case, $\eta = 50.0\%$; enhanced case, $\eta = 54.0\%$ per pass). Also the technological constraint on water content is considered. All model parameters other than the degree of CO conversion and the fuel production capacity are kept constant. The results obtained by solving the process and economic models are given in Table 7.7.

TABLE 7.6   Data for economic parameters

Parameter	Value	Parameter	Value
ccf (year^{-1})	0.20	t_{prod} $(\text{h} \cdot \text{year}^{-1})$	8000
$p_{C_{4+},\text{mass}}$ $\left(\$ \text{ kg}_{C_{4+}}^{-1}\right)$	1.00	$MW_{C_{4+}}$ $(\text{kg} \cdot \text{kmol}^{-1})$	184
$p_{C_{4+}} = p_{C_{4+},\text{mass}} MW_{C_{4+}}$ $\left(\$ \text{ kmol}_{C_{4+}}^{-1}\right)$	184	c_{syngas} $\left(\$ \text{ kmol}_{syngas}^{-1}\right)$	1.00
p_{offgas} $\left(\$ \text{ kmol}_{C_{1-3}}^{-1}\right)$	5.00	c_{CO_2} $\left(\$ \text{ kmol}_{CO_2}^{-1}\right)$	0.500
p_{H_2O} $\left(\$ \text{ kmol}_{H_2O}^{-1}\right)$	0.0200	η_{nom} $(-)$	0.500
$I_{nom}^{(conv)}$ $(\text{M}\$)$	300	λ_{nom} $(-)$	1.00
$I_{nom}^{(sep)}$ $(\text{M}\$)$	200	$\varphi_{nom}^{(*)}(\text{kmol} \cdot \text{s}^{-1})$	8.00
$I_{nom}^{(mix)}$ $(\text{M}\$)$	50	$q(-)$	0.600

Note that a cost penalty for CO_2 emissions is included.
*conv., sep., mixer.

TABLE 7.7   Performance results for a FT conversion process

	Economy				
Contributions to EP$_3$ ($\$ \text{ h}^{-1}$)	$\eta = 50.0\%$	$\eta = 54.0\%$	Investments $(10^6 \text{ US } \$)$	$\eta = 50.0\%$	$\eta = 54.0\%$
$P_{products}$	35,539	35,405	$I^{(conv)}$	251	231
C_{feed}	9,199	9,075	$I^{(sep)}$	142	135
C_{CO_2}	61.7	61.7	$I^{(mix)}$	41.9	40.5
$C_{cap,h}$	10,884	10,195	I_{cap}	435	407
EP$_3$	15,393	16,072	Ann. Cap. Cost	87.1	81.6
Ecology			Technological constraint:		
Carbon atom efficiency	0.759	0.770	Water fraction in stream 3	0.201	0.217

The economic performance, expressed by the economic potential (EP_3), accounts for the cost of capital investment. The ecological performance is the carbon atom efficiency as defined in Equation (7.14). The water fraction in the effluent from the FT conversion block reactor(s) is taken as a technological constraint specified in Equation (7.15).

When comparing the outcomes of the enhanced conversion case with the base case results, one sees that the higher CO conversion allows for slightly lower syngas intake to still make the same amount of liquid product (C_{4+}). This reduction results in smaller flows through the process and thus in smaller equipment. Consequently, the costs of feed and investments go down for all process units: conversion, separator, and mixer. It is seen from the data in Table 7.4 that the economic performance increases with conversion but less than proportional. The carbon atom efficiency improves relative to the base case but only marginally. The technological constraint (water fraction) is nearly hitting its upper bound (0.22; see Equation (7.15)). This implies that CO conversion cannot be increased much beyond 54% for this production rate. The conclusion of this evaluation is that economic performance and carbon atom efficiency both become somewhat better when moderately increasing the CO conversion. The ceiling for CO conversion is reached by hitting a technological constraint.

7.10 INTEGRATING PROCESS UNITS INTO THE FUNCTIONAL NETWORK

The design levels discussed so far (1–3) have resulted in optional networks of required processing functions jointly forming the process, as well as in targets for the functional duties. Design level 4 deals with process integration of heat, solvents, and water, between process units mutually and with the utility system. Then, at level 5, the design focus will shift to the internal structuring of process units with the associated equipment engineering aspects. This activity requires much insight into the nature of various existing specific processing technologies, equipment types, and the integration of a process unit into the process. The next part of this book extensively covers processing technologies for biomass as well as design aspects of process units. Development and design of process units must take place within the overall frame of a process. As a prelude to the integration of units into a process during design, one must ask a pertinent question:

> What kind of information does a process designer need about a process unit to be able to properly integrate it into the functional network?

The range of conditions under which a process unit must be able to operate must lie within its window of feasible operation. In other words, can the window(s) of feasible operation of a process unit be matched with the windows of other units in the network? Figure 7.13 shows some windows of unit variables that require attention when integrating a unit into a process design. The boundaries of the windows of a process unit are ideally provided as mathematical inequality constraints in a design model of a process unit. We will now briefly discuss the conditions represented in Figure 7.13.

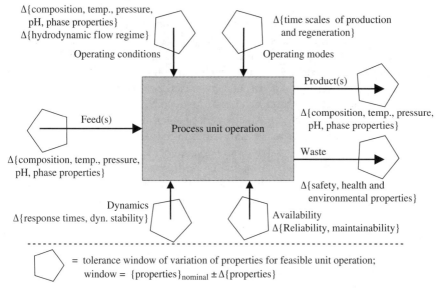

FIGURE 7.13 Windows of feasible operation of a processing technology.

1. *Chemical Species in Feeds and Products*
   First of all, the "knockout" items need to be considered: which species cannot be tolerated at all by the specific processing technology? Such species are, e.g., specific catalyst poisons or aggressive chemicals that attack membrane or containment materials. Which changes in composition, temperature, pressure, and pH can be tolerated? In case of multiphase streams, one needs tolerances with respect to changes in phase ratio and the typical size of the dispersed phase (droplets, bubbles, particles).

2. *Waste and SHE Aspects*
   Safety, health, and environmental (SHE) issues should be mentioned as well as the expected variability in waste, both in quantity and quality.

3. *Operating Conditions*
   Concerning the internal physical state of the process unit, a process designer needs to know the allowable ranges for temperature, pressure, and pH. For a multiphase system, the minimum and maximum phase ratios are required, as well as the preferred hydrodynamic flow regime with restrictions to stay in that regime.

4. *Operating Modes*
   Many process units run in different operating modes over time, covering manufacturing (either continuous or batchwise), regeneration of the active agents, and cleaning. Designers need to know the lifetime expectancy of active agents (e.g., catalysts) as well as the duration of nonmanufacturing modes, like regeneration and cleaning, with the associated procedures and means.

5. *Time Scales and Dynamics*

The dynamic behavior of a process must match the dynamics of the supply chain and be robust against external disturbances. If demand and supply change with, e.g., a typical time constant of one week, a process should be able to respond to these changes within a much shorter time (say, 1 day). For processes delivering power to the power grid, the response time to load changes can be much shorter. It is important to know the dynamics of a process unit, particularly the response times of its outputs to changes in inputs as well as the occurrence of any inverse responses. In addition, also from a safety point of view, it is essential to have information on the (dynamic) stability of a process unit. For example, how far are normal operating conditions removed from *runaway* conditions, where positive feedback mechanisms in the unit, such as heat release by chemical reactions, may cause temperature and pressure to escalate to dangerous levels?

6. *Availability*

Availability is the fraction of time over a specified time horizon a process unit is capable of fulfilling its nominal processing functions. This availability depends both on the reliability of the equipment (mean time to failure) and the maintenance efforts to repair a unit that broke down (mean time to repair). The reliability of a unit or of its components must be known to determine the availability of the entire process. Acceptance in a design of new process equipment with improved processing characteristics also requires sufficient reliability.

7.11 APPLICATION POTENTIAL

Process design requires understanding domain contents (e.g., physical, chemical, and equipment knowledge) as well as a procedure to structure the design decision making. The next chapters in this book offer mainly domain contents. However, the "problems" section at the end of some chapters will also ask questions about some underlying design decisions and trade-offs in developing process unit concept(s), in accordance with the design approach presented in this chapter. For instance, Chapter 17 on the use of syngas for the production of synthetic transportation fuels presents design related questions on FT synthesis.

CHAPTER SUMMARY AND STUDY GUIDE

This chapter introduces a top-down hierarchical, multilevel approach to decision making in process design. The design approach starts with the way of embedding a process into a supply chain, as represented by an input–output diagram. Then, the process can be split in subprocesses with intermediate feeds and products, and for each subprocess, a network of processing functions is generated and presented in a function block diagram. The next design step is to turn the network with processing functions into a process block diagram showing the required processing units. To ensure a successful integration of process units into a flow sheet, some specifications are formulated regarding the required design information and models for these units. At each design level, designers can follow the same sequence of generic steps, involving setting of design specs, synthesis of structure, analysis of physical behavior, and evaluation of performance. FT synthesis of liquid fuels from syngas is presented as a process example.

KEY CONCEPTS

Process design: decision making with a host of choices and many constraints
Decision making: hierarchical by distribution over multiple levels (7) of increasing detail
Generic engineering design steps, applicable at each level
Resulting matrix for conceptual process design
Iterative passes over engineering design steps and levels: often necessary to improve process designs
Input–output and functional block diagrams
Molar balances for a process block diagram with an economic potential function, solved for design analysis and evaluation

SHORT-ANSWER QUESTIONS

7.1 Who are the key stakeholders for a plant design for the conversion of lignocellulosic biomass into chemicals, heat, and power? Make a list of these stakeholders and give several performance criteria examples that each stakeholder will consider important for the design. It is recommended to use this format when making the list: (A) Stake holder category A, (A1) Stakeholder A1, Performance Criterion A1a, Performance Criterion A1b, and so on.

7.2 Give examples of the internal multiscale structures in a process (from enterprise down to the molecule level).

7.3 In the synthesis step of a design, one must identify suitable building blocks and connect them to a network. Give some examples of different types of building blocks and connections for a thermochemical process. What is the kind of knowledge that you would look for to get an understanding of the relevant design features of a block?

7.4 Describe the difference between the analysis of a process, using models, and evaluation of the performance of a process.

7.5 In the hierarchical decomposition approach to process design, the design decision variables are portioned in a hierarchy of smaller clusters (levels). Decisions that are taken at the top level consider those decision variables that have a high impact on the economic and ecological performance indicators. Give a list of these design variables and explain why these have a high impact on the economic performance indicators.

7.6 Explain the differences between the hierarchical design levels and the generic engineering design steps in the Conceptual Design Matrix.

How will the specification of a reactor product yield at the process network level set targets for design at the process unit (i.e., reactor equipment) level?

PROBLEMS

7.1 Pareto Trade-Off between Economy and Ecology in Design

Biomass (B) is converted to a hydrocarbon fuel product (P) and carbon dioxide (CO_2) in a well-mixed continuous-flow reactor, operating at steady state. The nominal chemical composition of B $= [-C_1O_{0.6}H_{1.4}-]_n$ and that of P $= [-C_1H_2-]_n$. The corresponding mass-based reaction stoichiometry is (in close approximation)

$$1 \text{ kg B} \xrightarrow{k} 0.425 \text{ kg P} + 0.575 \text{ kg CO}_2 \text{ with } k = 10^{-3} (\text{s}^{-1})$$

This reaction is the first order in the biomass amount; the symbol k represents the associated first-order reaction rate coefficient. The biomass is only partially converted as full conversion would require a reactor of infinite size and cost. The degree of conversion of the biomass (X) is defined as

$$X = \frac{F_B^{(in)} - F_B^{(out)}}{F_B^{(in)}} \text{ with } 0 \le X < 1$$

The biomass intake rate ($F_B^{(in)}$) and the volume flow (φ_V) are kept constant:

$$F_B^{(in)} = 1 \ (\text{kg}\cdot\text{s}^{-1}); \ \varphi_V = 4.0 \times 10^{-3} \ (\text{m}^3\cdot\text{s}^{-1})$$

The degree of conversion is an independent design variable. The dependent process variables are

$$\text{Outflow of B (kg}\cdot\text{s}^{-1}): \quad F_B^{(out)} = (1-X)F_B^{(in)}$$

$$\text{Outflow of P (kg}\cdot\text{s}^{-1}): \quad F_P^{(out)} = 0.425 \, X F_B^{(in)}$$

$$\text{Outflow of CO}_2 \ (\text{kg}\cdot\text{s}^{-1}): F_{CO_2}^{(out)} = 0.575 \, X F_B^{(in)}$$

$$\text{Reactor volume (m}^3): \quad V_R = \frac{\varphi_V X}{k(1-X)}$$

The process performance is evaluated by means of two different indicators. The ecological indicator is the *carbon atom efficiency* (ε_C). This efficiency is the fraction of carbon atoms in biomass B ending up in product P; carbon left in the unconverted biomass and in CO_2 is "lost":

$$\varepsilon_C = \frac{w_{c,P}F_P^{(out)}}{w_{c,B}F_B^{(in)}} \quad \text{with } w_{c,P} = \frac{12}{12+2\cdot1}; \; w_{c,B} = \frac{12}{12+0.6\cdot16+1.4\cdot1} \quad (-)$$

The *economic performance* is expressed by means of a profit function (P). The profit function is made up of earnings from product sales minus the costs of feed and wastes and the capital charge (cost of capital investment per unit of time). There are trade-offs in this design. A higher conversion increases product output as well as reactor size and cost. The carbon efficiency benefits from a high conversion. The company for which you design this reactor aims for a better balance between ecological and economic performances.

The profit per hour is given by

$$P = \left(F_P^{(out)}p_P - F_B^{(in)}c_B - F_B^{(out)}c_W - F_{CO_2}^{(out)}c_{CO_2}\right) 3600 - C_{cap,h} \quad (\$ \; h^{-1})$$

in which the capital charge, $C_{cap,h}$, is given by

$$C_{cap,h} = \frac{ccf}{t_{prod}}I_{cap} \quad (\$ \; h^{-1})$$

with the capital investment, I_{cap}, given by

$$I_{cap} = I_0 \left(\frac{V}{V_0}\right)^q \quad (\$)$$

Data are given in Table 7.8.

TABLE 7.8 Data for the evaluation of the economic performance

	Value
Price of fuel product, p_P ($ kg^{-1})	0.500
Price of biomass feed, p_B ($ kg^{-1})	0.050
Cost of biomass waste, c_W ($ kg^{-1})	0.010
Cost of CO_2 emissions, c_{CO_2} ($ kg^{-1})	0.005
Capital charge factor (annual), ccf (−)	0.200
Production hours, t_{prod} (h·year^{-1})	8000
Reference investment, I_0 (M$)	6
Reference capacity (for $X_0 = 0.8$), V_0, (m^3)	16.0
Capacity (=volume) scaling factor, q(−)	0.600
Validity range of investment correlation (m^3)	$1 \leq V \leq 200$

Evaluate performance functions ε_C and EP over the conversion interval $0.3 < X < 0.975$ and make a Pareto plot (EP vs. ε_C). What is the design point $\{X^*, \varepsilon_C^*, EP^*\}$ that you would advise to the management as being the optimum and why?

7.2 Analysis of Economic Potentials for Fischer–Tropsch Conversion

Each level of design has its own economic potential function. The definition of economic potential for the input–output level (1) is given by Equation (7.3), along with a feasible range for the profitability margin (0.50–0.75). Taking that range as criterion, has the Fischer–Tropsch base case design presented in Section 7.9 a satisfactory profitability margin? The economic results in Table 7.4 can be used for making an assessment.

The prices of hydrocarbons can be volatile over time. How much (by which fraction) can the product prices approximately drop before the lower limit on the profitability margin is reached while assuming the feed and waste costs remain constant?

7.3 Sensitivities in the Design of the Fischer–Tropsch Conversion

Consider the process block for the Fischer–Tropsch conversion with syngas recycle in Figure 7.12. The physical and economic models in Section 7.9 can be used with the base case feed composition Equation (7.12) and design conditions Equation (7.17). Determine and report for the next four design sensitivity cases with a one-at-a-time change in a design variable:

1. Increase in CO conversion (η: 0.50 $\Rightarrow$ 0.54)
2. Increase in separation efficiency (λ: 1.00 $\Rightarrow$ 1.04)
3. Increase in production rate ($F_{5,C_{4+}}$: 9.20 $\Rightarrow$ 10.0)
4. Increase in recycle fraction (r: 0.90 $\Rightarrow$ 0.94)

the following quantities on an hourly basis:

- The extents of the reactions
- The species flows in the reactor outlet stream (3) and the process outlet stream (5)
- The total molar flow rate of feed stream (1)
- The economic potential (EP_1)
- The capital cost
- The economic potential (EP_3)
- The carbon atom efficiency
- The molar fraction of water in the effluent of the conversion unit (stream 3)

Explain the differences in economic and carbon atom efficiency performances and the reactor effluent water fraction, relative to their base case values.

PROJECTS

P7.1. Take a recently published conceptual design of a biorefinery and analyze this design by going through the first three levels of the design template and generating input–output and functional block diagrams with associated mass balances and performance evaluations.

INTERNET REFERENCES

tinyurl.com/yocw7q

http://en.wikipedia.org/wiki/Pareto_efficiency

tinyurl.com/mfeymzk

http://www.aspentech.com/products/aspen-plus.aspx

tinyurl.com/nhw4msc

http://www.aspentech.com/products/aspen-icarus-process-evaluator.aspx

REFERENCES

Biegler LT, Grossmann IE, Westerberg AW. *Systematic Methods of Chemical Process Design.* Upper Saddle River (NJ): Prentice Hall; 1997.

DACE. *DACE Prijzenboekje.* 29th ed. Den Haag (the Netherlands): SDU Uitgevers; 2012.

Douglas JM. *Conceptual Design of Chemical Processes.* New York: McGraw-Hill; 1988.

Gerrard AM. *Guide to Capital Cost Estimation.* 4th ed. England: IChemE; 2000.

Grossmann IE, Westerberg AW. Research challenges in process systems engineering. AIChE J 2000;46:1700–1703.

Olofsson I, Nordin A, Söderlind U. Initial review and evaluation of process technologies and systems suitable for cost-efficient medium-scale gasification for biomass to liquid fuels. Technical report for University of Umeå. Umeå (Sweden): Energy Technology & Thermal Process Chemistry, University of Umeå; 2005.

Peters MS, Timmerhaus KD, West RE. *Plant Design and Economics for Chemical Engineers.* 5th ed. New York: McGraw-Hill; 2003.

Smith R. *Chemical Process Design and Integration.* Hoboken (NJ): John Wiley & Sons, Inc.; 2005.

Stuart PR, El-Halwagi MM, editors. *Integrated Bio-Refineries, Design, Analysis, and Optimization.* Boca Raton (FL): CRC Press; 2012.

Tassoul M. *Creative Facilitation.* Delft (the Netherlands): VSSD; 2009.

PART III

BIOMASS CONVERSION TECHNOLOGIES

8

PHYSICAL PRETREATMENT OF BIOMASS

Wiebren de Jong

Department of Process and Energy, Energy Technology Section, Faculty of Mechanical, Maritime and Materials Engineering, Delft University of Technology, Delft, the Netherlands

ACRONYMS

ar	as received basis
daf	dry and ash-free basis
HGI	Hardgrove grindability index
LHV	lower heating value
SRF	short rotation forestry
TOP	torrefaction and pelletizing process
VOC	volatile organic compound

SYMBOLS

A	area across which heat/mass transfer takes place	$[m^2]$
c_p	specific heat capacity	$[J{\cdot}kg^{-1}{\cdot}K^{-1}]$
C	constant in Equation (8.1)	various
D	diffusion coefficient	$[m^2{\cdot}s^{-1}]$
d	diameter	$[m]$
d_p	particle diameter	$[m]$
E	specific energy consumption	$[kJ{\cdot}kg^{-1}]$ or $[kWh{\cdot}kg^{-1}]$

Biomass as a Sustainable Energy Source for the Future: Fundamentals of Conversion Processes,
First Edition. Edited by Wiebren de Jong and J. Ruud van Ommen.
© 2015 American Institute of Chemical Engineers, Inc. Published 2015 by John Wiley & Sons, Inc.

f	driving force for drying ($= w - w_e$)	[kg]
h	mass-specific enthalpy	[J·kg^{-1}]
h	heat transfer coefficient	[W·m^{-2}·K^{-1}]
h_{fg}	heat of vaporization	[J·kg^{-1}]
k	mass transfer coefficient	[m·s^{-1}]
L	size of screen opening	[mm]
Le	Lewis number ($= Sc/Pr$)	[–]
M	moisture content	[wt%]
MW	molecular weight	[kg·mol^{-1}]
m	mass	[kg]
M_{cu}	cumulative fraction of particles by mass $<d_p$	[–]
p	pressure or partial pressure	[bar]
Pr	Prandtl number ($= \nu/a = \eta\, c_p/\lambda$)	[–]
$\dot{Q}$	heat flow supplied or extracted	[W]
R	drying rate per unit area	[kg·m^{-2}·s^{-1}]
Re	reynolds number ($= \rho vd/\eta$)	[–]
R_u	universal gas constant	[J·mol^{-1}·K^{-1}]
Sc	Schmidt number ($= \nu/D$)	[–]
T	temperature	[K]
t	time	[s]
V	volume	[m^3]
v	velocity	[m·s^{-1}]
$\dot{W}$	power	[W]
w	water content on a mass basis	[kg]
w_i	mass fraction of particles with size $d_{p,i}$	[–]
ε	voidage	[–]
η	dynamic viscosity	[Pa·s]
ν	kinematic viscosity	[m^2·s^{-1}]
ρ	density	[kg·m^{-3}]
ϕ	relative humidity	[–]
φ_m	mass flow	[kg·s^{-1}]
χ	porosity	[–]
ω	humidity ratio	[–]

Subscripts

a	air
app	apparent
ar	as received basis
b	biomass
c	constant-rate period, critical
cv	control volume
d	dry

e	equilibrium
f	fluid
f	falling-rate period related
i	initial
moist	moisture
p	particle
w	water
wb	wet bulb

Superscripts

sat	saturated

8.1 INTRODUCTION

Biomass and waste materials are available in widely varying physical appearances. Often, such sources contain substantial amounts of water and are available in the form of relatively large particles when having been harvested or obtained in other collection processes. Just to illustrate its wide spectrum of morphologies, biomass can have a tough, fiber-like structure or have a sticky, paste-like structure.

The majority of energy conversion processes that are dealt with in detail in Part III of this book need small particles for supply to the reactors, as large particles hamper feeding and fuel conversion extents due to the usually limited available residence times. Thus, particle size reduction is usually needed. Also, biomass is usually relatively wet, with typical moisture contents between 30 and 60 wt% (as received (ar)). In most processes, this cannot be accepted—with the exception of digestion processes requiring water and hydrothermal processes—because a high moisture content is at the expense of processing efficiency due to the lower effective heating value of the fuel. Therefore, moisture needs to be removed, which is a highly endothermic process. Often, due to usage of fertilizers and differences in rainfall and harvesting characteristics with time, the ash content of biomass is highly varying, which may impact its handling, in particular in thermochemical conversion processes.

In view of the aforementioned considerations, physical pretreatment of biomass is crucial, and in practice, many of the problems encountered in the processing of biomass can be traced back to improper, nonoptimized physical pretreatment. Physical pretreatment serves the following purposes:

- Storage ability improvement
- Reduction of the content of harmful species, e.g., removal of hard stones and adherent soil but also ash reduction (e.g., alkali salts)
- Size reduction
- Moisture content reduction
- (Volumetric) energy density increase

- Feedstock homogenization or smart blending
- Tailoring to a specific feeding system

As this chapter only deals with physical pretreatment, (thermo)chemical treatment is excluded; pretreatment processes such as torrefaction and steam explosion are dealt with in Chapters 12 and 13, respectively.

8.2 HARVESTING AND TRANSPORT

The first step in the chain from grown biomass-to-energy conversion is the process of harvesting. The ratio of the energy content of the biomass for further processing to that of the energy input (also called the *energy ratio balance*) may be significantly impacted by the harvesting technique, time (due to impact of weather conditions), and subsequent transportation parameters, such as distance and methods. In this section, harvesting of woody forest products, agricultural crops, grass, and aquatic biomass is discussed.

8.2.1 Wood

8.2.1.1 Primary Harvesting During wood harvesting in forestry, either labor-intensive manual chain sawing is practiced or use is made of a feller buncher, which is a forestry vehicle made for the purpose of grasping and holding whole trees with a hydraulically driven system while at the same time cutting them with an inbuilt chainsaw.

8.2.1.2 Extraction from the Forest When wood has been cut in the forest, it must be transported to a so-called landing (gathering location); this is done using *forwarders*. These machines enable the heavy transport of multiple collected logs, stems, and small trees using a trailer. When wood needs to be harvested from mountainous or hilly landscapes, *cable haulers* are used, as vehicles with wheels or tracks cannot safely move through such steep forests. The cable installation is mounted using a tractor or truck, both using a tower with drum winch.

Skidders are vehicles aimed at pulling harvested stems or even whole trees to the landing place. They usually lift biomass from one end, and as a result, soil and stones may contaminate the harvested material. Once the whole trees and stems have been delivered to the landing place, processors are used to remove limbs from the stem wood, cut off tops, and crosscut (also called "section") the stem wood to convenient lengths. The remaining smaller logs are then transport ready for delivery to a sawmill.

For compact, more energy-dense transport, the harvested wood product is usually baled; the wood is compressed, bundled, and tied together. The energy content of (forest residues) bales has been reported to be about 2.56 MWh $\cdot$ t^{-1} on a wet basis with a moisture content of 45 wt% (Van Loo and Koppejan, 2004).

For commercial pulp and paper logging application, also debarking (see, e.g., Hatton, 1987; Nurmi and Lehtimäki, 2011) takes place, which means that the outside part of the wood, the bark, is removed by machines.

When trees are harvested younger and more frequently compared to "conventional" forestry—and this also holds for woody energy crop plantations (e.g., the fast-growing plant species *Miscanthus sinensis*)—one speaks of short rotation forestry (SRF). Here, machines are applied that fell, bunch, and process single stems or "coppice" regrowth. A detailed overview of techniques is given by Sims (2002).

8.2.2 Agricultural Crops: Cereals' Straw

For the harvest of rice, wheat, or other cereals with a resulting by-product straw (or stover in case of corn), in most modern farming systems, a so-called combine is used. This machine combines cutting and threshing (removal of grains from the rest of the crop including stalks). A combine is a self-propelled harvesting machine that cuts the whole grain plant and separates the grain kernels from the remaining straw material. This straw is left in windrows spread throughout the harvested field. It is ensured that part of the straw (or stover) is left in the field for soil conditioning; also, to prevent taking in too much soil, not the whole plant is cut. After natural drying, the straw is usually picked up by a mechanical baling machine (baler), which compresses the straw in packs or bales. Some combines are equipped with cutting mechanisms for the straw enabling a speedup of straw degradation in the field. In areas where straw is rather collected for other uses, these cutters are generally not applied.

The main method for collecting the straw in the field is baling. Baling comprises collecting, exerting pressure on the loose biomass, and tying it together with a cord; these bales can be wrapped in plastic in order to prevent rewetting. Baling is a common method used to improve the characteristics of agricultural residues for transport, storage, and further handling (Maciejewska et al., 2006). The size of the bales as well as their density depends on the machine used, and the final density depends on the type of biomass feedstock and the machine. Transportation of bales in general is substantially cheaper than transportation of raw biomass, and in some cases, the gain is up to 50% (Van Loo and Koppejan, 2004). The bales are produced in various shapes, densities, dimensions, and weights. Bales can be square or round and their final size depends on the machine used. The density of the bales, for a given machine, depends on the type of biomass.

A Canadian study conducted by the Composites Innovation Centre showed that the harvesting and baling cost of wheat straw, when done by farmers themselves, for very large bales of 544 kg with a density of $160 \, \text{kg} \cdot \text{m}^{-3}$, is approximately $5 \, \text{€} \cdot \text{t}^{-1}$ ($7 \, \text{€} \cdot \text{t}^{-1}$ with collection included), while for custom round bales, of 408 kg, the harvesting and baling cost is $9.5 \, \text{€} \cdot \text{t}^{-1}$ ($12.8 \, \text{€} \cdot \text{t}^{-1}$ with collection included). Because harvesting and baling is done by farmers, labor costs have not been taken into account. Generally, the bigger the bale, the lower the baling cost and the cost of subsequent transportation of the bale (Composites Innovation Centre [CIC], 2008).

8.2.3 Grass

Grasses are usually harvested by cutting with mowers. In larger-scale agricultural practice, mowers are driven by tractors. The mown grass is left on the field for drying. Regular tedding ensures faster drying by exposing the grass to airflow and solar irradiation more evenly. Subsequently, by raking, lanes of dried grass are formed, which can then be optimally harvested as hay using a baling machine. Alternatively, grass can be ensiled, which needs less drying, but still the moisture content must be reduced as the optimal moisture content for ensiling grass is between 30 and 50 wt%. Too wet biomass leads to significant degradation during the silage process, associated also with smell issues. Another problem is that when pockets of biomass are stored that are too wet, self-heating and autoignition may occur, forming a serious hazard potential and leading to possibly severe economic loss.

Taller grass species (e.g., *Miscanthus giganteus sinensis*) need a forage chopper. This machine does not only cut off the stems but also chops the material in one continuous operation.

8.2.4 Aquatic biomass

A relatively novel development in the field of the production of biomass-to-energy carriers is the utilization of grown wet biomass plant material, particularly algae. Algae can be divided into two subclasses: microalgae and macroalgae (also termed "seaweeds").

8.2.4.1 Microalgae Microalgae can be cultivated in open or closed systems. Concerning open systems, one can discriminate both large open unmixed ponds and open mixed ponds. An unmixed pond is relatively simple and offers the advantage that no supplementary input of energy is required. This type of cultivation system can be used, e.g., for wastewater treatment. Mixed ponds are characterized by a high mixing degree that ensures exposure of the algae inventory to light, nutrients, and gases (in particular CO_2), causing higher growth rates of up to a factor of 10 higher than in unmixed ponds. Most commonly used is the *raceway pond*, which is a shallow circuit with a depth between 20 and 35 cm, ensuring adequate sunlight exposure while the inventory is stirred using paddle wheels (Benemann, 2008). A raceway pond is usually lined with plastic or cement, though this adds to its cost. Some of the main advantages of this concept are relatively low investment costs, simple operation, and easy scale-up (to the order of magnitude of hectares). Drawbacks compared to a closed system are the low volumetric productivity, little control of culture conditions, vulnerability to contamination, and high evaporative losses (Ugwu et al., 2008). Currently, raceway ponds are used for commercial generation of *Spirulina, Dunaliella salina, Chlorella vulgaris*, and *Haematococcus pluvialis* (Benemann, 2008).

Closed systems are called photobioreactors, and they exist as tubular, flat-panel, and vertical column reactors. The most pronounced difference with open systems is the separation from open outside air so that gas exchange can be better controlled and contamination prevented. Different algae require rather tailor-made solutions for their reactor design. Tubular configurations dominate, with diameters of typically 10 cm to facilitate light irradiation. A disadvantage, in particular when scaling up such systems, is that mass transfer may be poor, resulting in the buildup of O_2 and CO_2, which reduces the productivity of algae cultivation (Ugwu et al., 2008). Another disadvantage is the relatively high capital cost.

Also, hybrid concepts exist, where a photobioreactor acts as a kind of breeder of algae to a prespecified concentration; in a second step, the then densely populated algae are fed to an open system for larger-scale production.

Harvesting of microalgae, which consist of bioparticles in suspension, means separating them from the aqueous production medium. Challenges in harvesting follow from the relatively low algae concentration and the small sizes. Four harvesting methods can be distinguished (Bruton et al., 2009):

1. Sedimentation
2. Filtration
3. Flotation
4. Centrifugation

It should be noted that in addition flocculation is often necessary as pretreatment of the grown algae to improve harvesting yields. Flocculation is a process in which dispersed particles settle out of suspension in the form of flocks or flakes induced by the addition of some chemical. The process of flocculation is mostly used in freshwater and waste-water. Saltwater prevents the formation of stable colloidal suspensions, thus making flocculation ineffective.

Question: Why can stable colloidal suspensions not be formed in saltwater?

Sedimentation is a simple technique that can be applied for microalgae showing a naturally high sedimentation rate. Apart from flocculation, also the use of ultrasonic sound waves is a proven technique. The capital and operation costs of sedimentation are low. A disadvantage is that it is time intensive.

Filtration is a commonly used method in industry and can be kept simple or made more complex as, e.g., vacuum and pressure filtration systems, which are associated with higher costs. The biggest challenge in filtration is finding the optimal trade-off between very small pores that cause plugging and larger pores that do not filter out all the algae.

In contrast with sedimentation, flotation results in algae floating on the water surface from which the biomass can then be skimmed. Some microalgae strains float intrinsically, but flotation may be assisted by bubbling air through the grown slurry or by modification of the surface tension of the algae particles using chemicals for improvement of the adherence to the air bubbles. Advantages are that capital and operating costs can be kept comparatively low. A disadvantage is the low separation efficiency especially in shallow-depth ponds.

Centrifugation is an intensified sedimentation process with a much higher efficiency. The capital and operation costs are also much higher. Despite the high costs, centrifugation currently is the preferred method for harvesting due to the ease of cleanup and sterilization.

8.2.4.2 Macroalgae

8.2.4.2 Macroalgae Although presently it is possible to grow macroalgae in open ponds on shore, this is still uncommon. Offshore cultivation is much more interesting for macroalgae. Macroalgae can grow either attached to a solid or free-floating in water. Four different systems for cultivation can be distinguished (Reith et al., 2005):

1. Longlines
2. Ladders
3. Grids
4. Rings

A combination with power-producing wind parks is a highly attractive option as this forms a combined renewable energy supply system that requires no land.

The ladder system is a variation of the longline system in which multiple longlines are placed on top of each other. The advantages of these systems are the low capital costs and simple installation. Disadvantages are the mechanical damages that occur due to rough conditions prevailing at sea and difficulties in controlling process parameters such as temperature, nutrients, and light.

The grid system or net-style system is one in which algal spores are seeded onto nets that are then fixed to some kind of support system. At low tide, the nets are above the water line and are thus exposed to the atmosphere. This tidal variation enhances growth conditions and inhibits fouling by other organisms.

The ring system is assembled and inoculated on shore after which it is transported out to sea. The ring can be applied at depths of 5–8 m; a further refinement of the system still has to occur. The ring is transported back onto shore for harvesting.

When the macroalgae are attached to a solid, it is necessary to cut them, whereas free-floating algae can simply be collected. The most dominant harvesting practice is the collection of the total biomass of each plant, while a much rarer technique is cutting the plant above the holdfast to enable regeneration. The different techniques involved in macroalgae harvesting are discussed in the following.

The first distinguishing factor in harvesting is that it can be done either manually or using mechanical systems. Manual harvesting has been used since the preindustrial age, and the equipment involved is quite limited and may consist of a diving apparatus, knives, sacks for algae collection, and possibly boats for shore access.

Mechanical systems should ensure sustainable life and minimal damage to the ecosystem. They can be applied using specially equipped boats. Four categories of mechanical systems that are looked into for the collection of macroalgae are:

1. Drag rake
2. Cutter blade
3. Suction harvester
4. Scoubidou

The drag rake is a tool based on the hand rake and functions by dragging the system over the sea or ocean floor, thereby collecting solid-attached weeds. The rakes are then retrieved to the boat in which the macroalgae are collected, using haulers or winchers. After this, the whole process can be repeated.

In the cutter-blade system, a reciprocating cutter blade is lowered to a controlled height where it cuts the algae. The algae are then picked up by a conveyor belt and loaded onto the harvesting ship. This technique can be applied for both solid-attached algae and free-floating algae.

The suction harvester is a system consisting of pipes equipped with a bladed impellor that simultaneously draw up and cut seaweeds, which are then collected onto the ship or in a bag trailing behind the ship. This also is a technique that can be applied for both solid-attached algae and free-floating algae.

The Scoubidou system is a technique by which the seaweeds are twisted around a rotating hook, breaking the holdfasts through traction. The seaweed is then collected onto the ship by rotating the hook in the reverse direction. The Scoubidou system is applied to attached macroalgae species.

A fifth mechanical system for algae harvesting is directly linked with fishery and is performed by nets cast from and pulled by ships. This technique is used for free-floating seaweeds, which are thus simply collected by pulling in the nets.

8.3   STORAGE

Storing harvested material forms an intermediate step between harvesting and further physical and chemical processing of biomass. This creates a time bridge between these steps, which enables continuous supply. Although at first sight storing biomass does not seem to be a pretreatment process, its conditions partly determine the quality characteristics of the biomass for further processing. Storage process conditions can affect the moisture content of biomass, its energy value, and its dry matter content, and as such, it is related with physical biomass pretreatment (Maciejewska et al., 2006). Important safety and environmental aspects of storage are the moisture content and particle size distribution, degradation associated with loss of mass and energy content by microbiological activity (Hunder, 2005) causing self-heating, and emission of greenhouse gases (in particular CH_4, N_2O) (Wihersaari, 2005). In general, a higher original moisture content of the biomass causes higher losses of dry matter. Therefore, one targets at a moisture reduction to approximately 20 wt% before storage. In order to prevent as much as possible the loss of dry matter, airtight storage or storage of material not yet reduced in size is recommended; the smaller the size of biomass, the worse is its ventilation (Maciejewska et al., 2006). Wihersaari (2005) studied the release of CH_4 and N_2O from 6 months' wood chip storage. This amounted to 58 kg $CO_{2eq} \cdot MWh^{-1}$ fuel in case the wood chips were predried to 40 wt% moisture content; this figure increased to 144 kg $CO_{2eq} \cdot MWh^{-1}$ for biomass that was delivered fresh with a 60 wt% moisture content.

Risks associated with storing biomass in general are (see, e.g., Kaltschmitt et al., 2001):

- Loss of material by bacterial activity
- Fungi growth
- Self-heating and ignition
- Stench
- Renewed wetting and absorption
- Agglomeration/structure change by means of frost
- Demixing and loss of fines
- Release of water from a pile
- Dust issues with explosion risks

Systems for storing biomass can be subdivided into two types: outdoor and indoor storage. Outdoor storage is accomplished either without using cover or by using a simple cover (e.g., plastic) or below a shed. For indoor storage, a barn can be used, or storage can take place in a bunker or silo.

The type of biomass packing morphology mainly determines the storage solution. As occupation of space is costly, often, a precompaction is chosen, e.g., baling or pelletizing and loading big bags with the pellets. We will not deal with bulk material handling and transportation in detail as we focus on physical and chemical transformation processes. The reader is referred to general textbooks in this area (e.g., Mason, 1988; McGlinchey, 2008).

8.4   WASHING

Sometimes, it is worthwhile to extract mineral matter from the harvested biomass to prevent issues related to thermochemical processing. Especially for biomass containing high concentrations of both alkali metal ions (Na^+ and K^+) and chlorine, which is the case in particular for herbaceous biomasses, such as grasses and straw (Arvelakis and Koukios, 2002), thermochemical processing can become cumbersome due to agglomeration/sintering phenomena (fluidized beds), slagging and fouling of heat exchanger surfaces, as well as corrosion and erosion (see Chapters 9 and 10 for a more detailed description of these phenomena). This washing procedure (also called leaching) has been studied and suggested by many authors (e.g., Arvelakis et al., 2001; Jenkins et al., 1996, 1998; Turn et al., 1997) to be an efficient, quick, and cheap way to substantially diminish the ash content of a biomass material by eliminating alkali metals (K, Na), chlorine, and sulfur in the first place and secondly calcium and also minerals extraneously added to the biomass (such as clays), resulting in a material with improved ash thermal behavior. Leaching can be performed with water, which is most simple, or with acid and ammonia. Water ensures leaching of alkali sulfates, chlorides, and carbonates, while ammonia leaches out Mg, Ca, K, and Na, and aqueous HCl (as example of an acid) leaches carbonates and sulfates of alkaline earth and other elements. The use of ammonia and acids is usually not recommended as pretreatment process for bioenergy supply due to their costs. Simple water leaching in a controlled processing step is also associated with additional investments and operating costs mainly as a result of the after drying needed.

Herbaceous harvests are usually left exposed to the weather with occasional rainfalls; this already diminishes in particular the alkali and chlorine species with an associated reduction of deposition and corrosion in boilers achieved (Van Loo and Koppejan, 2004). On the other hand, qualities of biomass obtained in these cases can be highly varying as weather conditions vary and biomass partly degrades.

8.5 SIZE REDUCTION

Size reduction is applied to modify the particle size distribution of a biomass feedstock in order to obtain a larger fraction of finer sizes to comply with logistics and conversion technology demands. Particle size reduction leads to an increase of available specific surface area and a reduction of cellulose crystallinity and degree of cellulose polymerization (Kratky and Jirout, 2011; Zhang et al., 2007). It also generally leads to a denser product. These effects are needed for improved heat and mass transfer characteristics in subsequent processing steps, leading to, e.g., reduced processing time in biomass digestion and higher yields in hydrolysis. For the very heterogeneous biomass sources, size reduction is complex. Unlike coal and most mineral matter, only a minor fraction breaks down under crushing forces as biomass is often fibrous and tenacious. Most of these materials deform, stretch, or are simply compressed by crushing forces, so that shearing, ripping, and cutting actions are needed for size reduction (Niessen, 2010). Another complication may be that once grinded, biomass particles tend to stick together (see Van der Burgt in Knoef, 2005).

Drawbacks of size reduction are that the amount of energy required may be high depending on particle size targeted at. Furthermore, in herbaceous biomass, the morphology and (hard) silica content lead to extensive wear and tear of machinery. Also, size reduction often leads to a redistribution of particle sizes and creates comparatively fine fractions.

The main techniques used for size reduction of biomass are chunking, chipping, crushing, and milling/pulverization.

8.5.1 Chunking

When large biomass parts become available after harvesting, e.g., by tree cutting, chunkers are used to downsize them to a coarse size range of 50–250 mm.

8.5.2 Chipping and Shredding

Chipping or shredding can be applied when the biomass or waste material is relatively tough and possibly wet and downsizing to a relatively coarse size range of 25–50 mm is sufficient. This technology is also used for predownsizing of particles to enhance further size reduction characteristics (needed particle size distribution and energy requirement). Chipping and shredding devices are standard in use for the maintenance of, e.g., parks and gardens and make use of fast-rotating cutting blades mounted on the face of a flywheel. Large wood chippers are frequently equipped with grooved rollers for (reversible) material gripping in the throat of their feed funnels (tinyurl.com/bo9lcb). The following types exist:

- *High-torque roller*: This type of "shredder" uses low-speed grinding rollers, is driven with an electric motor, and is characterized by quiet, dust-free, and self-feeding operation.
- *Drum*: Drum chippers are among the oldest available and have a fast-rotating drum powered by an engine usually by means of a belt. The drum is mounted parallel to a hopper and spins toward the output chute. Modern types can handle biomass material with a diameter of 150–500 mm.
- *Disk*: Knives are mounted on a steel disk, and reversible hydraulically power wheels draw material from the hopper into the disk area, which is mounted perpendicularly to the fed material; the knives of the spinning disk cut the biomass into chips, which are thrown out toward the chute. Industrial-grade types can have disk sizes as large as 4 m; similar biomass sizes as in the drum type can be handled, and disk-type chippers produce more uniformly shaped and sized chips.

Shredding using rotary action has been shown to result in different specific energy consumption values for different types of biomass and derived products; e.g., paper is much more difficult to be shred than grass; shredding paper to ~40 mm size was shown to require an energy of 15.2 kWh $\cdot$ t_{od}^{-1} in a first pass and 7.6 kWh $\cdot$ t_{od}^{-1} in a second pass for further size reduction to 25 mm; for switchgrass, these figures were 8.2 and 4.1 kWh $\cdot$ t_{od}^{-1}, respectively (Schell and Hardwood, 1994).

8.5.3 Crushing

Materials that are comparatively hard and brittle are reduced in size by making use of their nondeformation nature, i.e., these materials rather break than bend. Crushers that are suitable for such materials are jaw crushers, but also roller mills and hammer mills can be used.

8.5.4 Milling and Pulverization

Mills of many different types can be used to grind biomass into the finest particle size classes. In the past or still in (very) traditional practices, mills for producing flour from grains (wheat) were driven either manually (e.g., mortar and pestle), with animals (horse mills), by wind power (windmills), or water (water mills). For the processing of coals on an industrial scale, ball mills are usually applied. In this concept, a horizontal or somewhat inclined rotating vessel is filled with balls up to a charge of approximately 30%. These balls grind the material by friction and impaction by their tumbling (Nag, 2008).

The ball-milling technique is also used to characterize the grindability of coals by the so-called Hardgrove grindability index (HGI). The HGI is a standardized test that has been developed empirically and is commonly applied to various coal classes to describe their uniformity and ease of pulverization. The traditional test involves grinding a sample *mass* of coal (50 g with an initial size between 0.6 and 1.18 mm) in a special milling apparatus, which crushes the coal using steel spheres at a specified load and number of revolutions (60) (British Standards Institution [BSI], 1995). The mass of the resulting pulverized coal fraction that passes through a 75 μm sieve determines the HGI. Thus, a higher HGI indicates a sample that is more easily grinded. The method first uses a calibration based on four known coals. Bridgeman et al. (2010) modified this traditional standardized test for (torrefied) biomass by using a fixed *volume* (50 cm^3) instead of a fixed mass (i.e., 50 g) and calibrated their results using coals of known HGI. The results of this modification are promising with respect to further standardization for heterogeneous biomass.

Table 8.1 gives an overview of the different techniques that are used for milling of biomass. Based on its high size reduction ratio and adequate control of the particle size range with comparatively good cubic particle shape, the hammer mill is widely used (Mani et al., 2004). Knife mills have shown successful functioning for shredding forages for different crops and machine conditions (Ige and Finner, 1976). Sometimes, disk mills are used to produce very fine particles, but input feed is needed from either knife mills or hammer mills (Hoque et al., 2007).

Generally, the energy requirements increase with increasing moisture content and decreasing final particle size distribution and also with increasing rotational speed of most equipment. Spliethoff (2010) showed that the energy consumption increases strongly with decreasing particle size (see Figure 8.1). The moisture content also has a significant impact on energy consumption as material with higher moisture content is tougher; for milling of straw with a moisture content of 30 wt% and a sieve size of 2 mm, the energy demand was more than 8% of its heating value (Spliethoff, 2010).

The energy needed for size reduction by milling can generically be expressed as (Ghorbani et al., 2010)

$$E = C \int_1^2 \frac{\mathrm{dL}}{\mathrm{L}^n} \qquad \text{(Eq. 8.1)}$$

with E being the specific energy consumption (kJ · kg^{-1}), C a constant, dL the differential size (dimensionless), and L the screen opening size (mm). Different models have been proposed in the literature. Bond (1961) assumed a value of 3/2 for n, and Rittinger assumed that the process is basically shearing and thus that the energy requirement is proportional to new surface created, so that n = 2. Finally, Kick assumed the energy requirement to be a function of the common dimension of the material only, and thus n = 1 (Henderson and Perry, 1976).

TABLE 8.1   Overview of milling techniques applied for biomass

Milling technique	Main forces	Advantages	Disadvantages
Ball mills	Shear Compression	Simple construction Easy to operate Cheap	Effectiveness for downsizing to very small sizes is relatively low Relatively high energy utilization
Vibro energy mills like ball mill but with vibration	Shear Compression	More effective in crystallinity reduction Lower end particle sizes compared to ball mills	More expensive than ball mills Relatively high specific energy utilization More complex construction
Knife mills (Kratky and Jirout, 2011)	Shear	Used for wide range of (tough) biomass Lower specific energy consumption (compared to hammer mills)	Only relatively dry biomass (<15 wt% moisture content)
Hammer mills	Compression	Cheap as they are standard equipment (though more expensive than ball mills) Easy to operate	Relatively high energy utilization (especially compared to knife mills)
Two-roll mills	Compression Some shear	Easy to operate, simple construction	Moderately fine distribution only
Disk mills and wet-disk mills	Shear	Processes wet streams, so no drying needed Finely ground product	High specific energy consumption for milling process
High shear—cavitation-based machines	Cavitation, shear	Very effective for fine grinding	Need upstream size reduction For smaller feed sizes only

These three approaches lead to the following equations for energy consumption:

$$\text{Bond}: E = C_B \left(\frac{1}{\sqrt{L_2}} - \frac{1}{\sqrt{L_1}} \right) \qquad \text{(Eq. 8.2)}$$

$$\text{Rittinger}: E = C_R \left(\frac{1}{L_2} - \frac{1}{L_1} \right) \qquad \text{(Eq. 8.3)}$$

$$\text{Kick}: E = C_K \ln \left(\frac{L_2}{L_1} \right) \qquad \text{(Eq. 8.4)}$$

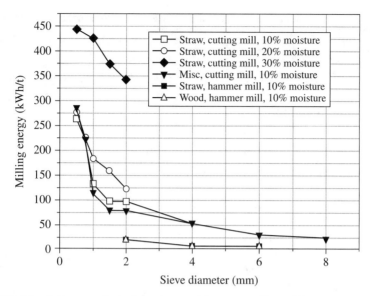

FIGURE 8.1 Influence of sieving size on biomass-specific milling energy. (Source: Reproduced with permission from Spliethoff (2010). © Springer.)

8.6 PARTICLE SIZE CHARACTERIZATION

Particle size distributions can be determined using standard sieve series. A sample of ground biomass with known mass is loaded on top of a stack of such sieves; these are closely held together by a fixation system and positioned on a vigorously vibrating bottom plate. After shaking for several minutes, the stack is dismantled and the accumulated mass is determined for each tray. Table 8.2 gives an example of particle size distributions of pellets crushed using a two-roll mill (De Jong, 2005). The highest value indicated in the d_p range per sieve is the sieve size. Sometimes, the sieve sizes are expressed in the Tyler number or mesh size.

Now, the question is how to determine an accurate average particle size based on such results. One approach might be just to calculate the ordinary weight average particle size, as is shown in Equation (8.5). In this equation, each $d_{p,i}$ value is determined as the average value of the indicated size range per sieve tray (d_p average in Table 8.2). The value for the top sieve is determined by taking the previous sieve range, here no. 17, and consider the same size range for the final one:

$$\overline{d_p} = \sum_{i=1}^{n} w_i d_{p,i} \qquad (Eq.\ 8.5)$$

Such a formulation is not very practical in view of its application for which area (transfer phenomena, surface reactions) and/or volume (pneumatic particle transport, fluidization) characteristics are of importance. Therefore, other formulations have

TABLE 8.2   Examples of analyses of particle size distributions determined by sieving

Sieve	d_p range (µm)	d_p average (µm)	*Miscanthus* (wt%)
1 (Bottom)	0–53	26.5	0.25
2	53–90	71.5	0.22
3	90–125	107.5	0.27
4	125–180	152.5	0.61
5	180–250	215.0	0.72
6	250–300	275.0	0.81
7	300–425	362.5	1.41
8	425–500	462.5	1.05
9	500–600	550.0	0.95
10	600–710	655.0	1.66
11	710–850	780.0	2.04
12	850–1400	1125	7.47
13	1400–2000	1700	7.04
14	2000–3150	2575	15.60
15	3150–4000	3575	16.38
16	4000–4750	4375	13.72
17	4750–5600	5175	10.54
18 (Top)	5600 +	6025	19.26

been put forward. These are based on area averaging Equation (8.6) and volume averaging Equation (8.7):

$$\overline{d_{p,\,area}} = \sqrt{\sum_{i=1}^{n}\left(w_i d_{p,i}^2\right)} \qquad \text{(Eq. 8.6)}$$

$$\overline{d_{p,\,volume}} = \sqrt[3]{\sum_{i=1}^{n}\left(w_i d_{p,i}^3\right)} \qquad \text{(Eq. 8.7)}$$

For the purpose of selecting the optimal average particle size for use in applications such as fluidized bed combustion (Chapter 9) and gasification (Chapter 10), the surface–volume average has been introduced (De Souza-Santos, 2004):

$$\overline{d_{p,\,area/volume}} = \frac{1}{\sum_{i=1}^{n}\left(\dfrac{w_i}{d_{p,i}}\right)} \qquad \text{(Eq. 8.8)}$$

Alternatively, the Sauter mean particle size is used, a volume–surface average (often applied in small-particle entrained flow processes):

$$\overline{d_{p,\,SM}} = \frac{\displaystyle\sum_{i=1}^{n} w_i d_{p,i}^3}{\displaystyle\sum_{i=1}^{n} w_i d_{p,i}^2} \qquad \text{(Eq. 8.9)}$$

An alternative for the characterization of particle size distributions forms laser diffraction measurement; the treatment of this characterization technique—and others—is considered to be out of the scope of this book (the reader is referred to, e.g., Barth and Winefordner, 1984; Merkus, 2008; Pankewitz, 2006; Syvitski, 2007).

Results of hammer milling of different types of biomass, corn stover, wheat straw, and switchgrass, show that the particle size distribution can be described by a Rosin–Rammler distribution:

$$M_{cu} = 1 - e^{\left(-\frac{d_p}{a}\right)^b} \qquad\qquad \text{(Eq. 8.10)}$$

in which M_{cu} is the cumulative fraction of particles (wt%) with a size smaller than d_p, d_p is the particle size (assumed equal to the nominal sieve aperture size), a is the size parameter (also called the Rosin–Rammler geometric mean diameter; this is the size at which 63.2 wt% of the particles are smaller than d_p), and b is the distribution parameter (also called the Rosin–Rammler skewness parameter).

Question: How would you determine b for a given size distribution?

8.7 SCREENING AND CLASSIFICATION

In order to separate possible adhering contaminants from the biomass and to generate a product stream with reasonably uniform particle size distribution, sieving and classification are needed. The biomass stream may contain contaminants such as sand, ferrometals (e.g., nails), nonferrometals, plastics, etc.

Iron (and other magnetic metals) can be separated by either permanent magnets or electromagnets, which can be integrated in a belt conveyor construction. The content of sand can be reduced by separating it from the biomass flow with, e.g., a horizontal vibrating sieve. Sieving technologies for biomass comprise disk sieving systems (flexible stacked sieves driven as a belt), plate sieves, and rotating (trommel) sieves.

Wind sifting can be used to remove light contaminants, such as plastics.

8.8 METHODS OF MOISTURE REDUCTION

Most applications of biomass in energy production systems require drying to a more or lesser extent. For example, for fluidized bed combustion, a moisture content of about 60 wt% still results in acceptably low heat losses; this is common practice in sludge combustion to reduce the volume of this waste. In contrast, in the production of wood pellets, the moisture content must be reduced to approximately 10–15 wt%. This difference in the required reduction of the moisture content also determines which technologies for dewatering can be used.

Moisture contents in widely differing types of biomass can be reduced by different techniques. These are natural drying, mechanical drying, and thermal drying.

Drying processes can be carried out in batch or continuous mode. The extent to which water can be removed depends on how the water is bound to the biomass. For very wet biomass sludge, Colin and Gazbar (1995) distinguished the contained water into different categories depending on its behavior during mechanical dewatering, namely, free water, bound water removable by moderate mechanical strain, bound water removable by maximal mechanical strain, and finally bound water not removable mechanically.

8.8.1 Natural Drying

The most straightforward method of moisture content reduction is drying in the open air. Dry, fresh air is used, occurring as natural weather condition during the harvesting process. The process can be enhanced, e.g., by raking in case of agricultural residues. This drying type is preferably utilized before transportation takes place, at least under the condition that the costs of transportation are not limited by volume but by mass. Costs are low, but the process is relatively slow. The final result depends mainly on the initial moisture content and the drying time. After harvest, straw and the entire cereal plants may still show moisture contents up to 40 wt%, but this value can be reduced to below 20 wt% within 2–3 days of drying on the field, provided the weather conditions are good (Hartman and Strehler, 1995). Sometimes, the choice of the season for harvesting is crucial to obtain a minimal moisture content harvest; *Miscanthus*, e.g., is preferably harvested in spring when only the stalks are still standing without leaves and nonwoody tops after the winter period, resulting in about 20 wt% moisture content (Lewandowski et al., 2000).

8.8.2 Mechanical Drying

Drying based on using mechanical work is applicable for compressible biomass and can be accomplished in several ways. The mechanical dewatering process itself may consume a large amount of energy and have high maintenance requirements, which must be weighed against the usually substantial reduction in drying energy. Mechanical drying can be performed as a batch or continuous process. Here, we only consider (larger-scale) continuous operation.

In the *screw press*, biomass is slowly compressed by means of a housing that is conically decreasing in diameter or by means of a screw with a decreasing "chamber" length. In both configurations, the biomass volume is decreased, by which the pressure increases and water is drawn off. Only limited dewatering is possible, so application of a screw press is more suitable for fiber-like biomass than for sludge, which contains comparatively large amounts of water. It can be used, e.g., to extrude green juice from grass that contains not only moisture but also potentially useful nutrients/chemicals that in this way can be retrieved (see also Chapter 15 for biorefinery concepts).

Other types of mechanical presses include *belt filter* presses, V-type presses, ring presses, and *drum presses* (Roos, 2008). A belt filter press accommodates the wet biomass between two permeable belts, which are moving over and under rollers, thereby squeezing the moisture out. A drum press comprises a perforated drum with a revolving press roll inside it that presses material against the drum. This kind of press has been used with a wide range of biomasses.

A *roller press* is another apparatus for mechanical dewatering; in a way, it is similar to a screw press. In a roller press, however, the pressure increase is accomplished by two cylindrical rolls of which one or two are electrically driven with a small gap between them through which the biomass is transported, simultaneously pressing out the water. Advantages of a roller press are that the pressure can be set relatively accurately and, more importantly, the energy use is comparatively low. In addition, the equipment is relatively simple and cheap to produce. There are some disadvantages though. Supply of biomass to the equipment is not very difficult for biomass with longer stalks, but for, e.g., grass, this is a rather challenging task. This issue can be solved by inclining the rolls and using a vertically mounted supply system so that gravity helps the flow. Alternatively, a press can be placed in front of the roller to ensure that biomass enters the system. Water removal in this way is limited, although it is somewhat improved (to ~50 wt% order of magnitude) by placing a few units in series.

Centrifuges can also be applied; these dewater based on centrifugal forces. An example is the bowl centrifuge, in which the material enters a conical, spinning bowl in which solids accumulate on the perimeter.

Example 8.1   Energy balance calculation for a biomass roller press for dewatering

Wet biomass is pressed through a roller press; a schematic is shown in Figure 8.2.

The biomass contains 75 wt% moisture (ar basis). Drying takes place by once-through pressing until a moisture content of 65 wt% (ar basis) is reached. Set up the mass and energy balances. Assume steady-state operation and adiabatic compression of the plug and neglect changes in potential and kinetic energy related to the biomass plug. Biomass enters the roller press at 25°C and leaves the press at a temperature of 26°C; assume that the dried biomass has the same temperature as the

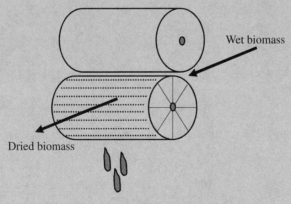

FIGURE 8.2   Schematic of the roller press.

moisture leaving in liquid form (no evaporation is assumed to take place). What is the specific work input of the roller press (in $kJ \cdot kg^{-1}$ dry fuel input)?

The c_p of dry biomass is assumed to be constant at $1200 \, J \cdot kg^{-1} \cdot K^{-1}$, and the c_p of water is assumed to be constant at $4180 \, J \cdot kg^{-1} \cdot K^{-1}$.

Solution

First, set up a macroscopic mass balance for the input biomass plug (stream 1), where splitting of the dry biomass part and the moisture part starts:

$$\varphi_{m,1,b,ar} = \varphi_{m,1,b,dry} + \varphi_{m,1,moist}$$

$$\varphi_{m,1,moist} = \left[\frac{M_{ar}}{100 - M_{ar}}\right]_1 \varphi_{m,1,b,dry}$$

with M_{ar} being the wt% of moisture on an ar basis.

The mass balance over the dewatering press then becomes

$$\varphi_{m,1,b,dry}\left(1 + \frac{M_{ar,1}}{100 - M_{ar,1}}\right) = \varphi_{m,2,b,ar} + \varphi_{m,2,moist}$$

$$= \varphi_{m,1,b,d}\left(1 + \frac{M_{ar,2}}{100 - M_{ar,2}}\right) + \varphi_{m,2,moist}$$

$$\varphi_{m,2,moist} = \left\{\frac{100}{100 - M_{ar,1}} - \frac{100}{100 - M_{ar,2}}\right\}\varphi_{m,1,b,d} = \frac{100(M_{ar,1} - M_{ar,2})}{(100 - M_{ar,1})(100 - M_{ar,2})}\varphi_{m,1,b,d}$$

The energy balance for the roller press is

$$\frac{dE_{cv}}{dt} = 0 = \dot{Q}_{cv} - \dot{W}_{cv} + \varphi_{m,1,b,d}\left\{h_{1,b} + \left(\frac{v_{1,b,d}^2}{2}\right) + gz_{1,b,d}\right\}$$

$$+ \varphi_{m,1,moist}\left\{h_{1,moist} + \left(\frac{v_{1,moist}^2}{2}\right) + gz_{1,b,d}\right\}$$

$$- \varphi_{m,2,b,d}\left\{h_{2,b,d} + \left(\frac{v_{2,b}^2}{2}\right) + gz_{2,b}\right\}$$

$$- \varphi_{m,2,moist}\left\{h_{2,moist} + \left(\frac{v_{2,moist}^2}{2}\right) + gz_{2,moist}\right\}$$

Based on the assumptions, the kinetic and potential energy terms are not taken into account as changes are negligible. Thus, with heat supply going to zero due to adiabatic compression for the power, we derive

$$\dot{W}_{cv} = \varphi_{m,1,b,d}h_{1,b,d} + \frac{M_{ar,1}}{(100-M_{ar,1})}\varphi_{m,1,moist}h_{1,moist}$$

$$-\varphi_{m,1,b,d}h_{2,b,d} - \frac{M_{ar,2}}{(100-M_{ar,2})}\varphi_{m,1,b,d}h_{2,moist}$$

$$-\frac{100(M_{ar,1}-M_{ar,2})}{(100-M_{ar,1})(100-M_{ar,2})}\varphi_{m,1,b,d}h_{2,moist}$$

So,

$$\frac{\dot{W}_{cv}}{\varphi_{m,1,b,d}} = (h_{1,b,d}-h_{2,b,d}) + \frac{M_{ar,1}}{(100-M_{ar,1})}(h_{1,moist}-h_{2,moist})$$

$$= -\left\{\Delta h_{b,d} + \frac{M_{ar,1}}{(100-M_{ar,1})}\Delta h_{moist}\right\}$$

$$\frac{\dot{W}_{cv}}{\varphi_{m,1,b,d}} = -\left\{c_{p,b,d}\Delta T + \frac{M_{ar,1}}{(100-M_{ar,1})}c_{p,moist}\Delta T\right\}$$

$$= -\{1200 \times 1 + 3 \times 4180 \times 1\}\ J \cdot kg^{-1} = -13.7\ kJ \cdot kg^{-1}$$

The minus sign indicates that work is done on the biomass.

Question: What would be the (minimum amount of) energy needed to evaporate the same amount of water, now removed mechanically?

8.8.3 Thermal Drying

Drying using (waste) heat is a more energy-intensive process than the other pretreatment techniques for biomass. This form of dewatering has been investigated widely and is commonly used in industrial practice, e.g., in the food processing industry where drying is needed to prevent degradation but also for easy handling/packaging and for giving the product the right flow characteristics (not sticking). For most bioenergy purposes, the requirements of thermal drying are less strict. Important aspects are energy consumption, emission of volatile organic compounds (VOCs) and dust, and, most importantly, safety (fire and dust-explosion hazards when drying with oxygen-containing gas).

There are different ways to classify existing drying technologies (see for a thorough treatise, e.g., Mujumdar, 2006). An important distinction can be made between *direct* and *indirect drying*. Direct drying takes place with a flow of hot air, steam, or flue gas in direct contact with the wet biomass. Indirect drying prevents this contact, but heat is transferred through a casing by conduction.

Direct convection dryers can be divided into the following types:

- Belt conveyer
- Flash (pneumatic flow) dryer
- Fluidized bed
- Rotary drum
- Spray type
- Tray type

Indirect conduction dryers can be of the following types:

- Drum
- Steam jacket rotary drum
- Steam tube rotary drum
- Tray type

The selection of a dryer depends on the physical form of the biomass feedstock (particle size distribution and morphology), the heat sensitivity of the material, the required throughput capacity, the turndown ratio, and the pre- and postdrying operations. Table 8.3 shows a dryer selection guide that has been adapted from Mujumdar (2006).

Before going into details on drying characteristics, we need to deal with some basics of moist air thermodynamics, i.e., the field of psychrometrics. Important definitions used in this field are given in the following.

Humidity, also called the humidity ratio, is defined as

$$\omega = \frac{m_w}{m_a} = \frac{(MW_w p_w V/R_u T)}{(MW_a p_a V/R_u T)} = \frac{MW_w p_w}{MW_a p_a} \approx 0.622 \frac{p_w}{p_a} = 0.622 \frac{p_w}{p-p_w} \qquad \text{(Eq. 8.11)}$$

When air is saturated, one speaks of the humidity of saturated air, ω^{sat}. The (percentage) relative humidity is defined as

$$\phi = \frac{p_w}{p_w^{sat}} \times 100 \qquad \text{(Eq. 8.12)}$$

This term is distinguished from the percentage humidity:

$$\frac{\omega}{\omega^{sat}} = \frac{\{MW_w p_w / MW_a (p-p_w)\}}{\{MW_w p_w^{sat} / MW_a (p-p_w^{sat})\}} = \phi \frac{(p-p_w^{sat})}{(p-p_w)} \qquad \text{(Eq. 8.13)}$$

When a gas flow is used for evaporating water from (the surface of) a wet substance, vaporization takes place provided that the air is not completely saturated with water (at its dew point). The vaporization causes the temperature of the water to drop, and

TABLE 8.3 Selection of dryer types

Feedstock nature →	τ (min)	Liquids: slurry	Liquids: paste	Solids: cakes	Solids: powder	Solids: granules	Solids: pellets	Solids: fibers
		Convection dryers						
Belt	10–60					X	X	X
Flash	0–0.1			X	X	X		X
Fluid bed	10–60	X	X	X	X	X	X	
Rotary	10–60			X	X	X	X	X
Spray	0.1–0.5	X	X					
Tray	10–60 60–360 (batch)			X	X	X	X	X
		Conduction dryers						
Drum	0.1–0.5							
Steam-jacketed rotary	10–60	X	X					
Steam tube rotary	10–60			X	X	X	X	X
Tray	60–360 (batch)			X	X	X	X	X

Source: Adapted from Mujumdar (2006).

the driving force for heat supply from the gas to the wet surface is enhanced. The so-called wet-bulb temperature, T_{wb}, can then be reached at equilibrium, realizing that heat transfer to the vaporizing surface must equal mass transfer times heat of vaporization. The rate of heat transfer is given by

$$\dot{Q} = hA(T - T_{wb}) \qquad \text{(Eq. 8.14)}$$

The vaporization rate follows from

$$\varphi_{m,w} = kA \frac{MW_w \left(p_w^{sat} - p_w \right)}{R_u T} \approx kA \frac{MW_a}{R_u T} \left\{ (\overline{p - p_w})(\omega^{sat} - \omega) \right\} \approx kA\rho_a(\omega^{sat} - \omega) \qquad \text{(Eq. 8.15)}$$

The associated heat flow then is

$$\dot{Q} = kA\rho_a(\omega^{sat} - \omega)h_{fg} \qquad \text{(Eq. 8.16)}$$

Necessarily, the heat flows given by Equations (8.14) and (8.16) are equal at equilibrium; thus,

$$(T - T_{wb}) = (\omega^{sat} - \omega) \frac{k \rho_a h_{fg}}{h} \qquad \text{(Eq. 8.17)}$$

Thus, T_{wb} is practically only dependent on the temperature of the surroundings and the humidity of the air.

The analogy of Chilton and Colburn concerning heat and mass transfer under turbulent conditions and transfer from "fixed walls" leads to an expression for h/k:

$$\frac{Nu}{Re\,Pr^{0.33}} = \frac{Sh}{Re\,Sc^{0.33}} \Leftrightarrow \frac{h}{k} = \frac{\lambda}{D} \left(\frac{Pr}{Sc}\right)^{0.33} = \rho c_p \left(\frac{Sc}{Pr}\right)^{0.67} = \rho c_p Le^{0.67} \qquad \text{(Eq. 8.18)}$$

The physical properties in this expression are related to the surrounding air above the vaporizing medium.

Question: Derive the expression for h/k yourself with intermediate steps.

The process of drying can be described as a sequence of transport phenomena that occur on and within a drying solid particle. Drying of different biogenic materials usually occurs in two or more distinct stages. This is related to the fact that initially free moisture is available at the drying surface, resulting in a virtually constant drying rate. This drying phase is also called the *constant-rate period*, in which the material dries from an initial water content w_i to the critical water content w_c. For further drying, bound water from capillaries/interstitial spaces is to diffuse to the surface making it *just* wet, which causes retardation of the process. This phase in drying is called the *first falling-rate period*, and in this phase, the water content decreases from w_c to a lower value but above the equilibrium water content, w_e, which is material dependent. Upon further drying, the transport process from within the drying particle can no longer ensure a wetted surface and then the process becomes not so much dependent on outer conditions, but rather on internal molecular diffusion processes that depend on the characteristics of the fuel. This final drying phase is called the *second falling-rate period*; the process continues till w_e is reached and then the driving force for water removal has become zero.

Figure 8.3 shows a schematic of the drying rate behavior of some biogenic materials, illustrating different drying characteristics.

During the "constant-rate period," the drying time t_c can be expressed as

$$t_c = \frac{w_i - w_c}{R_c A} \qquad \text{(Eq. 8.18)}$$

with R_c being the constant drying rate per unit area.

Question: Derive an expression for R_c.

During the first falling-rate period, it can be assumed that the rate of drying is linearly dependent on the actual free moisture content $(w - w_e)$, expressed as follows:

$$-\frac{1}{A}\frac{dw}{dt} = K(w - w_e) = Kf \Rightarrow -\frac{1}{KA}\int_{w=w_c}^{w=w} \frac{dw}{(w - w_e)} = \int_{t=0}^{t=t_f} dt \qquad \text{(Eq. 8.19)}$$

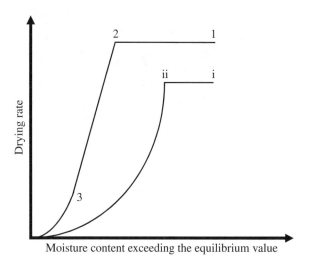

FIGURE 8.3   Schematic of the drying rate of some granular materials; 1–2 and i–ii represent a constant drying rate regime; 2–3 is the "first falling-rate" drying regime.

So,

$$\frac{1}{KA}\ln\left[\frac{w_c - w_e}{w - w_e}\right] = t_f \Rightarrow t_f = \frac{1}{KA}\ln\left(\frac{f_c}{f}\right) \qquad \text{(Eq. 8.20)}$$

The total drying time is $t_c + t_f$, and as the drying rate in the constant-rate period is equal to the rate of drying at the start of the falling-rate period, $R_c = Kf_c$, and therefore,

$$t = t_c + t_f = \frac{(w_i - w_c)}{KAf_c} + \frac{1}{KA}\ln\left(\frac{f_c}{f}\right) = \frac{1}{KA}\left\{\frac{(f_i - f_c)}{f_c} + \ln\left(\frac{f_c}{f}\right)\right\} \qquad \text{(Eq. 8.21)}$$

Associated with thermal drying are emissions of VOCs; these consist of, e.g., terpenes, alcohols, or organic acids. In case of an integrated combustion process, these gases can be incinerated to prevent VOC emissions. Their formation depends on biomass properties, resin contents, storing time, and operational aspects of the drying process: temperature, drying medium, required final moisture content, residence time, and particle size distribution of the dried biomass (Svoboda et al., 2009).

8.9   COMPACTION TECHNOLOGIES

Several of the disadvantages of raw biomass are a result of its low bulk density and the resulting low volumetric energy density. Therefore, in order to reduce transportation and storage and handling costs and to significantly improve its fuel characteristics, biomass is most often compacted. Compaction (or densification) has already been applied for a long time in food, feed, and pharmacy industrial sectors. Before giving an overview of the technologies, we first discuss some definitions of density.

The apparent density, ρ_{app} (also called bulk density), of a heap of biomass is the mass of all particles forming it divided by the total volume of the heap. This contains a substantial void fraction, ε, not occupied by particles. Its relation to the particle density, ρ_p, is as follows:

$$\rho_{app} = \rho_p(1-\varepsilon) \qquad \text{(Eq. 8.22)}$$

The particle density in its turn is related to the true material density but lower in value as the particle has a certain porosity, χ:

$$\rho_p = \rho_{true}(1-\chi) \qquad \text{(Eq. 8.23)}$$

Wetted particles, containing a fluid (e.g., moisture) of density ρ_f, have a density of

$$\rho_w = \rho_p + \chi\rho_f \qquad \text{(Eq. 8.24)}$$

In order to increase the apparent density of biomass, two techniques exist (we exclude baling as this is described as part of the harvesting phase in which it takes place), namely, briquetting and pelletizing.

An excellent overview of all densification techniques is provided by Tumuluru et al. (2011), but for the production of solid bioenergy carriers, the aforementioned technologies are most prominent.

Other advantages associated with the increase of the apparent density of biomass are:

- The energy density is increased.
- Handling of biomass is simplified (storage and logistics).
- Easier dosing is possibly realized depending on which technique is used.
- Lower water reabsorption can be realized.
- Pressing biomass can be combined with adding additives for improving downstream processes.
- Less dust is released.
- The stability is increased, preventing biological decay.
- A customer-tailored commodity product may be created.

8.9.1　Briquetting

Briquettes are commonly fabricated in an elongated press with extrusion of the biomass realized by a plunger or a screw (see Figure 8.4 for a schematic). The strain

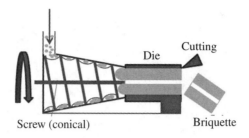

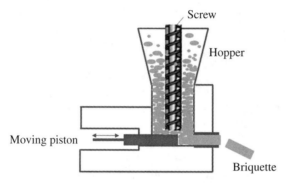

FIGURE 8.4 Some technologies for briquette production; screw press (top) and piston press (bottom).

that is released by the extrusion process is cut into well-shaped chunks, the briquettes. As briquettes are mainly applied in domestic energy supply and not so much industrially, we do not go into more details here.

8.9.2 Pelletizing

Pelletizing technology was introduced in 1880 for the purpose of producing cattle feed; since then, it increasingly shifted to fuel production for the (domestic) energy market, and now, pelletizing for biofuel production is a large-scale commercial activity. Pellets have become a rather common fuel type for domestic and industrial firing. In Europe, the two leading countries in this business are Finland and Sweden. Pellet production exceeded 1930 kt in 2007 (Hirsmark, 2002; Sikanen et al., 2009; Uslu et al., 2008). In 2008, the top four producers (in quantity) globally were the United States (4.1 Mt), Germany (3 Mt), Sweden (2.2 Mt), and Canada (2.1 Mt) (tinyurl.com/l9y3omo).

The major aim of the pelletizing process is to increase the energy density of the biomass and to achieve a considerably smaller specific volume, so that the pelletized biomass is more efficient in terms of storage, shipping, and converting it to heat and power or chemicals. The overall process comprises of pretreatment (size reduction,

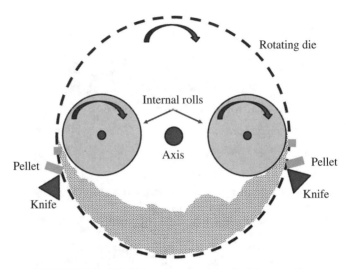

FIGURE 8.5   Working principle of a pelletizing machine.

drying), the pelletizing itself, and posttreatment (cooling, packaging). During the major processing phase of pelletizing, biomass is pressed through a press matrix and shaped into extrudates; also a ring matrix can be used. Figure 8.5 shows the working principle of a pelletizing machine.

Common dimensions of the pellets produced are diameters of 6–8 mm, though also larger diameter pellets are manufactured, and lengths varying from 3 to 50 mm (tinyurl.com/pq2f6us). The process demands a feedstock consisting of small particles with diameters preferably lower than ~6–8 mm (Samson et al., 2005) and moisture levels lower than 10–15 wt%, though a piston press can handle feedstock of a moisture content of up to 20 wt% (Uslu, 2005). Pelletizing is carried out with peak temperatures of about 150°C, because lignin starts softening at 100°C and then acts as a cellulose binder allowing the feedstock to fuse to form pellets. The moisture content of the feedstock is a crucial parameter, because biomass that is either too wet or too dry can cause a significant increase of the pressure required (Reed and Bryant, 1978). Pelletizing can be combined with torrefaction as is done in the development of the torrefaction and pelletizing process (TOP) technology, which is described in Chapter 12.

Intake fresh biomass generally has a moisture content of 50 wt%, but before the pelletizing step, this is reduced to ~15 wt%. The moisture content of the final product is 10 wt%. Since no significant volatilization occurs during the process, the total mass loss is due to water evaporation. Pelletizing typically has a thermal efficiency of 94% (94% of the initial energy remains in the pellet product), and the net efficiency, taking into account the energy consumption of the process, was calculated by Uslu to be around 87% (Uslu, 2005).

8.10   SEQUENCING THE PRETREATMENT STEPS

In the previous sections, we have discussed quite a broad spectrum of physical pretreatment techniques. Not all of these are needed in all cases; this depends on the quality characteristics of the biomass as well as on the foreseen downstream energy conversion processes that are described in the subsequent chapters including their pretreatment requirements. The final question then is what should be the sequence of the pretreatment steps.

Generally, some kind of storage follows upon the harvesting to create a buffer as direct downstream processing is usually not possible. However, storage poses restrictions on the moisture content in view of bacterial/fungal activity, so that drying to ~20–25 wt% moisture content is usually applied before storing the biomass. Usually, drying is preceded by a size reduction step to enhance the drying process. Sieving takes place after size reduction (when needed in view of the downstream processing). Leaching with water, sometimes used to improve the combustion characteristics of biomass, is in general followed by a drying step for thermochemical conversion. Compacting follows after a sufficient drying step has been applied to realize a stable product (bale/briquette/pellet).

CHAPTER SUMMARY AND STUDY GUIDE

This chapter provides an overview of physical biomass pretreatment techniques. It is important to appreciate the great heterogeneity of biomass and its subsequent processing technologies both giving rise to the need of such treatments. Various methods of harvesting are described, followed by storage and its impact on product qualities. Washing is needed for biomass with a high alkali and chorine content that is used as a feedstock for thermochemical conversion technologies. Size reduction is treated, which often precedes a drying step. Mechanical drying and thermal drying and its fundamentals are discussed next. Finally, different compaction techniques are dealt with, leading to a stable solid fuel product enabling more cost-effective transportation and other logistic handling.

KEY CONCEPTS

Necessity of physical pretreatment
Harvesting biomass and (local) transportation
Storage and its implications for biomass quality
Washing to ensure ash reduction
Size reduction (characterization and implication for downstream processing)
Moisture reduction via mechanical and thermal technologies
Compaction for production of stable intermediate solid products

SHORT-ANSWER QUESTIONS

8.1 What techniques are available for harvesting switchgrass?

8.2 Survey the literature concerning which part of the straw is (to be) left on the fields.

8.3 Why would debarking be advantageous when the wood will be burned for energy recovery?

8.4 Mention the pros and cons of indoor storage of grass in a barn.

8.5 Give an example of smart blending in a biomass-to-energy supply chain; which purpose(s) can it serve?

8.6 Why does a higher original moisture content of biomass cause higher losses of dry matter?

8.7 What is accomplished by biomass washing? For which types of biomass do you think this is relevant? List advantages and disadvantages of washing as a pretreatment technique.

8.8 Which particle properties change when size reduction is applied?

8.9 List advantages and disadvantages of direct comilling of biomass with coal. Is this always possible?

8.10 Which size reduction equipment would you select to mill sugarcane to enable pulverized fuel combustion?

8.11 How can particle size distributions be determined? Which average particle sizes can you name? Why are they needed?

8.12 In practice, is screening always needed for biomass processing?

8.13 Miao et al. (2011) show a correlation between the bulk density of biomass and the screen size as a power law expression. Why is the correlation characterized by a negative sign in the power of the characteristic size?

8.14 What are compaction techniques for straw? Is size reduction needed for the devices involved?

8.15 What are advantages and disadvantages of compacting biomass?

8.16 What are limits in the application of mechanical dewatering of biomass?

8.17 List advantages and disadvantages of direct and indirect drying techniques.

8.18 Describe the pros and cons of direct drying using flue gas and steam.

8.19 Is microwave heating an economically viable option for large-scale drying of biomass for heating purposes?

8.20 Which factors cause emissions of VOCs to increase when using direct drying techniques?

PROBLEMS

8.1   Estimate the fossil carbon footprint of 100 km truck transport (diesel consumption 1 L per 10 km) for the transportation of 30 tonnes of wood chips (moisture content 25 wt%) from storage location to processing unit. How does the fossil energy utilization compare to the energy content (LHV basis) of the biomass transported?

8.2   Under relatively wet conditions, storage of wood might lead to a greenhouse gas emission load of 144 $CO_{2eq} \cdot MWh^{-1}$ (Wihersaari, 2005); how does this compare to emissions of a coal-fired power station or a gas-fired combined cycle power station? What is your conclusion about the importance of appropriate storage conditions?

8.3   A certain type of straw (with 15 wt% moisture content) is submerged in water at a mass ratio of water to straw of 4. The straw initially contains 0.8 wt% Cl (ar basis). When taken out of the water, the straw is very wet with a moisture content of 80 wt% but a Cl content of only 0.03 wt%. What is the partitioning coefficient of Cl when this is defined as the mass fraction of Cl in the water phase divided by the Cl mass fraction in the wet biomass before submerging? What do you assume?

8.4   Kratky and Jirout (2011) have summarized literature concerning milling of biomass and its associated specific energy consumption. For the reported values for hardwood, try to fit the trends in terms of the presented models of Bond, Rittinger, and Kick. What do you conclude? Give an estimation of the relative energy use for the reported finest milling (to 1.6 mm) compared to the heating value of the hardwood.

8.5   In Table 8.2, a particle size distribution is given for crushed *Miscanthus* pellets. Determine the mass average, area average, volume average, area–volume average, and Sauter mean diameter. Compare the values. What do you conclude?

8.6   A community intends to use their sewage waste (using urine separation) to derive energy from it. The material is very wet, however, and even with urine separation, only about 29 wt% dry matter can be obtained. About 700 households produce 125 kg $\cdot$ day^{-1} of feces matter, and the combustion process that is to be applied requires 80 wt% dry matter material.

   a. How much water must be removed per second in a continuous process?

   b. In case no heat losses occur, what would be the energy needed for this drying process? Neglect the heating up of the feces, but only consider the water heat up and evaporation at atmospheric conditions.

   c. Would the assumption in (b) lead to a large difference in the calculated heat supply?

8.7 Very wet manure containing only 10 wt% dry matter (the rest can be considered to be water) is spray dried in a continuous-flow chamber. Small droplets of 1 mm radius with a density of 1000 kg · m^{-3} are introduced in the spray chamber with a velocity of 5 m · s^{-1}. A large flow of hot air is introduced at 130°C in cross-flow mode and has a relative humidity of 25%. Calculate the temperature of the droplets in the initial drying phase. Use steam tables and property tables for air for the calculations.

Can the Chilton–Colburn analogy be used in this case?

8.8 For a wet sludge of algae residue, a spray dryer is operated at steady state. The slurry, containing 30 wt% solids, enters the spray dryer through a spray head. Dry air enters at 177°C and 1 atm and moist air exits at 85°C and 1 atm, with a relative humidity of 21% and a volume flow of 310 m^3·min^{-1}. Dried particles are separated from the moist airstream.

Calculate: (i) The volume flow of dry air that enters in m^3·min^{-1}
(ii) The mass flow of leaving dried particles in kg·min^{-1}

8.9 A wood drying system consists of boiler in which dried wood (with 25 wt% moisture content) is burned with ambient air (25°C) and hot water is generated. The flue gas available at 130°C (immediately after the boiler) is used to dry the wood from 55 wt% moisture content in an integrated manner. The wood enters the dryer at 25°C. The initial LHV of the wood is 9.5 MJ·kg^{-1} (ar) and the wet wood feed rate is 250 kg·h^{-1}. Consider a combustion system in which combustion with 25% excess air is applied. Use the wood composition for pellets (daf basis) presented in Table 2.3. The boiler and dryer may be assumed to have no heat losses. The c_p value of wood can be assumed to be constant at 1200 kJ·kg^{-1}·K^{-1}.

a. Which type of dryer do you prefer for this system and why?
b. What must be the capacity of the air fan (in kg·h^{-1})?
c. What is the amount of heat transferred per unit of time to the water system in the boiler?
d. What is the end temperature of the flue gas after the dryer?
e. Is the temperature at the dryer exit above the dew point of the water?

PROJECTS

P8.1 Visit a plant where biomass (also, e.g., food/feed) is processed, and identify the different pretreatment techniques. Which techniques are most energy consuming and how much energy is consumed?

P8.2 Use a lab-scale ball mill (or develop one) and determine the HGI for some types of biomass. Study the impact of the moisture content and thermal pretreatment conditions (torrefaction; see Chapter 12).

INTERNET REFERENCES

tinyurl.com/bo9lcb
http://en.wikipedia.org/wiki/Woodchipper

tinyurl.com/l9y3omo
http://www.canbio.ca/upload/documents/canada-report-on-bioenergy-2010-sept-15-2010.pdf

http://tinyurl.com/pq2f6us
http://www.bioenergyconsult.com/tag/binders/

REFERENCES

Arvelakis S, Koukios EG. Physicochemical upgrading of agroresidues as feedstocks for energy production via thermochemical conversion methods. Biomass Bioenergy 2002;22:331–348.

Arvelakis S, Vourliotis P, Kakaras E, Koukios EG. Effect of leaching on the ash behavior of wheat straw and olive residue during fluidized bed combustion. Biomass Bioenergy 2001;20:459–470.

Barth HG, Winefordner JD. *Modern Methods of Particle Size Analysis.* New York: John Wiley & Sons; 1984.

Benemann JR. *Opportunities and Challenges in Algae Biofuels Production.* Singapore: Algae World, 2008.

Bond FC. Crushing and grinding calculations. Br Chem Eng 1961;6(6):378–385.

Bridgeman TG, Jones JM, Williams A, Waldron DJ. An investigation of the grindability of two torrefied energy crops. Fuel 2010;89:3911–3918.

British Standards Institution [BSI]. BS 1016–112:1995 (ISO 5074:1994) Methods for analysis and testing of coal and coke—part 112. Determination of Hardgrove grindability index of hard coals. London: BSI; 1995.

Bruton T, Lyons H, Lerat Y, Stanley M, Rasmussen MB. *A Review of the Potential of Marine Algae as a Source of Biofuel in Ireland.* Dublin: Sustainable Energy Ireland; 2009.

Colin F, Gazbar S. Distribution of water in sludges in relation to their mechanical dewatering. Water Res 1995;29(8):2000–2005.

Composites Innovation Centre [CIC]. Straw procurement business case: final report. Manitoba BioProducts Working Group, Prairie Practitioners Group Ltd. Winnipeg: Manitoba Agriculture, Food and Rural Initiatives; 2008.

De Jong W. Nitrogen compounds in pressurised fluidised bed gasification of biomass and fossil fuels [PhD thesis]. Delft (the Netherlands): Delft University of Technology; 2005.

De Souza-Santos ML. *Solid Fuels Combustion and Gasification: Modeling, Simulation, and Equipment Operation.* New York/Basel: Marcel Dekker Inc.; 2004.

Ghorbani Z, Masoumi AA, Hemmat A. Specific energy consumption for reducing the size of alfalfa chops using a hammer mill. Biosyst Eng 2010;105:34–40.

Hartman H, Strehler A. Die Stellung der Biomasse im Vergleich zu anderen erneuerbaren Energieträgern aus ökologischer, ökonomischer und technischer Sicht (in German). Schriftenreihe Nachwachsende Rohstoffe, Band 3. Münster-Hiltrup: Landwirtschaftsverlag; 1995.

Hatton JV. Debarking of frozen wood. Tappi J 1987;70(2):61–66.

Henderson SM, Perry RL. *Agricultural Process Engineering*. 3rd ed. Westport: AVI Publishing Co., Inc.; 1976.

Hirsmark J. Densified biomass fuels in Sweden: country report for the EU/IBDEBIF project [MSc]. Uppsala: Swedish University of Agriculture Sciences Department of Forest Management and Products; 2002.

Hoque M, Sokhansanj S, Naimi L, Bi X, Lim J, Womac A. Review and analysis of performance and productivity of size reduction equipment for fibrous materials. American Society of Agricultural and Biological Engineers (ASABE) Annual International Meeting. Minneapolis, MI: ASABE. p 18. 2007. Paper nr 076164.

Hunder M. Some aspects of wood chips' storage and drying for energy use. BioEnergy 2005: International Bioenergy in Wood Industry Conference and Exhibition. Jyväskylä (Finland); 2005. p 257–260.

Ige MT, Finner MF. Optimization of the performance of the cylinder type forage harvester cutterhead. Trans ASAE 1976;19(3):455–460.

Jenkins BM, Bakker RR, Wei JB. On the properties of washed straw. Biomass Bioenergy 1996;10(4):177–200.

Jenkins BM, Baxter LL, Miles TR. Combustion properties of biomass. Fuel Process Technol 1998;54(1–3):17–46.

Kaltschmitt M, Hartmann H, Hofbauer H. *Energie aus Biomasse: Grundlagen, Techniken und Verfahren*. Berlin: Springer-Verlag; 2001.

Knoef H. *Handbook Biomass Gasification*. Enschede: BTG Biomass Technology Group; 2005.

Kratky L, Jirout T. Biomass size reduction machines for enhancing biogas production. Chem Eng Technol 2011;34(3):391–399.

Lewandowski I, Clifton-Brown JC, Scurlock JMO, Huisman W. Miscanthus: European experience with a novel energy crop. Biomass Bioenergy 2000;19:209–227.

Maciejewska A, Veringa HJ, Sanders J, Peteves SD. Co-firing of biomass with coal: constraints and role of biomass pretreatment. Petten (the Netherlands): JRC; 2008. Report nr EUR 22461 EN.

Mani S, Tabil LG, Sokhansanj S. Grinding performance and physical properties of wheat and barley straws, corn stover and switchgrass. Biomass Bioenergy 2004;27:339–352.

Mason JS. *Bulk Solids Handling*. New York: Chapman and Hall; 1988.

McGlinchey D. *Bulk Solids Handling*. Oxford/Ames: Blackwell Publishers; 2008.

Merkus HG. *Particle Size Measurements: Fundamentals, Practices, Quality*. Dordrecht (the Netherlands): Springer; 2008.

Miao Z, Grift TE, Hansen AC, Ting KC. Energy requirement for comminution of biomass in relation to particle physical properties. Ind Crops Prod 2011;33(2):504–513.

Mujumdar AS, editor. Handbook of Industrial Drying. New York: CRC Press; 2006.

Nag PK. *Power Plant Engineering*. 3rd ed. New Delhi: Tata McGraw-Hill Education Private Ltd; 2008.

Niessen WR. *Combustion and Incineration Processes: Applications in Environmental Engineering*. Andover: CRC Press (Taylor & Francis Group); 2010.

Nurmi J, Lehtimäki J. Debarking and drying of downy birch (*Betula pubescens*) and Scots pine (*Pinus sylvestris*) fuelwood in conjunction with multi-tree harvesting. Biomass Bioenergy 2011;35:3376–3382.

Pankewitz A. Particle size analysis form lab to line: survey of state-of-the-art technologies. Powder Handl Process 2006;18(6):374–377.

Reed T, Bryant B. Densified biomass: a new form of solid fuel. Golden, CO: Solar Energy Research Institute (SERI), US Department of Energy Division of Solar Technology; 1978.

Reith JH, Deurwaarder EP, Hemmes K, Curvers APWM, Kamermans P, Brandenburg W, Zeeman G. Bio-offshore: grootschalige teelt van zeewieren in combinatie met offshore windparken in de Noordzee (in Dutch). Petten: ECN; 2005. Report nr ECN-C-05-008.

Roos CJ. *Biomass Drying and Dewatering for Clean Heat and Power.* Olympia (WA): Northwest CHP Application Center; 2008.

Samson R, Mani S, Boddey R, Sokhansanj S, Quesada D, Urquiaga S, Reis V, Lem CH. The potential of C4 perennial grasses for developing a global BIOHEAT industry. Crit Rev Plant Sci 2005;24(5–6):461–495.

Schell DC, Hardwood C. Milling of lignocellulosic biomass: results of pilot-scale testing. Appl Biochem Biotechnol 1994;45/46:159–168.

Sikanen L, Mutanen A, Röser D, Selkimäki M. Pellet markets in Finland and Europe: an overview. Northern Periphery Programme 2007–2013. Joensuu: North Karelia University of Applied Sciences; 2009.

Sims REH. *The Brilliance of Bioenergy.* London: James & James (Science Publishers) Ltd; 2002.

Spliethoff H. *Power Generation from Solid Fuels.* Berlin/Heidelberg: Springer Verlag; 2010.

Svoboda K, Martinec J, Pohorely M, Baxter D. Integration of biomass drying with combustion/ gasification technologies and minimization of emissions of organic compounds. Chem Pap 2009;63(1):15–25.

Syvitski JPM. *Principles, Methods and Application of Particle Size Analysis.* New York: Cambridge University Press; 2007.

Tumuluru JS, Wright CT, Hess JR, Kenney KL. A review of biomass densification systems to develop uniform feedstock commodities for bioenergy application. Biofuels Bioprod Biorefin J 2011;5:683–707.

Turn SQ, Kinoshita CM, Ishimura DM. Removal of inorganic constituents of biomass feed-stocks by mechanical dewatering and leaching. Biomass Bioenergy 1997;12(4):241–252.

Ugwu C, Aoyagi H, Uchiyama H. Photobioreactors for mass cultivation of algae. Bioresource Technology 2008;99:4021–4028.

Uslu A. Pre-treatment technologies and their effects on the international bioenergy supply chain logistics. Techno-economic evaluation of torrefaction, fast pyrolysis and pelletisation [MSc. Thesis]. Utrecht: Utrecht University; 2005.

Uslu A, Faaij APC, Bergman PCA. Pre-treatment technologies, and their effect on international bioenergy supply chain logistics. Techno-economic evaluation of torrefaction, fast pyrolysis and pelletisation. Energy 2008;33:1206–1223.

Van Loo S, Koppejan J, editors. *Handbook of Biomass Combustion and Co-Firing.* Prepared by task 32 of the implementing agreement on bioenergy under the auspices of the international energy agency. Enschede: Twente University Press; 2004.

Wihersaari M. VTT processes, evaluation of greenhouse gas emission risks from storage of wood residue. Biomass Bioenergy 2005;28:444–453.

Zhang W, Liang M, Lu C. Morphological and structural development of hardwood cellulose during mechanochemical pretreatment in solid state through pan-milling. Cellulose 2007;14:447–456.

9

THERMOCHEMICAL CONVERSION: DIRECT COMBUSTION

ROB J.M. BASTIAANS AND JEROEN A. VAN OIJEN

Department of Mechanical Engineering, Combustion Technology Section, Eindhoven University of Technology, Eindhoven, the Netherlands

ACRONYMS

BFB   bubbling fluidized bed
CFB   circulating fluidized bed
ESP   electrostatic precipitator
SCR   selective catalytic reduction
SNCR   selective noncatalytic reduction

SYMBOLS

A, B	constant	[–]
B	Spalding number	[–]
Bi	Biot number	[–]
C	integration constant	[Depending]
c_p	specific heat capacity	[$J \cdot kg^{-1} \cdot K^{-1}$]
D	diffusion coefficient	[$m^2 \cdot s^{-1}$]
D	diameter	[m]
d	diameter	[m]

Biomass as a Sustainable Energy Source for the Future: Fundamentals of Conversion Processes, First Edition. Edited by Wiebren de Jong and J. Ruud van Ommen.
© 2015 American Institute of Chemical Engineers, Inc. Published 2015 by John Wiley & Sons, Inc.

h	enthalpy	$[J \cdot kg^{-1}]$
h	heat transfer coefficient	$[W \cdot m^{-2} \cdot K^{-1}]$
h_{fg}	heat of vaporization	$[J \cdot kg^{-1}]$
K	evaporation constant	$[m^2 \cdot s^{-1}]$
k	reaction rate coefficient	$[s^{-1}]$
k_g	mass transfer coefficient	$[m \cdot s^{-1}]$
m	mass	$[kg]$
ṁ	mass flow rate	$[kg \cdot s^{-1}]$
r	radial coordinate	$[m]$
r_p	particle radius	$[m]$
Re	Reynolds number	$[-]$
s	source term	$[kg \cdot m^{-3} \cdot s^{-1}]$
Sc	Schmidt number	$[-]$
Sh	Sherwood number	$[-]$
T	temperature	$[K]$
Th	Thiele modulus	$[-]$
t	time	$[s]$
u, v	velocity	$[m \cdot s^{-1}]$
V	volume	$[m^3]$
Y_i	mass fraction of species i	$[-]$
ε	porosity	$[-]$
λ	thermal conductivity	$[W \cdot m^{-1} \cdot K^{-1}]$
ν	kinematic viscosity	$[m^2 \cdot s^{-1}]$
ρ	density	$[kg \cdot m^{-3}]$
τ	tortuosity	$[-]$
ξ	normalized radial coordinate	$[-]$
ψ	normalized concentration	$[-]$

Subscripts

0	initial value
1,2	integration constant numbers
boil	at boiling condition
d	diameter, droplet
eff	effective
f	flame
fg	evaporation
g	gas
m	mass
p	particle
q	evaporation
r	in radial direction
s	at the surface
∞	at infinity

9.1 INTRODUCTION

Combustion is the most direct process to convert biomass into usable energy. Hence, it is the most widespread method of biomass conversion. Scientific understanding is relatively well advanced but in several cases also limited mainly due to the complex nature of the process and the coupling of many physical processes. In this chapter, we describe some fundamental theory about combustion within a framework of practical applications.

The purpose of the combustion process is to release the energy of the biomass. This energy is stored in the form of chemical bonds. It is done by means of oxidation. Also, conversion in sensible heat, carried by the hot flue gases, is needed. This heat can be used directly for heating purposes (e.g., room heating), or it can be transferred to a working medium in a heat exchanger. The working medium then drives a heat engine to create mechanical energy for power production.

Most biomass combustion systems are based on technologies for the combustion of solid fuels, coal in particular. Coal combustion accounts for almost 50% of the world's electricity production, and the involved technologies are quite mature. Biomass is usually fired or cofired with coal in fixed bed, fluidized bed, or entrained flow systems. These reactors are discussed in Section 9.3.

In combustion processes, both homogeneous and heterogeneous reactions play a role. Homogeneous reactions are reactions taking place in the gas phase. Heterogeneous reactions occur at interfaces between solids and gases. Heterogeneous combustion processes of biomass and solid fuels in general are more complex than homogeneous combustion of gaseous fuels. Solid fuels, such as biomass, are composed of different fractions of minerals and organic matter, i.e., hydrocarbons. The combustion of biomass consists of four processes:

1. Drying
2. Pyrolysis
3. Combustion of volatile matter
4. Combustion of the residual char

Drying occurs by evaporation of moisture contained in the biomass and uses heat released by the combustion processes. Pyrolysis is the thermal decomposition (devolatilization) process as a consequence of the heating up of the fuel. The energy necessary to heat up the fuel to a temperature that is sufficiently high to induce ignition is mostly transferred by convection and radiation from the combustion zone. The drying, pyrolysis, and combustion processes do not necessarily run one after the other but may overlap, depending on the firing mode. In the following, these processes, as elements of solid fuel combustion, are discussed in more detail.

9.2 FUNDAMENTAL CONVERSION PROCESSES

9.2.1 Drying

Biomass particles may contain a large amount of water in different forms, either adhered to the particle surface, inside the particle pores, or also chemically bound to the solid matrix. When a fresh biomass particle enters a combustion system, it will first heat up by conduction, convection, and radiation. When the temperature of the particle rises above 100°C, the water it contains starts to evaporate, leaving the biomass particle by convection and diffusion through the pores of the solid matrix.

A flexible combustion system should be capable of drying solid fuels with different moisture contents. In grate or fluidized bed firing systems, the reactor can be fed with moisture-containing fuels without further treatment. The fuel in the case of a pulverized fuel firing is predried in order to ensure a fast combustion process within the available short residence time. A maximum moisture content is acceptable for combustion to be self-sustainable. This implies that the heat produced by combustion should be larger than the heat losses plus the heat needed for water evaporation.

9.2.2 Pyrolysis

The thermal decomposition of biomass and the formation of gaseous products during heating of the biomass particles are termed devolatilization or pyrolysis. Pyrolysis is an endothermic process by definition occurring in the absence of oxygen. Devolatilization of the solid biomass starts at temperatures above 200°C, by the thermal cracking of compounds. Formation of tars, liquids, and gaseous products occurs at temperatures of up to 600°C. The mixture of volatile gases is composed of carbon monoxide (CO), carbon dioxide (CO_2), methane (CH_4), and other light hydrocarbons such as C_2H_6, C_2H_4, and C_2H_2. Tars are complex hydrocarbon compounds with an organic structure similar to that of the base fuel. These tars evaporate from the biomass at temperatures between about 400 and 600°C. Tars condense at temperatures below 200°C forming a sticky layer at the walls of the furnace and other equipment.

Further heating up to temperatures of more than 600°C results in the conversion of the solid intermediate into char. This leads to a split-off of carbon monoxide and hydrogen. With increasing temperature, secondary cracking reactions cause the formation of light gas components such as hydrogen and carbon monoxide and also soot from the tar compounds. The fraction of the individual volatile components and the history of their release depend on the fuel type, the final temperature, and the heating rate. With increasing heating rate, the devolatilization maxima of the components shift toward relative later moments and thus higher temperatures. Higher end temperatures lead to a larger amount of volatile matter. The content of volatile matter determined at the high temperature and heating rate of an entrained flow reactor may amount to 1.1–1.8 times the amount detected in a proximate analysis at a relatively low temperature and heating rate.

The volatile matter content of biomass is in general much higher than that of coal. Therefore, during the combustion of biomass, the pyrolysis stage lasts a relatively long period compared to coal. For biomass, the pyrolysis stage may take 50% of the total conversion time, while for coal, it is often negligible compared to the time for char burnout to be completed. In addition, the heat released by the combustion of volatile matter relative to the heat released by char combustion is much higher for biomass than for coal, which could cause problems in the furnace design.

9.2.3   Combustion of Volatile Matter

The homogeneous combustion of the volatile components is characterized by a very high reaction rate. The burning time of volatile matter is essentially determined by the release rate of volatiles and their mixing with air. Volatile components formed on the surface of the biomass particles diffuse into the bulk gas. Therefore, the concentration of volatiles decreases with an increase of the distance to the biomass particle. In contrast, the temperature and oxygen concentration increase with an increase of the distance to the particle. At a certain distance from the particle, volatile matter and oxygen are present at stoichiometric concentrations. This leads to stabilization of the combustion and formation of a flame enveloping the particle. Under laminar flow conditions, the diameter of this flame is about three to five times the diameter of the particle. In pulverized fuel combustion and under more turbulent flow conditions, the volatile matter combustion processes of the individual particles combine, forming a large gas flame enclosing groups of particles. In fluidized bed operation, the flame cannot form inside the bed but only above the bed surface.

9.2.4   Combustion of Char

The porous structure that remains after the volatile matter is released from the particle consists almost only of carbon and ash and is called char. The carbon is oxidized with oxygen, at a sufficiently high particle surface temperature, resulting in carbon monoxide, carbon dioxide, and water vapor.

The reaction rate of the heterogeneous combustion of char is orders of magnitude lower than the homogeneous combustion of volatile matter. The total combustion time is therefore determined by the rate of char combustion. Together with the pyrolysis stage, it is decisive for the design of biomass firing systems.

The course of char combustion of a single particle is shown in Figure 9.1. Inside the particle or on the surface, the heterogeneous oxidation of carbon takes place with the oxidants oxygen, carbon dioxide, and water vapor:

$$2C + O_2 \rightarrow 2CO \qquad \text{(RX. 9.1)}$$

$$C + CO_2 \rightleftarrows 2CO \text{ (Boudouard reaction)} \qquad \text{(RX. 9.2)}$$

$$C + H_2O \rightleftarrows CO + H_2 \text{ (heterogeneous water – gas reaction)} \qquad \text{(RX. 9.3)}$$

In practice, reactions (RX. 9.2) and (RX. 9.3) are also virtually irreversible. Today, it is often assumed that first only conversion to carbon monoxide takes place directly on the particle, either by incomplete combustion (RX. 9.1) or by gasification ((RX. 9.2)

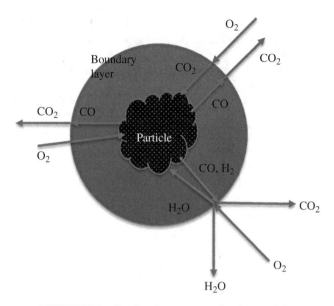

FIGURE 9.1 Combustion process of a char particle.

and (RX. 9.3)), and that CO will further react to CO_2. There is, however, still some uncertainty on this point. A gaseous atmosphere forms around the char particle, consisting of the combustion products CO and H_2 and the oxidants O_2, CO_2, and H_2O. The oxidants have to diffuse to the particle surface through the boundary layer, and, in the opposite direction, the combustion products diffuse from the particle to the environment. In the surrounding boundary layer, the following homogeneous oxidation reactions take place:

$$2CO + O_2 \rightarrow 2CO_2 \qquad (RX.\,9.4)$$

$$2H_2 + O_2 \rightarrow 2H_2O \qquad (RX.\,9.5)$$

In these homogeneous reactions, the products of the heterogeneous oxidation (CO and H_2) are oxidized into the final combustion products CO_2 and H_2O.

9.3 PARTICLE CONVERSION MODES

After the heterogeneous pyrolysis process, the resulting gas species can react further homogeneously in the gas phase as described in Section 9.2.3. However, depending on the pyrolysis reaction conditions, there will also be a certain amount of char left, which can also react further in the presence of an oxidizing agent.

Combustion of solid fuels is a very complicated process. The chemical reactions occurring are often not known. This especially is the case when there are both homogeneous (gas phase) reactions and heterogeneous reactions (at the surface of a solid fuel). Moreover, diffusion of the oxidizer to the outer surface of the solid fuel, where the surface can oxidize, has to be taken into account, as well as diffusion of the fuel through the pores where both homogeneous and heterogeneous reactions may take place. Also, the solid fuel will not only vaporize, but depending on the fuel, it might melt and eventually pyrolyze as well.

In order to understand the conversion process in a quantitative way, one could analyze one-dimensional systems like a propagating front. Studies are carried out on a planar traveling conversion wave or spherical particles, employing this strategy. The plane configuration is relevant for fixed bed conversion as explained by Van Kuijk et al. (2008), whereas the model of spherical particles applies to larger-scale fluidized beds and entrained flow reactors. Here, we will look at the combustion of the remaining (porous) solid particles that are composed of carbon, with mass fraction Y_C, and inert ashes that are not converted. It is assumed that reactions exclusively take place within the particle.

Figure 9.2 shows three simplified modes for the conversion of char. In these models, the char is converted either at the outside surface or inside the particle. Also, a practical particle conversion situation in which an ash layer builds up is considered in this figure. Models with more details and fewer assumptions can be found in the literature (see, e.g., Goméz-Barea and Leckner, 2010). A general discussion on this subject can be found in, e.g., Levenspiel (2012).

It is often assumed that the chemistry of this combustion process can be modeled with a single reaction:

$$C + O_2 \rightarrow CO_2 \tag{RX. 9.6}$$

At high temperatures, the observed behavior does not correspond to the reaction model shown in the equation above; it is found that the mass fraction of oxygen

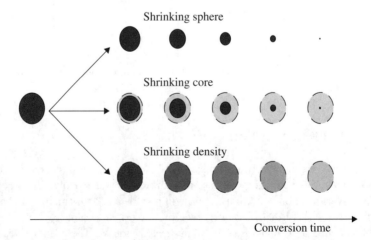

FIGURE 9.2 Simplified char particle combustion modes. (Source: Reproduced with permission from Thunman and Leckner (2007). © H. Thunman.)

has decreased to zero some distance away from the carbon surface. However, CO_2 can then act as an oxidizing agent and react with the surface carbon to form CO. This CO will be transported away from the surface by convection, and at some distance, it will be converted with O_2 to form CO_2. Thus, high-temperature combustion of carbon particles can be modeled using a two-step process:

$$C + CO_2 \rightleftharpoons 2CO \qquad (RX.\,9.7)$$

$$2CO + O_2 \rightarrow 2CO_2 \qquad (RX.\,9.8)$$

with the first reaction occurring at the surface and the second in the gas phase.

9.3.1 Shrinking Density Model

Here, the shrinking density model of Figure 9.2, in which reaction takes place inside the particle, resulting in a density decrease, is considered. If the particle is assumed to be isothermal, the mass conservation equation for oxygen in spherical coordinates can be written as

$$\frac{\partial}{\partial t}\left(\varepsilon \rho_g Y_{O_2}\right) + \frac{1}{r^2}\frac{\partial}{\partial r}\left(r^2 \varepsilon u\, \rho_g Y_{O_2}\right) = \frac{1}{r^2}\frac{\partial}{\partial r}\left(r^2 \rho_g D_{\mathit{eff}}\frac{\partial Y_{O_2}}{\partial r}\right) + s \qquad (Eq.\,9.1)$$

where ε is the particle porosity, ρ_g is the gas density, u is the gas velocity, and D_{eff} is the effective diffusion coefficient. The first term on the left hand side is the temporal change, and the second term represents convection of oxygen. On the right hand side, the first term represents diffusion of oxygen, and the second term, s, is the source term that describes the chemical consumption of oxygen due to conversion. The relation between the effective diffusion coefficient inside a char particle and the molecular diffusion coefficient D can be estimated from the porosity of the particle:

$$D_{\mathit{eff}} = \frac{\varepsilon}{\tau}D \sim \varepsilon^2 D \qquad (Eq.\,9.2)$$

This is an approximate equation where the inverse of the tortuosity τ is assumed to be almost equal to the porosity. So the transport equation for oxygen Equation (9.1) can be used for the situation inside and outside the particle where the effective diffusion coefficient is the molecular diffusion coefficient outside the particle and the porosity modulated diffusion coefficient inside the particle. Furthermore, we assume that the system is in quasi-steady state and that the carbon present in the char is converted to CO_2. The latter assumption (which is a simplification since carbon also can be converted to CO) makes the reaction equimolar, and no gas flow will therefore be present inside or outside the char particle. The equation to be solved is then reduced to a balance containing only a diffusion term and a source term. Assuming first-order kinetics, the source term can be written as

$$s = -k_{O_2}\rho_g Y_{O_2} \qquad (Eq.\,9.3)$$

Introducing normalized quantities for the coordinate and for the mass fraction of oxygen, $\xi = r/r_p$ and $\psi = \frac{Y_{O_2}}{Y_{O_2, \infty}}$ where r_p is the particle radius and $Y_{O_2, \infty}$ is the oxygen mass fraction far outside the boundary layer. After substitution of the equation for the source term Equation (9.3) and some manipulation, the conservation Equation (9.1) becomes

$$\frac{1}{\xi^2} \frac{1}{r_p} \frac{\partial}{\partial \xi} \left(\xi^2 D_{eff} \frac{Y_{O_2, \infty}}{r_p} \frac{\partial \psi}{\partial \xi} \right) - k_{O_2} Y_{O_2, \infty} \psi = 0 \qquad \text{(Eq. 9.4)}$$

which can be rewritten as

$$\frac{1}{\xi^2} \frac{\partial}{\partial \xi} \left(\xi^2 \frac{\partial \psi}{\partial \xi} \right) - Th^2 \psi = 0 \qquad \text{(Eq. 9.5)}$$

where

$$Th = r_p \sqrt{\frac{k_{O_2}}{D_{eff}}} \qquad \text{(Eq. 9.6)}$$

is a dimensionless number, called the "Thiele modulus." The Thiele modulus tells us whether the reaction is controlled by internal diffusion or the (intrinsic) reaction rate. If the Thiele modulus is large ($\gg 1$), the conversion of the char particle is controlled by diffusion, while if the Thiele modulus is small ($\ll 1$), the conversion is controlled by chemical reaction. With the introduction of the Thiele modulus, the general solution ψ of the equation is

$$\psi = \frac{1}{\xi} \left(A e^{Th\xi} + B e^{-Th\xi} \right) \qquad \text{(Eq. 9.7)}$$

The boundary condition in the center of the particle is a symmetry condition, which means that the first derivative of ψ is equal to zero. From this boundary condition, it follows that $A = -B$.

The boundary condition at the surface of the particle is that the molar flow of oxygen through the boundary layer is equal to the diffusion of oxygen from the surroundings into the particle:

$$-D_{eff} \frac{\partial Y_{O_2}}{\partial r} \bigg|_{r=r_p} = k_g \left(Y_{O_2} - Y_{O_2, \infty} \right) \qquad \text{(Eq. 9.8)}$$

where k_g is the mass transfer coefficient. Equation (9.8) can be written as

$$-D_{eff} \frac{Y_{O_2, \infty}}{r_p} \frac{\partial \psi}{\partial \xi} \bigg|_{r=r_p} = k_g \left(Y_{O_2, \infty} \psi - Y_{O_2, \infty} \right) \qquad \text{(Eq. 9.9)}$$

which can be written in dimensionless form as

$$-\frac{\partial \psi}{\partial \xi} = \frac{r_p k_g}{D_{eff}}(\psi - 1) = Bi_m(\psi - 1) \tag{Eq. 9.10}$$

which defines Bi_m, the Biot number for mass transfer. The Biot number is the ratio of external mass transfer to internal mass transfer by diffusion. It can be considered as a measure of the thickness of the boundary layer through which the molecules have to diffuse to reach the particle. In contrast to the Thiele modulus, the Biot number relates to the condition external to the particle. The Biot number has the same form as the Sherwood number, which for isolated spheres in a gas flow can be estimated from the Reynolds and Schmidt numbers as

$$Sh = \frac{2 r_p k_g}{D} = 2 + 0.6 Re^{1/2} Sc^{1/3} \tag{Eq. 9.11}$$

with $Re = \frac{2 r_p u}{\nu}$ and $Sc = \frac{\nu}{D}$.

After substitution of A with $-B$, deriving the equation for ψ, and substituting the result into the boundary condition at the surface of the particle, the constant A can be written as

$$A = -B = \frac{Bi_m}{(Bi_m + Th - 1)e^{Th} + (Th + 1 - Bi_m)e^{-Th}} \tag{Eq. 9.12}$$

The normalized oxygen mass fraction can now be plotted for different values of the Thiele modulus and the Biot number in order to investigate the behavior of the char combustion. To better visualize the influence of different parameters, with Equations (9.2) and (9.11), the Biot number is rewritten as

$$Bi_m = \frac{r_p k_g}{D_{eff}} = \frac{D}{D_{eff}} \frac{Sh}{2} = \frac{Sh}{2\varepsilon^2} \tag{Eq. 9.13}$$

As the minimum of the Sherwood number Equation (9.11) is 2, the Biot number always must be larger than 1. For a porous particle, e.g., char, the porosity is often around 0.5 or higher, which gives a Biot number around 4 or less. The normalized mass fraction of oxygen, plotted against the dimensionless particle radius, for different Biot number is shown in Figure 9.3. As can be seen in the figure, the Biot number influences the mass fraction of oxygen inside the particle. This is an expected result since a high Bi_m means that there is no external diffusion resistance; the rate of mass transfer of oxygen from the surroundings to the surface of the particle is much higher than the rate of mass transfer of oxygen into the interior of the particle.

The normalized mass fraction of oxygen, plotted against the dimensionless particle radius for different Thiele numbers, is shown in Figure 9.4. The result from the variation of the Thiele modulus is more interesting; it visualizes under which conditions

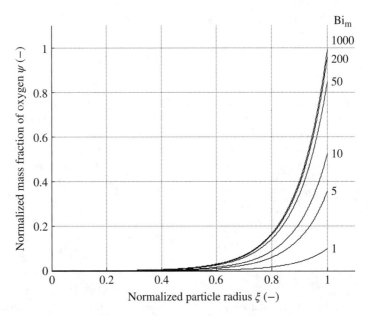

FIGURE 9.3 Normalized mass fraction of oxygen as a function of the normalized particle radius, left: for different Bi_m numbers at $Th = 10$.

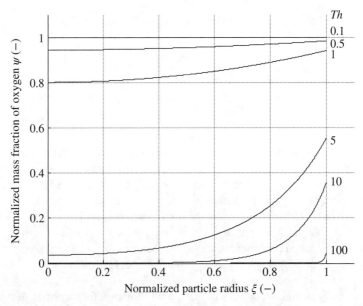

FIGURE 9.4 Normalized mass fraction of oxygen as a function of the normalized particle radius, for different values of Th at $Bi_m = 5$.

the different simplified char combustion models of Figure 9.2 can be used. The shrinking density model, discussed earlier, assumes that the reactions take place uniformly inside the char particle, which is true for low values of the Thiele modulus (<1) as can be seen from Figure 9.4. An alternative simplified model for char combustion is the shrinking sphere model, or when an ash layer is built up outside of the reacting core, the char burns like a shrinking core and the shrinking core model can be used. The assumption for the shrinking sphere and the shrinking core model is that the reactions take place in a narrow area close to what is defined as the reaction surface. Figure 9.4 shows that this assumption is valid for high values of the Thiele modulus (>10). More about the use of the dimensionless form in combustion and in fluid dynamics is given by Thunman and Leckner (2007).

9.3.2 Shrinking Sphere Model

The theory of surface combustion of isolated fuel droplets goes back to 1953 with a publication by Spalding (1953). This theory is used as a basis for the present analysis, supplemented by information as found on the website of Professor Dryer from Princeton University (tinyurl.com/o7cacsn). In this section, the general theory of surface conversion is explained by first looking at simple evaporation of a droplet. At the end of the chapter, in Problem 9.5, the extension to combustion of fuel droplets is treated.

Because of the simplicity of analyzing evaporation processes, we derive the model for evaporation in order to understand the approach. In this derivation, the so-called Spalding number pops up, which can be arranged in such a way that other processes like surface combustion and combustion with a standoff distance can be treated as well. A good reference is the book of Turns (2000). Models of solid-sphere combustion under the three simplified conditions of shrinking sphere, shrinking core, and shrinking density behavior are widely available in the literature.

We consider an evaporating, spherical droplet that is heated up to its boiling temperature and assume it to be present in a quiescent surrounding with an even higher temperature sufficiently far away from it. In this case, the droplet acts as a heat sink, associated simultaneously with the release of vapor to the surroundings. By this mechanism, the diameter of the droplet decreases at a certain rate. Hereby, the lifetime (or burnout time) of the droplet, t_d, is defined as the time at which its diameter reaches zero, starting from an initial diameter of $d_d (t = 0) = d_{d,0}$. It is found that the lifetime of such a droplet is proportional to a rate constant, $1/K$ (K is called the evaporation constant), times the initial diameter squared, $t_d = d_{d,0}^2/K$. The constant is determined by the ratio of available heat transported to the droplet and the heat of vaporization of the substance in the droplet, which is referred to as the Spalding number or transfer number. Therefore, the governing law is called the d^2-law. This law can be used for a set of shrinking processes like evaporation, surface gasification, surface combustion, and sheath combustion. It is important to stress that the principal change in the description of these processes is the definition of the Spalding number and therefore the rate constant K.

The basic derivation in case of vaporization proceeds as follows. We start with the observation that in a steady conversion, the mass flow rate, $\dot{m} = 4\pi r^2 \rho_g v_r$, remains the same in the gas surrounding the particle (continuity):

$$\frac{\partial\left(r^2 \rho_g v_r\right)}{\partial r} = 0 \qquad \text{(Eq. 9.14)}$$

Now, the energy equation can be written as a balance equation for convection and diffusion, which in terms of temperature becomes

$$\frac{\partial}{\partial r}\left(r^2 \rho_g v_r T\right) - \frac{\partial}{\partial r}\left(r^2 \rho_g D \frac{\partial T}{\partial r}\right) = 0 \qquad \text{(Eq. 9.15)}$$

We can use continuity, $r^2 \rho_g v_r = \dot{m}/4\pi$, to arrive at

$$\frac{\dot{m}}{4\pi}\frac{\partial T}{\partial r} - \frac{\partial}{\partial r}\left(r^2 \rho_g D \frac{\partial T}{\partial r}\right) = 0 \qquad \text{(Eq. 9.16)}$$

and assume for the Lewis number, being $Le = \lambda/(c_p \rho_g D)$, $Le = 1$. Herein, λ is the thermal conductivity and c_p the specific heat capacity. Then Equation (9.16) becomes

$$\frac{\dot{m}}{4\pi}\frac{c_p}{\lambda}\frac{\partial T}{\partial r} - \frac{\partial}{\partial r}\left(r^2 \rho_g D \frac{\partial T}{\partial r}\right) = 0 \qquad \text{(Eq. 9.17)}$$

Introducing $Z = c_p/(4\pi\lambda)$ and integrating give

$$r^2 \frac{\partial T}{\partial r} = Z\dot{m}T + C_1 \qquad \text{(Eq. 9.18)}$$

Integrating once again results in a solution for the temperature

$$T(r) = \frac{-C_1}{Z\dot{m}} + C_2 \exp\left(-\frac{Z\dot{m}}{r}\right) \qquad \text{(Eq. 9.19)}$$

Boundary conditions can be formulated at the droplet surface and far away from the droplet, $T(r = r_s) = T_{boil}$ and $T(r \to \infty) = T_\infty$. Substituting these solutions, it is easy to derive

$$C_2 = \frac{T_\infty - T_{boil}}{1 - \exp\left(-\frac{Z\dot{m}}{r_s}\right)} \qquad \text{(Eq. 9.20)}$$

and $C_1 = (C_2 - T_\infty)Z\dot{m}$, leading eventually to the temperature distribution

$$T(r) = \frac{(T_\infty - T_{boil})\exp\left(-\frac{Z\dot{m}}{r}\right) - T_\infty \exp\left(-\frac{Z\dot{m}}{r_s}\right) + T_{boil}}{1 - \exp\left(-\frac{Z\dot{m}}{r_s}\right)} \qquad \text{(Eq. 9.21)}$$

Using this temperature profile, we know the (conductive) heat transfer to the surface of the droplet, which should balance the heat of vaporization:

$$4\pi\lambda r_s^2 \frac{\partial T}{\partial r}\Big|_{r=r_s} = \dot{m}h_{fg} \qquad (Eq. 9.22)$$

in which h_{fg} is the heat of vaporization. Taking the derivative of the temperature at the particle surface,

$$\frac{\partial T}{\partial r}\Big|_{r=r_s} = \frac{Z\dot{m}}{r_s^2}\left(\frac{(T_\infty - T_{boil})\exp\left(-\frac{Z\dot{m}}{r_s}\right)}{1 - \exp\left(-\frac{Z\dot{m}}{r_s}\right)}\right) \qquad (Eq. 9.23)$$

the mass flow rate becomes

$$\dot{m} = \left(\frac{4\pi\lambda r_s}{c_p}\right)\ln(B_q + 1) \qquad (Eq. 9.24)$$

with $B_q = c_p(T_\infty - T_{boil})/h_{fg}$ being the Spalding number B for heat transfer (q) by evaporation. If, e.g., combustion is considered, we also have to take into account the release of heat by the conversion. This can be easily taken care of by involving this quantity in the definition of the Spalding number (this is shown in Problem 9.4).

Now, the lifetime of a droplet is determined from the mass flow rate according to

$$\frac{\partial m_d}{\partial t} = -\dot{m} \qquad (Eq. 9.25)$$

with the mass of a droplet as a function of the diameter being $m_d = \rho_d V_d = \rho_d \pi d_d^3/6$. For the change of the diameter as a function of time, we can write $\frac{\partial d_d^3}{\partial t} = 3d_d^2 \frac{\partial d_d}{\partial t}$ and thus

$$\frac{\partial d_d}{\partial t} = -\frac{4\lambda}{\rho_d c_p d_d}\ln(B_q + 1) \qquad (Eq. 9.26)$$

In the expression for the lifetime of a droplet, the diameter squared is generally used, so we write Equation (9.26) as

$$\frac{\partial d_d^2}{\partial t} = \frac{-8\lambda}{\rho_d c_p}\ln(B_q + 1) = -K \qquad (Eq. 9.27)$$

which shows that the diameter squared decreases linearly with time. This equation is important since it is the definition of the evaporation constant K. Integration yields

$$\int_{d_{d,0}^2}^{d_d^2} dd_d^2 = -\int_0^{t_d} K\,dt \qquad (Eq. 9.28)$$

and we find that

$$d_d^2(t) = d_{d,0}^2 - Kt \qquad \text{(Eq. 9.29)}$$

The lifetime of the droplet is determined at a droplet diameter of zero, so

$$t_d = \frac{d_{d,0}^2}{K} \qquad \text{(Eq. 9.30)}$$

The present derivation for the lifetime was made for an evaporation process. The same approach can be followed for film combustion or combustion with a flame with a standoff distance, and we can determine the burnout time. The only parameter that has to be changed is the Spalding number to include the actual exact heat fluxes associated with the process. For this, the reader is referred to Turns (2000).

Example 9.1   Calculation of a fuel's droplet lifetime

Consider a droplet of octane, C_8H_{18}, with a diameter of 50 µm, which is at its boiling temperature in an atmospheric environment at a temperature of 1000°C. The boiling point of octane is 125°C, its heat of vaporization h_{fg} is 125.7 kJ·kg^{-1}, and its specific heat capacity c_p is 2.238 kJ·kg^{-1}·K^{-1}. The density of liquid octane is 703 kg·m^{-3}, and the diffusivity into the surrounding medium, $\rho_g D$, can be taken as 10^{-5} kg·m^{-1}·s^{-1}. Notice the typical magnitudes of the quantities and their dimensions.

   How long does it take for this octane droplet to evaporate completely?

Solution
First, we need to evaluate the rate of vaporization. For that, the Spalding number for vaporization has to be calculated:

$$B_q = \frac{c_p(T_\infty - T_{boil})}{h_{fg}} = \frac{2.238(1000 - 125)}{125.7} = 15.1$$

The rate of vaporization then becomes (Equation (9.27) with $\lambda/c_p = \rho_d D$)

$$K = 8\rho_d D \frac{\ln(B_q + 1)}{\rho_d} = 8 \times 10^{-5} \frac{\ln 16.1}{703} = 3.2 \times 10^{-7} \, m^2 \cdot s^{-1}$$

Thus, the lifetime of the droplet is Equation (9.30)

$$t_d = \frac{d_{d,0}^2}{K} = \frac{(50 \times 10^{-6})^2}{3.2 \times 10^{-7}} = 0.0079 \, s$$

This time is of the same order as the time scale of combustion so that would be a good match. However, the mixing of the fuel with air and possible droplet group behavior will complicate the exact interaction seriously. Furthermore, a complete analysis including a combustion front would give more information.

In most cases, the phenomena take place under turbulent flow conditions. Under simplified laminar conditions, there is no mutual influence of droplets or particles, but in many practical cases, there is. Much research is being performed regarding these turbulent cases. For a start in this particular subject, the reader is referred to the work of Chiu et al. (1982). Their famous diagram indicates different regimes like single-droplet combustion, internal group combustion, external group combustion, and external sheath combustion, depending on the so-called group-combustion number G.

9.4 COMBUSTION SYSTEMS

Biomass combustion systems comprise fuel preparation and supply, fuel and combustion air transport and distribution, a combustion system for releasing the heat from the fuel in a furnace, and flue gas cleaning. Typical systems utilized for combustion of solid fuels are:

- Grate firing/fixed bed combustion
- Fluidized bed firing
- Pulverized fuel firing/entrained flow combustion

In the top of Figure 9.5, a schematic overview of the different firing systems is displayed. Furthermore, Figure 9.5 shows the characteristic gas and solid fuel flow velocities, pressure losses, and heat transfer coefficients for each of the different types of firing systems. The representation in the bottom graph of Figure 9.5 is not entirely correct since this assumes the bed to expand linearly with the gas velocity, which is certainly not true.

9.4.1 Grate Firing/Fixed Bed Combustion

In a grate firing system, the biomass lies in a stagnant bed on a moving grate (see Figure 9.6). The fuel burns with the combustion air, which is blown through the grate bars from below. At sufficiently low gas velocities, single biomass particles, with sizes of up to 30 mm, remain in the fuel layer on the grate. In that case, negligible quantities of solids are ejected from the bulk. Because of the limited size, grate furnaces are only used in the case of small capacities such as for small industrial and thermal power plants. Grate firing is the preferred system for ballast-containing fuels such as wastes or biomass.

9.4.2 Fluidized Bed Firing

In fluidized bed firing, the solid fuel whirls and burns staying in a fluidized gas–solid suspension. The fluidizing medium provides the oxygen, for the oxidation, of the fuel

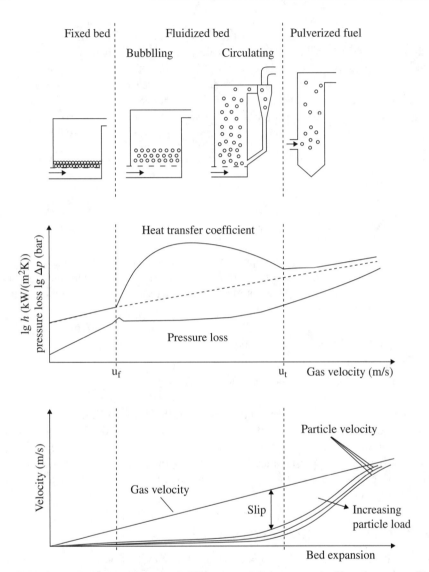

FIGURE 9.5 Distinctive features of different combustion systems. **Top: heat transfer coefficient and pressure loss as a function of gas velocity. Bottom: velocity as a function of the relative bed expansion**. (Source: Reproduced with permission from Görner (1991). © Springer Science + Business Media.)

as well. In a bubbling fluidized bed (BFB), relatively low gas velocities are applied, and consequently, only fine-grained ash is ejected from the fluidized bed after burnout and abrasion of the solid fuel. Coarse-grained ash, however, accumulates in these kinds of fluidized beds and, therefore, has to be removed. As a result of the higher flow velocities of air and combustion gases in the circulating fluidized bed (CFB),

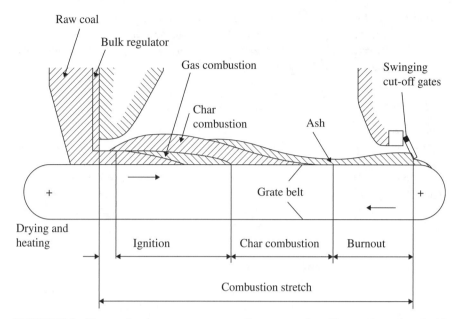

FIGURE 9.6 The combustion process on a traveling-grate stoker. (Source: Reproduced with permission from Adrian et al. (1986). © TÜV Media GmbH – TÜV Rheinland Group.)

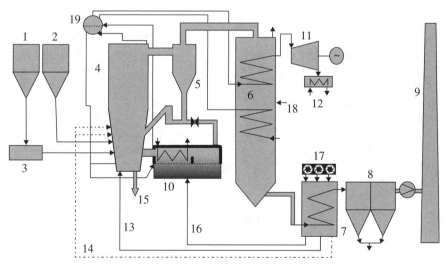

1. Coal bunker
2. Limestone bunker
3. Coal crusher
4. CFB combustor
5. Cyclone
6. Convective pass

7. Air preheater
8. Electrostatic filter
9. Stack
10. Fluid bed heat exchanger
11. Turbo generator
12. District heating

13. Primary air
14. Secondary air
15. Bottom ash
16. Fluid bed heat exchanger air
17. Blowers
18. Feed water
19. Drum

FIGURE 9.7 Schematic representation of a coal-fired CFB plant. (Source: Reproduced with permission from Koornneef et al. (2007). © Elsevier.)

the entire solid flow in the furnace is entrained and circulated. The CFB takes the total space of the furnace. In both the BFB and the CFB, the residence time of the solids in the furnace is markedly longer than that of the gas. Figure 9.7 presents an example of a CFB combustor integrated in a boiler system.

9.4.3   Pulverized Fuel Firing/Entrained Flow Combustion

In pulverized fuel firing systems, the particles are carried along with the air and combustion gas flow. Pulverized fuel and combustion air are injected into the furnace and mixed. As the solid fuel has been milled up to very small size and the flow velocities of the combustion gases are high, there is no slip velocity, and the residence times of the fuel particles and the gas are almost equal. The combustion of pulverized fuel is a rapid process, which is distributed over the entire furnace, so higher capacities are possible than in grate or fluidized bed firing systems.

9.4.4   Comparison of Different Firing Systems

Table 9.1 presents the advantages and disadvantages of the different combustion systems. The choice of the combustion system depends on the properties of the fuel and on the capacity for generating steam. Capacities for different solid fuel combustion systems offered on the market are shown in Table 9.2.

9.4.5   Power Generation

The heat produced by burning biomass is often transferred to a working medium, which is used subsequently in a heat engine to create mechanical energy for power generation. Steam turbines are the most widely spread technology for electricity generation in thermal power plants with capacities from 1 to several 100 MW. Other processes to generate power are based on the organic Rankine cycle, the externally fired combined cycle, or the Stirling engines. Biomass-fueled combustion plants normally have net electrical efficiencies (based on the lower heating value) between 25 and 30%. The main reason for this low efficiency is that the work produced by expansion of the gases, as they are combusted and heated up, is not utilized in standard combustion equipment such as boilers and furnaces.

9.4.6   Oxyfuel Combustion and Carbon Capture

Oxyfuel combustion is being developed for CFBs and also for pulverized coal plants. The main products of oxyfuel combustion are the usual carbon dioxide and water. Pulverized coal oxyfuel combustion burns coal in a mixture of recirculated flue gas and pure oxygen, rather than in air. The flue gas that is not recirculated is rich in carbon dioxide and water vapor, which makes it a good candidate for treatment by condensation of the water vapor and subsequent storage of the CO_2.

A driver for using optimized oxyfuel combustion power plants is the fact that they produce ultralow emissions. In addition, oxyfuel power cycles have the flexibility to use coal- or biomass-derived syngas containing both CO and H_2 for combustion in a

TABLE 9.1 Comparison of grate firing systems, fluidized bed systems, and pulverized fuel flow combustion systems

Grate firing systems	Fluidized bed firing systems (BFB and CFB)	Pulverized fuel firing systems
Advantages	Advantages	Advantages
• Little fuel preparation expenditure • Clear design • High degree of availability • Simple operation • Little auxiliary power requirements • Low NO$_x$ emissions • Partial desulfurization by limestone addition	• Little fuel preparation expenditure • Flue gas cleaning consists only of flue gas particle collection Disadvantages • High limestone demand for sulfur capture at excess air • Ash not usable without further preparation Advantages of CFB compared to BFB • Better burnout • Lower limestone demand for sulfur capture • Better emission values • No in-bed heating surfaces at risk of erosion • Better power control	• High degree of availability • Large capacities • High power density • Good burnout • Usable ash Disadvantages • High fuel preparation expenditures • Flue gas cleaning needed for particulates, SO$_2$, and NO$_x$
Disadvantages • High combustion losses of 2–4% by unburned carbon • High flue gas temperatures due to limited air preheating • Unsuitable for fine-grained fuels		

Source: Reproduced with permission from Spliethoff (2010). © Springer Science + Business Media.

TABLE 9.2 Output ranges of firing systems

Firing system	Output range (MW$_{th}$)
Entrained flow	40–2500
BFB	<80
Grate firing	2.5–175
CFB	40–750

Source: Reproduced with permission from Spliethoff (2010). © Springer Science + Business Media.

turbine power cycle. Temperatures are controlled by recycling water (or CO_2) in complete power systems. In both pulverized coal and power cycle applications, the current state of the art is such that oxyfuel combustion plants can be built but also existing plants can be retrofitted using existing technologies. However, such plants are not optimized due to a lack of data or proven computer models of oxyfuel combustors, boiler systems, and CO_2 recovery systems. Technology is in a continuously developing stage to model these systems with increasing accuracy. At present, several oxyfuel combustion facilities at various scales of power generation are in operation or being constructed around the world.

9.5 EMISSIONS

Combustion of biomass results in the emission of unwanted by-products. In this section, the emissions from biomass combustion and measures to reduce them are discussed. The emissions can be a result of complete and incomplete combustion.

9.5.1 Emissions from Complete Combustion

Carbon dioxide (CO_2) is the main combustion product, originating from the carbon content in the biomass, and anthropogenic CO_2 emissions are believed to be the main cause of global warming. However, combustion of biomass can be regarded as being highly CO_2 neutral with respect to the greenhouse effect as is explained in Chapter 1. This is the main environmental benefit of using biomass as a fuel.

Nitric oxides (NO_x) emissions from combustion applications are caused by oxidation of nitrogen originating from the fuel or the air. During the combustion process, mainly NO is formed (>90%), which is converted to NO_2 in the atmosphere. Nitric oxides react to form smog and acid rain. The three main NO formation mechanisms in combustion applications are (Glarborg et al., 2003; Hill and Douglas Smoot, 2000):

1. *Thermal NO mechanism*: At high temperatures (T > 1600 K), nitrogen in the air reacts with O radicals present in the postflame gases. This leads to the formation of NO and N radicals, which can further react with O_2 and OH radicals formed in the flame to NO. The initiating reaction step has a very high activation energy, which makes this formation mechanism very sensitive to temperature. The temperature in biomass combustion applications is often too low to form large quantities of thermal NO.

2. *Prompt NO mechanism*: Nitrogen in the air may also react with CH radicals in the flame. This reaction forms NCN, which is quickly converted to HCN and NH_3. If sufficient oxygen is available, NH_3 and HCN are mainly converted to NO via different routes. However, at fuel-rich conditions, NO will react with NH_3 and HCN, forming N_2. In applications of biomass combustion, the prompt NO formation mechanism has not been found to be of major importance.

3. *Fuel NO mechanism*: For nitrogen-bearing fuels such as coal and biomass, conversion of fuel-bound nitrogen to NO is often the main contribution to NO formation. The nitrogen is released from the fuel during pyrolysis, mainly as NH_3 and HCN. These components can subsequently be converted to NO or N_2 following the same pathways as in the prompt NO mechanism. Additionally, fuel nitrogen is retained in the char and is largely oxidized to NO in the char combustion phase but may subsequently be reduced to N_2 by a fast heterogeneous reaction with the char. The amount of fuel nitrogen retained in the char relative to the amount of fuel nitrogen released in the devolatilization phase is partially determined by the thermal exposure of the fuel.

Nitrous oxide (N_2O) emissions are also a result of the oxidation of fuel nitrogen. Although the N_2O emission levels measured in various biomass combustion applications are very low (below ppm level), N_2O emissions are worth mentioning because N_2O has a 310 times higher global warming potential than CO_2.

Sulfur oxides (SO_x) are a result of complete oxidation of sulfur originating from the fuel. Unlike coal, most biomass types hardly contain any sulfur. Mainly SO_2 (>95%) is formed. Not all fuel sulfur will be converted to SO_x; a significant fraction will remain in the ashes, while a minor fraction is emitted as salt (K_2SO_4) or H_2S at lower temperatures. Sulfur dioxide considerably contributes to air pollution, with an array of adverse respiratory effects including bronchoconstriction and increased asthma symptoms. In addition, SO_x is a precursor to acid rain and atmospheric particulates.

Part of the chlorine content in biomass can be released as **hydrogen chloride (HCl)**. On contact with water, this forms corrosive hydrochloric acid. Inhalation of the fumes causes severe respiratory problems. Not all chlorine is converted to HCl; the main fraction is retained in salts (KCl, NaCl) by reaction with K and Na. Although the chlorine content of wood is usually very low, significant amounts of HCl may be formed from biomass fuels containing higher amounts of chlorine, such as miscanthus, grass, and straw.

Particle emissions during complete combustion originate from fly ash, which is a result of entrainment of ash particles in the flue gas, and salts (KCl, NaCl, K_2SO_4), which are a result of reactions between K or Na and Cl or S. Other types of particle emissions are mentioned in the next section on emissions from incomplete combustion. Particle emissions from combustion applications, in general, have a negative effect on the human respiratory system and are carcinogenic. Furthermore, ash particles may have adverse effects on boiler operation due to agglomeration, slagging, and fouling.

Heavy metals (most importantly Cu, Pb, Cd, Hg) are present in all biomass fuels to some degree. In combustion applications, these elements remain in the ash or evaporate. They may also attach to emitted particles or be contained inside fly ash particles. Some heavy metals are toxic, and some are carcinogenic.

9.5.2 Emissions from Incomplete Combustion

When the biomass is not completely oxidized into its final products, intermediate reaction products leave the reactor as part of the flue gas.

Carbon monoxide (CO) is the final intermediate in the conversion of fuel carbon to CO_2. Oxidation of CO is the last elementary reaction step in this process. CO emissions can, therefore, be regarded as a good indicator of how complete the combustion is. If oxygen is available, the rate at which CO is oxidized to CO_2 depends primarily on temperature. Therefore, CO emission levels are minimal at an optimum air ratio: lower air ratios will result in fuel-rich regions, while higher air ratios will lead to lower combustion temperatures and shorter residence times, all resulting in increased CO emissions.

Methane (CH_4) is an important intermediate in the conversion of fuel carbon to CO_2 and fuel hydrogen to H_2O in biomass combustion. Hydrocarbons, in general, are earlier intermediates than CO and are, therefore, emitted at lower levels. Methane is usually mentioned separately from the other hydrocarbons since it is a direct greenhouse gas.

Volatile organic compounds (VOCs) are a group of species that include all hydrocarbons except methane, polycyclic aromatic hydrocarbons (PAHs), and other heavy hydrocarbons that condense and form particle emissions. They are all intermediates in the conversion of fuel carbon to CO_2 and fuel hydrogen to H_2O. VOCs have a negative effect on the human respiratory system. Furthermore, VOCs are indirect greenhouse gases because they are precursors of ozone in the atmosphere, which is a greenhouse gas.

PAHs are often mentioned separately from the other hydrocarbons due to their carcinogenic effects. PAHs that condense at relatively low temperatures (<200°C) are called tar (Milne et al., 1998). Tars evolve from primary tar, i.e., the tar that is released from the solid fuel by breaking of molecular bonds during pyrolysis. After release, these primary tars can grow larger in size by different kinds of reactions. Alternatively, when the residence time at elevated temperatures is sufficiently long, tars can also be cracked into smaller molecules, mainly gases. The fact that the tars condense is a major problem in biomass installations since the condensed tars foul downstream equipment and cause engine wear (by corrosion and erosion). The fouled installation parts need to be cleaned regularly, thus increasing operating costs.

Particle emissions from incomplete combustion consist of soot, char, and condensed heavy hydrocarbons (tar). Soot particles mainly consist of carbon and are formed in fuel-rich regions of the flame. Char particles are entrained in the flue gas due to their very low density, especially at high gas flow rates. Tar particles significantly contribute to the total particle emission level in small-scale biomass combustion applications such as woodstoves and fireplaces. Particle emissions have carcinogenic and negative respiratory effects.

Polychlorinated dibenzodioxins and furans (PCDD/F) are a group of highly toxic compounds. Dioxins are formed during combustion of fuels containing chlorine (Chagger et al., 1998). Ideally, a combustion process converts all carbon to CO_2 and all chlorine to HCl; however, in a temperature window between 400 and 700°C, carbon, chlorine, and oxygen react in the presence of a catalyst (Cu, present in fly ash generated) to form PCDD/F. The emissions of PCDD/F are highly dependent on the conditions under which combustion and flue gas cooling take place. Although herbaceous biomass fuels have high chlorine contents, their PCDD/F emissions are usually very low. This may be explained by their high alkali content, which leads to the formation of salts (KCl, NaCl) and thus to a lower level of gaseous chlorine for the formation of dioxins. In general, the PCDD/F emission level from biomass combustion applications using virgin wood as fuel is well below the health risk limit.

Ammonia (NH_3) is an intermediate in the conversion of fuel N to oxidized nitrogen-containing components. The fuel N is mainly released as NH_3 during devolatilization of biomass. At very low temperatures, this NH_3 is not completely oxidized and is emitted. Additionally, NO reduction measures utilizing NH_3 injection may contribute to the NH_3 emission level due to ammonia slippage.

9.5.3 Emission Reduction Measures

Reduction of harmful flue gas emissions can be achieved either by avoiding the formation of such substances (primary measures) or by removing these substances from the flue gas (secondary measures), which are described in the following sections.

9.5.3.1 Primary Measures for Emission Reduction
The goal of primary reduction measures is to prevent or at least reduce the formation of unwanted substances in the combustion chamber. Primary measures are mainly used to reduce NO_x emissions and emissions resulting from incomplete combustion. Since the latter emissions are mainly a result of a lack of oxygen, insufficient mixing of fuel and oxygen, too low temperatures, and too short residence times, primary measures often aim to improve these conditions. Several possible measures—often interrelated—are discussed below.

Modification of the Fuel Composition: One way to reduce harmful emissions is to decrease the amount of elements in the fuel that cause these emissions. The possibilities of decreasing the amount of specific elements in biomass are limited. However, the washing (leaching) of straw has been shown to reduce the amounts of chlorine and potassium significantly, leading to reduced dioxin and furan emission levels.

Modification of the Moisture Content of the Fuel: Biomass fuels often have a high moisture content. Fresh wood from forests may contain up to 60 wt% water. A high moisture content in the fuel reduces the heating value and makes it difficult to achieve a sufficiently high temperature in the combustion chamber. Drying is often too costly to make it economically feasible, unless waste heat from another process can be accessed at a very low cost (see Chapter 8 describing such drying techniques). Drying biomass in the open air by exposing it to the sun and the wind is a cheap and simple alternative.

Modification of the Particle Size of the Fuel: The fuel particle size is very relevant for the burnout time. The fuel size in biomass combustion applications may vary from whole wood logs to fine sawdust. Due to a large specific surface area, small particles burn much faster, resulting in more complete combustion. The size of large fuel particles can be reduced by a shredder or chipper. Special hammer mills can be used to create millimeter-sized particles for dust firing (see Chapter 8 for more details on particle size reduction). However, particle size reduction is only attractive if the benefits outweigh the additional investment and energy costs.

Improved Construction of the Combustion Application: In order to obtain optimal combustion with minimal emissions from incomplete combustion, one has to achieve:

- Sufficiently high combustion temperatures
- Sufficiently long residence times
- Optimal mixing of fuel gases and air, at varying loads

These factors are partly determined by the combustion technology and design of the furnace and partly by the operation of the combustion process. Process variables that can directly be adjusted typically are the amount of fuel fed into the furnace and the amount of primary and secondary combustion air supplied.

Staged-Air Combustion: It is widely applied in biomass combustion applications. It reduces both the emissions caused by incomplete combustion and NO emissions by separating devolatilization and gas-phase combustion, which leads to an improved mixing of volatiles and combustion air. In the first stage, primary air is added for devolatilization and gasification of the fuel. The oxygen-to-fuel molar ratio in this stage is smaller than one, resulting in a gas that consists of mainly CO, H_2, hydrocarbons, CO_2, and water. In the second stage, sufficient secondary air is provided to ensure complete combustion of this gas with low emission levels. The improved mixing of gas and air in the second stage reduces the amount of air needed, which leads to higher flame temperatures and better burnout. Air staging can also be used to reduce NO emissions from the fuel NO formation mechanism. The fuel gas contains NH_3 and HCN, which are converted to NO if sufficient O_2 is available. However, in fuel-rich conditions, NH_3 and HCN will react with NO, forming N_2. Therefore, the emissions of NO can be reduced by optimizing the excess air ratio in the first stage.

Staged-Fuel Combustion and Reburning: These are other possible methods to reduce NO_x emissions in biomass combustion applications. In the first stage of staged-fuel combustion, primary fuel is combusted at an air-to-fuel ratio larger than 1, leading to a flue gas with a relatively high NO concentration. In the second stage, secondary fuel is introduced into the flue gas without additional air supply. A fuel-rich condition is created in which the NO from the first stage reacts with NH_3 and HCN from the secondary fuel, reducing the NO levels in the same way as in staged-air combustion. An additional effect is that NO is converted back to HCN by reactions with HCCO (ketenyl) and CH_x radicals. This effect is called reburning. In the last stage, sufficient secondary air is supplied to achieve complete burnout.

9.5.3.2 Secondary Measures for Emission Reduction

Secondary measures are applied to remove emissions from the flue gas once it has left the combustion chamber. This concerns mainly emissions from complete combustion, in particular particle, NO_x, and SO_x emissions. Emissions of other components (HCl, heavy metals, PCDD/F) can be reduced by secondary measures as well, but they are not discussed here in detail.

Particle Control Technology: In most biomass combustion applications, particle emissions are significant, and secondary measures are needed to comply with emission limits. Since there are many different types of particles, there is also a wide range of particle control technologies. These include settling chambers, cyclones, electrostatic precipitators (ESPs), fabric filters, and scrubbers. Settling chambers are large compartments in which the gas velocity is reduced leading to sedimentation of the particles by gravity. Settling chambers have low collection efficiency and require much space, but they have simple designs and low investment and maintenance costs. In cyclones, the particles are separated based on the principle of centrifugal forces. The particle-laden gas is brought into a swirling motion by injecting it tangentially into a cyclone. Due to the centrifugal forces, the particles drift to the wall of the cyclone and then slide down into a container. The gas leaves the cyclone at the top of the cyclone. Cyclones have higher collection efficiency than settling chambers. To improve the collection efficiency, the centrifugal forces can be increased by reducing the diameter of the cyclone. To keep the same capacity, many small cyclones can be put in parallel, then called a multicyclone. The disadvantages of a multicyclone are the more expensive construction and the higher pressure drop, which leads to a higher energy consumption. At tolerable pressure loss, multicyclones can only remove particles larger than $10\,\mu m$. For the removal of finer particles, ESPs or fabric filters need to be used. In an ESP, the particles are first electrically charged and then exposed to an electric field, in which they are attracted to the collecting electrode. The collecting surface is cleaned periodically by vibration. The collected particles then drop into a container. ESPs have problems removing particles with a high electrical resistance, such as fly ash from dry straw combustion. Fabric filters are then preferred. In fabric filters, the flue gas has to pass through a tightly woven cloth consisting of special fibers. The filters are periodically cleaned by vibration or pressurized air. Bag filters can remove fine particles at high efficiency, but the cloths are often sensitive to high temperatures and fouling by condensing tars. For woody biomass, it is common practice to use cyclones for coarse separation and ESPs or fabric filters for fine separation. Another technology to remove particles is a scrubber. In scrubbers, a mist of small droplets is created, through which the flue gas flows. The particles collide with the droplets and are carried away by them. The droplets are collected at the bottom of the chamber resulting in a stream of wastewater that needs to be cleaned. The advantage of a scrubber is its ability to remove SO_x, NO_x, and HCl as well by gas absorption. Disadvantages are corrosion and erosion problems and the added cost of wastewater treatment.

NO_x and SO_x Control Technology: NO_x emission levels of biomass combustion applications are usually very low because of the small amount of fuel nitrogen in biomass (Spliethoff, 2010). For biomass fuels with relatively high nitrogen content (e.g., straw), primary measures are usually sufficient to meet the emission limits. Secondary measures for NO_x emission reduction are selective catalytic reduction (SCR) and selective noncatalytic reduction (SNCR). In both methods, a reducing agent—usually ammonia or urea—is injected in the flue gas to reduce NO to N_2. In SCR, platinum, titanium, or vanadium oxide catalysts are used in a

temperature range of 220–270°C for ammonia and 400–450°C for urea. SNCR processes have been developed to avoid the use of a catalyst. When the reducing agent is injected in the flue gas at a temperature between 850 and 950°C, the reactions can take place without catalyst. Both SCR and SNCR can reach about 90% NO_x reduction. Good mixing and temperature control, however, are essential; otherwise, ammonia is emitted or converted to NO.

SO_x emissions from biomass applications usually are also very low because of the low sulfur content of biomass (Spliethoff, 2010). Most of the sulfur is released into the gas phase and is captured in fly ash, which can be removed by particle control technologies. However, for some types of biomass, e.g., *miscanthus*, grass, and straw, emissions of SO_x can be significant. The common measure to remove SO_x from flue gas is scrubbing it with water containing finely ground limestone, $CaCO_3$, or lime, $Ca(OH)_2$. The SO_x dissolves in this slurry and reacts with the limestone or lime, producing calcium sulfite, $CaSO_3$, which is further oxidized to calcium sulfate, $CaSO_4$, a marketable industrial chemical. Limestone can also be injected directly in the combustion chamber for SO_x removal.

CHAPTER SUMMARY AND STUDY GUIDE

Combustion of biomass releases its energy in the form of heat, which can be used for direct heating or to drive a heat engine. The fundamental processes of biomass combustion involve drying, pyrolysis, and combustion of volatiles and char. These processes may occur sequentially or simultaneously depending on the biomass properties and heating conditions. Different particle conversion modes can be distinguished by analyzing the governing equations and considering the Thiele modulus and the Biot number for mass transport. At high values of the Thiele modulus, the conversion is diffusion controlled, and the chemical reactions occur at the surface of the particle. The conversion time in this shrinking sphere mode is analyzed and found to be proportional to the square of the initial diameter of the particle. The proportionality constant is closely related to the Spalding number, which is a measure of the amount of heat that is provided relative to the amount of heat that is needed for the conversion. The Spalding number can account for heats of vaporization, pyrolysis, gasification, and combustion.

Combustion is the most widespread technology for the conversion of biomass into energy. Many different combustion technologies exist, which can be categorized as fixed bed, fluidized bed, and entrained flow reactor technologies. The particle size of the biomass and the gas velocity of the medium mainly determine the level of fluidization. Although biomass combustion can be considered sustainable and highly CO_2 neutral, it is certainly not free from undesired emissions. These emissions may result from both complete and incomplete combustion. Improving the combustion efficiency of the reactor can prevent emissions resulting from incomplete combustion. Primary reduction measures aim to prevent the formation of emissions, while secondary measures (cyclones, filters, scrubbers, etc.) are used to remove formed undesired emissions from the flue gas.

KEY CONCEPTS

Thermal conversion
Shrinking sphere
Shrinking core
Shrinking density
Thiele modulus
Biot number
d^2-law
Spalding number
Combustion systems
Emissions from complete and incomplete combustion
Primary and secondary emission reduction measures

SHORT-ANSWER QUESTIONS

9.1 What are the main modes of particle conversion?

9.2 What dimensionless numbers are indicative for these modes?

9.3 Which model applies in film behavior?

9.4 Give three examples of emissions from complete combustion.

9.5 Give three examples of emissions from incomplete combustion.

9.6 What is the main NO formation mechanism in biomass combustion applications?

9.7 Name three different particle control technologies.

PROBLEMS

9.1 Draw the temperature profile that is assumed in the derivation of the d^2-law for evaporation inside and outside of a droplet as a function of the radius.

9.2 Draw the droplet diameter and its square as a function of time at a certain thermal load.

9.3 There are problems in a biomass power plant with the burnout time of the particles. A too large amount of the biggest particles needs to be separated in the exhaust system and fed into the reactor again. The distance of the reacting volume to the exhaust system is 20 m, and the main flow-through velocity is $40 \, \text{m·s}^{-1}$. The size of the biggest particles is 1 mm. Calculate the critical rate constant in $\text{m}^2 \cdot \text{s}^{-1}$ that is needed to convert the largest particles assuming a shrinking sphere model. What is the critical diameter of the biggest particle if it is required that it burns out in half the distance to the exhaust?

9.4 In Example 9.1, evaporation of boiling octane is considered. If we take combustion into account as well, we have to consider a laminar diffusive combustion front at some distance from the surface. Now, the definition for the Spalding number for this situation becomes

$$B_q = \frac{\Delta h_c / \vartheta + c_p (T_\infty - T_{boil})}{h_{fg}}$$

in which the enthalpy of combustion (Δh_c) and the oxidizer-to-fuel stoichiometric ratio (ϑ) have been added. For octane, the enthalpy of combustion is $\Delta h_c = 44{,}791 \text{ kJ·kg}^{-1}$, and the stoichiometric ratio is given by $\vartheta = 15.03$.
   Calculate the burnout time based on this information.

9.5 The Spalding theory can also provide a standoff distance and a burning temperature. This temperature can be used to calculate a new temperature in the far field and one can iterate the procedure. The standoff distance can be expressed as the ratio of the radii of the flame to the droplet, r_f/r_s, and becomes

$$\frac{\ln(1 + B_{0,q})}{\ln((1 + \vartheta)/\vartheta)}$$

Here, the index 0,q denotes the Spalding number for combustion and evaporation. Evaluate the standoff distance for Problem 9.4.

PROJECTS

P9.1 Visit a biomass-fired power plant and analyze the combustion system. Try to answer the following questions. What type of reactor is used? What is the average particle size? What is the residence time of the biomass particles in the reactor? Which emission reduction measures are taken? Try to set up a global mass and energy balance.

P9.2 In the analysis for determining the combustion mode, typical concentration profiles of oxygen are determined. Using these stationary profiles, try to find a method to estimate the time of total conversion for the different conversion modes.

P9.3 In the analysis of Project P9.2, a quasi-steady state is assumed. Is this correct? The fraction of time that the steady state exists should be a large part of the total conversion time for the steady-state solution to be meaningful. Add an unsteady-state term to the equations and solve them using numerical tools to see whether this is indeed the case.

INTERNET REFERENCES

tinyurl.com/o7cacsn

http://www.princeton.edu/~fldryer/nasa.dir/backgrnd.htm

REFERENCES

Adrian F, Quittek C, Wittchow E. *Fossil beheizte Dampfkraftwerke. Handbuchreihe Energie, Band 6*. Verlag TÜV Rheinland: Herausgeber T. Bohn. Technischer Verlag Resch; 1986.

Chagger HK, Kendall A, McDonald A, Pourkashanian M, Williams A. Formation of dioxins and other semi-volatile organic compounds in biomass combustion. Appl Energy 1988;60:101–114.

Chiu HH, Kim HY, Croke EJ. Internal group combustion of liquid droplets. Symp (Int) Combust 1982;19(1):971–980.

Glarborg P, Jensen AD, Johnsson JE. Fuel nitrogen conversion in solid fuel fired systems. Prog Energy Combust Sci 2003;29(2):89–113.

Goméz-Barea A, Leckner B. Modeling of biomass gasification in fluidized bed. Prog Energy Combust Sci 2010;36(4):444–509.

Görner K. *Technische Verbrennungssysteme. Grundlagen, Modellbildung, Simulation*. Berlin-Heidelberg/New York: Springer; 1991.

Hill SC, Douglas Smoot L. Modeling of nitrogen oxides formation and destruction in combustion systems. Prog Energy Combust Sci 2000;26(4):417–458.

Koornneef J, Junginger M, Faaij A. Development of fluidized bed combustion: an overview of trends, performance and cost. Prog Energy Combust Sci 2007;33(1):19–55.

Levenspiel O. *Tracer Technology: Modeling of the Flow of Fluids (Fluid Mechanics and Its Applications)*. New York: Springer; 2012.

Milne TA, Evans RJ, Abatzoglou N. Biomass gasifier "tars": their nature, formation, and conversion. Golden (CO): NREL; 1998. Report nr NREL/TP-570-25357.

Spalding DB. The combustion of liquid fuel. Symp (Int) Combust 1953;4(1):847–864.

Spliethoff H. *Power Generation from Solid Fuels*. Berlin-Heidelberg: Springer Verlag; 2010.

Thunman H, Leckner B. Char combustion and gasification. Thermo Chemical Conversion of Biomass and Wastes Nordic graduate school BiofuelGS-2. Göteborg (Sweden): Chalmers, November 19–23, 2007.

Turns S. *An Introduction to Combustion: Concepts and Applications*. 2nd ed. Boston (MA): McGraw-Hill; 2000.

Van Kuijk H, Van Oijen J, Bastiaans R, de Goey P. Reverse combustion: kinetically controlled and mass transfer controlled conversion front structures. Combust Flame 2008;153(3): 417–433.

10

THERMOCHEMICAL CONVERSION: (CO)GASIFICATION AND HYDROTHERMAL GASIFICATION

Sascha R.A. Kersten[1] and Wiebren de Jong[2]

[1]Sustainable Process Technology Group, Faculty of Science and Technology, University of Twente, Enschede, the Netherlands
[2]Department of Process and Energy, Energy Technology Section, Faculty of Mechanical, Maritime and Materials Engineering, Delft University of Technology, Delft, the Netherlands

ACRONYMS

ar	as received basis
BFB	bubbling fluidized bed
BTL	biomass-to-liquid
CC	carbon conversion
CFB	circulating fluidized bed
CGE	cold gas efficiency
CHP	combined heat and power
daf	dry and ash-free basis
db	dry basis
EF	entrained flow
FT	Fischer–Tropsch
IC	internal combustion
IGCC	integrated gasification combined cycle
PDU	Process Development Unit
PFBC	pressurized fluidized bed combustion

Biomass as a Sustainable Energy Source for the Future: Fundamentals of Conversion Processes,
First Edition. Edited by Wiebren de Jong and J. Ruud van Ommen.
© 2015 American Institute of Chemical Engineers, Inc. Published 2015 by John Wiley & Sons, Inc.

PNNL Pacific Northwest National Laboratory
SCGP Shell Coal Gasification Process
SCWG supercritical water gasification
SCW supercritical water
SNG substitute natural gas
TGA thermogravimetric analysis

SYMBOLS

A	area	[m^2]
a_s	specific surface area of a particle	[m^2·m^{-3}]
$a_{s,m}$	specific surface area (mass basis)	[m^2·kg^{-1}]
c_i	concentration of species i	[mol·m^{-3}]
CC	carbon conversion	[–]
CGE	cold gas efficiency	[–]
D	binary diffusion coefficient	[m^2·s^{-1}]
D_e	effective diffusion coefficient	[m^2·s^{-1}]
D_i	intraparticle diffusion coefficient	[m^2·s^{-1}]
d	diameter	[m]
E_a	activation energy	[J·mol^{-1}]
$\Delta_r H$	reaction enthalpy (change)	[J·mol^{-1}]
h	heat transfer coefficient	[W·m^{-2}·K^{-1}]
Kw	ion product of water	[mol^2·m^{-6}]
k	reaction rate coefficient	[depending on order]
k_g	mass transfer coefficient (gas phase)	[m·s^{-1}]
m	mass	[kg]
MW	molecular weight	[kg·mol^{-1}]
p	(partial) pressure	[Pa]
R_i	net production rate of species i	[mol·m^{-3}·s^{-1}]
R'_i	net production rate per unit mass of solid particle	[mol·kg$_{solid}$·s^{-1}]
R_u	universal gas constant (=8.3143)	[J·mol^{-1}·K^{-1}]
r	radius	[m]
SB	steam-to-biomass mass ratio	[–]
SB*	steam-to-biomass mass ratio including fuel moisture	[–]
T	temperature	[K]
t	time	[s]
X	(degree of) conversion	[–]
Y_i	mass fraction of compound/element i	[–]
α	distribution coefficient for C-oxidation to CO/CO$_2$	[–]
ε	voidage	[–]
η	particle's utilization factor in reactions	[–]
η_{HE}	heat exchanger efficiency	[–]

λ stoichiometric oxygen ratio (=1 for complete combustion)   [–]

λ thermal conductivity   [W·m^{-1}·K^{-1}]

ρ density   [kg·m^{-3}]

τ_p tortuosity of a particle   [–]

ϕ_n thiele modulus for nth-order reaction, def.: Equation (10.19)   [–]

φ_m mass flow   [kg·s^{-1}]

Subscripts

0   initial

b   reverse reaction (rate coefficient) or in the bulk phase (concentration)

f   forward reaction (rate coefficient)

g   grain

HE   heat exchanger

n   of an nth-order reaction

p   particle

s   at the surface

t   at time t

vap   vaporization

∞ after infinitely long time

10.1   WHAT IS GASIFICATION? A CHEMICAL AND ENGINEERING BACKGROUND

10.1.1   Introduction

Gasification is a thermochemical fuel conversion technology carried out at high temperature using a gaseous agent (mostly oxidizing, sometimes also reducing) to convert a liquid or solid fuel into a combustible product gas. When in this process air is used as the oxidizing gas, a product gas is generated that contains H_2, CO, CO_2, CH_4, and other higher molecular weight hydrocarbons together with H_2O and a bulk N_2 concentration. This gas is also called "producer gas" (Knoef, 2012; Milne et al., 1998). Other oxidizing gases that can be used to gasify a carbon-based resource are steam, (more or less pure) oxygen, CO_2, or steam/oxygen mixtures. The product gas mixture then generated is also known as (raw, uncleaned) biosyngas.

The technology can be integrated with many end uses of the produced gas. Gasification can be used for (sole) heat generation, (combined heat and) power production, as well as biofuel and chemical production (e.g., methanol, diesel). Figure 10.1 shows a generic schematic regarding the technology chain. This chapter deals with gasification fundamentals and reactor technologies.

10.1.2   Chemical Reactions

The process of biomass gasification is characterized by several different subprocesses (steps), namely, particle drying, pyrolysis, and gasification. Pyrolysis—also called

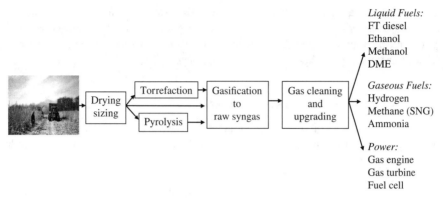

FIGURE 10.1 Scheme for thermochemical conversion of biomass via gasification to multiple products.

devolatilization—is the thermal decomposition of the main organic constituents into char, permanent gases, and heavier hydrocarbons (also called tars); details of this process are presented in Chapter 11. In the gasification step of the process, tars and char react further with already evolved gases to produce more gases. Gasification is a thermochemical conversion process and is performed at relatively high temperatures (typically in the range of 700–1500°C). The pressure can range from atmospheric pressure up to 7 MPa.

During gasification, numerous chemical reactions occur. The most significant of these are summarized in equations (RX. 10.1)–(RX. 10.13). In these reaction equations, the carbonaceous residue, usually called char, is considered to consist exclusively of carbon (C(s)) and tar, which is represented as C_nH_m:

$$C(s) + O_2 \rightarrow CO_2 \quad \Delta_r H^0{}_{298} = -394\,kJ\cdot mol^{-1} \tag{RX. 10.1}$$

$$C(s) + \tfrac{1}{2}O_2 \rightarrow CO \quad \Delta_r H^0{}_{298} = -111\,kJ\cdot mol^{-1} \tag{RX. 10.2}$$

$$C(s) + H_2O \leftrightarrow CO + H_2 \quad \Delta_r H^0{}_{298} = +131\,kJ\cdot mol^{-1} \tag{RX. 10.3}$$

$$C(s) + CO_2 \leftrightarrow 2CO \quad \Delta_r H^0{}_{298} = +173\,kJ\cdot mol^{-1} \tag{RX. 10.4}$$

$$C(s) + 2H_2 \leftrightarrow CH_4 \quad \Delta_r H^0{}_{298} = -75\,kJ\cdot mol^{-1} \tag{RX. 10.5}$$

$$CO + \tfrac{1}{2}O_2 \rightarrow CO_2 \quad \Delta_r H^0{}_{298} = -283\,kJ\cdot mol^{-1} \tag{RX. 10.6}$$

$$H_2 + \tfrac{1}{2}O_2 \rightarrow H_2O \quad \Delta_r H^0{}_{298} = -242\,kJ\cdot mol^{-1} \tag{RX. 10.7}$$

$$CO + H_2O \rightarrow CO_2 + H_2 \quad \Delta_r H^0{}_{298} = -41\ kJ\cdot mol^{-1} \tag{RX. 10.8}$$

$$CH_4 + H_2O \leftrightarrow CO + 3H_2 \quad \Delta_r H^0{}_{298} = +206\ kJ\cdot mol^{-1} \tag{RX. 10.9}$$

$$C_nH_m\,(tars) + nH_2O \rightarrow nCO + (n + m/2)\,H_2 \tag{RX. 10.10}$$

$$C_nH_m\,(tars) + nCO_2 \leftrightarrow 2n\,CO + m/2H_2 \tag{RX. 10.11}$$

$$C_nH_m\,(tars) \leftrightarrow nC + (n+m/2)\,H_2 \tag{RX. 10.12}$$

$$C_nH_m\,(tars) + (4n-m)/2H_2O \leftrightarrow nCH_4 \tag{RX. 10.13}$$

The supply of heat to accomplish the main endothermic gas-generating reactions can be provided *in situ* by exothermic partial oxidation (autothermal operation) or indirectly by integration of the reaction process with an external exothermic process, which transfers heat to the main gasification reactor.

10.1.3 Characteristic Process Parameters

Important for the generation of a good quality product gas is the ratio of oxidizer to fuel. A parameter that characterizes this is the stoichiometric oxygen ratio, λ (lambda value), which is defined as

$$\lambda = \frac{\text{external } O_2 \text{ supply}/\text{fuel supply}}{\text{stoichiometric } O_2 \text{ requirement}/\text{unit of fuel input (daf basis)}} \qquad \text{(Eq. 10.1)}$$

Different "oxidation regimes" can be distinguished for thermochemical biomass conversion depending on λ. Processes with $\lambda > 1$ refer to combustion, processes with $\lambda = 0$ to pyrolysis, and processes with $0 < \lambda < 1$ to gasification. In literature, one also finds the equivalence ratio (ER), which is the inverse of λ. Lower λ values signify higher concentrations of the targeted main gasification constituents, H_2 and CO, in the produced gas. What needs to be considered though is that with decreasing λ values, the endothermic reactions will increasingly prevail so that temperatures might decrease to values that are too low to sustain the gasification process. This can be overcome by external heat supply (indirect gasification), or the λ value can be increased so that heat is generated *in situ* by partial oxidation (autothermal gasification).

When steam is used as a component of the oxidizer flow supplied to the gasifier, the steam-to-biomass ratio plays an important role in determining the final gas composition. It can be defined in two ways, as presented below:

$$SB = \frac{\text{steam mass flow}}{\text{fuel feed rate}} \qquad \text{(Eq. 10.2)}$$

$$SB^* = \frac{\text{steam mass flow} + \text{fuel moisture flow}}{\text{fuel feed rate}} \qquad \text{(Eq. 10.3)}$$

Important output parameters for the evaluation of a gasification process are the carbon conversion (CC) and the cold gas efficiency (CGE), respectively. They are defined as

$$CC = \left(1 - \frac{\varphi_{m,C,\text{residue}}}{\varphi_{m,C,\text{feed}}} \right) \qquad \text{(Eq. 10.4)}$$

$$CGE = \left(\frac{\sum \varphi_{m,i} \cdot LHV_i}{\varphi_{m,\text{fuel}} \cdot LHV_{\text{fuel}}} \right) \qquad \text{(Eq. 10.5)}$$

Example 10.1   Calculation of CC and CGE for a gasifier

An air-blown, pilot-scale fluidized bed gasifier converts 277.8 kg·h⁻¹ of *Miscanthus* into product gas. The composition of this fuel on a dry mass basis is 48% C, 6% H, 42% O, 0.5% N, 3.5% ash; its moisture content on an as received (ar) basis is 10 wt%. The LHV of the fuel measured is 16 MJ·kg⁻¹ (ar basis). The used air stoichiometry value is $\lambda = 0.33$. The molecular weights of the elements are as follows: $MW_C = 12$ kg·kmol⁻¹, $MW_H = 1$ kg·kmol⁻¹, $MW_N = 14$ kg·kmol⁻¹, and $MW_O = 16$ kg·kmol⁻¹. Assume that the fuel-bound N is converted to NO in case of complete stoichiometric combustion.

Fly ash is tapped off from cyclone(s) and filters, and it contains 35 wt% carbon (pure) and 65 wt% mineral matter. Assume that all ash in the fuel goes to the fly ash fraction and that no carbon accumulation takes place in the bed. The gas composition is as follows: 4.5 vol.% methane, 15.5 vol.% carbon monoxide, 13.1 vol.% carbon dioxide, 11.6 vol.% hydrogen, 10.1 vol.% water, and 0.3 vol.% ammonia, while the remainder of the gas consists of nitrogen. The values of LHV of the gas constituents CH_4, CO, NH_3, and H_2 are, respectively, 50.016, 10.104, 22.5, and 119.97 MJ·kg⁻¹.

a. Calculate the carbon conversion, CC.
b. Calculate the cold gas efficiency, CGE.

Solution

a. Start with an inert balance (ash):

$$\varphi_{m,\text{fuel,dry}} m_{\text{ash,db}} = \left(1 - m_{C,\text{fly ash}}\right)\varphi_{m,\text{fly ash}}$$

$$\Rightarrow \frac{\varphi_{m,\text{fuel,ar}}\left(1 - m_{\text{moisture,ar}}\right)\cdot m_{\text{ash,db}}}{\left(1 - m_{C,\text{fly ash}}\right)} = \varphi_{m,\text{fly ash}}$$

Substituting the known data gives

$$\varphi_{m,\text{fly ash}} = 13.46 \text{ kg·h}^{-1}$$

Thus, the carbon flow as part of this fly ash mass flow is

$$\varphi_{m,C,\text{fly ash}} = 4.71 \text{ kg·h}^{-1}$$

This results in

$$CC = 96.1\%$$

b. To calculate the CGE, use Equation (10.5). Here, we know all the parameters in the denominator of the equation on an ar basis. Now, we only need to derive the mass flows of each species. First, we need to calculate the total mass flow of the gas produced. This can be derived from the fuel mass flow,

λ, and the given dry fuel composition basis. For complete combustion of the daf part of the fuel, one needs, according to the stoichiometry of the reaction below, x moles of air:

$$C_aH_bN_cO_d + x(O_2 + 3.76N_2) - > a\,CO_2 + \tfrac{1}{2}b\,H_2O + c\,NO + 3.76a\,N_2$$

Now, $x = a + \tfrac{1}{4}\,b + \tfrac{1}{2}\,c - \tfrac{1}{2}\,d$.

The stoichiometric air-to-fuel ratio (fuel on daf basis!) is

$$\left(\frac{Air}{Fuel}\right)_{stoichiometric} = \frac{4.76x}{1}\frac{MW_{air}}{MW_{fuel}}$$

We have $48/12\,mol = 4\,mol\,C$, $6/1 = 6\,mol\,H$, $0.5/14 = 0.036\,mol\,N$, and $42/16 = 2.63\,mol\,O$. The fuel then can be represented (daf) and normalized to 1 C per mol fuel by $CH_{1.5}O_{0.656}N_{0.0089}$. This leads to an air-to-fuel stoichiometric ratio of $4.76 \times (28.85/24.13) = 5.99$ kg air·kg^{-1} fuel (daf), which is on a dry basis (db) of 5.78 kg air·kg^{-1} fuel. With the given 277.8 kg·h^{-1} fuel (ar basis), we have 250 kg·h^{-1} dry feed flow. Now, the actually fed airflow, considering $\lambda = 0.33$, is then 476.5 kg air·h^{-1}. The total gas flow follows from an overall mass balance and is the sum of fuel and air minus the carbon-containing ash flow. This is $277.8 + 476.5 - 13.46$ kg·h^{-1} = 740.8 kg·h^{-1} product gas flow. Now, the only unknown is the product gas LHV. This is the sum of mass fractions of combustible gas compounds times their specific LHV. Given are volume fractions, which, with ideal gas law valid under these conditions, are equal to mole fractions (y_i). Mass fractions (Y_i) are calculated from these mole fractions (y_i) by

$$Y_i = y_i\frac{MW_i}{\overline{MW}}$$

Here, $\overline{MW}$ is the average molecular weight of the gas. The following mass fractions for the combustible species are obtained:

$$Y_{CH_4} = 0.028;\quad Y_{CO} = 0.17;\quad Y_{H_2} = 0.009;\quad Y_{NH_3} = 0.002.$$

The LHV of the mixture then directly follows, given the known LHV values for each species. This gives LHV = 3.42 MJ·kg^{-1}, so that

$$CGE = \frac{740.8 \times 3.42}{277.8 \times 16} \cdot 100\% = 57.0\%$$

10.1.4 A Thermodynamic Equilibrium Approach to Biomass Gasification

As a first approach to modeling biomass gasification processes, usually, a calculation of the chemical reaction equilibrium is performed so as to gain insight into which

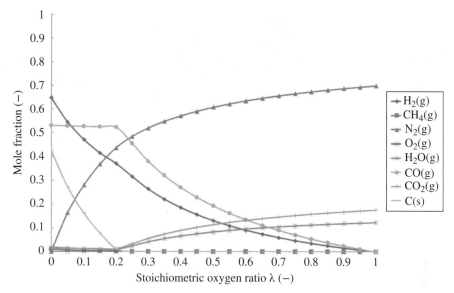

FIGURE 10.2 Chemical equilibrium for gasification of dry wood at T = 850°C and p = 0.1 MPa.

compounds are formed when full equilibrium would be reached (after infinite time at given temperature and pressure). This helps to understand trends in the formation of gaseous components as a function of the applied oxidizer stoichiometry and to elucidate which role catalysis might play, e.g., to reduce often unwanted hydrocarbon concentrations under the proposed process conditions.

Figures 10.2 and 10.3 illustrate a few typical thermodynamic equilibrium calculations based on air-blown gasification under adiabatic conditions. A representative temperature of 850°C has been chosen as well as two pressure levels, atmospheric pressure and a pressure of 2.0 MPa. As the dry biomass composition, a typical wood composition of $CH_{1.4}O_{0.6}$ was used, and the air-to-biomass stoichiometric ratio, λ, was varied between 0 and 1; the values of the heat of formation of the components (ΔH_f^0) were obtained from Zainal (2001). As components in the calculation procedure, the following ones were selected: H_2, CO, CO_2, CH_4, N_2, H_2O, O_2, and C(s). C(s) was taken as graphite. Minimization of the Gibbs energy was applied to calculate the chemical equilibrium. This procedure to calculate chemical equilibrium has already been explained in details in Chapter 5.

What can be observed is that when the value of λ decreases from 1 (the point of exact stoichiometric combustion), the composition of the gas changes in such a way that the complete oxidation products CO_2 and H_2O decrease in concentration and that the CO and H_2 concentrations increase. Furthermore, it can be observed that the values of the CH_4 concentrations remain very low under atmospheric conditions, but at increased pressure, they substantially increase. Important to note is that when decreasing λ, at a certain value of this parameter, solid carbon starts to form. This is called the "carbon boundary," and the value at which it occurs corresponds to the

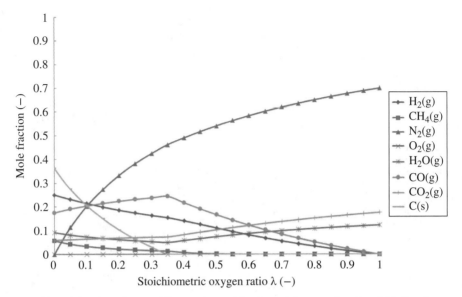

FIGURE 10.3   Chemical equilibrium for gasification of dry wood at T = 850°C and p = 2.0 MPa.

value of the highest CGE (see, e.g., Prins et al. (2003)). Regarding the graphs, below the carbon boundaries $-\lambda \sim 0.2$ for $p = 0.1$ MPa and $\lambda \sim 0.35$ for $p = 2.0$ MPa, CC is not complete anymore, which can be seen from the increased values of solid C predicted. This therefore leads to an optimum in the amount of energy in the gas compared to the thermal energy input.

10.1.5   Kinetic and Mass Transfer Aspects of Biomass Gasification

For gasification at temperatures lower than $\sim 1000°C$ without the presence of species in the reactor that (significantly) act as catalysts, usually chemical equilibrium is not attained, because practical residence times, in particular of the gas phase, are too short. In particular, the deviation from equilibrium appears from substantially higher hydrocarbon species concentrations (methane, light olefins, aromatic compounds) than predicted. Also, nitrogen compounds such as ammonia (NH_3) and HCN are present in the gas in significant amounts, despite the fact that they are not found to such extent in chemical equilibrium calculations, as N_2 is more stable (De Jong, 2005; Leppälahti and Kurkela, 1991). In order to be able to more accurately predict gas yields and composition, a reactor model based on finite kinetic rates of the most important gasification reactions is needed for more detailed reactor and process design.

The subprocesses taking place during biomass gasification are drying (dealt with in Chapter 8), pyrolysis (see Chapters 2 and 11), char oxidation (a combustion process treated in Chapter 9), and char gasification. This last step is treated in this chapter in some depth. Biomass devolatilization leaves a highly porous char with a substantial specific surface area. The development of the porosity and surface area of a biomass particle depend on the heating rate. A high heating rate causes volatile biomass fragments to quickly and easily escape from the remaining porous solid organic structure. In this case, these mostly organic compounds do not have sufficient residence time within the charring structure to recondense (forming a secondary char). In contrast, a slow heating rate for devolatilization results in relatively high amounts of residual char.

Heterogeneous reactions in coal/biomass gasification include the reaction of char, the solid residue of pyrolysis, with gas-phase species (mainly O_2, H_2O, CO_2, and H_2). In this part, the main reactions are presented, and kinetic modeling approaches are described.

The main heterogeneous reactions playing a role in gasification can be represented as

$$C(s) + (1/\alpha)O_2(g) \rightleftarrows (2\alpha-2)/\alpha CO(g) + (2-\alpha)/\alpha CO_2 \qquad \text{(RX. 10.1a)}$$

$$C(s) + H_2O(g) \rightleftarrows CO(g) + H_2 \qquad \text{(RX. 10.3a)}$$

$$C(s) + CO_2(g) \rightleftarrows 2CO(g) \qquad \text{(RX. 10.4a)}$$

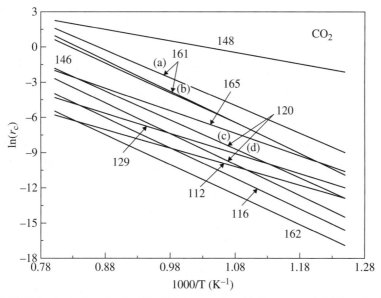

FIGURE 10.4   Arrhenius plot for CO_2–biochar reaction. (a) Cotton wood, (b) Douglas fir, (c) straw, and (d) spruce. (Source: Reproduced with permission from Di Blasi (2009). © Elsevier; numbers correspond to references in that publication.)

$$C(s) + 2H_2(g) \rightleftarrows CH_4(g) \qquad\qquad (RX.\ 10.5a)$$

with $1 < \alpha < 2$. The value of α in reaction (RX. 10.1a) is subject to some controversy in the literature. Some researchers suggest that this value cannot be found for the complex interacting phenomena taking place during gasification (Denn et al., 1979). According to Smith (1982), CO is the primary product ($\alpha = 2$) for coal. Laurendeau (1978) and Martens (1984) concluded that both CO and CO_2 are primary reaction products of this oxidation reaction. High temperature and low pressure favor CO formation. Jensen et al. (1995) showed that CO is the main primary product under pressurized fluidized bed combustion (PFBC) process conditions, but relatively low levels of 10–30% of primary CO_2 formation are possible. For graphite and coal char, Arthur (1951) obtained, under atmospheric conditions and suppressing further oxidation of gaseous CO using $POCl_3$, the following relation for the mole ratio CO/CO_2 (Take note that R_u is expressed in $cal \cdot mol^{-1}$ in this equation):

$$\frac{CO}{CO_2} = 10^{3.4} \exp\left(-\frac{12,400}{R_u T} \right) \qquad\qquad (Eq.\ 10.6)$$

At temperatures typical for fluidized bed operation, 700–1000°C, the CO/CO_2 ratio is between 4.1 and 18.8, so CO is the major component of initial carbon oxidation. Monson et al. (1995) give a relation for the molar ratio CO/CO_2 under pressurized bituminous coal char combustion conditions (verified in a drop tube reactor with a temperature range of 1000–1500 K and O_2 concentrations in the range of 5–21%):

$$\frac{CO}{CO_2} = 3 \times 10^8 \exp\left(-\frac{30,178}{T_s} \right) \qquad\qquad (Eq.\ 10.7)$$

At temperatures typical for fluidized bed operation, 700–1000°C and $1.02 \times 10^{-5} < CO/CO_2 < 1.52 \times 10^{-2}$, the coefficients in equation (RX. 10.1a) for CO and CO_2 are practically 0 and 1, respectively.

In Figure 10.4 and in Figure 10.5, a graphical overview of experimental correlations for the rate of char gasification reactions with CO_2 and H_2O is given, obtained from an excellent review of Di Blasi (2009). Only biochar data are shown here, and the given rates are based on the intrinsic kinetics only (no effect of surface development; see later in this section Equation (10.14)). The gasification reaction of char with H_2 is observed to be much slower than the char–H_2O/CO_2 reaction (Kosky and Floess, 1980) and is not further dealt with in this chapter. It can be observed that the reactivity values for both reactions show a wide range of values. Biomass heterogeneity plays an important role; certain ash elements can catalyze the reactions.

Char gasification reaction with CO_2 can be represented by the following mechanism, as pointed out by Barrio et al. (2001):

$$C_f(s) + CO_2 \underset{k_{1b}}{\overset{k_{1f}}{\rightleftarrows}} C(O) + CO \qquad\qquad (RX.\ 10.4b)$$

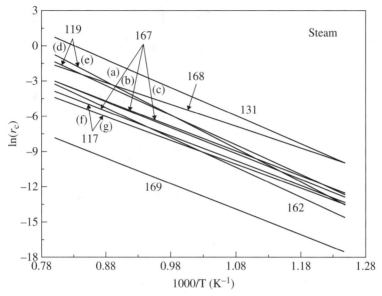

FIGURE 10.5   Arrhenius plot for H_2O–biochar reaction. (a) Straw, (b) poplar, (c) bark, (d) beech, (e) birch, (f) maple, and (g) pine. (Source: Reproduced with permission from Di Blasi (2009). © Elsevier; numbers correspond to references in that publication.)

$$C(O) \xrightarrow{k_3} CO + C_f(s) \qquad \text{(RX. 10.4c)}$$

In these expressions, C_f represents an available active carbon site and $C(O)$ an occupied site or, alternatively stated, a carbon–oxygen complex/transitional surface oxide. CO can have an inhibiting effect on the reaction rate, which consists of lowering the steady-state concentration of $C(O)$ complexes by increased reverse reaction 1b. The rate expression is of the Langmuir–Hinshelwood type:

$$r = \frac{k_{1f}p_{CO_2}}{1 + \dfrac{k_{1f}}{k_3}p_{CO_2} + \dfrac{k_{1b}}{k_3}p_{CO}} \qquad \text{(Eq. 10.8)}$$

Often, a further simplification to n^{th}-order kinetics is applied:

$$r = kp_{CO_2}^n \qquad \text{(Eq. 10.9)}$$

The char gasification reaction with H_2O is more complex than the char–CO_2 reaction, because more types of molecules are involved. Basically, there are two models of the reaction mechanism: the oxygen exchange model and the hydrogen inhibition model, as summarized by Barrio et al. (2001):

$$C_f(s) + H_2O \underset{k_{1b}}{\overset{}{\rightleftarrows}} C(O) + H_2 \qquad \text{(RX. 10.3b)}$$

$$C(O) \xrightarrow{k_3} CO + C_f(s) \qquad (RX.\ 10.3c)$$

$$C_f(s) + H_2 \underset{k_{4b}}{\overset{k_{4f}}{\rightleftharpoons}} C(H)_2 \qquad (RX.\ 10.3d)$$

$$C_f(s) + 1/2\,H_2 \underset{k_{5b}}{\overset{k_{5f}}{\rightleftharpoons}} C(H) \qquad (RX.\ 10.3e)$$

The oxygen exchange model is based on reactions (RX. 10.3b) and (RX. 10.3c), with the latter being irreversible. The traditional hydrogen inhibition model is an extension of the oxygen exchange model with reaction (RX. 10.3d). A second version of the hydrogen inhibition model consists of reactions (RX. 10.3b), (RX. 10.3c), and (RX. 10.3e). The rate expression relating to this mechanism is again of Langmuir–Hinshelwood type:

$$r = \frac{k_{1f}p_{H_2O}}{1 + \frac{k_{1f}}{k_3}p_{H_2O} + f(p_{H_2})} \qquad (Eq.\ 10.10)$$

with $f(p_{H_2}) = \left(\dfrac{k_{1b}}{k_3}\right) p_{H_2}$ [oxygen exchange model]

$f(p_{H_2}) = \left(\dfrac{k_{4f}}{k_{4b}}\right) p_{H_2}$ [hydrogen inhibition model, traditional]

$f(p_{H_2}) = \left(\dfrac{k_{5f}}{k_{5b}}\right) \sqrt{p_{H_2}}$ [hydrogen inhibition model, second version]

Often, a further simplification to n^{th}-order kinetics is applied:

$$r = k\, p_{H_2O}^n \qquad (Eq.\ 10.11)$$

It should be noted, however, that the heterogeneous reaction rate can be expressed in different ways. The char consumption rate during reaction can be directly measured, e.g., using a thermogravimetric analysis (TGA) (see Chapter 2), and the reaction rate can be expressed as

$$r_{char} = -\frac{1}{m}\frac{\partial m}{\partial t} = \frac{1}{1-X}\frac{\partial X}{\partial t} \qquad (Eq.\ 10.12)$$

with

$$X = \frac{m_t - m_0}{m_0 - m_\infty} \qquad (Eq.\ 10.13)$$

where m_0 and m_∞ are the char sample's initial and final mass values and

$$\frac{dX}{dt} = r_{char}(T, c_i) R_s(X), \quad \text{e.g.,} \quad \frac{dX}{dt} = k f(X) c_i^n \qquad (Eq.\ 10.14)$$

TABLE 10.1   Most common rate expressions for gas–solid reactions in different models

Model	Differential form $f(X) = \frac{1}{k}\frac{dX}{dt}$	Integral form $g(X) = kt$
Nucleation models		
Power law (P2)	$2X^{1/2}$	$X^{1/2}$
Power law (P3)	$3X^{2/3}$	$X^{1/3}$
Avrami–Erofe'ev (A2)	$2(1-X)[-\ln(1-X)]^{1/2}$	$[-\ln(1-X)]^{1/2}$
Avrami–Erofe'ev (A3)	$3(1-X)[-\ln(1-X)]^{2/3}$	$[-\ln(1-X)]^{1/3}$
Geometrical contraction models		
Contraction area (R2)	$2(1-X)^{1/2}$	$\left[1-(1-X)^{1/2}\right]$
Contraction volume (R3)	$3(1-X)^{2/3}$	$\left[1-(1-X)^{1/3}\right]$
Diffusion models		
1D diffusion (D1)	$\dfrac{1}{(2X)}$	X^2
2D diffusion (D2)	$[-\ln(1-X)]^{-2}$	$[(1-X)\ln(1-X)]+X$
Reaction-order models		
Zero order (F0/R1)	1	X
First order (F1)	$(1-X)$	$-\ln(1-X)$
Second order (F2)	$(1-X)^2$	$(1-X)^{-1}-1$
Third order (F3)	$(1-X)^3$	$0.5\left((1-X)^{-2}-1\right)$

Source: Khawam and Flanagan (2006).

Empirical expressions $f(X)$ and their integral form $g(X)$ commonly used for gas–solid reaction models are presented in Table 10.1.

For char reaction kinetics that is presented based on a per m^2 char surface, in order to arrive at the reaction rate in $mol \cdot s^{-1}$, one needs to multiply by the specific surface area of the char, a_s (in $m^2 \cdot m^{-3}_{char}$). This develops in time and with the extent of conversion:

$$a_s(X) = a_{s,0} \cdot f(X) \qquad \text{(Eq. 10.15)}$$

Here, the specific surface area changes with conversion as $f(X)$. There are different models describing this char area development.

In general, the char conversion process of reacting particles consists of several fundamental process steps (Di Blasi, 2009):

1. External mass and heat transfer from the bulk gas to the char's external surface layer
2. Internal mass and heat transfer through the ash layer created and the char particles
3. Pore diffusion and heat conduction inside the char particle
4. Surface chemical reaction taking place on the external and internal surfaces of the char particles

It is commonly agreed (Di Blasi, 2009; Hurt, 1998; Winter et al., 1997) that three main regimes can be distinguished during solid particle conversion. These are based on the Thiele modulus (a parameter indicating the ratio of the overall reaction rate to the internal diffusion rate) and the effectiveness factor (the ratio of the actual reaction rate to that which would occur if all the surface throughout the internal pores were exposed to the gaseous reactant at the same conditions as that existing at the external surface of the particle):

Regime I. Kinetic control prevails when the reaction occurs at low temperature with small char particles. Under these conditions, the value of the Thiele modulus is small and the effectiveness factor approaches unity. As a consequence, conversion occurs throughout the particle, which shows changes in density, but its particle size remains constant.

Regime II. Intraparticle mass transfer control takes place when the particle size is larger. This in combination with a low porosity of the particles leads to a limited gaseous reagent penetration into the char surface. Given these conditions, the Thiele modulus is much greater than unity, and the effectiveness factor much less than unity. Consequently, conversion occurs practically only at the particle's exterior surface. Therefore, the particle size decreases without much change in density.

Regime III. External mass transfer control, finally, occurs when the gas–solid reaction takes place at high temperature with even larger char particles present. In this situation, the reaction rate is proportional to the external surface of the particle and thus depends on its size.

Furthermore, heat transfer may also affect char conversion due to heat release and during combustion and heat absorption during gasification. This may enhance the temperature gradient between the surface and the core of the particle. Both mass and heat transfer effects are enhanced by high temperatures. To estimate the effects of mass and heat transfer during char reactions, various criteria are available in the literature. For instance, the Mears criterion (Mears, 1971) is widely used to estimate nonnegligible effects of external mass transfer Equation (10.16) and intraphase heat transfer Equation (10.18). The Weisz–Prater criterion (Fogler, 1999) is normally used to determine whether limitation by internal mass transfer occurs Equation (10.17). When these Equations (10.16–10.19) are satisfied, the effects of external mass transfer, internal mass transfer, and intraphase heat transfer effects can be neglected. In Example 10.2, alternative formulations for the first two criteria mentioned will be given.

Negligible external mass transfer:

$$\frac{-R_i' \rho_p (1 - \varepsilon_{bed}) r_p \mathrm{n}}{k_g c_{i,b}} < 0.15 \qquad \text{(Eq. 10.16)}$$

Negligible intraparticle mass transfer limitation (with η being the internal effectiveness factor of a particle and ϕ_1 the Thiele modulus):

$$\eta \phi_1^2 = 3(\phi_1 \coth \phi_1 - 1) << 1 \qquad \text{(Eq. 10.17)}$$

Negligible intraphase heat transfer limitation:

$$\left| \frac{-\Delta_r H\left(-R_i'\right)\rho_p\left(1-\varepsilon_{bed}\right)r_p E_a}{hT^2 R_u} \right| < 0.15 \qquad \text{(Eq. 10.18)}$$

For an nth order reaction the Thiele modulus is:

$$\phi_n = r_p \sqrt{\frac{k_n \rho_p a_{s,m} c_{i,s}^{n-1}}{D_e}} \qquad \text{(Eq. 10.19)}$$

NB in Equation (10.19), k_n is the rate coefficient of an nth-order reaction with the specific unit $[(m^3 \cdot kmol^{-1})^{n-1} \cdot (m \cdot s^{-1})]$; see the list of symbols for an explanation of the other parameters and their units.

Example 10.2   Use of a conversion model for gasification of a woody biomass particle in a TGA

Char derived from wood after a pyrolytic conversion process is a porous material. For modeling purpose, one can consider it to consist of round particles that themselves consist of many tiny, spherical grains. Such grains are surrounded by macropores and kept coherent in an inert matrix. The main pores can be assumed to have a diameter of 1 μm, and their tortuosity τ_p can be assumed to be 2. Consider the case that the char is converted with CO_2. So we have the reaction

$$CO_2(g) + C(s) \rightleftarrows 2CO(g)$$

which is simplified to

$$X(g) + Y(s) \rightleftarrows 2Z(g)$$

The process conditions are as follows: $T = 900°C$, $p = 1$ atm, and volume fraction of CO_2 is 0.10. Consider char as pure carbon (C) with a true density (thus only considering the organic part) of $2000 \, kg \cdot m^{-3}$. The particles are assumed to have a diameter of 1 mm. Assume binary diffusion of CO_2 in N_2 with the diffusion coefficient [unit: $m^2 \cdot s^{-1}$] being $D_{CO_2} - N_2 = 1.67 \times 10^{-5} \, (T/298)^{1.5}$ (Gomez-Barea et al., 2005):

a. Derive an expression for the specific surface area as a function of voidage. What is the relation with the degree of conversion?

b. Derive an expression for the degree of conversion as a function of time.

c. The conversion versus time behavior is as presented in Table 10.2. Plot the function derived in (b) versus time and derive the initial reaction rate, R_X (in $kmol \cdot m^{-3} \cdot s^{-1}$).

d. Have the experiments been biased by external mass transfer limitation?

e. Have the experiments been biased by internal mass transfer limitation?

TABLE 10.2   Reaction rate data for TG pyrolysis-based char with CO_2 reaction in a TGA (data are from Meng et al., 2012). AGROL wood gasification using 10 vol.% CO_2

t (min)	X (–)	t (min)	X (–)
0	0	10.1	0.1690
0.56	0.0111	10.66	0.1770
1.12	0.0221	11.22	0.1850
1.69	0.0326	11.78	0.1930
2.25	0.0424	12.34	0.2010
2.81	0.0522	12.90	0.2090
3.37	0.0621	13.46	0.2170
3.93	0.0719	14.02	0.2250
4.49	0.0811	14.58	0.2323
5.05	0.0904	15.14	0.2403
5.61	0.0996	15.70	0.2477
6.17	0.1088	16.26	0.2557
6.73	0.1174	16.83	0.2631
7.29	0.1266	17.39	0.2704
7.85	0.1352	17.95	0.2778
8.41	0.1438	18.51	0.2852
8.97	0.1524	19.07	0.2926
9.54	0.1604	19.63	0.2993

Solution

a. Particles consist of grains of size d_g. Now, in 1 m^3 of particles, there are n grains; thus,

$$n\left(\frac{\pi}{6}\right)d_g^3 = (1-\varepsilon)V \text{ with } V = 1 \text{ m}^3 \quad \Rightarrow \quad d_g^3 = \frac{6(1-\varepsilon)}{n\pi}$$

Then the specific surface area ($m^2 \cdot m^{-3}$) is (with the *n* particles making up 1 m^3 particles)

$$a_s = n\pi d_g^2 = (n\pi)\left(\frac{6}{n\pi}\right)^{2/3}(1-\varepsilon)^{2/3}$$

and the ratio with the original specific surface area at the start of the reaction is

$$\frac{a_s}{a_{s,0}} = \frac{(1-\varepsilon)^{2/3}}{(1-\varepsilon_0)^{2/3}} = \left[\frac{(1-\varepsilon)}{(1-\varepsilon_0)}\right]^{2/3}$$

Multiplying the numerator and denominator at the RHS of the equation with the true density (assuming this remains constant), the ratio of remaining mass at time t versus mass at t = 0 s is obtained:

$$\frac{a_s}{a_{s,0}} = \left[\frac{m_t}{m_0}\right]^{2/3} = (1-X)^{2/3}$$

This is the relation we were looking for.

b. The mass balance for the consumed char material, considering this being pure carbon, is given by

$$\rho_{T,Y}\frac{d(1-\varepsilon)}{dt} = -\nu_Y R_X MW_Y$$

where R_X is the rate of carbon dioxide consumption in $kmol\cdot m^{-3}\cdot s^{-1}$:

$$R_X = R_X'' a_s = k_1'' c_{X,g} a_s = k_1'' c_{X,g} a_{s,0}\frac{(1-\varepsilon)^{2/3}}{(1-\varepsilon_0)^{2/3}}$$

$$\Rightarrow \frac{d(1-\varepsilon)}{dt} = \frac{-\nu_Y MW_Y k_1'' c_{X,g} a_{s,0}}{\rho_Y}\left[\frac{1-\varepsilon}{1-\varepsilon_0}\right]^{2/3}$$

$$\Leftrightarrow \frac{d\left\{\frac{(1-\varepsilon)}{(1-\varepsilon_0)}\right\}}{dt} = \frac{-\nu_Y MW_Y k_1'' c_{X,g} a_{s,0}}{\rho_Y(1-\varepsilon_0)}\left[\frac{1-\varepsilon}{1-\varepsilon_0}\right]^{2/3}$$

$$\Leftrightarrow \frac{d(1-X)}{dt} = \frac{-\nu_Y MW_Y k_1'' c_{X,g} a_{s,0}}{\rho_Y(1-\varepsilon_0)}(1-X)^{2/3}$$

With the appropriate initial conditions, we can integrate this equation:

$$\int_{X=0}^{X=X}\frac{d(1-X)}{(1-X)^{2/3}} = \frac{-\nu_Y MW_Y k_1'' c_{X,g} a_{s,0}}{\rho_Y(1-\varepsilon_0)}\int_{t=0}^{t=t}dt$$

$$\Rightarrow 3\left[(1-X)^{1/3}\right]_0^X = \frac{-\nu_Y MW_Y k_1'' c_{X,g} a_{s,0}}{\rho_Y(1-\varepsilon_0)}(t-0)$$

$$\Leftrightarrow (1-X)^{1/3} - 1 = \frac{-\nu_Y MW_Y k_1'' c_{X,g} a_{s,0}}{3\rho_Y(1-\varepsilon_0)}\cdot t$$

$$\Leftrightarrow (1-X)^{1/3} = 1 - \frac{\nu_Y MW_Y k_1'' c_{X,g} a_{s,0}}{3\rho_Y(1-\varepsilon_0)}\cdot t = 1 - \frac{t}{\tau} \text{ with } \tau = \frac{3\rho_Y(1-\varepsilon_0)}{\nu_Y MW_Y k_1'' c_{X,g} a_{s,0}}$$

This is the conversion versus time relation we needed to derive.

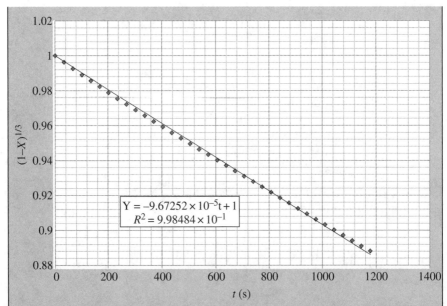

FIGURE 10.6 Plot of $(1 - X)^{1/3}$ versus time for the TGA char + CO_2 reaction of Example 10.2.

c. The plot is given in Figure 10.6.
   The requested maximum initial reaction rate per unit volume is

$$R_X = k_1'' c_{X,g} a_{s,0} = \frac{3\rho_Y(1 - \varepsilon_0)}{\tau MW_Y \nu_Y}$$

τ follows from Figure 10.6, from the slope of the straight line:
   $\tau = 1/9{,}67252 \times 10^{-5}$ s; substituting the other properties, with $\nu_Y = 1$, it follows that $R_X = 1.55 \times 10^{-2}$ kmol·m^{-3}·s^{-1}.

d. A check on lack of external mass transfer limitation can be made via the following criterion:

$$\frac{k_1 d_p}{6 \, k_g} = \frac{-R_X \pi d_p^3}{6 \, k_g \bar{c}_x \pi d_p^2} \ll 1$$

The nominator is the first-order reaction rate, and the denominator represents maximum mass transfer (with maximal driving force between bulk concentration and surface concentration).
   Substitution of the parameter values in the aforementioned equation leads to the quantity on the left-hand side of the equation to be 9.5×10^{-3}, which is much smaller than 1. This is OK.

e. A check on the lack of internal mass transfer limitation can be made by realizing that the mole flow of CO_2 (X) to the char particle (Y) must be equal to the inner particle transport (mole flow, Φ_A) and that the relative difference between bulk and inner concentration of X must be less than 5%. In this case, it has been derived (see Westerterp et al., 1988) that

$$\frac{-\Phi_X}{(1-\varepsilon)V_r D_i c_{X,i}}\,\frac{\delta^2}{} = \frac{|R_X|\left(\frac{d_p}{6}\right)^2}{D_i c_{X,i}} \leq 0.05$$

Now, R_X was calculated under subproblem (c). $c_{X,i}$—when the criterion is fulfilled–is within 5% of c_X (the bulk concentration), so we can substitute c_X. The only unknown left then is D_i, the effective internal diffusion coefficient. This can be calculated via

$$\frac{1}{D_i} = \frac{\tau_p}{\varepsilon}\left(\frac{1}{D} + \frac{1}{D_K}\right) \text{ with } D_K = \left(\frac{d_{pore}}{3}\right)\sqrt{\frac{8R_u T}{\pi\,MW_X}}$$

D_K is the Knudsen diffusion coefficient. The binary diffusion coefficient D at 900°C is $1.30 \times 10^{-4}\,m^2 \cdot s^{-1}$. We take ε_0 as a conservative value for the voidage. Filling out the relation for the Knudsen diffusion coefficient, we arrive at $D_K = 2.50 \times 10^{-4}\,m^2 \cdot s^{-1}$. Thus, the internal effective diffusion coefficient is $2.92 \times 10^{-5}\,m^2 \cdot s^{-1}$. c_X is calculated based on the assumption that ideal gas law holds:

$$c_X = \frac{y_X p}{R_u T} = 1.04 \times 10^{-3}\,kmol \cdot m^{-3}$$

Now, the aforementioned criterion can be evaluated given the calculated values, and it reaches a value of 0.014, which is OK.

10.2   A SHORT HISTORY OF GASIFICATION

Already around 1850, there was a considerable gasification industry in Europe. In those early days, the gas was used for lighting and industrial heating and as fuel for the internal combustion (IC) engine (power generation). So-called "town gas" generation was mostly based on coal. This was a two-step process of consisting of thermal cracking of solid fuel and subsequent steam blowing of the remaining char in a fixed bed reactor. A real breakthrough in gasification technology was the Siemens gasifier (1861). The Siemens gasifier was the first continuous gasification reactor and had spatially separated combustion and gasification sections. All early gasifiers were air-blown fixed bed reactors with a maximum temperature in the gasification zone of ca. 900°C. Around the WWII era, a large part of the cars and trucks was powered by gas produced by built-in fixed bed wood and waste gasifiers. Winkler introduced the first alternative to the fixed bed gasifier in 1926 by developing a (relatively) low-temperature fluidized bed gasifier. The advantages of a fluid bed over a fixed bed were claimed to be the ability to accept all types of coal, smaller-sized coal, and more ash removal flexibility. The availability of oxygen supply on plant scale (von Linde) and advances made in the manufacturing of high-pressure vessels set off the development of high-pressure oxygen-blown gasification. The Lurgi dry ash (1936) process was the first oxygen-blown moving bed gasifier. Like the Siemens gasifier, the Lurgi gasifier was operated at temperatures below 1000°C in order to prevent ash melting. This system is, though in slightly modified form, still in operation (e.g., by Sasol). In 1938, the Koppers-Totzek entrained flow (EF) gasifier came into commercial use. The Koppers-Totzek gasifier produced synthesis gas (CO and H_2) containing no tars and methane on a continuous basis at ca. 1850°C and atmospheric pressure from oxygen-entrained coal. In relation to these developments, a start was made in Germany with the production of Fischer–Tropsch (FT) diesel from, a.o., wood-derived synthesis gas. This was all done because of the scarcity of liquid fuels and to become independent of imported oil. After the war, the interest in this technology rapidly declined because of the increasing availability of cheap crude oil. At the end of the 1940s and in the early 1950s, Texaco and Shell developed technologies for the production of the synthesis gas by oil gasification. These were EF reactors with top-mounted burners (atomizers) operated in the downflow mode. Operating pressures up to 8.0 MPa and temperatures in the range of 1250–1500°C were used. Apart from Texaco and Shell, also Lurgi developed oil gasification technology, known as multipurpose gasification.

Nowadays, most oil gasifiers are part of a refinery and are used for polygeneration of power, H_2/synthesis gas, and steam. As a result of the oil crisis of the early 1970s, coal gasification was taken up again. It was again Texaco and Shell (together with Krupp-Koppers) that developed EF high-pressure (2–7 MPa) and high-temperature (>1400°C) coal gasification.

The past decades have shown a gradually increased momentum in the development of biomass gasification related to the need for green transportation fuels accompanied with decreased oil dependency, a minimum impact on net CO_2 emissions, and no competition with food and feed production and land usage.

10.3 (CO)GASIFICATION TECHNOLOGIES FOR DRY BIOMASS

Gasification of biomass, also together with other fuels, e.g., coals, can be carried out using a variety of chemical reactors, still a number of which are under development at different commercial companies and research institutes. Since the first (controlled) attempts regarding thermochemical gasification of biomass, a number of reactor designs have evolved that are considered as suitable. These reactors can be classified according to the transport processes occurring (Brown, 2011):

- Fixed beds/moving beds: countercurrent updraft, cocurrent downdraft, and cross-draft
- Fluidized beds: bubbling, circulating, and dual connected
- EF reactors

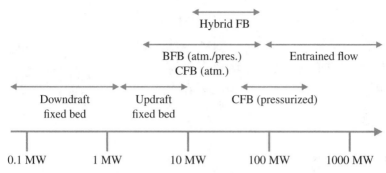

FIGURE 10.7 Different gasification technologies for biomass and their scales.

TABLE 10.3 Some examples of biomass gasification process implementation

Downdraft fixed bed	Updraft fixed bed	Bubbling fluidized bed (BFB)	Circulating fluidized bed (CFB)	Hybrid fluidized bed	EF
Fluidyne (NZ)	Babcock and Wilcox Vølund (DK)	Carbona	VVBGC (SE)	MILENA/ ECN (NL)	Shell
Pyroforce (CH)	Condens Oy/VTT (FI)	Enerkem (CA)	Foster Wheeler (USA)	Repotec (AU)	NUON/ Vattenfall (NL)
TERI (IN)					Bioliq/KIT (DE)
Xylowatt (CH)					Chemrec (SE)

In each of these reactor types, biomass gasification can be performed, but one should realize that they are also a compromise between the product gas quality, conversion efficiency, suitability for handling feedstocks with varying physical and chemical properties, complexity and scalability of the design, complexity of operation, and investment costs. In this light, Figure 10.7 illustrates the relevant size ranges of application of the aforementioned classes of gasifiers. Related to this, Table 10.3 gives a few examples of current demonstrations and (semi)commercial gasification activities.

10.3.1 Small-Scale Gasifiers: Fixed and Moving Bed Gasifiers

For comparatively small-scale operation of up to the order of a few MW_{th}, fixed bed (or moving bed) gasification technology is suitable. There are three basic types of reactor configuration in this class: downdraft, updraft, and cross-draft

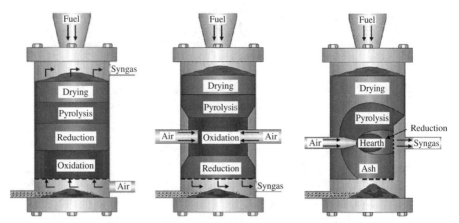

FIGURE 10.8 Diverse fixed bed reactor types: updraft (left), downdraft (mid), and cross-draft (right). (Source: Adapted with permission from Olofsson et al. (2005). © Umeå University (Sweden).)

(see Figure 10.8). Typically, such gasifiers are coupled to diesel or gas engines; systems of $100 - 200$ kW$_e$ with an approximate electrical efficiency of between 15 and 25% are commercially available on the market. High costs and the need for gas cleaning and careful operation have prevented application in large numbers. Some systems are being applied more or less successfully in, e.g., rural areas of India, Indonesia, and China (Stassen, 1995).

The *downdraft gasifier* is a cocurrent reactor, in which the fuel and the gasification agent move in the same direction, and it is presented in Figure 10.8 (mid). Different distinct reaction zones are present in this reactor type: a drying zone in which moisture is evaporated from the biomass fuel while slowly moving downward in the direction of the pyrolysis zone; in that zone, the biomass is decomposed into chars, tars, and gases. A part of the pyrolysis gas/vapor products is burned in the combustion zone below. The heavier (polyaromatic) hydrocarbon species, called tars, are cracked there due to the high temperature, which is a result of the overall exothermic reactions in the combustion zone. Thus, the produced gas exiting from below the last gasification zone is relatively clean, i.e., with low tar concentrations. In the gasification zone, remaining char is converted using the steam and carbon dioxide in the gas from the combustion zone above that has run out of oxygen. This gasifier configuration is uncomplicated and practically mostly reliable and has been proven for quite some biomass fuels, such as relatively dry blocks or lumps with a low ash content and containing a low portion of fine and coarse particles (Bridgwater, 1995). The physical limitations of the diameter in relation to the particle size mean that there is an upper limit to the capacity of this configuration of around $500 \, \text{kg} \cdot \text{h}^{-1}$ (Bridgwater, 1995).

Another alternative fixed bed gasifier design is the *updraft gasifier*, which is a countercurrent reactor in which the fuel and the gasification agent flow in opposite directions. In Figure 10.8 (left), a schematic representation of an updraft gasifier is given. In the updraft gasifier, the biomass is dried in the drying zone by the downward generated hot producer gas. Below this zone, the biomass is pyrolyzed, and the char generated there moves further down the reactor. On the other hand, higher molecular weight tar vapors and gases follow the updraft flow pattern within the reactor. Only a fraction of the tars condenses on the drying biomass particles; a substantial amount of the tars stays in the gas and leaves the reactor as product gas. The generated char slowly moves down toward the combustion zone in which it is converted by heterogeneous gasification reactions. As a result of this process configuration, the gas produced in an updraft gasifier is characterized by a relatively high tar and hydrocarbon content, which leads to a comparatively high heating value of the gas. However, this producer gas requires substantial cleaning before it can be processed further in, e.g., prime movers for power generation. The updraft gasifier has found application in, e.g., the Lurgi dry ash gasifier, developed in the 1930s, with subsequent installation of about 150 gasifiers since. It is a pressurized gasifier, and the oxidizer introduced in the bottom is a mixture of steam and oxygen. Temperatures in the bottom combustion part are moderated by steam supply ($\sim$1100°C) so that ash is still removed in non-molten form using a rotating grate and lock hopper in series. The produced raw synthesis gas is quenched with recycle water to remove tars. A water jacket surrounds the reactor, generating part of the steam used in the gasifier.

Finally, in the category of fixed/moving bed gasifier configurations, the *cross-draft gasifier* is described. In this reactor configuration, the fuel and gasification agent flow in a direction mainly perpendicular to each other, which is shown in Figure 10.8 (right). The cross-draft gasifier is only suitable for the gasification of charcoal. The temperatures in the combustion zone can rise to 1500°C, and a point of concern is that the reactor wall material has to endure the high temperature in the reactor for long-running durations. Like the updraft gasifier, the cross-draft gasifier has a low tar conversion. Therefore, a high-quality charcoal has to be used.

For large-scale biosyngas generation, these gasifiers are not attractive due to their relatively small scale. Two other classes of reactors, the fluidized bed and EF gasifiers, can be scaled up to the large sizes needed for biosyngas production in view of transportation fuel production.

10.3.2 Large-Scale Gasifiers: Entrained Flow Gasifiers

EF gasifiers are designed for the large-scale operation ($\gg$ 100 MW$_{th}$) on varying fuels. Such scales allow more economical production of biofuels for transportation or for power production using advanced, efficient cycles such as the integrated gasification combined cycle (IGCC: gas turbine with downstream steam turbine operating on the hot flue gas of the turbine).

The operating temperatures in these EF gasifiers range from approximately 1200 to 1500°C. They are mostly operated using varying coal types. EF gasifiers generally use fuel in the form of a powder or slurry. The fuel is then mixed with steam, or steam and oxygen, and gasified in a flame. When using biomass as fuel, it must either be ground to a powder or in some cases pyrolyzed to gas, oil, and coke; the latter again needs to be ground or converted to slurry. Manufacturers of "slagging" gasifier types are Shell, Texaco, Krupp-Uhde, Future Energy (formerly Noell/Babcock Borsig Power), ConocoPhillips (E-gas technology, formerly Destec/Dow), Mitsubishi Heavy Industries (MHI), Hitachi, and Choren (formerly UET, now Linde-patented, Germany).

Coal gasification was the initial application of the EF type of gasifier, in the 1950s by Koppers. Koppers-Totzek slagging gasifiers operating at atmospheric pressure for syngas production aimed at the production of ammonia (NH_3). The gasifier is characterized by coal and oxygen being fed via pairs of opposed burners inside niches on the side parts of a typically egg-shaped reactor. At the top part of the reactor, the produced raw syngas leaves the reactor at a maximum temperature of about 1500°C and is rapidly cooled down (quenching) using water to a temperature of about 900°C; it is then cooled down further in a syngas cooler in which steam is produced. Quenching is crucial to prevent sticky slag droplets entering and depositing on the heat exchanger. A steam jacket as reactor wall protects the steel from the high temperatures. The produced slag mainly leaves the reactor via a bottom hole.

The PRENFLO and Shell Coal Gasification Process (SCGP) are processes that have been developed later and are pressurized versions (3.0 – 4.0 MPa) of the Koppers-Totzek process with a higher efficiency. They maintain the feature of diametrically opposed burners inside side niches of the vertically mounted cylindrical reactor. A gas quench has replaced the water quench.

Also, in these processes, membrane walls are applied instead of a steam jacket to protect the pressure shell from too high temperatures. These walls are used for the generation of high-pressure saturated steam for additional power production. Aimed at power production, the developments led to the erection of a 2000 t·day^{-1} SCGP unit in Buggenum (the Netherlands) and a 3000 t·day^{-1} PRENFLO unit in Puertollano (Spain). Both gasifiers form the heart of an IGCC unit for electricity production. The power-producing company NUON has demonstrated cofiring of coal with chicken manure in the Buggenum plant up to about 30 wt% biomass share using the (modified) Shell process-based 253 MW$_e$ plant for electricity production (see, e.g., Van Dongen and Kanaar (2006)). The concept and reactor technology can be extended to polygeneration of heat, power, and syngas-based chemicals.

The General Electric (GE)-owned Texaco gasification process is also of the EF type, with the main difference being that the fuel is fed as a water slurry, whereas Koppers-Totzek- and Shell-based coal gasifiers use steam. Future Energy GmbH developed a down-fired EF reactor. Instead of burners inside of the reactor, this gasifier only consists of a single burner in the top part, which results in a simpler and lower-cost reactor. Also, the control of this single burner is easier than the other types mentioned with a resulting lower capital and operational cost. The German company Choren developed a three-stage gasification process involving the following subprocesses:

1. Low-temperature pyrolysis/gasification
2. High-temperature gasification
3. Endothermic EF gasification

The first process step comprises the carbonization of relatively dry biomass (moisture contents varying up to 15 – 20 wt%) using partial oxidation with air/oxygen at mild temperatures of 400 – 500°C. In this stage, volatiles and char are formed as products. The second process step comprises the understoichiometric combustion of the generated volatiles with oxygen and/or air in a combustion chamber that operates above the melting point of the fuel's ash to turn it into a hot gasification medium. The third stage, finally, consists of char grinding to a pulverized fuel that is blown into the hot gasification medium created in the EF reactor of the bottom part of the second stage. The pulverized fuel here reacts in an endothermic process that generates a raw syngas. Further treatment by cooling and cleaning follows. The syngas can be used for different purposes; nowadays, the focus is on diesel fuel production, called SunDiesel.

10.3.3 Large-Scale Gasifiers: Fluidized Bed Gasifiers

The fluidized bed reactor has been one of the workhorses for large-scale coal conversion since it was first patented by Fritz Winkler in 1922 and commercialized in 1926 (see, e.g., Howard (1983) and Kunii and Levenspiel (1991)). A number of basic designs of fluidized bed gasifiers have been developed since these early days. Since the oil crises in the 1970s, biomass use in such reactors has been investigated. Such biomass-based fluidized bed reactors have been targeted at the midscale thermal capacities of 10 MW_{th} to large-scale capacities of more than 100 MW_{th}.

An inert or catalytic bed material of small solid particles is used to facilitate heat and mass transfer throughout the reactor. The bed is kept in a fluidized state by blowing a gasification agent through it, lifting the bed against gravity by drag force. Herewith, the turbulence in the bed creates an even temperature distribution in the bed. Thus, no different distinct reaction zones appear in the bed, like in the case of fixed bed gasifiers. The prevailing temperatures in a fluidized bed are usually in the range of 700–900°C, and the reactors are operated with gauge pressures between 0 and 7.0 MPa. Even within the relatively low operating temperature window, bed sintering is a common problem when biomass with a high ash content is used. The alkali components in the ash, facilitated by chlorine, show the tendency to form low-melting eutectics with silica sand, which is the most common bed material. These eutectics cause bed agglomeration and bed sintering, which can lead to loss of fluidization (Bartels, 2008; Fryda et al., 2008).

The most common types of fluidized beds are the BFB and the CFB. Also, hybrid forms or interconnected fluidized bed reactors are encountered in modern designs. Figure 10.9 shows the basic reactor configurations (Olofsson et al., 2005).

In a BFB, the gasification agent is blown through the bed in such a way that it forms bubbles within the bed zone. The gas velocity is significantly above the minimum fluidization velocity and below the maximum terminal velocity (typical values range from ~0.5 to 2 m·s^{-1}), so that the bed material largely remains in the reactor.

In the BFB design configuration, the majority of the gasification reactions take place in the dense fluidized bed part. Some reactions, especially thermal cracking and reforming reactions, the water–gas shift reaction, and the gasification of entrained small particles, continue in the freeboard above the bed. The carbon conversion in the

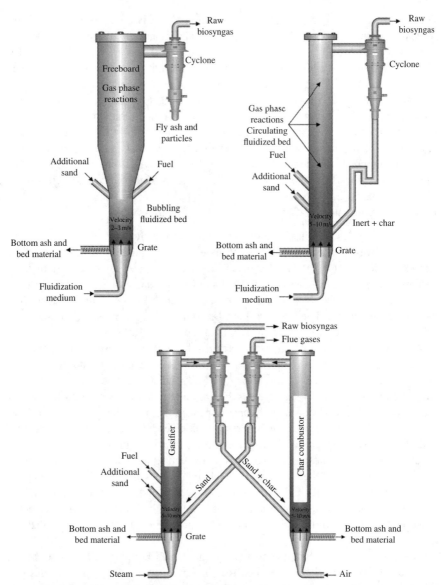

FIGURE 10.9 Configurations of fluidized bed reactors: BFB (top left), CFB (top right), and interconnected (indirect) fluidized beds (bottom). (Source: Adapted with permission from Olofsson et al. (2005). © Umeå University (Sweden).)

process is well above 90%, due to the high residence time of the biomass particles and the residual conversion when entrained to the freeboard. The tar content of the gas is in between the tar content of the downdraft and updraft gasifiers. An example of a BFB design is the steam-/oxygen-blown gasifier developed by IGT, now commercially manufactured by Carbona.

In a CFB gasifier, the oxidizing gas is introduced into the reactor with such a high velocity that large amounts of solids are entrained with the product gas. The entrained solids are separated from the gas in a cyclone and recycled back to the gasifier. The CFB has a high CC efficiency (typically >95%), because of the recycling of the bed material. The raw product gas generated by a CFB gasifier has a relatively high dust content. An example of this technology is the 18 MW_{th} Värnamo (Sweden) pressurized gasifier (see, e.g., Ståhl and Neergaard (1998)); this gasifier was operated for demonstration in the 1990s and restarted for a short period in 2007, but hereafter it was mothballed.

Besides the BFB and the CFB, there are also designs of gasification installations based on interconnected beds. These reactors make use of indirect, or allothermal, gasification, in which the heat to drive the endothermic char gasification reactions is generated by a separate combustion process. An excellent overview is given by Corella et al. (2007). A recent example is the fast internal circulating fluidized bed (FICFB) reactor concept developed at Vienna University (Hofbauer et al., 2002; Pfeifer and Hofbauer, 2008), which has become commercially available on the market (see Table 10.4). This advanced gasifier type has been designed as a steam-blown BFB in which biomass is gasified and char is carried over to an interconnected CFB combustor, which provides heat for the endothermic reactions in the BFB reactor via internal transportation of hot bed material. Another process using interconnected beds is

TABLE 10.4 Some of the main fluidized bed biomass gasification demonstrations

Project/demo	Gasifier, thermal input	Product gas use	Current status
Güssing (Austria)	Interconnected FICFB, ~8 MW_{th}	Combined heat and power (CHP) (Pfeifer and Hofbauer, 2008), Demo-FT (Weber et al., 2010) Demo-substitute natural gas (SNG)	Running
Oberwart (Austria)	Interconnected FICFB, ~8 MW_{th}	CHP	Running
Villach (Austria)	Interconnected FICFB, ~15 MW_{th}	CHP	Commissioned
Skive (Denmark)	BFB 28 $MW_{th,max}$	CHP	Running
NSE Biofuels, Varkaus (Finland)	CFB (Foster Wheeler), ~12 MW_{th}	Lime kiln, Demo-FT	Commissioned
HVC, Alkmaar (the Netherlands)	Interconnected MILENA, 10 MW_{th}	SNG (Van der Meijden et al., 2009)	Development
AMER (the Netherlands)	CFB, 84 MW_{th}	Cofiring (Willeboer, 1998)	Running
Värnamo (Sweden)	Foster Wheeler, CFB (air blown) 18 MW_{th}	CHP (Rensfelt and Gardmark, 2010)	Mothballed, after last demo in 2007

the SilvaGas process, which was initially developed at the Battelle Columbus labs (United States) and is characterized by two interconnected CFBs of which one is a steam-blown biomass gasifier and the second a char combustor storing heat into bed material that is circulated to the gasifier reactor. Finally, there is the MILENA concept (van der Meijden et al., 2009), developed at ECN (the Netherlands), in which the central part is a riser in which steam-blown biomass gasification takes place; the surrounding annular space has been designed for combusting the char from the core part. Table 10.4 shows an overview of large-scale fluidized bed biomass gasification demonstrations.

Factors playing decisive roles in the selection of employing a certain reactor design for biomass gasification are:

- Scale of the energy conversion process
- Feedstock flexibility (particle sizes and fuel composition)
- Sensitivity to the amount of ash and its composition
- Tar generation characteristics

The scale of operation is most probably the primary criterion. Small, decentralized systems will benefit from a simple and cheap reactor that is easy to control and maintain. Here, the feedstock is probably well defined based on local conditions. On the other hand, a biomass-to-liquid (BTL) plant or maybe even a biorefinery, where the gasifier is only one of the units of operation, benefits from the larger scale of the reactor in terms of its thermal efficiency and economies of scale.

The feedstock flexibility is another aspect. The structural appearance of biomass is often fibrous and tough, and consequently, it is difficult to cut or pulverize. Therefore, it is not desirable to reduce the biomass in size too much, because of the adverse effect on the energy efficiency of the whole process. Additionally, raw biomass is not dry, but contains varying amounts of moisture. Considering the preceding text, the gasification reactor should be able to cope with the changes in fuel supply characteristics, which are both physical and chemical in nature.

Next to the moisture and volatile fraction, biomass also contains inorganic matter, usually referred to as ash. This ash-forming matter can be variable in composition (see, for instance, Chapter 2). In general, wood has low ash contents with calcium and silica as the main constituents, but agricultural residues may contain appreciable amounts of alkali metals (potassium, sodium) and chlorine, which may pose challenges to fluidized bed reactors (agglomeration) as well as EF gasifiers (slagging and fouling).

While ash-related issues may lead to difficulties in process operation and unscheduled maintenance stops, the tar produced in the gasifier may affect the downstream equipment in a negative way, resulting in the need for extensive downstream gas treatment and upgrading. "Tar" is an umbrella term for various kinds of larger hydrocarbons produced during gasification. There are diverse definitions of the term, but there is a broadening global consensus in defining biomass tar as organic contaminants with a molecular weight larger than benzene (Brown et al., 2009; Van Paasen and Kiel, 2004), comparable to a European standard—CEN/TS 15439—for quantifying tars (CEN, 2006). Tar formation is a commonly encountered issue in lower-temperature (e.g., fluidized bed) gasification processes (Corella et al., 2006; Kinoshita et al., 1994; Milne et al., 1998; Van Paasen and Kiel, 2004). Although the main issues related to tar are formed by condensation in the equipment downstream the gasifier, which operates at lower temperatures (typically below 500°C), tar also significantly contributes to the heating value of the product gas. Therefore, gas cleaning is needed with respect to this class of contaminants.

10.3.4 Comparison of Large-Scale Gasification Technologies for Dry Biomass

An overview of the characteristics of different dry biomass-based gasifier technologies for the larger-scale applications—so excluding the fixed and moving bed types, as these are less relevant for large-scale syngas generation—is given in Table 10.5. The major challenges for fluidized bed gasification are in gas cleaning for removal of particles and tars in particular, as well as agglomeration prevention. The main challenges for EF gasifiers are biomass pretreatment to obtain finely sized fuel and slagging/fouling of heat transfer surfaces.

10.3.5 Some Other Types of Gasifier Concepts

Apart from fixed/moving bed, EF, and fluidized beds, there are also comparatively novel reactor designs under development. One example is the cyclone gasifier. It combines reaction and separation, intensifying the latter. Fuel is fed via different tangential inlets using steam. Design and evaluation have been performed for atmospheric and pressurized operation. In order to realize the pressurized design, the cyclone is surrounded by a pressure vessel. Produced gas leaves swirling at the top of the reactor; the remaining solids are separated by centrifugal and gravity forces (Syred et al., 2004).

Another relatively new concept is the vertical vortex gasifier, which has been derived from the downdraft moving bed gasifier. In this reactor, preheated air is used as an oxidizer, and it is fed tangentially at the top of the reactor, so that a swirl is generated. By the strong circulating flow toward the wall, another upflow in the center part is generated; here, pyrolysis gases generated from the top of a charring bed are sucked upward toward the entering airstream(s). Then these tar-containing volatiles are converted at temperature of about 1100°C. The heat generated in the top part of the reactor radiates to the top part of the pyrolyzing bed, which enhances the process of fuel conversion.

More design alternatives, separating in varying ways the different subprocesses of gasification, have been suggested and tested. The Blue Tower is one such a concept. It is an indirect staged reforming process with a heat-carrying system to transfer heat

TABLE 10.5   Comparison of different large-scale gasifier types

Advantages	Disadvantages
(Bubbling) fluidized bed	
Flexible feed rate and composition	Operating temperature limited by ash melting temperature
High-ash fuels acceptable	High tar and fine-particle content in gas
High volumetric capacity	Possibility of high-C content in fly ash
Temperature distribution	
CFB	
Easy to scale up to medium/large scale	Corrosion and attrition problems
Medium operating temperature (about 850°C)	Potential for agglomeration
Double (indirect) fluidized bed	
Oxygen not required	More tar due to lower bed temperature
Temperature distribution	Difficult to operate under pressure
	Scalability more limited than common BFB/CFB
EF	
Very low tar and CO_2	Severe feedstock size reduction required
Flexible feedstock	Complex operational control
Exit gas temperature	Carbon loss with ash
	Higher exergy loss due to high gasifier temperature
	Ash slagging

from a combustion zone to a pyrolysis zone and a reforming zone. The transportation is mechanical and the heat carriers, alumina (Al_2O_3) granules, are transported to its coolest stage before being heated by flue gases that originate from char combustion. The char is reformed by steam into a product rich in hydrogen and carbon dioxide. The steam amount used ensures a high hydrogen concentration. The gasifier is able to operate on various fuels.

Another way to split the drying/pyrolysis, combustion, and reforming zones has been realized in the heat-pipe reformer. Here, the heat is not transferred mechanically, but using heat pipes. These are tubes with a medium (e.g., liquid Na) inside that takes up heat on one side via evaporation and gives off heat via condensation at the end of the pipe (Karellas et al., 2008).

Novel ways of intensification of the gasification process that result in cleaner product gases are nowadays coming up. In this context, plasmas are used to convert biomass or waste into clean synthesis gas. Plasma is considered as the fourth aggregation state next to solid, liquid, and gas; an example is lightning. Organic material is decomposed to combustible gas and (partially) molten mineral matter at extreme temperature and radical and charged reactive species concentrations in the plasma. There are basically two types of plasma torches: transferred and nontransferred torches. In the transferred torch type, an electric arc is established between torch-tip and metal or slag phase in the reactor bottom part, or the conductive reactor wall lining. In nontransferred torches, the position of the arc is within the torch, and plasma is established in the gas flowing through the device.

10.4   GASIFICATION IN AN AQUEOUS ENVIRONMENT: HYDROTHERMAL BIOMASS CONVERSION

10.4.1   Introduction

Although wet biomass has a very low overall heating value, still products with a high heating value can be extracted from it by applying advanced conversion processes. Gasification in hot compressed water is considered a promising technique to convert such wet streams into medium calorific gas. The produced gas is rich in either hydrogen or methane depending on the operating conditions and applied catalysis. By practicing countercurrent heat exchange between the feed stream and the reactor effluent (see Figure 10.10), high thermal efficiencies can be reached despite the low dry matter content of the feedstock. Apart from this essential energetic benefit, the unique properties of hot compressed water are believed to be beneficial for the desired chemistry. It promotes ionic reaction pathways over radical routes, leading to less char formation. Generally, organic molecules are reactive in pressurized water at temperatures above 250°C.

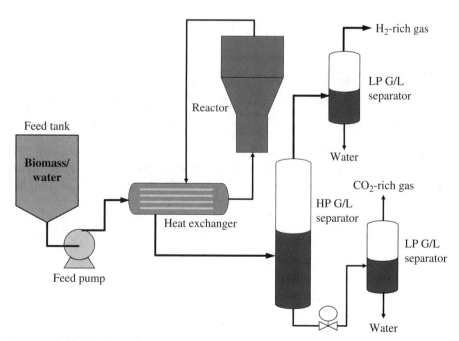

FIGURE 10.10   Simplified process scheme of the PDU for SCWG in Enschede, the Netherlands.

TABLE 10.6   Composition of the gas produced in the PDU (see Figure 10.10) by SCWG of 5 wt% glycerol in water with and without addition of NaOH. $p = 27.0$ MPa; T = 580°C

Component	Concentration (vol.%)	Concentration (vol.%) 0.0075 wt% NaOH
Hydrogen (H_2)	29	60
Carbon monoxide (CO)	30	0.5
Carbon dioxide (CO_2)	13	21
Methane (CH_4)	15	12
Ethene/ethane	11	5
Propene/propane	2	1
LHV [MJ m_n^{-3}]	21	15

Biomass gasification in supercritical water (SCW) (600°C, 30.0 MPa) is a novel process under development after the pioneering work of, e.g., Modell and Antal and coworkers since the mid-1980s (Antal et al., 2000). In the laboratory and in pilot-plant work that has been carried out so far, stirred autoclaves and tubular reactors are used frequently in connection with a shell and tube heat exchanger. Not much attention has been paid to practical design aspects of the supercritical water gasification (SCWG) process (including heat exchanger and reactor). Figure 10.10 shows a possible flow sheet of the process, and typical results for glycerol are listed in Table 10.6.

10.4.2   Supercritical Water Properties

The properties of SCW are quite different from those of the normal liquid or steam at atmospheric pressure. For instance, at the critical point, the density is around 300 kg·m^{-3} (vs. 1000 kg·m^{-3} for liquid water at ambient conditions), the dielectric constant is 5 (vs. 80 for liquid water at ambient conditions), and the ion product Kw = [H^+] [OH^-] is 10^{-11} (vs. 10^{-14} for liquid water at ambient conditions). The behavior of water near and above the critical point has been studied extensively. In fact, all relevant physical properties have been determined experimentally and have been tabulated by the US National Institute of Standards and Technology (NIST) (tinyurl.com/9s23f). The most striking feature of SCW is the possibility to manipulate and control its properties around the critical point by tuning the temperature and pressure. In the vicinity of the critical point where the ion product is high (10^{-11}), the H^+ concentration is about 30 times higher than at ambient conditions, offering increased opportunities for acid-catalyzed reactions. As an important consequence of the change in the dielectric constant, SCW behaves like a nonpolar solvent and exhibits a high solubility for nonpolar organic compounds like benzene. Gases like oxygen, nitrogen, carbon dioxide, and methane are completely miscible in SCW. On the contrary, the solubility of inorganic salts like NaCl is decreased to very low values.

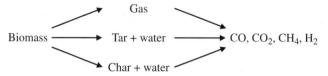

FIGURE 10.11 Global mechanism for biomass conversion under hydrothermal conditions.

10.4.3 Chemistry and Thermodynamics

Figure 10.11 shows a global mechanism for biomass conversion under hydrothermal conditions. Biomass gasification in SCW is supposed to be the result of thermal decomposition reactions, followed by homogeneous reactions (water–gas shift and the methanization) of the resulting gases.

In general, the overall reaction equation (for glucose as model compound) can be written as

$$uC_6H_{12}O_6 + vH_2O \rightarrow wCO_2 + xCH_4 + yCO + zH_2 \qquad \text{(RX. 10.14)}$$

The desired reaction of complete reforming to hydrogen is

$$C_6H_{12}O_6 + 6H_2O \rightarrow 6CO_2 + 12H_2 \qquad \text{(RX. 10.15)}$$

Maximal methane yields are obtained by

$$C_6H_{12}O_6 \rightarrow 3CH_4 + 3CO_2 \qquad \text{(RX. 10.16)}$$

Figure 10.12 presents results of thermodynamic calculations for supercritical gasification of $C_6H_{12}O_6$, which have been obtained with a model based on Gibbs free energy minimization (see Chapter 5). Such calculations have a limited quantitative value in case the reactions involved are too slow to reach equilibrium, but they may be useful in predicting trends and the results desired upon application of an appropriate catalyst.

Complete gasification of the organic feedstock is thermodynamically possible for both the proposed low- and high-temperature gasification processes. Actually, dry matter concentrations of up to 50 wt% do not have thermodynamic or stoichiometric limitations (see Figure 10.12a) regarding the conversion. Thermodynamic equilibrium calculations predict that high-temperature gasification produces a hydrogen-rich gas (at least for dry matter contents of less than 10 wt%), while at low temperature a methane-rich gas results (see Figure 10.12b and c). This shift in the product distribution is also observed experimentally.

For low-temperature gasification, the content of dry matter in the feed does not influence the product distribution to a large extent; beyond 5 wt%, the yields are almost unaffected (see Figure 10.12b). In contrast, at higher temperature, there is a continuous varying product distribution ranging from nearly pure hydrogen for very low weight percentages of dry matter to a mixture of ca. 50 mol% hydrogen and 50 mol % methane for high organic fractions in the feed. Once above 15 MPa, the operating

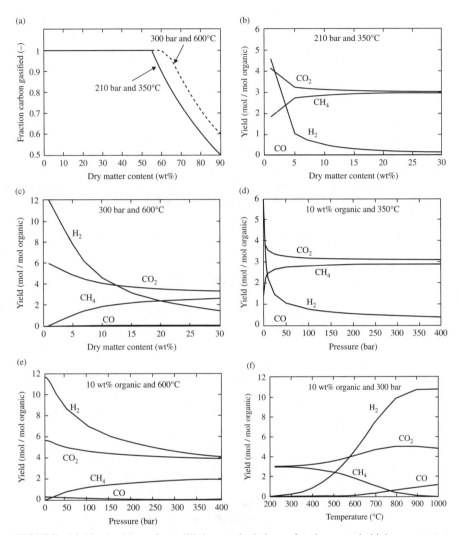

FIGURE 10.12 Results of equilibrium calculations for low- and high-temperature gasification of $C_6H_{12}O_6$ in hot compressed water.

pressure does not influence the product distribution to a large extent (see Figure 10.12d and e). It is worthwhile to note that, from a thermodynamic point of view, high-temperature gasification should be carried out at the lowest possible pressure to achieve maximal hydrogen yields (see Figure 10.12e).

Figure 10.12f shows that, according to thermodynamics, there is strong shift from methane toward hydrogen and carbon monoxide upon increasing the temperature. Methane-rich gas can be produced up to temperatures of approximately 500°C, while higher temperatures favor the production of hydrogen.

10.4.4   Reported Results for Low-Temperature and High-Temperature Gasification: Catalyst Selection and Development

To achieve complete conversion of the feed into gases, a catalyst is always required in a hydrothermal gasification process. Moreover, that catalyst plays an important role in steering the product distribution. In the following part, gasification in hot compressed water is discussed for low (250 – 400°C) and high (>550°C) temperature separately.

10.4.4.1 Low-Temperature Gasification   In an extensive research program that started in the 1980s, the Pacific Northwest National Laboratory (PNNL, United States) developed a catalytic process for the destruction of organic waste at approximately 350°C while producing a methane-rich gas. Tests were carried out at laboratory and pilot scale focusing on both catalyst and process development. Ruthenium on rutile titania, ruthenium on carbon, and stabilized nickel catalysts showed the highest activity and the best stability. With these catalysts, nearly 100% gasification of model components (1–10 wt% in water) was achieved, while without catalyst the extent of gasification is very limited at this temperature. The gas produced consisted of nearly only CH_4 and CO_2, as dictated by the overall thermodynamic equilibrium. The catalytic process was carried out in a series of fixed bed reactors. When using feedstock materials with the tendency to produce char/coke, a continuous stirred tank reactor (CSTR) was required before the fixed bed to soften the feed and to prevent the buildup of solids. Pilot plant runs using complex feeds such as potato waste and manure were carried out. The required liquid hourly space velocity (LHSV) was in the range of 1.5–3.5 m_n^3 feed·m_{cat}^{-3}·h^{-1}. For a waste disposal process, these LHSVs are acceptable, but for the production of gaseous energy carriers from biomass, the activity is too low. Researchers at PSI (Peterson et al., 2008) reported high extents of gasification and equilibrium methane yields for concentrated (up to 30 wt%) wood sawdust slurries using Raney nickel as catalyst at 400°C. For complete gasification, 90 min reaction time was required in their batch reactor.

How the catalysts enhance the extent of gasification at these low temperatures has not been completely clarified. Either they accelerate the rate of the gasification reaction relative to the rate of polycondensation/polymerization reactions, or they are able to gasify the formed polymers, or a combination of both. Obviously, these catalysts catalyze all gas-phase component reactions because a good agreement was found between the observed gas composition and the gas composition dictated by thermodynamic equilibrium. Reported problems with respect to the catalysts are poisoning by trace components such as sulfur, magnesium, and calcium and the growth of the active metal crystals during operation (sintering). A general problem of the near-supercritical and supercritical region is that it enhances leaching of the catalytic active phases and degeneration of the support. Hot compressed water is a good solvent for most organic chemicals and thus especially useful to keep coke precursors dissolved. Further, if coke is formed on the surface of the catalyst, the high H_2O concentration helps in keeping it clean via gasification. In accordance with that, it was found that coke formation on the catalyst surface is only a minor problem.

Researchers from the university of Wisconsin-Madison have reported interesting catalysis around 230°C for the production of hydrogen-rich gas from small oxyge- nated hydrocarbons. They were able to decrease the methane formation rate via C–O bond cleavage and methanation (hydrogenation) while maintaining high rates of C–C bond cleavage and the water–gas shift reaction for hydrogen production. They used a Pt catalyst and nickel catalyst promoted with tin. High hydrogen yields were obtained for methanol, ethylene glycol, and glycerol. However, with sorbitol and glu- cose as feedstock, already significant amounts of methane were being produced next to hydrogen. In an embryonic stage, the methodology of decelerating methane- producing reactions at catalytic sites while keeping a high rate of catalytic hydrogen production seems promising to produce hydrogen-rich gas at conditions for which overall chemical equilibrium dictates methane-rich gas, viz., at subcritical temperature and at the combination of high temperature and high concentration of organics. In this concept, it will be important to decrease homogeneous reactions to undesired by- products (oil/char/CH$_4$) and to increase the reaction rate. This is quite a challenge for both catalyst and reactor design.

10.4.4.2 High-Temperature Gasification   For high temperatures (>500°C), alka- lis have been proposed as catalysts. However, recovery of alkalis from the process may be a problem, because alkalis hardly dissolve in SCW. One of the pioneers in this field, Antal et al. (2000), has reported that leading the effluent of their empty tube reactor over a fixed bed of activated carbon derived from coconut increased the extent of gasification from 0.7 to 1.0. Despite the successful use of this activated carbon as a catalyst on laboratory scale, it may not be the catalyst finally selected for the process. Two important reasons for this are: (i) the catalytic activity nor its decline is under- stood, and (ii) the rate of charcoal gasification is slow but certainly not zero, leading to consumption of the catalyst.

Researchers at the University of Twente used the Ru/TiO$_2$ catalyst of PNNL and found complete gasification of glucose (1–17 wt% solutions) at 600°C and approxi- mately 60 s residence time (Kersten et al., 2006). The produced gas was at chemical equilibrium. The reaction is much faster at 600°C compared to 350°C, which is ben- eficial for the size of the reactor. However, no information is yet available concerning the stability of catalysts in the high-temperature-range SCW. Gas produced by gasi- fication in hot compressed water typically has a (very) low CO content. Gasification in hot compressed water produces either methane- or hydrogen-rich gas in combination with CO$_2$. This makes the product gas unsuitable as synthesis gas for FT synthesis. The hydrogen can be used for hydrogenation of bioliquids (see Chapter 11).

10.4.5 Reactor and Process Design Aspects

1. *High Temperatures.* High temperatures of above 600°C are needed in the reac- tor to allow significant noncatalytic thermal degradation of the biomass mole- cules and subsequent cracking or gasification of the intermediate-sized fragments to the small molecules H$_2$, CO, CH$_4$, and CO$_2$. Obviously, by apply- ing catalysis, the required operating temperature can be reduced. It is, however,

not evident that at these lower temperatures also high concentrations of organic matter can be treated. This subsection is focused mainly on gasification, without or with catalysis, at about 600°C.

2. *High Pressures.* It is crucial for the process that the heat content of the reactor effluent is utilized as far as possible to preheat the feedstock stream (mainly water) to reaction conditions (see Figure 10.10). However, heat exchange between these streams is not practical at low pressure, because of the high heat of evaporation at nearly isothermal and isobaric conditions. Heating of the feedstock stream to the desired gasification temperatures in a heat exchanger without evaporation requires operation at high pressures. This is the true incentive of the high pressures involved in wet gasification. The efficiency of the heat exchange in relation to the applied pressure can be calculated from the heat balance for a countercurrent shell and tube heat exchanger. The result is presented in Figure 10.13, in which the heat exchanger efficiency is plotted as a function of the operating pressure and the available area per unit throughput. In case of an infinite surface area (no overall transport limitations), the efficiency is given by

$$\eta_{HE} = 1 - \frac{\Delta H_{vap}}{H_{hot,\,in} - H_{cold,\,in}} \qquad \text{(Eq. 10.20)}$$

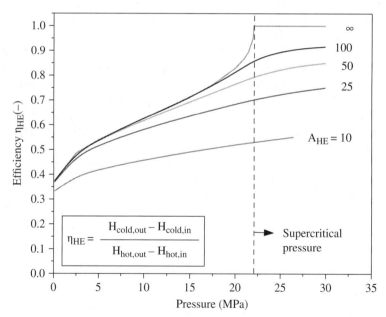

FIGURE 10.13 Calculated efficiencies of a water–water countercurrent heat exchanger plotted versus the operating pressure for different surface areas; AHE = area (m^2) per unit throughput (kg·s^{-1}). The flow rates (kg·s^{-1}) on both sides were assumed to be equal. U = 1000 W·m^{-2}·K^{-1}). Inlet conditions: T$_{hot,in}$ = 600°C, and T$_{cold,in}$ = 25°C.

The steep asymptotic approach to 100% efficiency above 20 MPa is a result of the sharp decrease of ΔH_{vap} to zero beyond that pressure. For heat exchangers with a finite surface area, the effect of the operating pressure is less pronounced. In practice, a hundred percent transfer of the available heat in the reactor effluent to the feedstock stream is impossible. In fact, efficiencies of approximately 75% are typical for liquid–liquid shell and tube heat exchangers. For such an efficiency, the operating pressure should be about 20 MPa in case of 50 m^2 per kg·s^{-1} throughput or 30 MPa in case of 25 m^2 per kg·s^{-1} throughput (see Figure 10.13).

3. *Additional Heat.* Mass and energy balance calculations show that in most cases additional heat input is required for the reactor. This heat can be supplied externally and/or by *in situ* generation of oxidation heat. For the latter, oxygen has to be added to the process. *In situ* oxidation is preferable over external heating because of efficiency and construction reasons. However, it is then crucial to do this selectively, i.e., without consuming any desired products. This is discussed further under point 4.

4. *Incomplete Conversion of the Biomass.* This is something that has to be taken into account in reactor design considerations. Thermal decomposition of real biomass will definitely result in carbon formation, either simultaneously with the production of tars (condensable vapors) and gaseous compounds or as a result of secondary reactions of tar. It was found that even under the high water pressure prevailing in the SCWG process, steam gasification of char is still extremely slow. This result excludes recycling of the carbon from the water outlet stream or applying very long residence times, as a reactor design option. *In situ* oxidation of the carbon indeed seems interesting, because of the corresponding heat production inside the reactor. Besides, in case of catalyst addition (discussed under point 5), regeneration is required to burn off any carbon deposited on the catalyst surface. While considering *in situ* oxidation of the carbon, it should be appreciated that oxygen introduced into the SCWG reactor will react preferentially with gaseous products and dissolved organic molecules while leaving the carbon largely unconverted. Reactor staging, or creating separate combustion and gasification zones inside a single reactor, may be a solution. Alternatively, a secondary wet-oxidation reactor can be used to clean the effluent water from the SCWG reactor.

5. *Catalysis.* This has been suggested to lower the required gasification temperatures for complete conversion and to improve the selectivity to either H_2 or CH_4 in case SNG is aimed at. For reactor design, the application of a heterogeneous catalyst has complicating consequences, e.g., regeneration is required (see following text). On the other hand, it also creates the opportunity for additional removal of minerals that are deposited together with the coke on the catalyst surface. Moreover, particle circulation can be used to prevent the heat exchanger from coking and plugging. The formation of ash and coke occurs in a temperature range of roughly 200–400°C and possibly mainly in a confined region of the tubular heat exchanger.

6. *Corrosion and Fouling.* Corrosion is a serious problem at high temperatures, especially if oxygen and sulfur are present. This is one of the reasons for developing catalysts that can reduce the required gasification temperatures. With respect to the selection of the construction material for a commercial plant, a two-barrier solution could be adapted from the technology of SCW oxidation that has been developed earlier. Up to now, it is recommended to construct the essential parts from a custom-made alloy like Inconel 625.

A problem of general nature in SCWG is the required heat exchange between the reactor outlet and inlet streams. Heating of the biomass slurry in a heat exchanger is likely to cause fouling/plugging problems because the thermal decomposition starts already at approximately 250°C producing oily products (tars) and, more seriously, polymers (char). Another likely cause of fouling is the production and accumulation of minerals from the biomass ash. Ash removal from the reactor as deposits on inert or catalyst particles circulating through the reactor system seems a better option than entraining the ash with the reactor effluent. For feedstock with higher than 10 wt% dry matter, there is evidence that deposition of salts inside the reactor occurs. Removal of these deposits together with circulating particles may prevent blockage and corrosion/erosion problems.

7. *Feeding.* A special problem in the process development of SCWG is the feeding. The nature of biomass feedstocks for SCWG varies from dilute waste streams of organics dissolved in water to heavy slurries of biomass in water. In case the starting material is coarse or fibrous, the original biomass should be ground and mixed with water to make a pumpable slurry. For instance, verge grass, wine-grape residues, and municipal waste fractions must be treated in that way. Although pumps for light slurries of fines are commercially available, they have been hardly tested for biomass feedstocks. High-pressure pumping is required for heavy viscous streams, and sometimes, the cement pump, known from building with concrete, is proposed as a possible solution.

10.5 GAS CLEANING FOR BIOMASS GASIFICATION PROCESSES

10.5.1 Introduction: Requirements for Gas Cleaning

The product gas generated by the different biomass gasification processes is usually far from being applicable in downstream conversion processes. Different downstream product gas utilization processes impose a variety of requirements for removal of impurities from the gas. This involves a multistep approach. The most relevant classes of impurity species are:

- Particulate matter
- Tars (polyaromatic hydrocarbons (PAHs), two-ring and higher ring PAHs, such as naphthalene, phenanthrene, and chrysene)

- Sulfur species (e.g., H_2S, COS, thiophenes, mercaptans)
- Chlorine species (e.g., HCl)
- Alkali and other trace elements (e.g., KCl, KOH, NaCl)
- Nitrogen compounds (e.g., NH_3, HCN, aromatic nitrogen compounds such as pyrroles and pyridines)

Feedstock composition as well as gasification process layout and conditions determine the concentration of such species in the raw product gas. Also, certain compounds of these classes have impact on the effective emission of other species, e.g., particulate matter can capture alkali and trace species (e.g., Hg compounds) depending on the concentration and on the process conditions.

An overview of typical gas cleaning demands concerning the aforementioned species classes for different prime movers (engines) and biofuel production processes is presented in Table 10.7 (Knoef, 2012). Concerning high-temperature fuel cell application, gas cleaning requirements are extensively dealt with in Chapter 16 and in Aravind and De Jong (2012).

Different end uses of the raw synthesis gas generated demand different gas treating sections. A generic scheme of such part of a BTL technology chain is presented in Figure 10.14. The dashed circle points to the opportunity of integration of particle removal and tar/hydrocarbon removal. Leibold et al. (2008) have suggested an even further simplification by addition of sorbent materials for removal of H_2S and HCl

TABLE 10.7 Overview of important gasification product gas applications and compound restrictions

| Contaminant | Application | | | |
	IC engine	Gas turbine	Methanol synthesis	FT synthesis
Particulate matter (soot, dust, char, ash)	$<50 \, mg \cdot m_n^{-3}$ (PM10)	$<30 \, mg \cdot m_n^{-3}$ (PM10)	$<0.02 \, mg \cdot m_n^{-3}$	n.d.*
Tars • Condensables • Inhibitory species (class 2 heteroatoms, benzene, toluene, xylenes)	$<100 \, mg \cdot m_n^{-3}$		$<0.1 \, mg \cdot m_n^{-3}$	$<10 \, ppb$ $<1 \, ppm$
Sulfur species (H_2S, COS)		$<20 \, ppm$	$<1 \, mg \cdot m_n^{-3}$	$<10 \, ppb$
Nitrogen species (NH_3, HCN)		$<50 \, ppm$	$<0.1 \, mg \cdot m_n^{-3}$	$<20 \, ppb$
Alkali compounds		$<24 \, ppb$		$<10 \, ppb$
Halides (mostly HCl)		$1 \, ppm$	$<0.1 \, mg \cdot m_n^{-3}$	$<10 \, ppb$

(Source: Adapted with permission from Knoef (2012). © Biomass Technology Group (BTG).)
* n.d., not detectable.

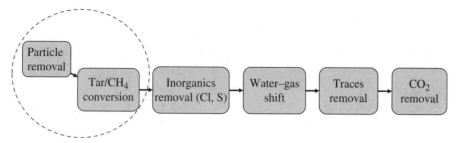

FIGURE 10.14 Gas cleaning and upgrading for biomass gasification in a BTL chain.

upstream the particle removal unit so that the indicated third step is also integrated in the first cleaning step.

The following subsections deal with the different cleaning strategies with respect to the aforementioned classes of compounds to be removed.

10.5.2 Particle Cleaning Techniques

Ash, attrited bed material and carbonaceous solids as well as fine droplets in gasification product gas make up the particulate matter inventory to be cleaned up to certain values needed regarding emission restriction laws and the downstream application. Gasifiers emit such particles in a typical range of approximately 0.1–100 μm. Major inorganic compounds are alkaline earth metal compounds (Ca, Mg oxides), silica, alkali species, and iron compounds. Minor inorganic species constitute, e.g., Zn, Pb, Cu, etc. depending on the biomass source gasified. One usually discriminates PM2.5, particles with diameters up to 2.5 μm, and PM10, which is particulate matter with diameters up to 10 μm.

Now, gas cleaning can be carried out at low (ambient) temperature (LT), intermediate temperature (IT, also called warm gas cleaning, up to ~350°C), and high temperature (HT, >> ~350°C). Table 10.8 summarizes available particulate matter cleanup equipment.

10.5.3 Tar Compound Cleanup

Tars form a challenging class of organic aromatic compounds. They have been defined in widely differing terms. In the European tar measurement standard (CEN/TS 15439 (CEN, 2006)) or the tar guideline (tinyurl.com/pwo46l4), tar is defined as follows: "Generic (unspecified) term for the entity of all organic compounds present in the producer gas excluding gaseous hydrocarbons (C1 through C6). Benzene is not included in tar." The primary issue is that they can condense and deposit, thereby clogging sampling lines, fouling heat exchanger surfaces. Upon decomposition, they can form soot, which in downstream combustion applications leads to higher emissions of particulate matter and CO. Only gasifiers in which the

TABLE 10.8   Particulate matter cleaning equipment

Equipment	Temperature level (LT, IT, HT)	Minimum dp size captured (μm)	Level of gas cleanup	References
Settling chamber	HT	>> 50	Low	Jacob and Dhodapkar (1997)
Impingement separator	HT	>10 – 20	Low	Towler and Sinnott (2012)
Cyclone	HT/IT	>10	Moderate	
Multicyclone	HT/IT	>5	Moderate	
Rotating particle separator	HT/IT	>0.1	High	Brouwers (1997)
Electrostatic filter	IT (/HT)	>2	High	
Scrubber	LT	0.5 – 10	Moderate to high	
Baghouse (fabric) filter	IT	>0.2	High	
Ceramic filter	HT	>0.1	High	Heidenreich (2013)
Metal filter	HT	>0.1	High	

product gas passes a hot zone of temperatures higher than approximately 1200°C have limited tar issues due to much lower concentrations. Acceptable limits to tar content in gasification product gas depend on the final application of this gas. Already during combustion in gas engines and gas turbines, tars may cause serious operational problems. In more advanced end-use applications, a virtually tar-free gas is required. Catalyst deactivation in downstream processing of the product gas should be avoided, and a limit of $2 \, gtar \cdot m_n^{-3}$ has been proposed before additional catalytic reforming (Aznar et al., 1998). A literature review by Milne et al. (1998) indicates that limiting values of less than $50\text{–}500 \, mg \cdot m_n^{-3}$ are recommended for compressors, $50\text{–}100 \, mg \cdot m_n^{-3}$ for IC systems, and $5 \, mg \cdot m_n^{-3}$ for direct-fired industrial gas turbines.

Tar quantity and species composition are determined by several gasification process parameters: type of feedstock, gasification conditions (temperature, pressure, oxidizer-to-fuel ratio), and gasifier type. Fluidized bed gasifiers can easily produce up to $\sim 10 \, gtar \cdot m_n^{-3}$; an EF gasifier produces almost no tars as it operates at high temperatures (>1200°C).

Strategies to decrease tar concentration levels in biomass gasification-derived product gases can be classified into primary methods (*in situ* removal within the gasifier) and secondary measures (downstream the gasifier). This is depicted in Figure 10.15.

Table 10.9 presents different catalytic materials used for tar reduction. These can be distinguished into two major classes, natural rock mineral matter and synthetic catalytic materials.

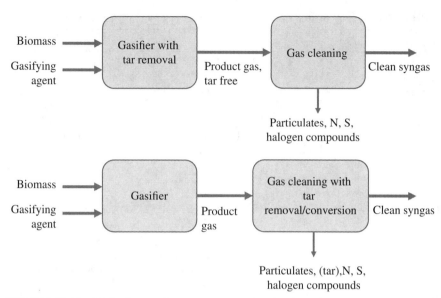

FIGURE 10.15   Methods to reduce tar concentration levels during biomass gasification: primary methods (top) and secondary methods (bottom).

10.5.3.1   Primary Tar Reduction Methods   Secondary methods would not be needed in case of application of ideal primary methods (Devi et al., 2003). When gasification process conditions, such as temperature, gasifying medium, stoichiometric oxygen ratio (λ), and residence time, are properly selected, this can reduce the amount of tars formed during gasification of biomass (Van Paasen and Kiel, 2004). High CC and diminished tar formation both are favored by a higher gasification temperature (Gil et al., 1999; Narvaez et al., 1996). For fluidized bed gasification, though, serious attention must be paid to the temperature in the denser bed zone so as to prevent sintering (Bartels, 2008). High values of λ enhance tar destruction, which, however, is at the expense of gas quality due to increased CO_2 formation in oxidation reactions. Values in the λ range of 0.25–0.3 have been reported to be optimal (Devi et al., 2003). Also, the gasifying medium has an impact on the composition of tars. In fluidized bed gasification tests, it has been shown (Gil et al., 1999) that steam addition leads to an increased share of phenolic species in comparison with the use of air as gasifying medium. There is also an impact of gasifier pressure on tar formation and distribution. Entire elimination of phenols was found by Knight (2000) in the case of pressurized gasification of wood chips.

The gasifier design also has an influence on tar production and destruction. The addition of a secondary oxidizer (Pan et al., 1999) and, for instance, the process concept of two-stage FICFB gasification combined with the use of tar reducing bed material, e.g., olivine and added Ni species (Pfeifer et al., 2004), are a few existing process design alternatives for enhancing fluidized bed gasification in terms of gas

TABLE 10.9   Tar reduction materials based on ECN report by Zwart (2009)

		Positive aspects	Negative aspects
Natural rock materials			
Calcined dolomite, magnesite, and limestone		Abundant Inexpensive	High attrition tendency Variable qualities
		Large tar conversions Disposable Investigated to large extend	Recarbonization with deactivation
Olivine		Attrition resistant Inexpensive	Lower tar reduction activity than dolomite
Clay minerals		Abundant inexpensive Disposal after usage less problematic	Loss of activity at higher T (>800 °C) Lower activity than dolomite
Iron and iron minerals		Abundant Inexpensive	Without H_2 rapid deactivation Lower activity than dolomite
Synthetic materials			
Transition metal catalysts	For example, Ni on carrier materials with promoters	Almost complete tar conversion ~900 °C, high activity Increase CO/H_2 yields Commercial catalysts available	Deactivation due to S and high tar content Relatively expensive Disposal
Activated alumina		High tar conversion, comparable to dolomite	Deactivation due to coking
Alkali metals		Natural production in gasifier	Agglomeration Lower activity compared to dolomite
FCC catalyst		Reasonably cheap Experience in industrial use	Deactivation by coking Lower activity compared to dolomite
Char		Inexpensive *In situ* generation High tar conversion	Consumed in reaction Varying qualities

composition and tar formation. Usually, when investigating a material as an in-bed catalyst for tar elimination and/or upgrading of the gas composition, it is used as an additive compound instead of a bed material.

In situ use of catalysts in the gasifier has been studied widely, and additives like dolomites (Corella et al., 1991; Delgado et al., 1996; Gil et al., 1999; Ising, 2002; Olivares et al., 1997), magnesites (Corella et al., 1991; Delgado et al., 1996; Siedlecki et al., 2009), limestones (Corella et al., 1991; Delgado et al., 1996), olivines (Aznar et al., 2007; Corella et al., 2004b; Devi et al., 2005a, 2005b), and Ni-based catalysts (Pfeifer et al., 2004) have found application. Table 10.9 refers to the natural rock materials as having relatively good activity. Generally, they are cheap, but show less favorable attrition behavior, and therefore, they impose higher capacity requirements to remove particles. Successful demonstration of this *in situ* measure has been shown using magnesite as bed material in the 18 MW_{th} CFB demonstration gasifier of Värnamo (Ståhl et al., 2004). *In situ* destruction seriously reduces the cleaning needed in downstream processing of product gas. In addition, catalytically active materials applied in the gasifier can also promote char gasification, modify the major product gas concentrations and decrease tar levels, and prevent agglomeration in the gasifier.

10.5.3.2 Secondary Tar Reduction Methods Secondary tar reduction techniques are accomplished using reactor units that are placed downstream the gasifier. Research and implementation have mainly focused on fixed bed reactors, although (circulating) fluidized beds have also been selected. An example of commercial use was technology offered by the (former) Swedish company TPS, using dolomite in a second CFB reactor (Rensfelt, 1997). A disadvantage is that such reactors show attrition issues as the mineral rock materials are soft. In order to comply with the rigorous requirements concerning tar concentration levels, combining primary measures and downstream reduction may be effective in view of keeping catalyst life-times acceptable. Dayton (2002), Abu El-Rub et al. (2004), and Yung et al. (2009) have presented excellent reviews on tar reduction by secondary measures. El-Rub has performed extensive research toward the use of biomass-derived char to reduce tar levels (Abu El-Rub et al., 2008). This material is active, cheap, and made *in situ* but shows variable quality and is consumed in the reactions, so it must be supplied continually.

Novel materials concern monoliths with impregnated metal catalysts (often Ni based) in relatively dusty syngas flows. Corella et al. (2004a) used them for product gas obtained from biomass gasification, but they concluded that the activity was only moderate and to be improved. They also indicate that upstream tar concentrations need to be reduced to approximately $2\,g \cdot m_n^{-3}$. Another issue indicated is the presence of sticky ash due to alkali eutectics. The monolith concept is applied to reduce tar in the CHP demonstration project in Skive (Denmark). Recently, in China, a new catalyst was tested (60 h run) with real gas from a fluidized bed biomass gasifier (scale, $150-300\,kg \cdot h^{-1}$) impregnated in a monolith that shows promising properties concerning tar conversion. Coking and sintering of the catalyst are prevented on the material, which consists of a NiO–MgO/γ-Al$_2$O$_3$/cordierite monolithic catalyst with a pore channel size of 7 by 7 mm. Moreover, such a monolith unit can function properly even under high dust load conditions ($\sim$330 g·m$_n^{-3}$) (Wang et al., 2011).

Particle gas cleanup can be combined with tar conversion, which is a novel development that holds good promise for process integration and intensification (Nacken et al., 2009).

10.5.4 Sulfur Species Cleanup

During different gasification processes, depending on the fuel-bound sulfur content of the fuel source, sulfur species are formed as part of the product gas. Most biomass types contain little to no sulfur ($0 - 2.0$ wt%). Biomass-derived feedstock such as municipal solid waste or sewage sludge, though, contains higher sulfur levels. The species formed during gasification are usually H$_2$S (major compound), COS, CS$_2$, and organic compounds like thiophenes and mercaptans. Up to 93–96% of the sulfur appears as H$_2$S in a typical gasification process (Higman and Van der Burgt, 2003). The demonstration-scale woody biomass gasifier in Güssing (Austria) shows sulfur compound levels of $\sim$150 ppm H$_2$S and $\sim$30 ppm organic sulfur (e.g., thiophene) (Weber et al., 2010). For emission control, not the strictest reduction measures are needed (yet), but rather these measures are needed for protection of downstream catalysts as they can be poisoned or materials can be corroded by traces of such compounds.

The overall thermal efficiency of a biomass gasification-based energy conversion system can be kept relatively high when hot gas desulfurization is applied, due to the fact that product gas cooling is not needed. Sulfur cleanup can be accomplished by using either primary methods (in the gasifier) or downstream capture, or a combination.

10.5.4.1 Primary Sulfur Species Concentration Reduction Methods   Like for tar species reduction, natural rock materials, limestone and dolomite, from various sources are among the most commonly used materials for sulfur capture in fluidized bed gasification applications. An extensive review has been presented by Meng et al. (2010). Natural rock materials have a widespread availability and are comparatively cheap. They show different compositions depending on geographic sourcing, which has significant effects on the final quality of the gas. Again, problems of attrition and incomplete conversion below the calcining temperatures have been shown for limestone. In dolomite, the higher Mg/Ca ratio causes an even higher attrition, but dolomite generally shows an increased reaction rate and reduced agglomeration behavior. Nowadays, also other Ca-based materials are being investigated that show both increased strength and absorption capacity (Akiti et al., 2002).

10.5.4.2 Downstream Sulfur Capture   Downstream of the gasifier, many different metal oxide materials can be utilized for high-temperature gas cleaning, with each having their own limitations and advantages. Key performance criteria for downstream sulfur sorbents are their sulfur sorption capacity and temperature at which good absorption takes place. Mixed oxides comprising a combination of the properties of various metals and supported metal oxides are the most promising for effective and stable sulfur capture, but depending on the optimal absorption temperature, other metal oxide sorbents may be used.

Among such materials, ZnO has the most favorable thermodynamic properties for H_2S capture in high-temperature gas cleaning systems. Vaporization, though, is an issue (Westmoreland and Harrison, 1976), and zinc migration and agglomeration limit the use of ZnO as sulfur sorbent to approximately 600°C. Therefore, zinc sorbents with higher stability comprising the inclusion of other metal elements have been developed. They include zinc copper ferrite that can ensure H_2S reduction to lower levels than zinc ferrite. Moreover, zinc ferrite doped with titanium shows higher stability under certain preparation conditions.

Copper oxide is also able to ensure H_2S concentration reduction from $\gg 1000$ ppm_v to sub-ppm_v levels. An issue regarding CuO is that metallic copper is easily formed by the reducing action of the product gas compounds H_2 and CO. This of course decreases the desulfurization efficiency.

Manganese oxides have demonstrated an attractive combination of high sulfur capture capacity and high reactivity even in a moderate temperature range, without the need for sorbent preconditioning or activation. The required process conditions better match biomass gasification and tar reforming temperatures. Manganese oxide sorbents, though, are prone to the formation of sulfates and also need a regeneration at very high temperature.

Also iron oxide, an abundant and relatively cheap material class, can desulfurize product gas, but its potential is slightly lower than the aforementioned compounds, mainly as a result of severe reduction as well as iron carbide formation at temperatures in excess of 550°C. The sulfidation or absorption step forms iron sulfide, which can be regenerated effectively by oxidation using air or nitrogen-diluted air at considerably lower temperatures than other metal sulfides. Furthermore, the sulfidation product, FeS_x, can react with SO_2 to form Fe_3O_4 and elemental sulfur, which is the most preferable route for SO_2 capture from the regeneration product gas.

As yet another material type, cerium oxide may be utilized for sulfur capture, and it can be regenerated well with reasonable elemental sulfur recovery. The extent of sulfur absorption increases when the temperature and CO/CO_2 ratio are increased. Cerium oxide, though, shows lower sulfur removal efficiency than ZnO. This is in particular the case at comparatively low temperatures (<700°C) in the range of hot gas cleaning. Compared to other sulfur capture materials, cerium oxide is far more expensive.

Additive elements, such as Ti, Al, Si, Zr, Co, Ni, and Fe, and promoters (Co, Ni, and Fe) are included in different metal oxide-based sorbents to enhance their sulfur capture capacity as well as their regeneration properties. Moreover, mesoporous materials and zeolites, composed of Al_2O_3, TiO_2, Fe_2O_3, and SiO_2, are used as support. Zinc oxide-based sorbents ($ZnFe_2O_4$, $ZnTiO_3$, Zn_2TiO_4, and Zn_3TiO_8), mixed and dispersed copper oxide-based sorbents (CuO/Al_2O_3, $CuO/Fe_2O_3/Al_2O_3$, CuO/Fe_2O_3, and CuO/MnO_2), manganese oxide-based sorbents ($MnO/\gamma - Al_2O_3$,

Mn–Cu, and Mn–Cu–V), iron oxide-based sorbents (Fe_2O_3/SiO_2), and cerium oxide-based sorbents (Ce/Mn) have been extensively studied. Some intrinsic problems using sulfur sorbents still need to be solved, even though many research works have been carried out to find suitable sorbents for different desulfurization processes.

Regarding the progress toward a low-cost, environmentally friendly, highly efficient, rapidly reacting, and regenerable sorbent with high sulfur capture capacity and durability, still, additional development work must be done, applying the material under real conditions. Optimization of supported or mixed metal sorbents from zinc, copper, and manganese is one of the options. Binary oxides and "promoted" binary oxides may have better attrition resistance and higher sulfidation equilibrium constants. Experimental and modeling studies focusing on the molecular structure of different sulfur capture materials could be useful for the design and development of new sulfur sorbents. Last but not least, mechanistic information for rational design and modification of such materials can be conventionally achieved with the aid of a variety of microscopic and spectroscopic characterization methods.

10.5.5 Chlorine Compound Removal

The chlorine content of biomass can vary significantly (Leibold et al., 2008) (see also Chapter 2). Chlorine can act as an alkali vaporization shuttle, in such a way that alkali-induced issues are occurring in the gasifier as well as in downstream equipment. Moreover, ash softening can take place in high-temperature gas cleaning processes due to the formation of low-melting compounds including in particular chlorides.

Chloride poisoning (HCl) may harm relevant (Cu- and Zn-containing) catalysts in the BTL chain via several parallel mechanisms (Twigg and Spencer, 2001):

i. Adsorbed chlorine atoms, formed by reaction, can block or modify catalytic sites.

ii. The low melting point and high surface mobility of Cu(I) chloride mean that even extremely small amounts of copper halide are sufficient to provide mobile species that accelerate the sintering of Cu catalysts.

iii. Poisoning of Cu catalysts by reduced sulfur compounds (e.g., H_2S) is worsened by traces of mobile Cu(I) chloride.

iv. ZnO, often present in Cu catalysts, reacts to form Zn halides, which also have low melting points, and causes further poisoning and sintering problems.

Therefore, chlorine-containing species need to be removed by sorbents. Natural rock materials, such as dolomite and limestone, already substantially reduce the HCl content. This component is absorbed and $CaCl_2$ is formed, which might contribute to stickiness of the ashes as it has a relatively low melting point.

10.5.6 Alkali and Trace Metal Cleaning

Many biomass types among the agricultural residues show high contents of alkali salts, in particular potassium based (straw, perennial grasses). When the temperature exceeds approximately 700°C, these (eutectic) salts are evaporated into the gas phase (Stevens, 2001). This gives rise to problems of deposition on downstream surfaces that have lower temperatures (below ~650°C). Herewith, aerosols with fine particle sizes (<5 μm) are formed, or the species condense directly on any surface, such as other particulate matter or the walls of process equipment. During biomass gasification, alkali vapors can be removed by cooling the hot producer gas to below 600°C to allow for condensation of the material into solid particulates (Stevens, 2001). These solids are subsequently removed using various dry or wet particle removal systems. In the design of these particle removal systems, not only the chemical behavior of the condensed alkali salt has to be taken into account but also the effect of tar condensation (Zwart, 2009). Ash stickiness, often resulting from lower-melting alkali species, is also impacting dust filtration as sticky ash is difficult to remove.

The capture of alkali species upstream of the gas cleaning section, and thus within the gasifier, is preferred. This also retards or prevents agglomeration in case of fluidized bed gasification and is possible using "alkali getters," which are clay minerals, e.g., kaolinite (Bartels, 2008).

10.5.7 Nitrogen Compounds

The main nitrogen compounds in gases produced by biomass gasification are ammonia (NH_3) and to a lesser extent hydrogen cyanide (HCN) and other species like aromatic nitrogen compounds (e.g., pyridine) and HNCO. These nitrogen compounds originate from fuel-bound nitrogen present in the feedstock (see, for instance, De Jong, 2005; Leppälahti and Koljonen, 1995). 0.2 wt% of nitrogen in sawdust led to concentrations of 300–400 ppm NH_3 in the fuel gas of a pressurized fluidized bed gasifier with fuel-bound nitrogen to NH_3 conversion values in the range of ~50–70% at a CC of >90% (Wang et al., 1999).

In IGCC concepts for (combined) power and heat production, a substantial part of the nitrogen species (NH_3, HCN, etc.) is converted into NO_x in the gas turbine combustors. NO_x is difficult to remove and highly undesirable as atmospheric pollutant.

Given the experience obtained in coal to FT diesel production (Sasol company), HCN is indicated to be less tolerable than NH_3; HCN exposure leads to the deactivation of the FT catalyst (Leibold et al., 2008; Olofsson et al., 2005). Thus, conversion or removal of NH_3 and HCN prior to exposure of the catalyst to these species is required. Regarding removal of NH_3, wet scrubbing can be applied, but this creates a liquid waste stream. An alternative is dry, hot gas cleaning. Catalysts used for tar conversion also show activity toward ammonia conversion in which harmless N_2 is formed. At temperatures approaching those of fluidized bed gasifiers (~900°C), dolomite,

(Ni-based) steam reforming catalysts (Wang et al., 1999, 2000), and Fe-based catalysts have all been shown to be able to largely convert NH_3 (Leppälahti and Koljonen, 1995; Leppälahti et al., 1991; Stevens, 2001). Carbon deposition may be a problem, although high steam contents can prevent this. A high H_2 concentration limits the conversion. A novel development is the application of monoliths to remove tar and NH_3 as applied by Corella et al. (2004a, 2005); these require less strict cleaning of the gas from particles, though sticky ashes may cause deposition and related activity decrease. Catalytic filtration using ceramic filters impregnated with a Ni-based catalyst can also decrease tar and NH_3 concentrations simultaneously; the configuration might also consist of an annular packed bed on the clean side of the ceramic candles.

When chlorine is present in the raw syngas, this might result in the formation of a condensable (fouling) species, NH_4Cl, which solidifies below 250–280°C and presents a fouling risk. Reaction of NH_3 with H_2S can result in the formation of ammonium (poly)sulfide, with a melting point below 150°C.

CHAPTER SUMMARY AND STUDY GUIDE

This chapter deals with gasification of biomass as a technology that is flexible regarding the production of a spectrum of products, such as heat, power, chemicals, and biofuels. Gasification is treated from different points of view: chemical thermodynamics, reaction kinetics of heterogeneous char-oxidizer reactions, and heat and mass transfer; reactor technologies are reviewed based on the different applications and properties of biomass and the qualities of the produced gas. A distinction is made in technologies for dry biomasses and wet biomass. In the last case, hydrothermal gasification is a promising, relatively new conversion technology making use of the water content in the fuel so that drying is not needed. Finally, in view of requirements of emissions and demands imposed by downstream equipment, gas cleaning options and devices are addressed. Of particular importance is the cleaning of biomass gasification product gas with respect to particles, tar, sulfur species, chlorine compounds, and alkali metal species, as well as trace elements.

KEY CONCEPTS

Differences between combustion and (hydrothermal) gasification of biomass
Chemical equilibrium; trends with actual oxidizer amount compared to stoichiometric combustion
Reaction kinetics; description of char conversion during gasification with H_2O and CO_2
Models for char surface development with conversion

Heat and mass transfer limitations and their verification via criteria for gasification
Selection criteria for different reactor types
Gas cleaning requirements and selection options for contaminant removal

SHORT-ANSWER QUESTIONS

10.1 What are advantages and disadvantages of the use of air as the oxidizing agent as compared to steam for biomass gasification for combined heat and power production?

10.2 Mention a number of different bio-based ways to produce synthesis gas.

10.3 CO_2 can be used as gasifying agent; which advantages and complications do you see for this use?

10.4 In Figure 10.1, indicate which by-products are formed (only the main products are shown in this figure). Can these be used, and if so, how?

10.5 Why are fixed/moving beds not suitable for large-scale synthesis gas generation based on biomass gasification?

10.6 Why does a model based on chemical equilibrium calculations predict higher CH_4 formation at higher pressure and typical fluidized bed gasifier temperatures of about 850°C?

10.7 Why does the "carbon boundary" correspond to obtaining the highest cold gas efficiency in gasification?

10.8 When a biochar particle is converted in Regime I, its conversion rate depends on its particle size. Right or wrong?

10.9 A high value of the Thiele modulus during a biochar particle conversion process corresponds to a Regime I conversion. Right or wrong?

10.10 How can you check experimentally whether gasification carried out in a fixed bed gasifier takes place in Regime III?

10.11 What does it mean when a particle's gasification process is limited by intraphase heat transfer?

10.12 Why would the cross-draft gasifier only be suitable for the gasification of charcoal?

10.13 Recap the pros and cons of the use of circulating fluidized bed gasifiers versus entrained flow gasifiers.

10.14 What are the main differences between bubbling and circulating fluidized bed gasifiers?

10.15 In a BFB gasifier, biomass fuel can be fed from the top or from the bottom part (in-bed feeding); what are the implications of these different ways of feeding for the gas composition?

10.16 In which way are reaction and separation combined in the gasifiers dealt with in this chapter? Can you think of further intensifications of these functions?

10.17 Explain the role of NaOH (aq.) addition in the SCW gasification process.

10.18 Which problems do you foresee when applying SCW gasification with seaweed (macroalgae) as feedstock?

10.19 Which oxidizer(s) could be used to *in situ* generate heat in SCW gasification?

10.20 Name a few advantages and disadvantages of gas cleaning at high temperature versus low (near ambient) temperature.

10.21 Often, a (multi)cyclone is positioned before a filter. Mention one advantage and one disadvantage of such a particle cleaning constellation.

10.22 Identify a few compounds in the product gas of the fluidized bed gasifier that are not shown in the gas composition of Example 10.1.

PROBLEMS

10.1 Repeat the chemical equilibrium calculation using a Gibbs energy minimization routine for the case of dry biomass gasification ($CH_{1.4}O_{0.6}$) at 1300°C and 2.0 MPa. What are the main differences with the results obtained at 850°C (see Figs. 10.2 and 10.3)?

10.2 Have a look at Table 10.6. Perform chemical equilibrium calculations for the given feedstock and process conditions. Which of the given compositions approaches the obtained results better?

10.3 For Example 10.1, verify the elemental balances for C, H, and O. Do they close to a reasonable extent?

10.4 Consider a perfectly stirred reactor to be used for dry wood gasification. The composition of wood is given as overall molecular formula: $CH_{1.4}O_{0.6}$. The reactor volume (V) is 4 m³, and the mass flow rate of wood is 2 kg·h⁻¹. Primary air consists of 23 wt% O_2 and is fed to the gasifier with a mass flow rate $\phi_{m,air}$. Oxygen in the air reacts with biomass in an idealized way so as to form only CO and H_2. The reaction is given as $CH_{1.4}O_{0.6} + |v_1|O_2 \rightarrow v_2CO + v_3H_2$

with v_i being the stoichiometric coefficients. The rate of consumption of O_2, $R_{O2,1}$, in [kmol·m⁻³·s⁻¹] is given by

$$R_{O2,1} = k_1\ Y_{O2}\ exp(-T_{a1}/T),$$

with Y_{O2} being the O_2 mass fraction, $k_1 = 10^7 \, kmol \cdot m^{-3} \cdot s^{-1}$, and $T_{a1} = 2.5 \times 10^4 \, K$. The reactor is operated at steady state and at isothermal conditions with $T = 1000 \, K$.

a. Calculate λ.

b. Suppose that just enough air is fed into the reactor for complete wood conversion into CO and H_2. Compute $\phi_{m,air}$.

c. Write down the conservation equations for total mass and O_2 (mass fraction), respectively.

d. Determine Y_{O2} in the reactor by solving the equations. N. B. $Y_{O2} > 0$, though just enough air is introduced in the reactor for complete conversion.

10.5 For rice husk char derived from slow pyrolysis, the following kinetic rate expression for the char–CO_2 reaction is presented by Di Blasi (2009):

$$\frac{dX}{dt} = \frac{9.87 \times 10^{12}}{T} \exp\left(-\frac{197}{R_u T}\right) p_{CO_2}[1-X]$$

Plot X versus t at 800°C and 1 atm for the case of conversion with pure CO_2 as gasifying agent. What changes when the surface factor $[1 - X]$ changes to $3(1 - X)^{2/3}$? Which models for surface area development are represented by these two factors?

10.6 Char gasification reactions are generally slow and the rate-limiting step during biomass gasification. Therefore, a good understanding of char reactivity kinetics is essential for the effective modeling and operation of gasification processes. Reaction data concerning char gasification using 10 vol.% CO_2 (C + $CO_2 \rightarrow 2CO$) are presented in the following table, where the weight (%) and the derived weight loss (%.min^{-1}) at three different temperatures, 900, 1000, and 1100°C, are given. The residual weights at 900, 1000, and 1100°C are 2.52, 1.931, and 2.141%, respectively.

a. Plot the curve of weight versus time at different temperatures, and explain the influence of temperature on char conversion. Assume that the overall reaction rate of char gasification can be expressed as

$$\frac{dX}{dt} = k_0 \exp\left(-\frac{E_a}{R_u T}\right) f(X) C_{CO_2}$$

b. Use two different models, $f(X) = (1-X)$ and $f(X) = 3(1-X)^{2/3}$, to calculate the activation energy (E_a) and pre-exponential factor (k_0).

c. Use the determined E_a and k_0 to recalculate the weight change (%) and then compare it with the experimental value. Which model is better and why?

Reaction time (Min)	Temperature 900°C		Temperature 1000°C		Temperature 1100°C	
	Weight (%)	Deriv. weight (%.min^{-1})	Weight (%)	Deriv. weight (%.min^{-1})	Weight (%)	Deriv. weight (%.min^{-1})
0	20.89	0.3808	21.24	1.261	20.36	1.726
0.56	20.59	0.641	20.47	1.377	19.42	1.645
1.12	20.22	0.6602	19.70	1.375	18.51	1.616
1.68	19.85	0.6716	18.93	1.376	17.61	1.599
2.24	19.47	0.6779	18.17	1.367	16.72	1.581
2.8	19.09	0.6772	17.40	1.361	15.84	1.568
3.36	18.71	0.6728	16.64	1.345	14.96	1.557
3.92	18.33	0.6696	15.89	1.346	14.09	1.547
4.48	17.96	0.6681	15.14	1.335	13.23	1.538
5.04	17.59	0.6596	14.39	1.320	12.37	1.525
5.60	17.22	0.6517	13.66	1.307	11.52	1.516
6.16	16.86	0.6451	12.93	1.299	10.68	1.496
6.72	16.50	0.6361	12.21	1.277	9.843	1.488
7.28	16.14	0.6267	11.50	1.266	9.014	1.472
7.84	15.79	0.6178	10.79	1.262	8.188	1.468
8.40	15.45	0.6086	10.09	1.246	7.369	1.452
8.96	15.11	0.5972	9.394	1.238	6.562	1.433
9.52	14.78	0.5897	8.704	1.235	5.768	1.402
10.08	14.45	0.5814	8.017	1.220	4.992	1.366
10.64	14.13	0.5714	7.334	1.223	4.246	1.294
11.2	13.81	0.5646	6.648	1.227	3.546	1.193
11.76	13.50	0.5537	5.961	1.227	2.924	1.010
12.32	13.19	0.5426	5.280	1.206	2.434	0.7071
12.88	12.88	0.5384	4.610	1.186	2.141	0.3438
13.44	12.59	0.5286	3.972	1.070		
14.00	12.29	0.5166	3.429	0.8759		
14.56	12.00	0.5138	2.981	0.7302		
15.12	11.71	0.5110	2.620	0.5557		
15.68	11.43	0.5023	2.361	0.3776		
16.24	11.15	0.4987	2.189	0.2435		
16.8	10.87	0.4919	2.074	0.1675		
17.36	10.6	0.4880	2.000	0.1014		
17.92	10.33	0.4844	1.957	0.05572		

10.7 For Example 10.2, check whether the criterion for negligible intraparticle heat transfer limitation is fulfilled. Take $\Delta_r H = 227.5$ kJ·mol^{-1} and $E_a = 220$ kJ·mol^{-1}.

Assume the gas properties are those of pure nitrogen at 900°C. These are $\lambda = 0.074$ W·m^{-1}·K^{-1}, $\rho = 0.2875$ kg·m^{-3}, and $c_p = 1.199$ kJ·kg·K^{-1}.

10.8 Supercritical water has a high density and is a highly nonideal gas state. At 24.0 MPa and 380°C, calculate the density according to ideal gas law. Also use the van der Waals equation of state to calculate density and finally the steam tables. Compare the values obtained. What is your conclusion?

10.9   Consider an aqueous feedstock with 10 wt% organics (take glucose for the calculations). This solution is gasified at 600°C and 25.0 MPa (abs.), and the reactor effluent is heat exchanged with the feed at an efficiency of 70%. Estimate the amount of energy that has to be provided to the reactor per kg of (wet) feed. You may assume that the reaction reaches equilibrium.

PROJECTS

P10.1   With access to a TGA analyzer, you can determine the reactivity of a typical BBQ (barbecue) char with CO_2. Under which conditions do you expect can the kinetics be determined without being biased by mass and heat transfer effects?

P10.2   Develop a process scheme for gasification of woody biomass to produce the biofuel dimethyl ether (DME). Set up mass and energy balances for a production plant of 40 kt·year^{-1}.

INTERNET REFERENCES

tinyurl.com/9s23f

http://www.nist.gov

tinyurl.com/pwo46l4

http://www.tarweb.net

REFERENCES

Abu El-Rub Z, Bramer EA, Brem G. Review of catalysts for tar elimination in biomass gasification processes. Ind Eng Chem Res 2004;43:6911–6919.

Abu El-Rub Z, Bramer EA, Brem G. Experimental comparison of biomass chars with other catalysts for tar reduction. Fuel 2008;87:2243–2252.

Akiti TT, Constant KP, Doraiswamy LK, Wheelock TD. A regenerable calcium-based core-in-shell sorbent for desulfurizing hot coal gas. Ind Eng Chem Res 2002;41:587–597.

Antal MJ, Allen SG, Schulman D, Xu X. Biomass gasification in supercritical water. Ind Eng Chem Res 2000;39:4040–4053.

Aravind PV, De Jong W. Evaluation of high temperature gas cleaning options for biomass gasification product gas for solid oxide fuel cells. Prog Energy Combust Sci 2012;38:737–764.

Arthur JR. Reactions between carbon and oxygen. Trans Faraday Soc 1951;47:164–178.

Aznar MP, Caballero MA, Gil J, Martín JA, Corella J. Commercial steam reforming catalysts to improve biomass gasification with steam-oxygen mixtures. 2. Catalytic tar removal. Ind Eng Chem Res 1998;37:2668–2680.

Aznar MP, Toledo JM, Sancho JA, Francés E. The effect the amount of olivine has on gas quality in gasification with air of different mixtures of biomass and plastic waste in fluidized bed. 15th European Biomass Conference and Exhibition. Berlin: ETA Florence; 2007. p 1176–1179.

Barrio M, Goebel B, Risnes H, Henriksen U, Hustad JE, Sorensen LH. Steam gasification of wood char and the effect of hydrogen inhibition on the chemical kinetics. In: Bridgwater AV, editor. *Progress in Thermochemical Biomass Conversion*. Oxford: Blackwell Science Ltd; 2001. p 32–46.

Bartels M. Agglomeration in fluidized beds: detection and counteraction [PhD thesis]. Delft: Delft University of Technology; 2008.

Bridgwater AV. The technical and economic feasibility of biomass gasification for power generation. Fuel 1995;74(5):631–653.

Brouwers JJH. Particle collection efficiency of the rotational particle separator. Powder Technol 1997;92:89–99.

Brown D, Gassner M, Fuchino T, Marechal F. Thermo-economic analysis for the optimal conceptual design of biomass gasification energy conversion systems. Appl Therm Eng 2009;29:2137–2152.

Brown RC. *Thermochemical Processing of Biomass: Conversion into Fuels, Chemicals and Power*. Chichester (UK): John Wiley & Sons; 2011.

CEN. Biomass gasification—tar and particles in product gases—sampling and analysis; 2006. Report nr CEN/TS 15439.

Corella J, Aznar MP, Delgado J, Martinez MP, Aragüés JL. Deactivation of tar cracking stones (dolomites, calcites, magnesites) and of commercial steam reforming catalysts in the upgrading of the exit gas from steam fluidized bed gasifiers of biomass and organic wastes. Stud Surf Sci Catal 1991;68:249–252.

Corella J, Toledo JM, Molina G. Calculation of the conditions to get less than 2 g tar/m_n^3 in a fluidized bed biomass gasifier. Fuel Process Technol 2006;87:841–846.

Corella J, Toledo JM, Molina G. A review on dual fluidized-bed biomass gasifiers. Ind Eng Chem Res 2007;46:6831–6839.

Corella J, Toledo JM, Padilla R. Catalytic hot gas cleaning with monoliths in biomass gasification in fluidized beds. 1. Their effectiveness for tar elimination. Ind Eng Chem Res 2004a;43:2433–2445.

Corella J, Toledo JM, Padilla R. Olivine or dolomite as in-bed additive in biomass gasification with air in a fluidized bed: which is better? Energy Fuels 2004b;18(3):713–720.

Corella J, Toledo JM, Padilla R. Catalytic hot gas cleaning with monoliths in biomass gasification in fluidized beds. 3. Their effectiveness for ammonia elimination. Ind Eng Chem Res 2005;44:2036–2045.

Dayton DC. A review of the literature on catalytic biomass tar destruction. Golden (CO): NREL; 2002. Report nr NREL/TP-510-32815.

De Jong W. Nitrogen compounds in pressurised fluidised bed gasification of biomass and fossil fuels [PhD thesis]. Delft: Delft University of Technology; 2005.

Delgado J, Aznar MP, Corella J. Calcined dolomite, magnesite, and calcite for cleaning hot gas from a fluidized bed biomass gasifier with steam: life and usefulness. Ind Eng Chem Res 1996;35(10):3637–3643.

Denn MM, Yu W-C, Wei J. Parameter sensitivity and kinetics-free modeling of moving bed coal gasifiers. Ind Eng Chem Res 1979;18(3):286–288.

Devi L, Craje M, Thune P, Ptasinski KJ, Janssen FJJG. Olivine as tar removal catalyst for biomass gasifiers: catalyst characterization. Appl Catal A Gen 2005a;294:68–79.

Devi L, Ptasinski KJ, Janssen FJJG. A review of the primary measures for tar elimination in biomass gasification processes. Biomass Bioenergy 2003;24:125–140.

Devi L, Ptasinski KJ, Janssen FJJG. Pretreated olivine as tar removal catalyst for biomass gasifiers: investigation using naphthalene as model biomass tar. Fuel Process Technol 2005b;86:707–730.

Di Blasi C. Combustion and gasification rates of lignocellulosic chars. Prog Energy Combust Sci 2009;35(2):121–140.

Fogler HS. *Elements of Chemical Reaction Engineering*. 3rd version. Upper Saddle River (NJ): Prentice-Hall, Inc.; 1999.

Fryda L, Panopoulos KD, Kakaras E. Agglomeration in fluidised bed gasification of biomass. Powder Technol 2008;181:307–320.

Gil J, Caballero MA, Martín JA, Aznar MP, Corella J. Biomass gasification with air in a fluidized bed: effect of the in-bed use of dolomite under different operation conditions. Ind Eng Chem Res 1999;38(11):4226–4235.

Gomez-Barea A, Ollero P, Arjona R. Reaction-diffusion model of TGA gasification experiments for estimating diffusional effects. Fuel 2005;84:1695–1704.

Heidenreich S. Hot gas filtration. Fuel 2013;104:83–94.

Higman C, Van der Burgt M. *Gasification*. Amsterdam (The Netherlands): Elsevier/Gulf Professional Publishing; 2003.

Hofbauer H, Rauch R, Loeffler G, Kaiser S, Fercher E, Tremmel H. Six years experience with the FICFB-gasification process. 12th European Conference and Technology Exhibition on Biomass for Energy, Industry and Climate Protection. Amsterdam (The Netherlands). Florence: ETA Florence; 2002. p 982–985.

Howard JR. *Fluidized Beds Combustion and Applications*. London/New York: Applied Science Publishers; 1983.

Hurt RH. Structure, properties, and reactivity of solid fuels. Symp (Int) Combust 1998;27(2): 2887–2904.

Ising M. Zur katalytischen Spaltung teerartiger Kohlenwasserstoffe bei der Wirbelschichtvergasung von Biomasse [PhD thesis]. Dortmund (Germany): Universität Dortmund; 2002.

Jacob K, Dhodapkar S, editors. *Gas-Solid Separations: Handbook of Separation Processes for Chemical Engineers*. 3rd ed. New York: McGraw-Hill; 1997.

Jensen A, Johnsson JE, Andries J, Laughlin K, Read G, Mayer M, Baumann H, Bonn B. Formation and reduction of NO_x in pressurized fluidized bed combustion of coal. Fuel 1995;76(11):1555–1569.

Karellas S, Karl J, Kakaras E. An innovative biomass gasification process and its coupling with microturbine and fuel cell systems. Energy 2008;33:284–291.

Kersten SRA, Potic B, Prins W, Van Swaaij WPM. Gasification of model compounds and wood in hot compressed water. Ind Eng Chem Res 2006;45:4169–4177.

Khawam A, Flanagan DR. Solid-state kinetic models: basics and mathematical fundamentals. J Phys Chem B 2006;110(35):17315–17328.

Kinoshita CM, Wang Y, Zhou J. Tar formation under different biomass gasification conditions. J Anal Appl Pyrolysis 1994;29:169–181.

Knight RA. Experience with raw gas analysis from pressurized gasification of biomass. Biomass Bioenergy 2000;18:67–77.

Knoef H. *Handbook Biomass Gasification*. 2nd ed. Enschede (The Netherlands): Biomass Technology Group; 2012.

Kosky PG, Floess JK. Global model of coal gasifiers. Ind Eng Chem Proc Des Dev 1980;19:586–592.

Kunii D, Levenspiel O. *Fluidization Engineering*. Boston (MA): Butterworth-Heinemann; 1991.

Laurendeau NM. Heterogeneous kinetics of coals char gasification and combustion. Prog Energy Combust Sci 1978;4:221–270.

Leibold H, Hornung A, Seifert H. HTHP syngas cleaning concept of two stage biomass gasification for FT synthesis. Powder Technol 2008;180:265–270.

Leppälahti J, Koljonen T. Nitrogen evolution from coal, peat and wood during gasification: literature review. Fuel Process Technol 1995;43:1–45.

Leppälahti J, Kurkela E. Behaviour of nitrogen compounds and tars in fluidized bed air gasification of peat. Fuel 1991;70:491–497.

Leppälahti J, Simell P, Kurkela E. Catalytic conversion of nitrogen compounds in gasification and combustion. Fuel Process Technol 1991;29:43–56.

Martens FJA. Freeboard phenomena in a fluidized bed combustor [PhD thesis]. Delft: Delft University of Technology; 1984.

Mears DE. Diagnostic criteria for heat transport limitations in fixed bed reactors. J Catal 1971;20(2):127–131.

Meng X, Benito P, de Jong W, Basile F, Verkooijen AHM, Fornasari G, Vaccari A. Steam-O_2 blown CFB biomass gasification: characterization of different residual chars and comparison of their gasification behavior with TG-derived pyrolysis chars. Energy Fuels 2012;26 (1):722–739.

Meng X, de Jong W, Pal R, Verkooijen AHM. In bed and downstream hot gas desulphurization during solid fuel gasification: a review. Fuel Process Technol 2010;91(8):964–981.

Milne TA, Abatzoglou N, Evans RJ. Biomass gasifier "tars": their nature, formation, and conversion. Golden (CO): NREL; 1998. Report nr NREL/TP-570-25357.

Monson CR, Germane GJ, Blackham AU, Smoot LD. Char oxidation at elevated pressures. Combust Flame 1995;100:669–683.

Nacken M, Ma L, Heidenreich S, Baron GV. Performance of a catalytically activated ceramic hot gas filter for catalytic tar removal from biomass gasification gas. Appl Catal B 2009;88 (3–4):292–298.

Narvaez I, Orio A, Aznar MP, Corella J. Biomass gasification with air in an atmospheric bubbling fluidized bed. Effect of six operational variables on the quality of produced raw gas. Ind Eng Chem Res 1996;35(7):2110–2120.

Olivares A, Aznar MP, Caballero MA, Gil J, Francés E, Corella J. Biomass gasification: produced gas upgrading by in-bed use of dolomite. Ind Eng Chem Res 1997;36:5220–5226.

Olofsson I, Nordin A, Söderlind U. Initial review and evaluation of process technologies and systems suitable for cost-efficient medium-scale gasification for biomass to liquid fuels. Technical report for University of Umeå. Umeå (Sweden): Energy Technology & Thermal Process Chemistry, University of Umeå; 2005.

Pan YG, Roca X, Velo E, Puigjaner L. Removal of tar by secondary air injection in fluidised bed gasification of residual biomass and coal. Fuel 1999;78:1703–1709.

Peterson AA, Vogel F, Lachance RP, Fröling M, Antal MJ, Tester JE. Thermochemical biofuel production in hydrothermal media: a review of sub- and supercritical water technologies. Energy Environ Sci 2008;1(1):32–65.

Pfeifer C, Hofbauer H. Development of catalytic tar decomposition downstream from a dual fluidized bed biomass steam gasifier. Powder Technol 2008;180(1–2):9–16.

Pfeifer C, Rauch R, Hofbauer H. In-bed catalytic tar reduction in a dual fluidized bed biomass steam gasifier. Ind Eng Chem Res 2004;43(7):1634–1640.

Prins MJ, Ptasinski KJ, Janssen FJJG. Thermodynamics of gas-char reactions: first and second law analysis. Chem Eng Sci 2003;58(13–16):1003–1011.

Rensfelt E. Atmospheric CFB gasification: the Greve plant and beyond. Gasification and Pyrolysis of Biomass. Stuttgart: CPL Press; 1997. p 139–159.

Rensfelt E, Gardmark L. VVBGC, Växjö Värnamo Biomass Gasification Centre AB: the VVBGC four-year-project. 18th European Biomass Conference and Exhibition: From Research to Industry and Markets. Lyon (France). Florence: ETA Florence; 2010.

Siedlecki M, Nieuwstraten R, Simeone E, de Jong W, Verkooijen AHM. Effect of magnesite as bed material in a 100 kW_{th} steam-oxygen blown circulating fluidized-bed biomass gasifier on gas composition and tar formation. Energy Fuels 2009;23:5643–5654.

Smith IW. The combustion rates of coal chars: a review. Symp (Int) Combust 1982;19: 1045–1065.

Ståhl K, Neergaard M. IGCC power plant for biomass utilisation, Värnamo, Sweden. Biomass Bioenergy 1998;15(3):205–211.

Ståhl K, Waldheim L, Morris M, Johnsson U, Gårdmark L. Biomass IGCC at Värnamo, Sweden: past and future. GCEP Energy Workshop; Apr 27, 2004. Palo Alto (CA): Stanford University; 2004.

Stassen HE. Small Scale Biomass Gasification for Heat and Power Production: A Global View. Washington, DC: World Bank; 1995.

Stevens DJ. Hot gas conditioning: recent progress with larger-scale biomass gasification systems. Update and Summary of Recent Progress. Golden (CO): NREL; 2001. Report nr NREL/SR-510-29952.

Syred C, Fick W, Griffiths AJ, Syred N. Cyclone gasifier and cyclone combustor for the use of biomass derived gas in the operation of a small gas turbine in cogeneration plants. Fuel 2004;83(17–18):2381–2392.

Towler G, Sinnott R. Chemical Engineering Design: Principles, Practice and Economics of Plant and Process Design. 2nd ed. Oxford: Butterworth-Heinemann; 2012.

Twigg MV, Spencer MS. Deactivation of supported copper metal catalysts for hydrogenation reactions. Appl Catal A 2001;212:161–174.

van der Meijden CM, Veringa HJ, Vreugdenhil BJ, van der Drift B. Bioenergy II: scale-up of the Milena biomass gasification process. Int J Chem React Eng 2009;7:A53.

Van Dongen A, Kanaar M. Co-gasification at the Buggenum IGCC power plant. Energetische Nutzung von Biomassen. Velen (Germany): DGMK; 2006. p 57–58.

Van Paasen SVB, Kiel JHA. Tar formation in a fluidised-bed gasifier: impact of fuel properties and operating conditions. Petten (The Netherlands): ECN; 2004. Report nr ECN-C-04-013.

Wang T, Li Y, Wang C, Zhang X, Ma L, Wu C. Synthesis gas production with NiO-MgO/ γ-Al2O3/cordierite monolithic catalysts in a pilot-scale biomass-gasification-reforming system. Energy Fuels 2011;25:1221–1228.

Wang W, Padban N, Ye Z, Andersson A, Bjerle I. Kinetics of ammonia decomposition in hot gas cleaning. Ind Eng Chem Res 1999;38:4175–4182.

Wang W, Padban N, Ye Z, Olofsson G, Andersson A, Bjerle I. Catalytic hot gas cleaning of fuel gas from an air-blown pressurized fluidized-bed gasifier. Ind Eng Chem Res 2000;39: 4075–4081.

Weber G, Potetz A, Rauch R, Hofbauer H. Development of process routes for synthetic biofuels from biomass (BTL). 18th European Biomass Conference and Exhibition: From Research to Industry and Markets. Lyon (France): ETA Florence; 2010:1829–1833.

Westerterp KR, Van Swaaij WPM, Beenackers AACM. Chemical Reactor Design and Operation. 2nd ed. New York: John Wiley & Sons; 1988.

Westmoreland PR, Harrison DP. Evaluation of candidate solids for high-temperature desulfurization of low-BTU gases. Environ Sci Technol 1976;10:659–661.

Willeboer W. The amer demolition wood gasification project. Biomass Bioenergy 1998; 15(3):245–249.

Winter F, Prah ME, Hofbauer H. Temperatures in a fuel particle burning in a fluidized bed: the effect of drying, devolatilization, and char combustion. Combust Flame 1997;108(3): 302–314.

Yung MM, Jablonski WS, Magrini-Bair KA. Review of catalytic conditioning of biomass-derived syngas. Energy Fuels 2009;23:1874–1887.

Zainal ZA. Prediction of performance of a downdraft gasifier using equilibrium modeling for different biomass materials. Energy Convers Manag 2001;42:1499–1515.

Zwart R. Gas cleaning downstream biomass gasification. Petten (The Netherlands): ECN; 2009. Report nr ECN-E-08-078.

11

THERMOCHEMICAL CONVERSION: AN INTRODUCTION TO FAST PYROLYSIS

STIJN R.G. OUDENHOVEN AND SASCHA R.A. KERSTEN

Sustainable Process Technology Group, Faculty of Science and Technology, University of Twente, Enschede, the Netherlands

ACRONYMS

DP	degree of polymerization
GC-MS	gas chromatography combined with mass spectrometry
ESP	electrostatic precipitator
FCC	fluid catalytic cracking
FID	flame ionization detector
HDO	hydrodeoxygenation
HDS	hydrodesulfurization
LHV	lower heating value
nbp	normal boiling point
TGA	thermogravimetric analysis/analyzer

SYMBOLS

B	permeability	$[m^2]$
Bi	Biot number	$[-]$
c_p	specific heat capacity	$[J{\cdot}kg^{-1}{\cdot}K^{-1}]$

Biomass as a Sustainable Energy Source for the Future: Fundamentals of Conversion Processes, First Edition. Edited by Wiebren de Jong and J. Ruud van Ommen.

d	diameter (of pore)	[m]
E_a	activation energy	[J·mol^{-1}]
g	gravitational acceleration	[m·s^{-2}]
ΔH	heat of reaction	[J·mol^{-1}]
h	mass-specific enthalpy	[J·kg^{-1}]
h	external heat transfer coefficient	[W·m^{-2}·K^{-1}]
h_c	convective heat transfer coefficient	[W·m^{-2}·K^{-1}]
k	apparent rate coefficient for mass loss	[s^{-1}]
k_0	pre-exponential factor	[s^{-1}]
m	mass	[kg]
$\overline{MW}$	mean molecular weight	[kg·mol^{-1}]
p	pressure	[Pa]
Q	heat flow	[W]
Py, Py$'$	pyrolysis numbers	[–]
R	characteristic length	[m]
R_u	universal gas constant (=8.3143)	[J·mol^{-1}·K^{-1}]
r_i	reaction rate of component i	[kg·m^{-3}·s^{-1}]
T	temperature	[K]
t	time	[s]
u	velocity in x-direction	[m·s^{-1}]
V	volume	[m^3]
v	velocity in y-direction	[m·s^{-1}]
ε	porosity	[–]
η	dynamic viscosity	[Pa.s]
λ	thermal conductivity	[W·m^{-1}·K^{-1}]
ν	stoichiometric coefficient	[–]
ρ	density	[kg·m^{-3}]
σ	Stefan–Boltzmann constant (=5.6704 × 10^{-8})	[W·m^{-2}·K^{-4}]
τ	residence time	[s]
χ	degree of biomass conversion defined by Equation (11.23)	[–]
ω	emissivity	[–]

Subscripts

ab	active biomass
b	biomass
c	char
g	gas
g + v	total volatiles
l	liquid
p	particle
r	reactor
s	solid
v	vapor

x   in x-direction
y   in y-direction
0   initial or reference conditions

11.1 INTRODUCTION

Fast pyrolysis is a thermochemical process to convert biomass into bio-oil that can be further upgraded or refined for the production of heat, electricity, (transportation) fuels, and chemicals. Lignocellulosic streams such as sawdust, forest thinning, straw, and empty fruit bunches are targeted feedstocks. At temperatures between 200 and 550°C in the absence of oxygen, biomass particles decompose into char, liquids (removed from the solid as vapors or as aerosols), and gases by a process known as pyrolysis. The liquid product, termed bio-oil or pyrolysis oil, is captured downstream of the reactor in condensers. When the pyrolysis is conducted at temperatures between 450 and 550°C, combined with a high heating rate of the biomass particles and rapid quenching of the produced vapors, maximal liquid production is achieved. This process is termed fast pyrolysis. Typically, small particles (millimeters) are used as feedstock, to achieve high heating rates. Processes conducted at lower temperatures (200–350°C), often using larger particles (centimeters), resulting in high char yields are typically known as slow pyrolysis or carbonization. Pyrolysis processes are typically operated at atmospheric pressure or below. In addition to its practical importance for the production of bio-oil, pyrolysis is also the initial chemical step in all biomass gasification and combustion processes.

Advocated advantages of bio-oil produced via fast pyrolysis over the bulky inhomogeneous biomass from which it originates are (i) the increased volumetric energy density ($\sim$5x); (ii) the ease of storage and transportation, which facilitates the decoupling of the refining location from biomass production; (iii) better processability; and (iv) the possibility to return minerals to the soil via the carbonaceous by-product. At the time of writing, several fast pyrolysis demonstration plants were in operation or being built, aiming at maturing the technology, maximizing oil production, and producing sufficient oil for application tests. Research is focusing on understanding the underlying processes at all relevant scales, ranging from the chemistry of the cell wall compounds to optimization of production plants, in order to produce better quality oils for targeted uses.

For wet biomass (>ca. 60% moisture), a process called hydrothermal liquefaction has been studied and operated at pilot scale. In this process, which is operated at high pressure (typically 100–300 bar), water is used as a solvent and reaction medium. Also, other solvents, such as ethanol, are considered. In this case, the process is called solvolysis. Fast pyrolysis, hydrothermal liquefaction, and solvolysis are all liquefaction processes: processes to convert biomass into a liquid. This chapter deals with the fast pyrolysis process, while the other liquefaction processes are briefly touched upon.

Fast pyrolysis is described at the level of chemistry and processes. Bio-oil properties are reported together with proposed applications. This chapter is intended as a first introduction to the subject; more general information on fast pyrolysis can be found in Basu (2010), Bridgwater (2008, 2012), Kersten and Garcia-Perez (2013), Mohan et al. (2006), and Venderbosch and Prins (2010).

11.2   A FIRST LOOK AT A LIQUEFACTION PROCESS

At the level of the mass balance, any liquefaction process can be described by

$$\text{Biomass} \rightarrow \text{solid residues} + \text{vapors} \{\text{organics} + H_2O\} + \text{permanent gases}$$

$$(RX. 11.1)$$

Solid residues are carbonaceous and contain a large amount of the ash-containing compounds present in the feedstock. Permanent gases are typically CO_2, CO, CH_4, and H_2. Water can already be present in the feed and is also produced in the reactions. Organics are multicomponent mixtures containing mostly oxygenated hydrocarbons. In the pyrolysis reactor, these organics are present as vapors and aerosols, while after condensation they are collected together with the water as a liquid, which can contain particulate fragments. During hydrothermal liquefaction and solvolysis, the vast majority of the organics remain in the liquid phase. Core elements in a liquefaction plant are (i) pretreatment (drying, cutting, etc.), (ii) reactor, (iii) heat production system, (iv) heat transfer to the reactor, (v) liquid recovery and collection, (vi) solid residue removal and utilization, and (vii) gas recovery and utilization. This archetype process is visualized in Figure 11.1.

Many of the proposed processes utilize the produced char and/or gases to generate the energy required for the process. In this way, no additional energy is necessary. In some hydrothermal/solvolysis processes, the liquid reactor effluent is recycled. The yield of the liquid organic product typically varies between 40 and 65 wt%. Yield is defined here as kg organic liquid produced per kg of the organic part of the biomass fed (i.e., dry ash free). Hereafter, all yields reported in this chapter are on dry ash free basis.

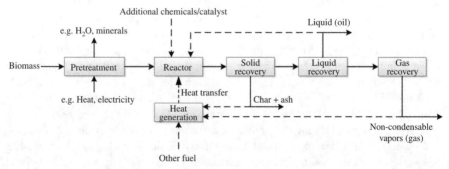

FIGURE 11.1   Archetype conceptual liquefaction process.

11.3   A FIRST LOOK AT FAST PYROLYSIS OIL

Figure 11.2 shows graduated cylinders containing pyrolysis oil, which is the common name of the liquid produced by the fast pyrolysis process, and the wood chips from which it is produced. The oil in the picture is produced from pinewood chips of ~1 mm, which are converted in a fluidized bed operated at 500°C, typical fast pyrolysis conditions. Both cylinders contain the same amount of energy, showing that during pyrolysis the volumetric energy content increased considerably. Pyrolysis oil is a brown, free-flowing liquid typically containing significant amounts of water.

The elemental composition and some properties of bio-oils are given in Table 11.1. Clearly, the elemental composition of pyrolysis oil resembles that of wood, showing that this bio-oil is liquefied biomass. Depending on the feedstock, its moisture content, and the applied condensation method, water is present in this liquid in concentrations ranging from 5 to 35 wt%. Higher concentrations of water cause phase separation, leading to the formation of an aqueous phase and an oil phase. Hydrothermal liquefaction tends to produce liquids that contain less oxygen compared to the feedstock (see Table 11.1). In Table 11.1, also, properties of heavy fuel oil (fossil origin) are given for comparison.

FIGURE 11.2   Pyrolysis oil (left) and pinewood from which it originates (right).

TABLE 11.1   Elemental composition and some characteristics of pinewood, fast pyrolysis oil (from pinewood), hydrothermal liquefaction oil, and heavy fuel oil

	Pinewood	Pyrolysis oil	Hydrothermal liquefaction oil	Heavy fuel oil
C (wt% db)	46.6	50–64	65–82	85
H (wt% db)	6.3	5–7	6–9	11
O (wt% db)	47.0	35–40	6–20	1
Water content (wt% ar)	9	5–35	3–6	0.1
Lower heating value (LHV) (MJ.kg^{-1})	17–19	16–19	25–35	40
Viscosity (cP at 20°C)	—	40–150	$\sim 10^4$	180
Density (kg.m^{-3})	570^a	1150–1250	1050–1150	900

Data taken from Kersten and Garcia-Perez (2013) and Knezevic (2009).
^{a} Particle density.

11.4   CHEMISTRY AND KINETICS OF PYROLYSIS

In this section, firstly, the chemistry of pyrolysis is introduced. Secondly, similarities and possible differences with hydrothermal liquefaction and solvolysis are briefly discussed.

11.4.1   Pyrolysis

Lignocellulosic biomass is a very complex material that consists mainly of cellulose, hemicellulose, and lignin (see Chapter 2). These building blocks together form the cell wall of lignocellulosic materials. When lignocellulosic materials are heated to temperatures over 200°C, reactions occur in the cell walls, which are known as primary reactions. These are depolymerization reactions resulting in liquids, vapors, and a solid residue. Upon subjecting lignocellulosic biomass to a temperature ramp, hemicelluloses will decompose first followed by cellulose, while lignin reacts over a broad temperature range. This behavior is illustrated in Figure 11.3, which shows thermogravimetric analysis/analyzer (TGA) (see Chapter 2 for details on TGA) curves of wood and its building blocks.

Because of the complex nature of the building blocks themselves (e.g., the monomers and structure of the hemicelluloses and lignin vary from biomass type to biomass type) and their interaction, many research works regarding the chemistry are carried out using model compounds. Model compounds can be the individual building blocks (e.g., cellulose), their monomers (e.g., levoglucosan) or decay products of the monomers (e.g., glycolaldehyde). Generally, depolymerization of cellulose is proposed to proceed via either an unzipping or a random chain cleavage mechanism (see Antal, 1982 for more information about the proposed cellulose decomposition models). In dedicated experiments at high heating rates and with a quick product quenching oligomer anhydrosugars with a degree of polymerization (DP) = 2 (cellobiosan) to DP = 9 (cellononasan) have been observed, pointing toward the dominance of the random chain mechanism (Piskorz et al., 2000).

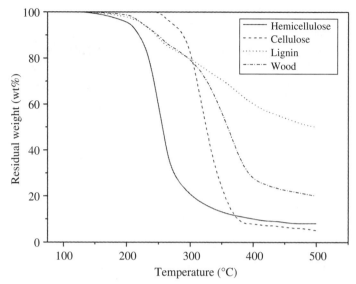

FIGURE 11.3 TGA of wood, cellulose, hemicellulose and lignin in nitrogen. A typical result is shown at ca. $10 \, K \cdot min^{-1}$. Data taken from Basu (2010).

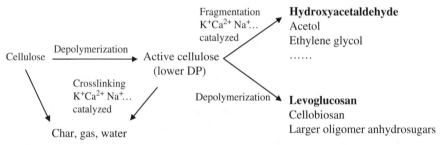

FIGURE 11.4 Cellulose depolymerization (modified Waterloo model). From Piskorz et al. (1989).

An often used reaction path scheme for cellulose pyrolysis is shown in Figure 11.4. In this scheme, solid cellulose decomposes firstly into an (liquid) intermediate, often termed active cellulose, with a lower DP. The active cellulose can depolymerize further into anhydrosugars or react via fragmentation (e.g., ring opening) or cross-linking reactions, which are catalyzed by alkali metals such a potassium, sodium, and magnesium (present in the ash).

For the pyrolysis of pure cellulose, the depolymerization reaction is rather selective with levoglucosan yields of up to 70% reported. However, it is known that while pyrolysis of cellulose can result in such high levoglucosan yields, a much smaller part of the cellulose fraction of lignocellulosic materials can be currently converted into sugars in fast pyrolysis reactors. Overall, it can be concluded that detailed chemical knowledge is available for pure cellulose, whereas for hemicelluloses, lignin, and real lignocellulosic biomass, this understanding is still in its early stages.

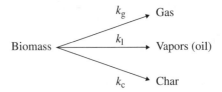

FIGURE 11.5   Global lumped component scheme for pyrolysis (of wood).

TABLE 11.2   Kinetic rate coefficients for wood pyrolysis

	Oak (Thurner and Mann, 1981)	Unspecified wood species (Chan et al., 1985)	Pine (Wagenaar et al., 1993)	Beech (Di Blasi and Branca, 2001)
k_c [s^{-1}]	7.4×10^5 $\exp(-107 \times 10^3/R_uT)$	1.1×10^7 $\exp(-121 \times 10^3/R_uT)$	3.1×10^7 $\exp(-125 \times 10^3/R_uT)$	3.3×10^6 $\exp(-112 \times 10^3/R_uT)$
k_l [s^{-1}]	4.1×10^6 $\exp(-113 \times 10^3/R_uT)$	2×10^8 $\exp(-133 \times 10^3/R_uT)$	9.3×10^9 $\exp(-149 \times 10^3/R_uT)$	1.1×10^{10} $\exp(-148 \times 10^3/R_uT)$
k_g [s^{-1}]	1.4×10^4 $\exp(-88.6 \times 10^3/R_uT)$	1.3×10^8 $\exp(-140 \times 10^3/R_uT)$	1.1×10^{11} $\exp(-177 \times 10^3/R_uT)$	4.4×10^9 $\exp(-153 \times 10^3/R_uT)$

For practical purposes, pyrolysis kinetics are often described based on so-called "global" lumped component schemes. The most used one includes three parallel first-order reactions to "gas", "vapors (which form oil after condensation)", and "char" (see Figure 11.5).

Measured kinetic rate coefficients for different wood types are given in Table 11.2, and Figure 11.6 (top) shows the liquid (oil) yield as a function of temperature for pyrolysis in the absence of internal and external heat transfer limitations predicted using these sets of kinetic data (see Section 11.5). It can be clearly seen that the predictions of the oil yield deviate considerably. This may be caused by errors in the measurement techniques, the interpretation model employed (probably too simple), and the effects of the biomass type used. In view of the actual complex reaction network, including catalytic effects of minerals, it is indeed unrealistic to expect that pyrolysis of woody biomass can be described accurately with such a simple kinetic model. The existing kinetic schemes may have value in describing trends and as mathematical description of rates and yields for the purpose of reactor design for a particular biomass stream. For the latter, it has to be realized that the kinetic coefficients have to be determined for every feedstock considered. Unfortunately, the existing kinetic schemes cannot predict the composition (and thus the quality) of the produced liquid. This would require models that provide a (much more) detailed description of the chemistry and, at the particle level, mass and heat transfer effects (see Section 11.5).

The volatile chemical species (vapors and aerosols) that are formed by depolymerization may undergo secondary reactions with the nascent char (containing

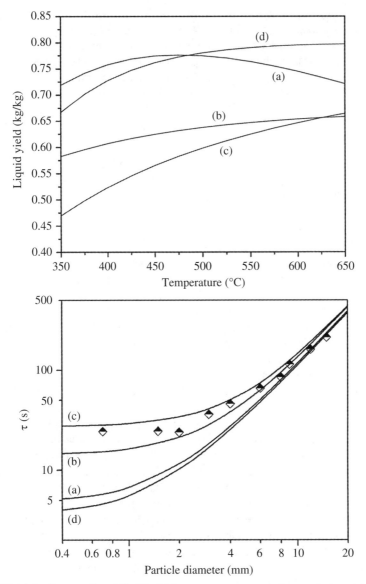

FIGURE 11.6 *Top*: predicted liquid yields versus temperature for pyrolysis not hindered by transfer limitations (data from Table 11.2). (Source: Reproduced with permission from Kersten et al. (2005). © American Chemical Society.) *Bottom*: predicted (single-particle model as described in Appendix 11.1) and measured and predicted pyrolysis times of pinewood at 99% conversion versus particle size at 500°C. a, Wagenaar et al. (1993); b, Chan et al. (1985); c, Thurner and Mann (1981); d, Di Blasi and Branca (2001). (Source: Reproduced (in adapted form) with permission from Wang et al. (2005). © American Chemical Society.)

minerals) along their diffusion path (intraparticle) out of the biomass particle, with other particles (interparticle), or in the vapor phase (homogeneous vapor-phase reactions). Kinetic relations have also been derived for the conversion of vapors to (permanent) gases. An example is the relation derived by Wagenaar (1994) for homogeneous cracking, which uses a first-order decay reaction with

$$k = 1.6 \times 10^5 \ \exp\left(\frac{-87.8 \times 10^3}{R_u T}\right) \tag{Eq. 11.1}$$

Obviously, the reaction rates of the secondary reactions will depend strongly on whether these are homogeneous or heterogeneous reactions. Generally, once the vapors are cooled down below 450°C, homogeneous cracking reactions become very slow.

11.4.2 Hydrothermal Liquefaction and Solvolysis

The simplest way of depicting hydrothermal liquefaction and solvolysis is by assuming that it starts with the primary pyrolysis reactions. The main difference is then that the reaction products are not being exposed to a vapor/gas phase, like in pyrolysis, but are dissolved and diluted in a solvent both inside and outside the biomass particle. Whether or not the solvent has part in the primary decomposition reactions is not unequivocally settled. Possible roles of water in the primary reactions could the supply of OH⁻ and H⁺ ions for catalysis and the participation in hydrolysis reactions. The solvent can be water or an organic solvent such as an alcohol, acid, or the liquefaction product itself. Typical reactor temperatures are in the range of 250–400°C with pressures of 100–250 bar (vapor pressure of water/solvent). An advantage of organic solvents over water is that they can have lower vapor pressures leading to a lower reactor pressure.

Figure 11.7 depicts the hydrothermal liquefaction process of wood shred in an isochoric capillary reactor. It can be clearly seen that the reaction products dissolve in water. After cooling the reactor effluent, two phases are present: an oil phase and an aqueous phase. At the same temperature, less char is found for hydrothermal liquefaction compared to pyrolysis. This might be explained by the rapid dilution (into the solvent) of char precursors. On the other hand, under hydrothermal liquefaction conditions, clearly, more decarboxylation takes place leading to an oil with a low(er) oxygen content but with a higher molecular weight (see Table 11.1).

Further readings concerning the kinetics and chemistry involved in fast pyrolysis are Antal (1982, 1985), Bradbury and Allan (1979), Chan et al. (1985), Cheng et al. (2012), Dauenhauer et al. (2009), Di Blasi (2008), Haas et al. (2009), Lin et al. (2009), Piskorz et al. (1989), Thurner and Mann (1981), and Wagenaar (1994).

11.5 PROCESSES AT THE PARTICLE LEVEL

Processes at the particle level are only discussed for fast pyrolysis of a wood particle. At particle level, wood pyrolysis is a multilevel (length and time scale) process (see Figure 11.8). Wood consists of cells (cavities) oriented in longitudinal direction; in a

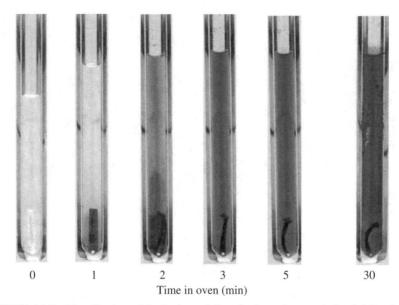

| 0 | 1 | 2 | 3 | 5 | 30 |

Time in oven (min)

FIGURE 11.7 Visualization of hydrothermal liquefaction of a wood shred (after 3 min, the temperature is ca. 340°C and the pressure ca. 150 bar). Figure adapted from Knezevic (2009).

rectangular wood particle of 1 mm × 1 mm (depth × width), 1500–3000 of such cells are present. Transport of gases, vapors, and aerosols proceeds predominantly via these cells in longitudinal direction. Heat transfer into the particle is a multidimensional process; however, for typical biomass particles with an aspect ratio (length over diameter or width) of three or larger, the dominant heat transfer direction is the diameter (width). Heat transfer into the particle can be described by the energy balance with the appropriate boundary conditions for heat transfer to the particle. This equation has to be solved together with mass balances for biomass, char, and gases/vapors/liquids in order to give predictions about product yields and conversion time. These parameters are most important for reactor design. Appendix 11.1 gives the equations of a two-dimensional single-particle model.

Figure 11.6 (bottom) shows predicted and measured pyrolysis times (time to reach 99% conversion in this case). Clearly, the predictions deviate considerably for small particles because for these particles the "uncertain" chemical kinetics are dominant in determining the conversion rate. For these particles, heat transfer to and within the particles is (much) faster than the chemistry as a result of which the reactions proceed at the temperature of the surroundings. For larger particles, the predictions become, more or less, independent of the kinetics because heat transfer inside the particle is the rate-limiting step. It can also be seen that in this regime the predictions fairly well match with the measured values. Hence, the properties determining heat transfer such as density, conductivity, and specific heat and their change during pyrolysis are known with sufficient accuracy to predict heat penetration. For particles larger than

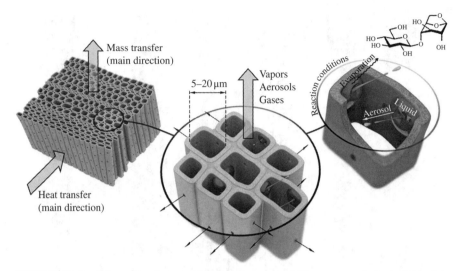

FIGURE 11.8 Schematic of the multilevel character of biomass pyrolysis (softwood is shown) ranging from a single cavity (cells) to a wood particle. (Source: Reproduced from Kersten and Garcia-Perez. Copyright (2013) with permission from Elsevier.)

TABLE 11.3   Bi, Py, and Py′ classification scheme from Pyle and Zaror (1984)

Model	Bi	Py	Py′
I	<1	∼0–10	>10
II	<1	>1	>1
III	>50	<10^{-3}	<<1

5 mm, the pyrolysis time might be estimated by calculating the time required to reach ∼95% of the temperature of the environment at the center of the particle by using averaged properties. The prediction of product yields is under all circumstances affected by the implemented reaction path scheme and kinetic data.

To estimate whether the rate of pyrolysis of a pyrolyzing particle is:

- Not hindered at all by heat transfer limitations (I)
- Controlled by external heat transfer (II)
- Controlled by internal heat transfer (III)

Biot (Bi), Py, and Py′ numbers can be used with the classification scheme shown in Table 11.3.

The dimensionless numbers are defined as

$$Bi = \frac{hR}{\lambda} \quad \text{(Biot number)} \tag{Eq. 11.2}$$

$$Py = \frac{\lambda}{k \rho c_p R^2} \quad \text{(pyrolysis number)} \qquad \text{(Eq. 11.3)}$$

$$Py' = \frac{h}{k \rho c_p R} \quad \text{(pyrolysis' number)} \qquad \text{(Eq. 11.4)}$$

The symbols used in these equations are explained in the general symbols list at the beginning of this chapter.

Besides heat transfer, also, mass transfer can play an important role at the particle level. As mentioned before, the actual pyrolysis reactions occur in the cell walls producing liquids, vapors, and a residual carbonaceous solid called char. The liquid intermediates formed during pyrolysis (see publication and movies of Haas et al., 2009) can remain in the cell at reaction conditions and are important because they accelerate dehydration and cross-linking reactions, which result in the formation of more char. Aerosols form thin liquid layers of reacting cellulose (see Teixeira et al., 2011). It is argued that their subsequent entrainment is an important mechanism for transporting heavy, nonvolatile products out of the biomass particle. Aerosols and vapors formed may undergo secondary reactions with the nascent char, containing (alkali) minerals, on their way out of the biomass particle. These secondary reactions could be cracking/depolymerization reactions leading to the formation of gases/light vapors or polymerization reactions leading to the formation of char, CO_2 and water. Inside the particles, secondary polymerization reactions leading to the formation of more char are dominant. It may be argued that if the transport distance of vapors/aerosols or the residence time of liquids in the particles decreases, this will lead to the production of more oil and less char. This indeed has been observed by making the transport distances very small by pyrolyzing only cell wall fragments and very rapid removal (e.g., under vacuum) of the (volatile) products (see the smallest particle sizes in Figure 11.9).

The heat and mass transfer effects described can be clearly observed by comparing oil and char yields of particles of different size (heating rate) as shown in Figure 11.9. Obviously, heat and mass transfer effects may interfere with attempts to unravel the reaction pathways and measure the intrinsic kinetics of the primary pyrolysis reactions.

More on the subject of heat and mass transfer and particle models can be found in the works of Di Blasi (1997, 2008), Di Blasi and Branca (2001), Hoekstra et al. (2012), Kersten et al. (2005), Pyle and Zaror (1984), Teixeira et al. (2011), Wang et al. (2005), and Westerhof et al. (2012).

11.6 A CLOSER LOOK AT PYROLYSIS OIL

Pyrolysis oil, produced by fast pyrolysis, is a complex mixture of water and a large variety of different oxygenated organic compounds, originating from mainly depolymerization and fragmentation reactions of the cell wall compounds. The composition of the pyrolysis oil largely depends on the biomass feedstock, process conditions, and pyrolysis oil handling. The compounds in the pyrolysis oil cover a large range of

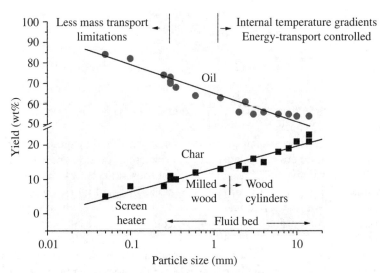

FIGURE 11.9   Effect of the particle size on the char and oil yields of pyrolysis of pinewood. Data from Wang et al. (2005), Westerhof et al. (2012), and Hoekstra et al. (2012).

TABLE 11.4   Chemical composition of pyrolysis oil from pinewood. Oil condensed at 20°C. Data from Westerhof et al. (2011)

	wt%
Extractives	0.2
Water	30
Lights, normal boiling point (nbp) < 154°C	10
Midboilers, nbp 154–300°C	9
Sugars	10
Water insoluble	16
Unidentified organics	30

molecular weights (up to 4000 g·mol^{-1}), boiling points, and functional groups. This results in difficulties in the identification and quantification of the compounds in the oil due to limitation of analytical techniques. For example, a normal gas chromatography combined with mass spectrometry (GC-MS) can only identify compounds with a boiling point < 250°C. Therefore, the composition of pyrolysis oil is often presented in lumped groups and determined by a combination of several analytical and solvent fractionating techniques. The composition of a typical pyrolysis oil produced from pinewood is shown in Table 11.4.

The extractives are obtained as a top layer on the pyrolysis oil and originate from the organic non-cell wall compounds such as fats, waxes, and proteins, which have a low oxygen content. The lights (e.g., formic acid, acetic acid, hydroxyacetaldehyde, acetol) and midboilers (mainly phenolic and furanic compounds) comprise typically around 20 wt% of the oil and are the fractions that can be relatively easily determined by GC-MS/flame ionization detector (FID) analysis. The sugars present in biomass originate from the hemicellulose and cellulose and are in the dehydrated form (anhydro). Besides the monosugars, also, large sugars (oligomers) are present in the oil. The water-insoluble fraction, often referred to as pyrolytic lignin, is the solid residue obtained after diluting the oil with cold water. It is generally accepted that this residue exists mainly of large lignin-derived molecules and is therefore rich in aromatics. A relatively large fraction of the organics in the oil cannot be identified with current analytical techniques. This fraction is soluble in water and has a relatively high boiling point. It is argued that this fraction consists mainly of cross-linked oligomer sugars (see, e.g., Kersten and Garcia-Perez, 2013).

Analysis of identical oil samples by different laboratories resulted in significant differences in the oil compositions (see Oasmaa et al., 2005 and Elliott et al., 2012); therefore, careful interpretation of the absolute values reported in the literature is required.

Since the chemical composition of pyrolysis oil varies significantly compared to that of fossil fuels, the properties differ from those of standard fuels. The most important differences are mentioned hereafter. The high oxygen and water contents result in a heating value of 14–19 MJ·kg^{-1}, which is only ~40% of the heating value of petroleum-derived fuels. Pyrolysis oil contains substantial amounts of organic acids, such as formic and acetic acid, resulting in a pH of 2–3. This low pH causes corrosion to common construction materials used in, e.g., engines and turbines. The compounds in pyrolysis oil contain reactive functional groups, which react further via polymerization reactions producing water, heavier compounds, and eventually char. This phenomenon occurs already at low temperatures (e.g., at room temperature during storage, called aging) or more rapidly when heated (e.g., excessive coke formation during distillation, referred to as coking). Typical pyrolysis oil is a single-phase mixture, but at high water content ($\sim > 35$ wt%), phase separation of the oil can occur affecting further processing of the oil. Phase separation of a single-phase oil can also occur due to aging during storage.

The earlier reported adverse properties of pyrolysis oil negatively influence the introduction of pyrolysis oil as a substitute for fossil fuels. Oil improvements can be achieved through actions before, during, or after the pyrolysis step. At present, research is focusing on oil quality improvement. It seems that the conditions for optimal oil quality differ from those for maximum oil yield. Obviously, the required oil quality depends on the targeted applications. Oil improvements might be lowering the acid content to reduce the burden in any downstream processing, increasing the content of fermentable sugars for biotechnology applications, and lowering the oxygen content and stability for corefining in crude oil refineries. These applications will be discussed later in this chapter.

More details on the analysis of the bio-oil can be found in Oasmaa and Meier (2005), Elliott et al. (2012), and Garcia-Perez et al. (2007). Concerning the bio-oil properties and its applications, more can be read in the works of Czernik and Bridgwater (2004), Elliott (2007), Jarboe et al. (2011), de Miguel Mercader et al. (2010), Oasmaa et al. (2005), Venderbosch et al. (2010), and Westerhof et al. (2010).

11.7 FAST PYROLYSIS PROCESSES

Figure 11.10 gives the flow sheets of four selected fast pyrolysis processes: Ensyn, Dynamotive, BTG-BTL, and Metso–Fortum. The core elements of a pyrolysis plant have already been introduced. In this section, these elements will be discussed in more detail based on the flow sheets of the four (proposed) demonstration units.

Before doing this, first, the overall mass and energy balances of pyrolysis of woody biomass will be discussed at input–output level starting from the reactor feed, which consists of biomass with ∼10 wt% moisture (see Table 11.5). The energy required to run the pyrolysis reactor (heating of the feed, the reaction enthalpy, plus the heat loss of the reactor to the surrounding) is not very well known; however, it can be estimated to be in the range of 1.5–2 MJ·kg^{-1} feed for relatively dry biomass (∼10 wt% moisture). Based on these numbers, it can be calculated that combustion of the produced char yields 4.8 MJ·kg^{-1} (0.16 × 30), which is more than enough to run the pyrolysis reactor. Combustion of only the product gases does not result in enough energy for running the reactor. Combusting the product gases together with the light part of the oil (vapors), however, can also deliver enough energy.

The core elements of a pyrolysis process are discussed below:

Pretreatment: Usually consists of (i) size reduction and homogenization to simplify feeding into the reactor and to obtain a high heating rate of the solids and (ii) drying to ∼10–20 wt% moisture.

Feeding system: Screw conveyers are most often used for larger-scale installations.

Reactor: Many concepts for pyrolysis reactors (auger, moving bed, rotary drum, Herreshoff furnace, fluidized bed, circulating fluidized bed, ablative, microwave) have been proposed and reviewed (see Bridgwater, 2012 and Garcia-Perez and Kruger, 2011 for more information). Most operational or planned larger-scale units are of the fluid bed type. These reactors can be scaled, are fuel flexible, and show good heat transfer characteristics. The fluidized beds recycle the gases produced within the process for fluidization in order to prevent the need for additional gas. Characteristics of the proposed reactors are discussed in more detail in Chapter 9. Important design parameters for pyrolysis reactors are:

- Operating temperature: 450–550°C.
- Residence time of the biomass: Should be long enough to achieve a high conversion of the biomass particles. The required conversion can be calculated by using single-particle models. For small particles of several millimeters, typically, tens of seconds are required, while for large particles (centimeters), minutes are required.
- Residence time and temperature of the vapors: The residence time should be minimized to ca. 2 s inside the reactor. Downstream, in the hot zone, cracking reactions are minimized below 400°C, allowing longer residence times.

ENSYN

Dynamotive

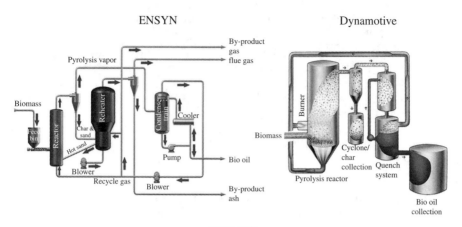

BTG-BTL

Metso-fortum

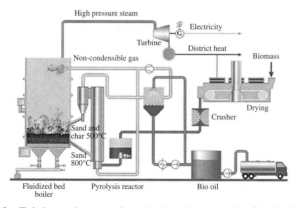

FIGURE 11.10 Existing and proposed pyrolysis units at a scale of 1–10 t.h^{-1}. The Metso–Fortum system is integrated with a boiler. Figures obtained (with permission) from the websites of the companies Ensyn, Dynamotive, BTG and Metso (see Internet References).

TABLE 11.5   Mass and energy balance of pyrolysis at
530°C. Input wood with 10 wt% moisture

	Yield [kg·kg^{-1}]	LHV [MJ·kg^{-1}]
Wood	—	16.5
Char	0.16	30
Oil	0.65	16.5
Gas	0.19	5

Data adapted from Westerhof et al. (2010).

- Biomass/char holdup: The vapor–char contact time should be minimized to limit cracking and polymerization reactions. The holdup of biomass/char in the reactor can be estimated by coupling a single-particle model with a biomass/char population balance.

Heat transport to the reactor: Can be done directly via a heat transport medium (e.g., sand, gas) that is cycled between the reactor and a hot utility (e.g., combustor) or indirectly via the reactor wall or heating pipes. Ensyn, BTG, and Metso circulate hot sand from a combustor (boiler) to the reactor, while Dynamotive circulates hot gas.

Energy (heat) generation: Is generally done by combustion of the produced char, gases, and/or light vapors. For detailed information about combustors, see Chapter 9.

Solid (char) removal: Is usually accomplished by cyclone systems. Deep removal of solids from the oil has to be done in the liquid phase by filtration.

Liquid recovery/collection: Typical systems used are electrostatic precipitators (ESPs) and (countercurrent) spray columns. The recovery of the liquid does not only involve condensation but also capturing of the aerosols. ESPs collect the aerosols by charging them utilizing an electric field, while in spray columns the aerosols are captured by contacting them with the spray. In spray columns, the oil produced can be used as spraying liquid. A first separation can be achieved by applying a series of condensers at decreasing temperature. In Figure 11.11, it can be seen that by operating the first condenser at 80°C, an oil can be obtained that contains only 1% of acetic acid (HAc), while the aqueous phase recovered in the second condenser contains 8–10 wt% HAc.

The websites of some companies give further information on pyrolysis processes (see Internet References). Recent scientific articles dealing with fast pyrolysis processes for further reading are Garcia-Perez and Kruger (2011), Hoekstra et al. (2012), Oudenhoven et al. (2013), Venderbosch and Prins (2010), and Westerhof et al. (2011).

Wet liquefaction is the pyrolysis process performed under aqueous hot, pressurized conditions. Here, only a brief reference is made to different research works in this specific domain; more can be read in Davis (1983), Goudriaan and Peferoen (1990), Knezevic (2009), Naber et al. (2005), and Venderbosch (2000).

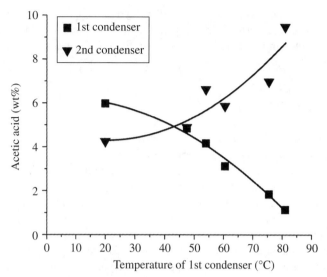

FIGURE 11.11 Staged condensation; acetic acid distribution over the first and second condenser. The reactor temperature is 480°C. The second condenser operates at 20°C. Spray columns are used. (Source: Reproduced with permission from Westerhof et al. (2011). © American Chemical Society.)

11.8 CATALYTIC PYROLYSIS

Catalytic pyrolysis has been considered to overcome the problematic characteristics of pyrolysis oil, as mentioned in the introduction, and to more selectively produce target compounds. Impregnating specific compounds such as H_2SO_4 and $ZnCl_2$ into the biomass particles in order control the primary reactions and the use of heterogeneous catalysts in the reactor to steer the secondary reactions have received significant attention in the last years. This section is not intended to give a detailed account of catalytic pyrolysis, but merely to mention its scope. Dickerson and Soria (2013) listed and reviewed the developments. Selective oxygen removal from the pyrolysis vapors, preferably by decarboxylation, aiming at increasing the heating value and decreasing the acidity of bio-oil, has been studied most. Deoxygenation reactions are catalyzed by acids, and the most studied ones are solid acids such as zeolites and clays. Aiming at the production of chemicals, yields of aromatics of ca. 20% (carbon basis) when using modified ZSM-5 zeolites have been reported (Cheng et al., 2012). All reported results on catalysis to steer the heterogeneous reactions show that low product yields and excessive coke and water formation are still items that need considerable attention.

The company KiOR (United States; see Internet References) developed a fluid catalytic cracking (FCC)-like catalytic pyrolysis process. Not much detailed information is available about this process. However, patent information available indicates the use of alkali and alkaline earth catalysts on mildly acidic support oxides such as

Boehmite alumina in a conventional fluid catalytic cracker, modified to suit solid biomass feedstocks. In 2012, the company claimed to have started up a $500\,t\cdot d^{-1}$ day facility in the United States (see US patent 2010,0105,970 A1).

11.9   OIL APPLICATIONS

The oil produced can be used locally (decentralized) for, e.g., the production of heat and power. It is also possible to transport the oils from many/several small-scale (decentralized) plants to central power production or a refinery facility. Figure 11.12 shows centralized and decentralized applications of bio-oil. Using liquefied biomass in large-scale existing fossil feed facilities enables the integration of biomass with the current fossil-based infrastructure. This route may speed up the introduction of biomass in the energy mix because capital costs are reduced (making use of existing infrastructure) and because existing products are made for existing markets. Liquefaction is often projected as a technology for relatively small capacities of 5–50 tonnes per day due to limited biomass availability and high transportation costs. Liquefaction can also be practiced at a larger scale, for instance, near a sugar mill where up to $10^6\,t\cdot y^{-1}$ of biomass can be available.

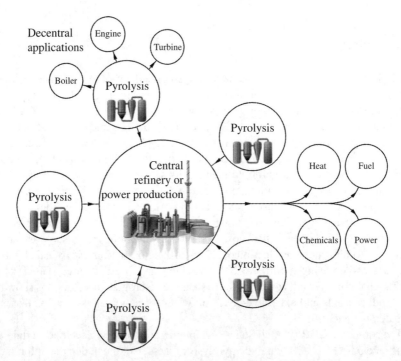

FIGURE 11.12   Decentralized and centralized applications of bio-oil. (Source: Reproduced from Kersten and Garcia-Perez. Copyright (2013) with permission from Elsevier.)

A short overview of the current status of processing pyrolysis oil in several applications to produce energy fuels and/or chemicals is given below (for more information, see, e.g., Czernik and Bridgwater, 2004 or Bridgwater, 2012):

Boilers: Successful combustion tests of pyrolysis oil in boilers and furnaces for the production of heat and power have been carried out. This is especially interesting for replacement of fuel oil by pyrolysis oil for heating purposes.

Engines: Direct combustion of pyrolysis oil in standard diesel engines suffers some difficulties in the ignition, corrosiveness, and coking. However, for larger (e.g., ship) and stationary diesel engines, it is expected that these problems can be overcome by engine modifications and blending the pyrolysis oil with alcohols.

Gasification: Gasification of solid biomass is already a proven technology to produce syngas. The advantages of the gasification of pyrolysis oil compared to the gasification of solid biomass are the increased volumetric energy density (transportation) and easier feeding. Using pyrolysis oil, high-temperature gasification (entrained flow) tests have been successful, while low-temperature gasification using a catalyst suffers from rapid catalyst deactivation.

Chemicals: Chemicals, including highly oxygenated ones, are currently almost always produced from fossil fuels. It may be simpler and energetically attractive to produce these chemicals from biomass. Chemicals such as acetic acid, acetol, glucose (levoglucosan), and phenols can be produced via further processing of the oil or its fractions or can be directly extracted from the oil. Separation technology will play a very important role in the production of chemicals from bio-oils. Improving the selectivity toward compounds more readily suitable for the production of fuels and chemicals would increase the value of the pyrolysis process. An interesting process concept to increase the selectivity toward anhydrosugars is via demineralization of the biomass, which can be done using the acids produced during the process, prior to pyrolysis (see Oudenhoven et al., 2013). Pyrolysis might become one of the technologies linking thermochemistry and biotechnology. More detailed studies of pyrolysis reactions using novel experimental and theoretical tools are likely to result in a substantial increase of the yield of anhydrosugars produced, which could make the production and fermentation of these sugars technically viable. New and more efficient microorganisms able to directly ferment anhydrosugars with higher resistance to bio-oil inhibitors are expected to be developed (see Kersten and Garcia-Perez, 2013).

Cofeeding in existing refineries: Recently, many studies have been published on cofeeding upgraded pyrolysis oil in crude oil refineries in selected processes such as FCC, hydrodesulfurization (HDS) or hydrocracking. This route makes use of available infrastructure and generates existing products such as gasoline and diesel for existing markets. At small, microunit scale, it has been shown that these upgraded oils can be corefined yielding fuels in the gasoline and diesel range. Upgrading involves deoxygenation and stabilization of pyrolysis oil, i.e., minimizing the tendency to form coke during refining (heating). This is achieved by removing the chemical functionality from the oil that is responsible for coke. Oxygen removal is required to enhance the miscibility with refinery streams and to (already) approach the composition of the target products. The remaining part of the oxygen is removed in the refinery. The upgrading process is a catalytic one and is operated at 250–400°C and 100–250 bar. For this process, (more stable) catalysts are being developed and tested, and the first stages of process development are in progress.

For more information, see Elliott (2007), de Miguel Mercader et al. (2010), and Venderbosch et al. (2010).

11.10 OUTLOOK

Liquefaction is an interesting thermochemical process for converting bulky inhomogeneous biomass into a liquid, which is easier to store, transport, and process. At the time of writing, only one commercial pyrolysis process was in operation. However, research is making good progress, and several demonstration plants are scheduled to come on stream in the near future. In the years to come, companies should mature the liquefaction technologies and develop the first large-scale applications of pyrolysis oil. The scientific community should advance at a higher level of integration of the different levels described in this chapter, allowing a much better control of the processes occurring in reactors and consequently of the composition of the oils produced. Improving the quality of the oils will be of paramount importance for the introduction of this technology in the market. Many applications of the oils are envisaged; in the authors' view, refinery feeds, precursors for chemicals, and pyrolytic sugars are the most promising.

APPENDIX 11.1 SINGLE-PARTICLE MODEL (BASED ON THE MODEL BY DI BLASI, 1997)

The single-particle model uses the Broido–Shafizadeh pyrolysis scheme (see Figure 11.13).

In this model, the biomass particle is described as an anisotropic, porous medium with different properties along the grain (permeability to gas flow and thermal conductivity are larger along the grain).

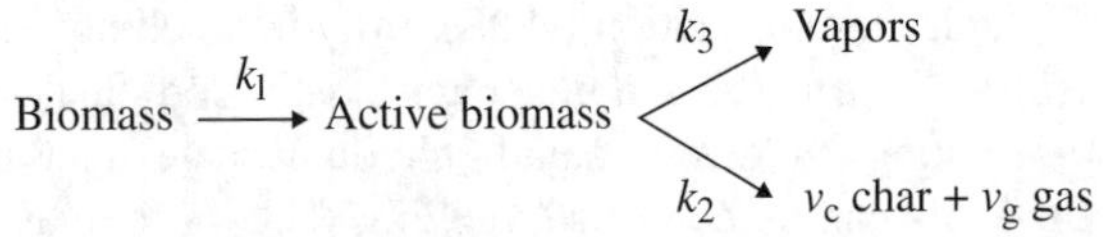

Biomass $\xrightarrow{k_1}$ Active biomass $\underset{k_2}{\overset{k_3}{<}}$ Vapors / v_c char + v_g gas

FIGURE 11.13 Broido–Shafizadeh scheme for biomass pyrolysis as presented by Bradbury and Allan (1979).

The physical processes described by the model include:

- Convective and radiative heat transfer from the surrounding fluid/environment to the surface of the biomass particle
- Radiative, convective, and conductive heat transfer inside the biomass particle
- Momentum transfer to account for nonzero pressure gradients and nonuniform velocity
- Variable properties (thermal conductivity, porosity, and permeability)
- Accumulation of volatile enthalpy and mass in the biomass pores
- Volatile convection and diffusion in the pores of both the virgin biomass and the charred region

The model assumes:

- No secondary vapor decomposition and gasification of char
- No particle shrinkage/swelling $\Rightarrow$ constant volume
- Symmetry

These assumptions result in the following equations:

$$\frac{\partial \rho_b}{\partial t} = -r_1 \tag{Eq. 11.5}$$

$$\frac{\partial \rho_{ab}}{\partial t} = r_1 - r_2 - r_3 \tag{Eq. 11.6}$$

$$\frac{\partial \rho_c}{\partial t} = \nu_c r_2 \tag{Eq. 11.7}$$

$$\frac{\partial (\varepsilon \rho_g)}{\partial t} + \frac{\partial (\rho_g u)}{\partial x} + \frac{\partial (\rho_g v)}{\partial y} = \nu_g r_2 + r_3 \tag{Eq. 11.8}$$

$$\left(\rho_c c_{p,c} + \rho_b c_{p,b} + \rho_{ab} c_{p,ab} + \varepsilon \rho_g c_{p,g} \right) \frac{\partial T}{\partial t} + \rho_g c_{p,g} u \frac{\partial T}{\partial x} + \rho_g c_{p,g} v \frac{\partial T}{\partial y}$$

$$= \frac{\partial}{\partial x} \left(\lambda_x^* \frac{\partial T}{\partial x} \right) + \frac{\partial}{\partial y} \left(\lambda_y^* \frac{\partial T}{\partial y} \right) + Q_r \tag{Eq. 11.9}$$

$$u = -\frac{B_x}{\eta} \frac{\partial p}{\partial x} \tag{Eq. 11.10}$$

$$v = -\frac{B_y}{\eta} \left(\frac{\partial p}{\partial y} + \rho_g g \right) \tag{Eq. 11.11}$$

$$\text{The ideal gas law} : p = \frac{\rho_g R_u T}{MW_g} \tag{Eq. 11.12}$$

$$\frac{V_s}{V_{s,0}} = \frac{(\rho_b + \rho_c + \rho_{ab})}{\rho_{b,0}} \qquad \text{(Eq. 11.13)}$$

$$k_i = k_{0,i} \exp\left(-\frac{E_{a,i}}{R_u T}\right) \quad i = 1...3 \qquad \text{(Eq. 11.14)}$$

$$r_1 = k_1 \rho_b \qquad \text{(Eq. 11.15)}$$

$$r_i = k_i \rho_{ab} \quad i = 2, 3 \qquad \text{(Eq. 11.16)}$$

$$\rho_b = \frac{m_b}{V} \qquad \text{(Eq. 11.17a)}$$

$$\rho_c = \frac{m_c}{V} \qquad \text{(Eq. 11.17b)}$$

$$\rho_g = \frac{m_g}{V_g} = \frac{m_g}{\varepsilon V} \qquad \text{(Eq. 11.17c)}$$

$$\varepsilon = \frac{V_g}{V} \qquad \text{(Eq. 11.17d)}$$

$$V_g = V - V_s \qquad \text{(Eq. 11.17e)}$$

$$
\begin{aligned}
Q_r = \; & k_1 \rho_b \left[\Delta h_1 + (T - T_0)(c_{p,b} - c_{p,ab})\right] \\
& + k_2 \rho_{ab} \left[\Delta h_2 + (T - T_0)(c_{p,ab} - \nu_c c_{p,c} - \nu_g c_{p,g})\right] \\
& + k_3 \rho_{ab} \left[\Delta h_3 + (T - T_0)(c_{p,ab} - c_{p,v})\right]
\end{aligned}
\qquad \text{(Eq. 11.18)}
$$

$$\lambda_x^* = \chi \lambda_{b,x} + (1-\chi)\lambda_{c.x} + \varepsilon \lambda_g + \sigma T^3 \frac{d_x}{\omega} \qquad \text{(Eq. 11.19)}$$

$$\lambda_y^* = \chi \lambda_{b,y} + (1-\chi)\lambda_{c.y} + \varepsilon \lambda_g + \sigma T^3 \frac{d_y}{\omega} \qquad \text{(Eq. 11.20)}$$

$$B_x = \chi B_{b,x} + (1-\chi)B_{c.x} \qquad \text{(Eq. 11.21)}$$

$$B_y = \chi B_{b,y} + (1-\chi)B_{c.y} \qquad \text{(Eq. 11.22)}$$

$$\chi = \frac{(\rho_{ab} + \rho_b)}{\rho_{b,0}} \qquad \text{(Eq. 11.23)}$$

The density of the gas phase in the model (combination of the tars and gases, since both are in the gas phase inside the particle during pyrolysis) is expressed in kg/m_{gas}^3, while the density of the char/biomass/active biomass is expressed in $kg/m_{particle}^3$.

The boundary conditions are as follows:

$$\text{At } t = 0$$

Initially, there is only biomass; the particle density depends on the type of biomass and moisture content.

At the particle-surrounding boundary,

$$\text{For } x : \lambda_x^* \frac{\partial T}{\partial x} = -\sigma\left(T^4 - T_r^4\right) - h_c(T - T_r) \quad p = p_0 \qquad \text{(Eq. 11.24)}$$

$$\text{For } y : \lambda_y^* \frac{\partial T}{\partial y} = -\sigma\left(T^4 - T_r^4\right) - h_c(T - T_r) \quad p = p_0 \qquad \text{(Eq. 11.25)}$$

At the center of the particle, there are no gradients due to symmetry.

CHAPTER SUMMARY AND STUDY GUIDE

This chapter describes the processes to generate bio-oil from biomass. Bio-oil can be used as a fuel in combustion or further upgraded for more sophisticated applications. For dry biomass, the process is termed pyrolysis; in case of wet biomass, reactions are carried out at subcritical liquid (water or solvent) conditions, and the process is called hydrothermal liquefaction or solvolysis. The kinetic background of biomass conversion for this process type is presented as well as reactor engineering fundamentals to obtain a better understanding of the process.

KEY CONCEPTS

Difference between pyrolysis and hydrothermal liquefaction/solvolysis

Chemistry

Reaction kinetics

Heat, mass, and momentum balances for the process

Different reactor configurations

SHORT-ANSWER QUESTIONS

11.1 a. Explain why the observation of larger oligomers during the pyrolysis of a cellulose particle contradicts an unzipping mechanism.
  b. Argue how these large oligomer anhydrosugars (DP = 9) leave the particle.

11.2 The Biot number is the ratio of the resistance of internal heat transfer to external heat transfer. Which processes do the Py and Py' numbers compare?

11.3 The concentration of acetic acid in pyrolysis oil varies between 1 and the 8 wt%. HAc is the most abundant acid in the oil. Can you explain the low (2–3) but hardly varying pH of the oil?

PROBLEMS

11.1 For the case where pyrolysis is not hindered by transfer limitations based on the scheme in Figure 11.5, derive:

    a. The equation for the mass loss (m = biomass + char) of a biomass particle as a function of time for a given temperature

    b. The equations for the product yields as a function of temperature

11.2 a. Derive an expression that relates the loss of vapors due the homogeneous vapor cracking reaction in the hot zone above a fluidized bed as a function of temperature and vapor residence time.

    b. Consider a fluidized bed pyrolysis reactor that contains ideally mixed biomass particles that are producing vapors. In the fluidized bed reactor, also, vapor cracking proceeds, which can be described by a first-order reaction in vapors. Derive an expression to calculate the fraction of the vapors that is cracked relative to the amount produced in the pyrolysis reactions as a function of temperature and residence time. You may assume that the gas/vapors show plug flow behavior.

11.3 Estimate the pyrolysis time for cylindrical biomass particles of aspect ratio L/D > 3 as a function of the diameter at a reactor temperature of 500°C. Use the estimation method discussed earlier.

11.4 Consider a process in which wet biomass is the feed. This biomass has to be dried to 10 wt% moisture before being fed to the reactor. Estimate the maximum intake moisture content that can be handled if the process is overall autothermal (no external energy input) and the required energy for drying and reaction is generated by combustion of produced char and gas. Use the figures from Table 11.5.

11.5 For the case where sand is circulated between a hot utility and the pyrolysis reactor, derive an equation that gives the required sand flow (in kg) per kg biomass fed as a function of the temperature differences between the pyrolysis reactor and the hot utility and the energy requirement of the reactor.

11.6 Calculate the amount (kg) of H_2 needed per kg product for complete deoxygenation of pyrolysis oil and discuss the economics of the process based on this number.

PROJECTS

P11.1 Perform TGA of woody biomass at a certain heating rate and derive the parameters of the first-order rate equation describing the weight loss. Examine whether these parameters can predict the measured weight loss at a considerably lower or higher heating rate. Discuss the results.

P11.2 Design a very simple laboratory pyrolysis system using glassware. Measure the oil yield of woody biomass. You will probably not obtain more than 50 wt% liquid yield. Discuss this deviation from the 65–70 wt% yield obtained in industrial reactors.

INTERNET REFERENCES

Companies

    BTG: www.btgworld.com/en/references/brochures/btg-btl-pyrolysis.pdf

    Dynamotive: www.dynamotive.com/technology

    Ensyn: www.ensyn.com/wp-content/uploads/2011/04/EC-Corp-PPT-April-2011.pdf

    KiOR: www.kior.com

    Metso: www.metso.com/News/Newsdocuments.nsf/Attachments/Bio-oilproductionplantto FortumpowerplantinJoensuu,Finland/$File/Fortum%20Joensuu%20press%20event% 202012-03-07-EN.pdf?OpenElement

REFERENCES

Antal MJ. Biomass pyrolysis: A review of the literature Part I—Carbohydrate pyrolysis. *Advances in Solar Energy*. New York: Springer; 1982. p 61–111.

Antal MJ, editor. *Biomass pyrolysis: A review of the literature Part 2—Lignocellulose pyrolysis*, Volume 2. New York: Springer; 1985.

Basu P. *Biomass Gasification and Pyrolysis: Practical Design and Theory*. Burlington, MA: Academic Press/Elsevier; 2010.

Bradbury AGW, Allan GW. A kinetic model for pyrolysis of cellulose. J Appl Polym Sci 1979;23(11):3271–3280.

Bridgwater AV, editor. *Fast Pyrolysis: A Handbook*, vols. 1–3. Newbury: CPL Press; 2008.

Bridgwater AV. Review of fast pyrolysis of biomass and product upgrading. Biomass Bioenergy 2012;38:68–94.

Chan W-CR, Kelbon M, Krieger BB. Modelling and experimental verification of physical and chemical processes during pyrolysis of a large biomass particle. Fuel 1985;64(11): 1505–1513.

Cheng Y-T, Jae J, Shi J, Fan W, Huber GW. Production of renewable aromatic compounds by catalytic fast pyrolysis of lignocellulosic biomass with bifunctional Ga/ZSM-5 catalysts. Angew Chem Int Ed 2012;51(6):1387–1390.

Czernik S, Bridgwater AV. Overview of applications of biomass fast pyrolysis oil. Energy Fuels 2004;18(2):590–598.

Dauenhauer PJ, Colby JL, Balonek CM, Suszynski WJ, Schmidt LD. Reactive boiling of cellulose for integrated catalysis through an intermediate liquid. Green Chem 2009; 11(10):1555–1561.

Davis HG. 1983. Direct liquefaction of biomass. Final report and summary of effort, 1977–1983, LBL-16243; DE83014066. Available at http://www.osti.gov/scitech/biblio/6374017. Accessed May 30, 2014, p 94.

Di Blasi C. A transient two-dimensional model of biomass pyrolysis. In: Bridgwater AV, Boocock DGB, editors. *Developments in Thermochemical Biomass Conversion*. London: Blackie Academic & Professional, p 147–160; 1997.

Di Blasi C. Modeling chemical and physical processes of wood and biomass pyrolysis. Prog Energy Combust Sci 2008;34(1):47–90.

Di Blasi C, Branca C. Kinetics of primary product formation from wood pyrolysis. Ind Eng Chem Res 2001;40(23):5547–5556.

Dickerson T, Soria J. Catalytic fast pyrolysis: A review. Energies 2013;6(1):514–538.

Elliott DC. Historical developments in hydroprocessing bio-oils. Energy Fuels 2007;21(3): 1792–1815.

Elliott DC, Oasmaa A, Meier D, Preto F, Bridgwater AV. Results of the IEA round robin on viscosity and aging of fast pyrolysis bio-oils: Long-term tests and repeatability. Energy Fuels 2012;26(12):7362–7366.

Garcia-Perez M, Chaala A, Pakdel H, Kretschmer D, Roy C. Characterization of bio-oils in chemical families. Biomass Bioenergy 2007;31(4):222–242.

Garcia-Perez M, Lewis T, Kruger CE. Methods for producing bio-char and advanced biofuels in Washington state. Part 1: Literature review of pyrolysis reactors. First Project Report. Ecology Publication Number 11-07-017. Pullman, WA: Department of Biological Systems Engineering and the Center for Sustaining Agriculture and Natural Resources, Washington State University; 2011. 137pp.

Goudriaan F, Peferoen DGR. Liquid fuels from biomass via a hydrothermal process. Chem Eng Sci 1990;45(8):2729–2734.

Haas TJ, Nimlos MR, Donohoe BS. Real-time and post-reaction microscopic structural analysis of biomass undergoing pyrolysis. Energy Fuels 2009;23(7):3810–3817.

Hoekstra E, Van Swaaij WPM, Kersten SRA, Hogendoorn KJA. Fast pyrolysis in a novel wire-mesh reactor: Decomposition of pine wood and model compounds. Chem Eng J 2012a;187:172–184.

Hoekstra E, Westerhof RJM, Brilman W, Van Swaaij WP, Kersten SRA, Hogendoorn KJA, Windt M. Heterogeneous and homogeneous reactions of pyrolysis vapors from pine wood. AIChE J 2012b;58(9):2830–2842.

Jarboe LR, Wen Z, Choi D, Brown RC. Hybrid thermochemical processing: Fermentation of pyrolysis-derived bio-oil. Appl Microbiol Biotechnol 2011;91(6):1519–1523.

Kersten S, Garcia-Perez M. Recent developments in fast pyrolysis of ligno-cellulosic materials. Curr Opin Biotechnol 2013;24(3):414–420.

Kersten SRA, Wang X, Prins W, van Swaaij WPM. Biomass pyrolysis in a fluidized bed reactor. Part 1: Literature review and model simulations. Ind Eng Chem Res 2005;44(23):8773–8785.

Knezevic D. *Hydrothermal conversion of biomass [PhD Thesis]*. Enschede: Twente University; 2009.

Lin Y-C, Cho J, Tompsett GA, Westmoreland PR, Huber GW. Kinetics and mechanism of cellulose pyrolysis. Int J Phys Chem C 2009;113(46):20097–20107.

de Miguel Mercader F, Groeneveld MJ, Kersten SRA, Way NWJ, Schaverien CJ, Hogendoorn JA. Production of advanced biofuels: Co-processing of upgraded pyrolysis oil in standard refinery units. Appl Catal B 2010;96(1–2):57–66.

Mohan D, Pittman CU, Steele PH. Pyrolysis of wood/biomass for bio-oil: A critical review. Energy Fuels 2006;20:848–889.

Naber JE, Goudriaan F, et al. Conversion of biomass residues to transportation fuels with the HTU® process. ACS Division of Fuel Chemistry, Preprints 2005;50(2):685. Available at http://www.scopus.com/record/display.url?eid=2-s2.0-32244439487&origin=inward&txGid= D1F0BC57C8E1E2DA752DC10586B5474A.kqQeWtawXauCyC8ghhRGJg%3a1. Accessed May 30, 2014.

Oasmaa A, Meier D. Norms and standards for fast pyrolysis liquids. Part 1: Round robin test. J Anal Appl Pyrolysis 2005;73(2):323–334.

Oasmaa A, Peacocke C, Gust S, Meier D, McLellan R. Norms and standards for pyrolysis liquids. End-user requirements and specifications. Energy Fuels 2005;19(5):2155–2163.

Oudenhoven SRG, Westerhof RJM, et al. Demineralization of wood using wood-derived acid: Towards a selective pyrolysis process for fuel and chemicals production. J Anal Appl Pyrolysis 2013;103:112–118.

Piskorz J, Radlein DSAG, Scott DS, Czernik S. Pretreatment of wood and cellulose for production of sugars by fast pyrolysis. J Anal Appl Pyrolysis 1989;16(2):127–142.

Piskorz J, Majerski P, Radlein D, Vladars-Usas A, Scott DS. Flash pyrolysis of cellulose for production of anhydro-oligomers. J Appl Polym Sci 2000;56(2):145–166.

Pyle DL, Zaror CA. Heat transfer and kinetics in the low temperature pyrolysis of solids. Chem Eng Sci 1984;39(1):147–158.

Teixeira AR, Mooney KG, Kruger JS, Williams CL, Suszynski WJ, Schmidt LD, Schmidt DP, Dauenhauer PJ. Aerosol generation by reactive boiling ejection of molten cellulose. Energy Environ Sci 2011;4(10):4306–4321.

Thurner F, Mann U. Kinetic investigation of wood pyrolysis. Ind Eng Chem Process Des Dev 1981;20(3):482–488.

Venderbosch RH. Hydrothermal conversion of wet biomass: A review, Novem GAVE-9919 report. Utrecht: Novem. p 134; 2000.

Venderbosch RH, Prins W Fast pyrolysis technology development. Biofuels, Bioprod Biorefin 2010;4(2):178–208.

Venderbosch RH, Ardiyanti AR, Wildschut J, Oasmaa A, Heeres HJ. Stabilization of biomass-derived pyrolysis oils. J Chem Technol Biotechnol 2010;85(5):674–686.

Wagenaar B. *The rotating cone reactor [PhD thesis]*. Enschede: Twente University; 1994.

Wagenaar BM, Prins W, et al. Flash pyrolysis kinetics of pine wood. Fuel Process Technol 1993;36(1–3):291–298.

Wang X, Kersten SRA, Prins W, van Swaaij WPM. Biomass pyrolysis in a fluidized bed reactor. Part 2: Experimental validation of model results. Ind Eng Chem Res 2005;44(23): 8786–8795.

Westerhof RJM, Brilman DWF, Van Swaaij WPM, Kersten SRA. Effect of temperature in fluidized bed fast pyrolysis of biomass: oil quality assessment in test units. Ind Eng Chem Res 2010;49(3):1160–1168.

Westerhof RJM, Brilman DWF, Oudenhoven SRG, Van Swaaij WPM, Kersten SRA, Garcia-Perez M, Wang Z. Fractional condensation of biomass pyrolysis vapors. Energy Fuels 2011;25(4):1817–1829.

Westerhof RJM, Nygård HS, van Swaaij WPM, Kersten SRA, Brilman DWF. Effect of particle geometry and microstructure on fast pyrolysis of beech wood. Energy Fuels 2012;26(4): 2274–2280.

12

THERMOCHEMICAL CONVERSION: TORREFACTION

JAAP H.A. KIEL[1,2], ARNO H.H. JANSSEN[1], AND YASH JOSHI[2]

[1]ECN, Biomass & Energy Efficiency, Petten, the Netherlands
[2]Department of Process and Energy, Energy Technology Section, Faculty of Mechanical, Maritime and Materials Engineering, Delft University of Technology, Delft, the Netherlands

ACRONYMS

LHV   lower heating value
NMR   nuclear magnetic resonance

12.1 INTRODUCTION

As compared to modern fossil hydrocarbons, raw biomass is an inferior fuel. It typically has a lower energy density; is hydrophilic (leading to a decreased heating value); is vulnerable to biodegradation (leading to problems in storage); is often tenacious and fibrous (leading to higher energy consumption during size reduction); and furthermore is highly heterogeneous. These properties impede the logistics (handling, transport, and storage) and large-scale end use (combustion, gasification, and chemical processing) of raw biomass. Torrefaction is the thermal treatment of biomass carried out typically between 240 and 320°C under nonoxidizing conditions aimed at

Biomass as a Sustainable Energy Source for the Future: Fundamentals of Conversion Processes,
First Edition. Edited by Wiebren de Jong and J. Ruud van Ommen.
© 2015 American Institute of Chemical Engineers, Inc. Published 2015 by John Wiley & Sons, Inc.

improving the physicochemical properties of biomass as a fuel or a feedstock. The term torrefaction is derived from the French verb *torréfier*, meaning roasting (typically that of coffee beans).

Ever since man could control fire, biomass has been harvested, collected, stored, traded and used as a fuel. However the technologies applied in combustion have become increasingly sophisticated in time, with a view to achieving higher fuel efficiency and reducing environmental impacts. To meet the increasingly stringent quality requirements of combustion processes, fuel pretreatment technologies have been developed. The severity of treatment (in terms of applied temperatures) depends on the intended application of the product. As an example, the most basic and oldest of these thermal treatment processes involves low-temperature sun drying of wood or dung cakes for combustion in order to meet domestic energy requirements. However, developments in metallurgy necessitated higher-temperature furnaces and reducing agents, resulting in the development of slow pyrolysis for producing charcoal from wood. Charcoal production was widespread until the nineteenth century, but later, charcoal was slowly replaced by coke (in metallurgy) or coal (as an industrial fuel) due to cost considerations. In general, the use of thermal pretreatment technologies imply energetic and economic costs, which become higher with greater improvements in fuel properties. For drying, the energy cost is the heat required in evaporating the moisture from the biomass. For charcoal production, in addition to the drying cost, there is a significant (up to 70–80%) dry mass loss with an associated energy content, leading to a much lower energetic efficiency. However, the resulting charcoal is undoubtedly a better fuel than dry wood. Torrefaction, conceptually, can be placed in between drying and charcoal production with respect to pretreatment severity. Unlike charcoal production, torrefaction does not aim at removing all the volatiles from biomass. Instead, it aims at maximizing the energy yield of the fuel while bringing about the necessary improvements in the fuel properties suiting logistics and end use.

Charcoal production at relatively low temperatures has been mentioned already in the 1800s (Percy, 1861), but the first reference to a torrefaction process is of more recent date; in the 1930s in France, research on gasification of fuels was performed. In the 1980s, a pilot plant was set up by Pechiney in Laval-de-Cere, France, for the production of torrefied material as a reducing agent in metallurgical applications (Bergman et al., 2005b). With increasing focus on biofuels in the twenty-first century, several start-ups in Europe and North America have initiated work on torrefaction, and several pilot/demo plants have been built.

12.2 FUNDAMENTALS OF TORREFACTION

The optimum choice of conditions (temperature and time) used in torrefaction depends on the nature of the input as well as on the application of the end product. In case of woody biomass, it may be assumed that the optimum conditions are reached with a mass loss of about 30 wt% of the dry solid matter. At this point, the torrefied wood typically contains 90% of the energy content (LHV basis) of the input (Bergman et al., 2005b).

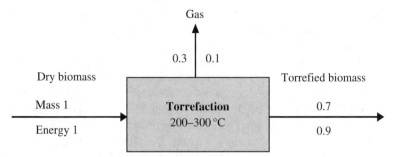

FIGURE 12.1   Typical mass and energy balances in torrefaction of woody biomass.

Figure 12.1 shows a simplified schematic of the torrefaction reaction starting with dry biomass as the input. The endo- or exothermicity of torrefaction is insignificant as compared to the chemical energy potentials of the inlet and outlet flows. Consequently, the heat of torrefaction is neglected in Figure 12.1. It can be seen that in comparison to the mass, a higher fraction of energy is retained in the torrefied biomass. Consequently, there is an increase in the heating value (or specific energy content on a mass basis) of the biomass. It can also be seen that the gas emanating from the torrefaction process carries away part of the energy of the incoming fuel. Several process schemes utilize the torrefaction gases for providing a substantial fraction of the heat input required for either drying the biomass or heating up of the dried biomass prior to torrefaction.

Biomass is primarily composed of organic matter, inorganic "ash", and moisture. In case of lignocellulosic biomass, such as wood or straw, the dominant polymeric structures in the organic matter are cellulose, hemicellulose, and lignin. Woody biomass contains 20–40 wt% hemicellulose, 40–60 wt% cellulose, and 10–25 wt% lignin. In reality, there are differences in composition, morphology, and decomposition behavior for each class of biological macromolecules (celluloses, hemicelluloses, and lignins) depending on their source. However, these (comparatively small) interclass distinctions are not discussed here (see Chapter 2 for more background information).

The regimes of physicochemical changes on account of torrefaction are shown in Figure 12.2. Upon heating biomass to temperatures of around 100°C, first, the physically bound water undergoes evaporation. As the temperature is increased further, light volatiles and chemically bound water start to evaporate. The dehydration and devolatilization reactions are endothermic, but condensation reactions may be initiated as well, which are exothermic. Hemicelluloses are the first to decompose, with their devolatilization starting at temperatures even below 200°C. Cellulose and lignin are included in the decomposition process at higher temperatures of typically 240–320°C. In this temperature range, also, extractives (e.g., resins, fats, and fatty acids), which are present in lignocellulosic biomass in small quantities, may evaporate.

The gases emanating from the decomposition of the biocompounds are referred to as *torrefaction gases*. The composition of torrefaction gases depends on the torrefaction conditions and extent. The gases are typically a combination of water vapor and carbon dioxide along with condensable organic volatiles such as acetic

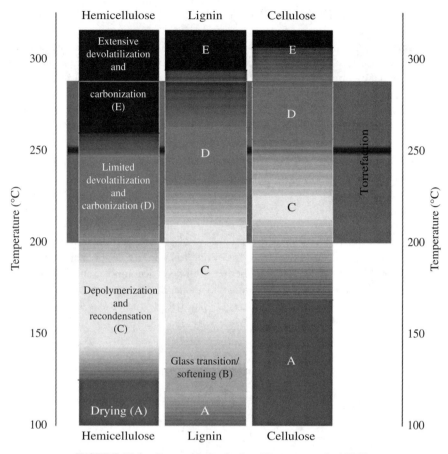

FIGURE 12.2 Stages of torrefaction (Bergman et al., 2005b).

acid, methanol, and formic acid. With increasing torrefaction severity, carbon monoxide is also present along with traces of phenols, furfural, and ammonia (Bridgeman et al., 2008). On a molecular level, nuclear magnetic resonance (NMR) studies suggest that torrefaction leads to removal of the acetyl group in hemicelluloses along with the cleavage of the aryl ether bonds in lignin and changes in the cellulose structure like depolymerization (Ben and Ragauskas, 2012; Melkior et al., 2012).

The macroscopic structure and bulk volume of the biomass are not affected greatly by torrefaction, and consequently, the volumetric energy density of the fuel is lower than that of the input. To overcome the reduction in energy density, the torrefaction process can be followed by densification of the biomass to form pellets or briquettes. Densification can be particularly important for economic viability in situations where the torrefied biomass is to be transported over long distances or where the inventory costs are high.

12.3   ADVANTAGES OF TORREFACTION

1. *Improvement in Lower Heating Value.* Torrefaction results in the loss of approximately 30% of the mass of the fuel (for woody biomass), primarily chemically bound water, and other light volatiles that do not contribute significantly (around 10%) to the energy content. Consequently, the specific energy content of the solid matter left behind after torrefaction is higher (Bridgeman et al., 2008; Patel et al., 2011; Pimchuai et al., 2010; Prins et al., 2006b; Rousset et al., 2011). A significant increase in the LHV is also due to drying of the biomass during the torrefaction process; the resulting increased hydrophobicity largely prevents the reabsorption of moisture.

2. *Improved Grindability.* For application in large pulverized fuel boilers and (dry feed) entrained flow gasifiers, biomass must be ground to (sub)millimeter sizes. Woody biomass typically consists of long tenacious fibers, the pulverizing of which requires a significant amount of energy. Torrefaction (in addition to dehydration) leads to shorter fibers due to cellulose depolymerization with less interconnections owing to the hemicellulose decomposition. This reduces the energy consumption in grinding and results in a more even particle size distribution and particles with increased sphericity (Arias et al., 2008; Bergman et al., 2005a; Phanphanich and Mani, 2011).

3. *Increased Hydrophobicity and Decreased Biological Activity.* Hemicellulose is the most hydrophilic compound in the biomass structure and is preferentially decomposed in the process of torrefaction. This is associated with the loss of free hydroxyl groups that can act as sites for hydrogen bonding, the absence of which lead to increased hydrophobicity (Chew and Doshi, 2011; van der Stelt et al., 2011).

Due to the sustained high temperatures, torrefaction leads to sterilization of the feedstock. Furthermore, the increased hydrophobicity and the removal of monosaccharides and hemicelluloses serve to prevent the recurrence of fungi/mold in storage. The elimination of biological activity in the feedstock prevents decomposition and loss of solid matter and also leads to limitation of biochemically induced self-heating of the biomass in stockpiles. The improved storage properties of torrefied biomass may enable outdoor storage, which can lead to significant savings.

Typical properties of torrefied wood pellets in comparison with other solid fuels are shown in Table 12.1.

12.4   TORREFACTION TECHNOLOGY

Many research groups are involved in the research and development of torrefaction technologies. Rather than developing entirely new reactor designs, most developers rely on reengineering of proven technologies developed for other applications, such as drying, pyrolysis, combustion, and ore roasting.

TABLE 12.1 Properties of *solid fuels*

	Wood chips	Wood pellets	Torrefied wood pellets	Charcoal	Coal
Moisture content (wt% wet basis)	30–55	7–10	1–5	1–5	10–15
LHV (MJ.kg^{-1})	7–12	15–17	18–24	30–32	23–28
Volatile matter (wt% dry basis)	75–85	75–85	55–80	10–12	15–30
Fixed carbon (wt% dry basis)	16–25	16–25	20–40	85–87	50–55
Bulk density (kg.L^{-1})	0.2–0.3	0.55–0.65	0.65–0.8	0.18–0.24	0.8–0.85
Volumetric energy density (GJ.m^{-3})	1.4–3.6	8–11	12–19	5.4–7.7	18–24
Hygroscopic properties	Hydrophilic	Hydrophilic	(Moderately) Hydrophobic	Hydrophobic	Hydrophobic
Biological degradation	Fast	Moderate	Slow	None	None
Milling requirements	Special	Special/ standard	Standard	Standard	Standard
Product consistency	Limited	High	High	High	High
Transportation costs	High	Medium	Low	Medium	Low

Torrefaction was primarily conceived as a technology for application in coffee processing. The philosophy of designing biomass torrefaction reactors, however, is conceptually dissimilar, since unlike coffee, biomass fuel is not a high value product. Also, the technology is targeted at processing biomass in quantities that are several orders of magnitude larger than the quantity of coffee beans roasted. However, just as in coffee beans roasting, good process control in terms of temperature, residence time, and mixing is imperative for achieving consistency in product quality. For a certain reactor configuration, variations in temperature generally have a larger impact on product quality than variations in residence time. Energetic efficiency is the key to economic viability and to the overall sustainability of torrefied biomass-based value chains. This necessitates a proper integration between the several heat sources and sinks available in the process. Although the torrefaction process itself is not a largely endo- or exothermic process, the predrying unit requires a substantial energy supply. Hence, proper design of the predrying unit is also of particular importance in achieving good overall torrefaction process efficiencies (especially in case the feedstock has a high moisture content).

Figure 12.3 gives an example of heat integration, where predrying of the biomass and torrefaction are carried out by utilizing three heat sources, namely, the torrefaction

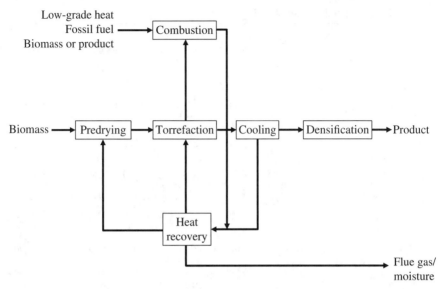

FIGURE 12.3   Example of heat integration in a torrefaction process.

gases, supplementary fuel combustion (either fossil or biomass), and heat recovered from cooling the solid product.

One of the primary choices to be made in designing the heat integration strategy is the means of heating biomass and cooling the torrefied product. It is possible to use the heat contained in the torrefaction gas or flue gas by bringing them into direct contact with the biomass. Alternatively, heat exchangers can be used to transfer the heat to an inert convective heating medium. Furthermore, the designer can choose the reactor type (e.g., fixed bed, fluidized bed, moving bed) as well as the reactor geometry (screw, conical, column). The mode of heat transfer can also be varied; in addition to convection, conductive and radiative (infrared, microwave) heat transfer modes have also been tested. While the design choices focus on achieving high efficiencies and low costs, they also aim at avoiding operational and safety complications emanating from, e.g., blockages, exothermicity, or condensation. For example, directly contacting hot flue gas with biomass for reasons of more efficient heat transfer or a strong overall exothermicity of the torrefaction reactions may lead to uncontrolled temperature excursions. Using controlled quantities of an inert convective medium instead of flue gas might be safer at the cost of a higher energy penalty. Recirculation of torrefaction gases for effective heat management might lead to condensation or polymerization of organic constituents leading to fouling or blocking of pipes.

Current torrefaction technology initiatives are based on different reactor concepts, some of which are illustrated in Figure 12.4.

The **rotating drum reactor** is widely used for drying and mixing of biomass and waste streams and can therefore relatively easily be adapted to the torrefaction

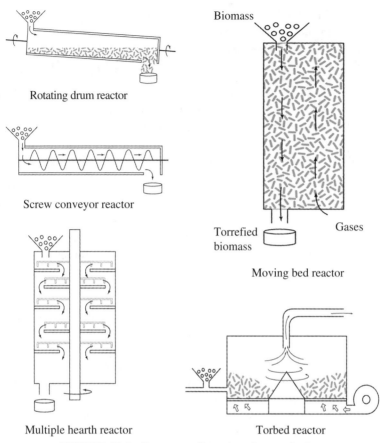

FIGURE 12.4   Reactor configurations for torrefaction.

process. The biomass can be heated directly or indirectly with steam or flue gas. The temperature and residence time can be controlled by varying the temperature of the heat supplying medium and the rotational velocity, length, and angle of the drum. Some residence time distribution generally is unavoidable, which may lead to a distribution in product quality. The tumbling of the relatively brittle torrefying material results in a relatively large fraction of fines. Rotary drum reactors have a limited scalability, necessitating a modular approach for higher capacities. Companies applying this type of reactor include TSI (United States), Torr-Coal (the Netherlands), and Andritz (Austria).

A **screw conveyor reactor** typically uses indirect heating of the biomass via the outer reactor wall and possibly via a hollow screw, which limits the heat transfer rates and makes scale-up difficult. Varying the screw speed can effectively control the residence time of the biomass due to the (near) plug flow of the solids. The first commercial torrefaction plant, operated in the 1980s by Pechiney in France, used a screw reactor. Similar reactor concepts are now applied by CENER (Spain), Biolake (the Netherlands), and BioEndev (Sweden).

The **multiple hearth furnace reactor** is well known for ore refining and smelting and consists of several layers with separate heating sources (usually gas burners). The biomass is fed at the top layer and driven by rotating arms to the center of the reactor from where it drops to the next layer, in which it is pushed to the perimeter and again falls down one layer, etc. Hot gas is fed at the bottom of the reactor or per layer through the side wall. The top layers are used for drying and the bottom layers for torrefaction. With a gas supply per layer, the temperature at each layer of the reactor can be independently set to control the torrefaction process. This reactor allows flexibility of operation due to accurate control concerning the residence time. Also, the relatively low agitation ensures limited production of dust. Among others, CMI-NESA in Belgium has been involved in applying this reactor technology. Wyssmont (United States) has been developing a slightly different design (the so-called TurboDryer), in which the trays move while the arms are stationary. Hot gas is circulated through the reactor using internal fans with the possibility of controlling the gas flow in each zone.

Recently, Andritz and ECN have successfully demonstrated a tray-type reactor design in Denmark. The design involves multiple stationary trays with rotating scrapers and separate gas loops for final drying and torrefaction. The reactor is operated at an elevated pressure level to improve gas–solid heat transfer and allow better scale-up.

The **moving bed reactor** is a packed column where biomass is continuously fed from the top and removed from the bottom of the reactor. Hot gas (flue gas or recirculated torrefaction gas) is fed from the bottom and heats up the biomass. This reactor type has a relatively high biomass loading (ratio of mass contained to reactor volume). The temperature distribution, however, can be uneven, and the pressure drop over the bed is strongly dependent on the particle size of the biomass. The slowly moving mass causes relatively long retention times. AREVA (formerly Thermya, France) is developing a torrefaction process based on this reactor type.

The **TORBED reactor** is a proven reactor concept for several applications including drying and combustion. It involves entraining the solid biomass particles in a high velocity swirling flow of gas. The "tor" in the name of the reactor does not refer to torrefaction, but to the resulting toroidal motion of the fluidized solid particles within the bed. The heat exchange in this reactor is very good and residence times are short (several minutes). The biomass input must consist of adequately small particles that can be fluidized in the bed. The reactor typically has a comparatively low degree of biomass holdup. This concept is applied by Topell (the Netherlands).

The **microwave reactor** is a relatively new development in which microwaves are used to heat up the biomass. Unlike conductive or convective heating, microwaves can penetrate surfaces and accurately deliver heat to target sites (principally water molecules), thus ensuring high heat transfer rates. Microwave sources, however, require electricity, a more expensive and higher-quality energy source than heat. Rotawave (United Kingdom) is currently in the process of demonstrating this technology.

12.4.1 Pelletizing and Briquetting

The energy density of the torrefied material can be improved by milling and subsequent densification through pelletizing or briquetting. A homogeneous product quality after torrefaction is essential for proper densification. For the production of durable pellets or briquettes, a binding agent may be necessary to act as the glue for the ground torrefied particles. Woody biomass contains lignin as a natural binding agent. However, the amount and structure of lignin still present after torrefaction depend on the feedstock and the torrefaction temperature and time. A less severe torrefaction leads to self-binding materials and durable pellets or briquettes. With high severity of torrefaction, the remaining lignin might not be sufficient, and the addition of a binding agent, e.g., starch, vegetable oil, molasses, glycerin, paraffin, or another lignin source, is necessary. When applying a binder, care should be taken that other product properties such as hydrophobicity are not adversely affected.

12.4.2 Safety Issues

The fire hazard already associated with fuel preparation is exacerbated by the low ignition temperatures of biomass volatiles. A sufficiently high concentration of biomass dust in air can furthermore lead to an explosion, which is a serious risk for torrefaction plant operators. This problem is largely associated with the raw biomass bunkers, as volatiles from torrefied biomass are less susceptible to spontaneous combustion. Before densification, the torrefied product is highly reactive, and self-heating and self-ignition can occur already at relatively low temperatures upon contact with ambient air. During pelletizing or briquetting, the product is heated because of the friction with the die wall, and hence, pelletizing presses are commonly water cooled. Furthermore, to ensure safe storage, the product must be cooled thoroughly following densification. Generally, densification considerably reduces the risk of self-heating and self-ignition.

12.5 TORREFACTION: AN ENABLING TECHNOLOGY

The favorable properties of torrefied (and densified) biomass enable the decoupling of biomass production and its end use in place, time, and scale. Torrefied biomass pellets or briquettes fit much better into existing solid fuel infrastructures and end-use technology. Moreover, these products may allow advanced trading schemes (comparable to oil and coal) and then have the potential, through standardization, of becoming commodity biomass fuels. The better fit in existing conversion technology in particular pertains to pulverized fuel combustion, which is used widely for high-efficiency power generation from coal. Large-scale cofiring of biomass with coal is seen as an easily implementable means of increasing the share of renewables in electricity production. However, pulverized fuel boilers (owing to the small residence times) require the fuel to be pulverized to (sub)millimeter sizes prior to combustion. Biomass feedstocks such as wood chips or agricultural residues cannot be milled using conventional coal grinding equipment due to their tenacious and fibrous nature.

Consequently, cofiring wood chips or agricultural residues in a coal-fired power plant necessitates installation of additional machinery, leading to additional capital costs. Furthermore, for biomass, the energy required for grinding is several times higher, leading to high operating costs. The most common practice, circumventing most grinding problems, is the use of wood pellets, which are essentially compacted sawdust. However, wood pellets require closed handling and storage facilities and, at high cofiring percentages, may reduce the rating (power output) of the power plant considerably. Torrefied biomass pellets or briquettes potentially do not require closed handling and storage nor special grinding equipment. Furthermore, due to their larger heating value, there will be no or only minor derating at high cofiring percentages.

The favorable properties in terms of grindability and energy density are even more important in dry feed oxygen-blown entrained flow gasification (especially with respect to pneumatic feeding), which is seen as a key technology in the future production of liquid biofuels and biochemicals. There also may be opportunities for torrefied biomass as a fuel in smaller-scale pellet boilers and stoves. Finally, torrefaction can potentially enable the utilization of a wide range of agricultural residues for the production of electricity, heat, fuels and chemicals from biomass.

12.6   THE FUTURE OF TORREFACTION

The initial main driver for torrefaction has been the increasing demand for solid biofuels in Western Europe emanating from legislation and incentives in favor of cofiring biomass in large power utilities. As the availability of clean woody biomass in this region is limited, North America serves as the principal source of imports. Consequently, several developers in Europe and North America are aiming to commercialize this technology, with several projects having made it to the pilot/demo phase. The continuing interest in torrefaction and its development relies on the sustained presence of these drivers, in addition to a further exploration of alternative raw biomass supply chains for which torrefaction seems to be a strong enabler.

More recently, interest in torrefied biomass as a renewable fuel has grown in other parts of the world, e.g., in Brazil, South Africa, and in several Asian countries including Malaysia, Indonesia, Japan, and South Korea. There is also a growing interest in the use of torrefied biomass for applications such as entrained flow gasification (for biofuel production) and smaller scale heat and power production.

Even as a lot is known about the chemistry of torrefaction reactions, process scale-up, optimization, and control will benefit from a more thorough fundamental understanding. Furthermore, in view of economics and sustainability, it is also imperative that the developed technologies focus on employing more efficient heat integration strategies.

Handling, storage, grinding, feeding, and conversion (e.g., combustion, gasification) of torrefied biomass need to be explored extensively in large-scale test campaigns to optimize product quality and inspire end user confidence. This requires that torrefied biomass be produced in adequately large quantities, which has only recently been accomplished with several larger demonstration plants starting to continuously and consistently produce good quality torrefied biomass. With several commercial plants coming online in the coming few years, the future of torrefaction seems as exciting as it is promising!

CHAPTER SUMMARY AND STUDY GUIDE

This chapter describes torrefaction as an emerging biomass pretreatment technology. Following the explanation of the necessity of thermal pretreatments, the chemical background of torrefaction is briefly discussed along with basic heat and mass balances involved therein. This is followed by a discussion of the properties of torrefied biomass as a fuel. The various reactors that can be used for carrying out torrefaction are then discussed along with their important features. The chapter concludes with an exposition of torrefaction as an enabling technology for the use of biomass fuels along with a future view to its development and implementation.

KEY CONCEPTS

History and necessity of thermal pretreatment

Fundamentals of torrefaction

Properties of torrefied fuels

Torrefaction reactors

Torrefaction: An enabling technology

SHORT-ANSWER QUESTIONS

12.1 What are the main constituents of woody biomass?

12.2 At what temperatures does each of these constituents decompose?

12.3 Name the (potential) heat sources and sinks in a typical torrefaction process.

12.4 Which step in the torrefaction process is the most energy consuming?

12.5 Name the main constituents of torrefaction gas.

12.6 Does the composition of torrefaction gas change with the extent of reaction? If so, in what way?

12.7 Name five potential uses of torrefied biomass.

12.8 Which properties of biomass are improved by torrefaction?

12.9   Name five types of reactors that have been developed for biomass torrefaction.

12.10 Which of the reactor designs seem most promising for scale-up (for production rates > 50 kt per year)?

12.11 What are the safety issues and risks involved in torrefaction processes?

12.12 What is the impact of the torrefaction temperature on the properties of the product biomass?

12.13 What role can densification (pelletizing/briquetting) play in the torrefaction process?

12.14 Name a few binder additives used in pelletizing of torrefied biomass. When is it necessary to use them?

12.15 Explain the effect of the particle size on the choice of reactor used in torrefaction.

PROBLEMS

12.1   Dry woody biomass can be represented with the molecular formula $CH_{1.4}O_{0.6}$. A sample of this biomass is torrefied to give a product gas with the following composition (on a mass basis): H_2O, 51%; acetic acid (CH_3COOH), 20%; CO_2, 26%; and CO, 3%.

12.2   Assuming typical values for the loss of solid mass and energy during torrefaction, calculate the LHV of the torrefied biomass and compare it with the LHV of the input biomass (you may use an equation based on the chemical composition of the solid fuel). Further, plot the compositions of both the raw biomass and products on a van Krevelen diagram.

12.3   The following scheme, depicted in Figure 12.5 (Prins et al., 2006a), uses pseudocomponents to model the weight loss kinetics of biomass in torrefaction.

The relations for the reaction rate coefficients (expressed in $kg.kg_{solid}^{-1}s^{-1}$) are as follows:

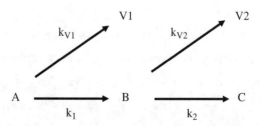

FIGURE 12.5   Torrefaction reaction kinetics scheme (adapted from Prins et al., 2006a). A, unreacted biomass; B, a moderately torrefied biomass; and C, a severely torrefied product. V1 and V2 represent a mixture of volatiles released in the conversion of A to B and B to C, respectively.

$$k_1 = 2.48 \times 10^4 \, \exp\left(\frac{-75,976}{R_u T}\right)$$

$$k_{V_1} = 3.23 \times 10^7 \, \exp\left(\frac{-114,214}{R_u T}\right)$$

$$k_2 = 1.10 \times 10^{10} \, \exp\left(\frac{-151,711}{R_u T}\right)$$

$$k_{V_2} = 1.45 k_2$$

where T is the temperature (K) and R_u is the universal gas constant (8.3143 $J.(mol.K)^{-1}$).

A particle of dry biomass (A) preheated to a temperature of 150°C enters a torrefaction reactor. It is heated at a rate of approximately 20°C per minute to a final torrefaction temperature of 270°C and is then held there for 20 min. Plot the weight loss of the particle and the mass of all pseudocomponents as a function of time.

PROJECTS

P12.1 The improvement in biomass quality depends on the severity of the thermal pretreatment. In increasing order of severity, treatments that can be used include drying, torrefaction, and charcoal production. Which of these treatments would be most appropriate for the following applications:

- Large-scale cofiring in pulverized coal-fired boilers

- Gasification for production of syngas

- Reduction of iron ores in a blast furnace

Work out each case in detail and justify your choices.

P12.2 One of the greatest advantages of torrefaction is the reduction in the energy required in grinding the biomass. Conceptualize and design an apparatus that can measure the energy required in grinding.

REFERENCES

Arias B, Pevida C, Fermoso J, Plaza MG, Rubiera F, Pis JJ. Influence of torrefaction on the grindability and reactivity of woody biomass. Fuel Processing Technology 2008;89(2):169–175.

Ben HX, Ragauskas AJ. Torrefaction of Loblolly pine. Green Chemistry 2012;14(1):72–76.

Bergman PCA, Boersma AR, Kiel JHA, Prins MJ, Ptasinski KJ, Janssen FJJG. Torrefaction for entrained flow gasification of biomass. Report nr ECN-C-05-067. Petten: Energy Research Centre of the Netherlands (ECN); 2005a.

Bergman PCA, Boersma AR, Zwart RWR, Kiel JHA. Torrefaction for biomass co-firing in existing coal-fired power stations—BIOCOAL. Report nr ECN-C-05-013. Petten: Energy Research Centre of the Netherlands (ECN); 2005b.

Bridgeman TG, Jones JM, Shield I, Williams PT. Torrefaction of reed canary grass, wheat straw and willow to enhance solid fuel qualities and combustion properties. Fuel 2008;87(6): 844–856.

Chew JJ, Doshi V. Recent advances in biomass pretreatment – torrefaction fundamentals and technology. Renewable and Sustainable Energy Reviews 2011;15(8):4212–4222.

Melkior T, Jacob S, Gerbaud G, Hediger S, Le Pape L, Bonnefois L, Bardet M. NMR analysis of the transformation of wood constituents by torrefaction. Fuel 2012;92(1):271–280.

Patel B, Gami B, Bhimani H. Improved fuel characteristics of cotton stalk, prosopis and sugarcane bagasse through torrefaction. Energy for Sustainable Development 2011;15(4): 372–375.

Percy J. *Metallurgy: The Art of Extracting Metals from their Ores*. London: J. Murray; 1861.

Phanphanich M, Mani S. Impact of torrefaction on the grindability and fuel characteristics of forest biomass. Bioresource Technology 2011;102(2):1246–1253.

Pimchuai A, Dutta A, Basu P. Torrefaction of agriculture residue to enhance combustible properties. Energy & Fuels 2010;24(9):4638–4645.

Prins MJ, Ptasinski KJ, Janssen JG. Torrefaction of wood. Part 1. Weight loss kinetics. Journal of Analytical and Applied Pyrolysis 2006a;77(1):28–34.

Prins MJ, Ptasinski KJ, Janssen JG. Torrefaction of wood. Part 2. Analysis of products. Journal of Analytical and Applied Pyrolysis 2006b;77(1):35–40.

Rousset P, Aguiar C, Labbe N, Commandre JM. Enhancing the combustible properties of bamboo by torrefaction. Bioresource Technology 2011;102(17):8225–8231.

van der Stelt MJC, Gerhauser H, Kielb JHA, Ptasinskia KJ. Biomass upgrading by torrefaction for the production of biofuels: a review. Biomass & Bioenergy 2011;35(9):3748–3762.

13

BIOCHEMICAL CONVERSION: BIOFUELS BY INDUSTRIAL FERMENTATION

MARIA C. CUELLAR AND ADRIE J.J. STRAATHOF

Department of Biotechnology, BioProcess Engineering Group, Faculty of Applied Sciences, Delft University of Technology, Delft, the Netherlands

ACRONYMS

ABE	acetone–butanol–ethanol
ADP	adenosine diphosphate
ATP	adenosine triphosphate
FAAE	fatty acid alkyl ester(s)
FAEE	fatty acid ethyl ester(s)
FPP	farnesyl pyrophosphate
HMF	5-hydroxymethylfurfural
IPP	isopentenyl pyrophosphate
LCA	life cycle assessment
MEP	methylerythritol phosphate
MEV	mevalonate
TAG	triacylglycerols

Biomass as a Sustainable Energy Source for the Future: Fundamentals of Conversion Processes, First Edition. Edited by Wiebren de Jong and J. Ruud van Ommen.
© 2015 American Institute of Chemical Engineers, Inc. Published 2015 by John Wiley & Sons, Inc.

SYMBOLS

c_p	specific heat capacity	$[J.kg^{-1}.K^{-1}]$
G	Gibbs free energy	$[J.mol^{-1}]$
H	enthalpy on mass basis	$[J.kg^{-1}]$
h_{fg}	enthalpy of evaporation	$[J.mol^{-1}$ or $J.kg^{-1}]$
H	enthalpy on mole basis	$[J.mol^{-1}]$
MW_i	molecular weight of species i	$[kg.mol^{-1}]$
Y	mass yield	$[kg.kg^{-1}]$
τ	residence time	$[s]$

Subscripts

c	combustion
f	formation
glc	glucose
prod	product(s)
r	reaction

Superscript

max	maximum

13.1   INTRODUCTION

13.1.1   Fermentation: Biological Background

Living organisms are complex structures of mostly organic compounds. Building up such structures from smaller molecules ("anabolism") requires energy. Photosynthetic organisms obtain this energy from sunlight. Other organisms convert part of the chemical compounds available to them into compounds with a lower Gibbs energy ("catabolism") and use this difference in Gibbs energy to drive the anabolic reactions. The biochemistry of the cell is strictly organized for this; specific enzymes catalyze only the necessary reactions. Generally, the catabolic reactions are such that the available Gibbs energy is not liberated as heat but used to convert adenosine diphosphate (ADP) with inorganic phosphate (P_i) into adenosine triphosphate (ATP). The anabolic reactions are usually driven forward by coupling them to the reverse conversion (ATP to ADP and phosphate). Thus, ATP is the carrier of Gibbs energy equivalents at the biochemical level.

Carbohydrates or other compounds available to living organisms can be completely oxidized to CO_2 and H_2O when sufficient O_2 is present. This combustion-like catabolism generates much ATP and allows fast growth. However, in the absence of O_2 or any other electron donor such as sulfate or nitrate, some ATP can still be generated by certain rearrangements of carbohydrate molecules. These often lead to a more oxidized product (such as CO_2) and a more reduced product, such as ethanol (C_2H_5OH). Figure 13.1 shows the catabolic reactions in a yeast cell when carrying out ethanol fermentation from glucose or fructose (both $C_6H_{12}O_6$). The overall catabolic reaction in this case is

$$C_6H_{12}O_6 + 2ADP + 2P_i \rightarrow 2C_2H_5OH + 2CO_2 + 2ATP \qquad \text{(RX. 13.1)}$$

The formed ATP remains intracellular and is converted back to ADP and P_i by anabolic reactions, which lead to conversion of carbohydrates into cells.

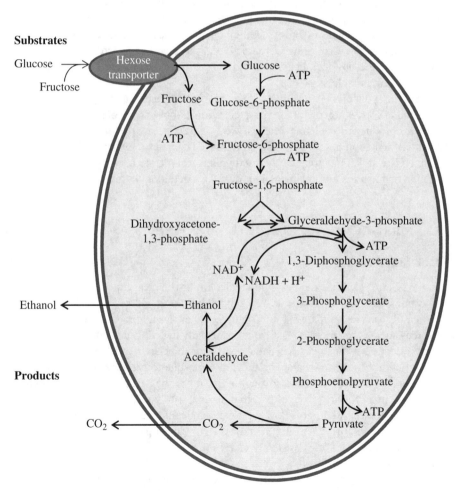

FIGURE 13.1 Formation of ethanol in a yeast cell. Only the catabolic reactions are shown, but thousands of other enzymatic reactions occur in the cell. "ATP" implies its synthesis from ADP and phosphate, but for brevity, these are left out.

A simplified elemental formula per carbon mole of a baker's yeast cell is $CH_{1.748}N_{0.148}O_{0.596}S_{0.0018}$ (Lange and Heijnen, 2001). Although the one-carbon mole formula of carbohydrates is not very different (CH_2O), modest consumption of additional nutrients and modest formation of side products are required to close the mass balance of the anabolic reactions.

13.1.2   Outline of This Chapter

Reduced products excreted by some microbial cells upon fermentation of carbohydrates can be isolated for use as biofuel. This approach has been very successful for ethanol production, which therefore is the default biofuel produced by industrial fermentation. Section 13.2 treats a traditional process for producing ethanol. Such a so-called 1st-generation process consumes carbohydrate sources that might otherwise be used as food. Therefore, 2nd-generation ethanol processes that consume nonfood carbohydrate sources are being developed, and these are described in Section 13.3. This includes the physical, chemical, and enzymatic process steps that are required to liberate fermentable sugars.

Industrial production is also investigated for many other compounds resulting from fermentative conversion of 1st- or 2nd-generation carbohydrate sources. Section 13.4 treats the production of 1-butanol, and Section 13.5 treats diesel-like compounds. Which fermentation products should be produced remains a matter of debate, but in Section 13.6, stoichiometric and thermodynamic criteria are used to facilitate such a selection on the basis of fundamental knowledge. An outlook is given in Section 13.7.

13.2   FIRST-GENERATION BIOETHANOL PROCESSES

13.2.1   Ethanol

Ethanol (CH_3CH_2OH) is the best known alcohol. It is a clear liquid that mixes in all proportions with water. It boils at $79°C$ and solidifies at $-114°C$.

Ethanol can be prepared from petrochemical resources or from renewable resources such as sugar; in the last case, it is usually called "bioethanol." Its fermentative production from sugars has been known for thousands of years.

Distillation of ethanol–water mixtures at atmospheric pressure does not lead to pure ethanol (although it has a lower boiling point than water) but to a so-called azeotropic mixture of 95.6 vol.% ethanol and 4.6 vol.% water, because their molecules interact and do not show thermodynamically ideal behavior. Pure ethanol was first obtained in 1796. Applications were developed as solvent, industrial intermediate (e.g., for ethyl esters), and as fuel. The T-Ford was designed for bioethanol use.

Between about 1950 and 2000, petrochemical resources were used for the production of ethanol, but now, the production of bioethanol is again cheaper than the production of this petrochemical ethanol. The most important reason for the return of bioethanol is the need for an automotive fuel from renewable resources. Such bioethanol has to be virtually free of water in order to prevent engine problems. Since some decades, Brazilian car engines are suited for use of 100% ethanol and for blends with gasoline, while cars and tank stations in other countries usually cannot yet deal with 100% ethanol but only with blends. Ethanol, however, is by no means an ideal automotive fuel. Combusting 1 kg liberates less energy than combusting 1 kg gasoline does, because the oxygen atom in ethanol has no combustion value but makes up a relatively large part of the mass of ethanol. Besides, ethanol cannot replace diesel.

13.2.2 Fermentation

As feedstock for ethanol fermentation, various carbohydrates are used:

- Sucrose (from sugarcane in Brazil/India, sugar beet in Europe, etc.)
- Starch (from corn in United States/China, wheat in Europe, cassava in Thailand, etc.)
- Lignocellulosic sugars (from straw and other "wastes," everywhere according to many future scenarios as shown in Section 13.3)

Criteria for the choice of feedstock are:

- Price
- Availability
- Convenience of handling
- Stability
- Effect on productivity of the process including product isolation
- Societal issues (food vs. fuel discussion)

Enzymes and sometimes also acid catalysts are used to hydrolyze polysaccharides and oligosaccharides to fermentable monosaccharides. Baker's yeast (*Saccharomyces cerevisiae*) is the default microorganism used for ethanol fermentation and has the natural advantage of excreting the sucrose-hydrolyzing enzyme invertase. Other advantages of baker's yeast are its fast and easy growth, its potential to virtually completely convert a range of C_6 monosaccharides into ethanol and CO_2, and its high microbial safety. Therefore, ethanol fermentations are traditionally carried out using baker's yeast. Still, alternatives are considered (other types of microorganisms or recombinant strains of *S. cerevisiae*) because baker's yeast does not naturally ferment some important polysaccharides and C_5 monosaccharides and does only ferment up to 50°C, while a higher temperature could lead to energy savings in some process types.

13.2.3 The Brazilian Bioethanol Process

There are many varieties of bioethanol processes, mainly due to feedstock differences. Here, we treat a typical sugarcane process such as performed in Brazil. A sugarcane

TABLE 13.1   Average composition of the millable part of sugarcane according to local data

Component	wt%
Sucrose	13.74
Glucose/fructose	1.70
Minerals	0.62
Other water-soluble components	5.09
Fiber	13.21
Water	65.64

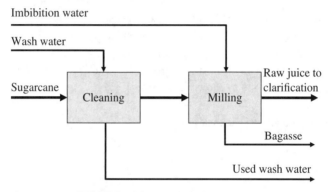

FIGURE 13.2   Milling of sugarcane.

composition is given in Table 13.1. Sucrose, glucose, and fructose are fermentable to baker's yeast.

A representative process for ethanol production from sugarcane is summarized in Figures 13.2, 13.3, 13.4, 13.5, 13.6, and 13.7. Six parts are distinguished: milling, clarifying, concentrating, fermenting, purifying, and cogenerating. These parts are connected in many ways. For simplicity, only material flows are shown and no energy flows (via steam, cooling water, and electricity). In Table 13.2, a summary of technical terms in the bioethanol process can be found.

Clearly, there are other outputs besides ethanol. These either result in treatment costs (vinasse, ash) or can be sold as coproduct (fusel oil, filter cake). The main coproduct is the electricity that is produced by cogeneration as shown in Figure 13.7; it is more than sufficient to cover the plant's requirement. Actually, we are dealing with a typical biorefinery, where the agricultural feedstock crop, in this case sugarcane, is separated and converted into various fractions, in which each needs to have a clear destination. The huge scale would lead to problems even if only a mere 0.1% of the feedstock had no destination.

The sugar plants are located in areas with cane fields. Transport costs for cane are minimized in this way and also sugar losses because the cane cells consume sugars during transport and storage. Calculations show that truck transport in excess of

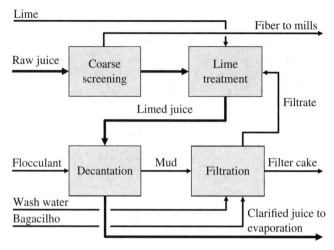

FIGURE 13.3 Clarification. Suspended particles are removed from the juice.

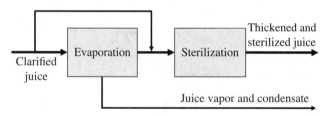

FIGURE 13.4 Evaporation and sterilization.

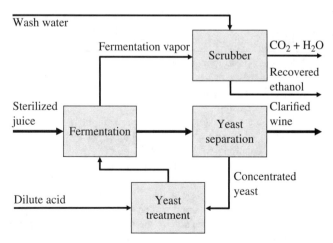

FIGURE 13.5 Fermentation. Ethanol is not only recovered from the liquid but also from the off-gas. The yeast is recycled.

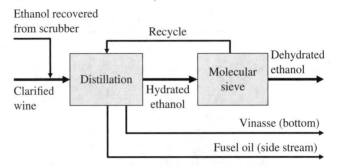

FIGURE 13.6   Purification. A series of distillations yields 96 vol.% ethanol, and molecular sieve adsorbs the remaining 4 vol.% water.

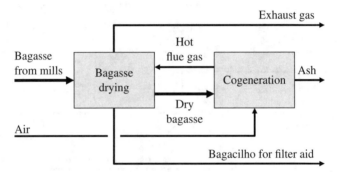

FIGURE 13.7   Cogeneration of electricity. Before incineration, the wet bagasse is dried using the hot flue gas. Note that the diagram just gives the mass flows but that electricity is the most important product of the cogeneration.

TABLE 13.2   Glossary for the bioethanol process

Ash	Minerals (salts) remaining after combustion
Bagacilho	Dried bagasse
Bagasse	Remainder of sugarcane plant after pressing juice, mainly polysaccharide from cell walls
Filter aid	Inorganic powder (e.g., silica) which, on a filter, prevents clogging
Fusel oil	Mixture of C_3 to C_5 alcohols and other volatile side products of fermentation
Imbibition	Replacement of interstitial liquid by water
Lime	Calcium oxide (CaO) or calcium hydroxide (Ca(OH)$_2$)
Molecular sieve	Material with nanopores in which water fits (e.g., aluminosilicate crystals)
Mud	Aqueous mass containing ~20 wt% undissolved polymer
Sucrose	Plain sugar, a disaccharide of the monosaccharides glucose and fructose
Vinasse	Water fraction (impure) from ethanol distillation
Wine	Aqueous ethanol solution, centrifuged after fermentation

TABLE 13.3 Calculated material balance of a bioethanol factory

Feed stream	kg.s^{-1}	Exit stream	kg.s^{-1}
Sugarcane	139.96018	Ethanol	9.95870
Imbibition water	38.39679	Fermentation: CO_2 + H_2O vapor	10.07730
Wash water	363.17538	Used wash water	364.52082
Air	204.39191	Exhaust gas	247.87209
Lime	0.12559	Ash	0.66098
Flocculant	0.04616	Filter cake	5.69991
Dilute acid	25.92310	Vinasse	125.83721
Fermentation additives[a]	0.24126	Fusel oil	0.61930
Dilution water[a]	24.12937	Yeast bleed[a]	0.55403
Dilution water[a]	2.26117	Condensate[a]	34.25009
Total in	798.6509	Total out	800.0504

[a] For simplicity, these streams are not indicated in Figures 13.1–13.6. The condensate is composed of water and can be recycled as wash water. The yeast bleed prevents accumulation of dead yeast, and in a steady-state process, this bleed equals the growth of the yeast.

50 km might easily become uneconomical. Then, the available cane within, e.g., 50 km determines the maximum plant size. Of course, the produced ethanol needs to be transported, but storage and transport in tanks are easy, and ethanol occupies only 2% of the volume of the sugarcane. The shipping costs, from Brazil to Europe, are acceptable. The factories run during the harvest season, which may be half a year.

13.2.4 Material Balances of the Process

In 2005, in collaboration with Brazilian experts, a sugarcane process has been quantified with respect to size and composition of all process streams. That was a conceptual process design and assumed to be representative for a modern plant. Table 13.3 shows its overall material balance. In the calculated flow sizes, the number of significant digits shown is much larger than the accuracy of measurements of flows would have been, if these were available. However, keeping so many digits in the calculations facilitates checks for consistency with subprocess mass balances such as shown later.

Most striking are the huge quantities of water needed for washing the sugarcane and air for the incineration of the bagasse, also resulting in a large amount of exhaust gas.

Another observation is that only 1 kg ethanol is produced from 14 kg sugarcane; the other 13 kg ends up as side products, mainly water, as the water content of the cane is ~65 wt%.

13.2.5 Material Streams in the Fermentation

The fermentative conversion is the heart of the process. It occurs at 34°C and atmospheric pressure. Table 13.4 gives the material streams for the fermentation, including their composition.

TABLE 13.4 Calculated material streams (kg.s^{-1}) for the fermentation of Figure 13.5 (where the smallest stream, "fermentation additives," has been left out for simplicity)

Component	Juice in	Fermentation additives	Recycled yeast	Unclarified wine	Fermentation vapor
Acetic acid			0.00160	0.17153	
Yeast			4.86066	5.18276	
CO_2			0.00842	0.12458	9.81202
Ethanol			0.27260	10.21233	0.12587
Glucose/fructose	2.24468		0.00255	0.27432	
Glycerol			0.00548	0.58929	
Impurity	0.58625		0.00525	0.56369	
Isoamyl alcohol			0.00144	0.15510	
Minerals	0.06240		0.00059	0.06299	
Succinic acid			0.00039	0.04184	
Sucrose	18.32557				
H_2SO_4			0.00110	0.00110	
H^+			0.00004	0.00406	
Antifoam		0.09728	0.00091	0.09820	
Disperging agent		0.02594	0.00024	0.02619	
Soda		0.11804	0.00111	0.11914	
Water	70.19729		43.80449	112.94909	0.11026
Total	91.41620	0.24126	48.96687	130.57619	10.04815

    The described fermentation is continuous, although batch fermentation is more common. Here, we only consider the steady state of the continuous process. The sugars are converted almost completely by baker's yeast. Products besides ethanol are CO_2, glycerol, succinic acid (butane-1,4-diacid), isoamyl alcohol (3-methyl-1-butanol), and more yeast.

    During fermentation, sucrose is first hydrolyzed into D-glucose and D-fructose by an extracellular enzyme:

$$C_{12}H_{22}O_{11} + H_2O \rightarrow 2C_6H_{12}O_6 \qquad \text{(RX. 13.2)}$$

Glucose and fructose are isomeric C_6 sugars ($C_6H_{12}O_6$). The C_6 sugars are taken up by the cells. The fermentation itself runs without O_2 to prevent conversion of sugars into H_2O and CO_2. Instead of liberating a large amount of chemical energy, the yeast only liberates a small amount by formation of ethanol:

$$C_6H_{12}O_6 \rightarrow 2C_2H_5OH + 2CO_2 \qquad \text{(RX. 13.3)}$$

Section 13.1.1 shows the complete catabolic reaction. The involved phosphate compounds remain intracellular and do not play a role in the material balance of the fermentation. The other compounds pass the cell membrane. Less than 2 kg m^{-3} CO_2 dissolves in water, so most escapes the solution as gas bubbles, with small amounts of evaporated water and ethanol. Almost all ethanol remains in solution and achieves a concentration of about 8.5 vol.% (=7.8 wt%) at the end of the fermentation.

The yeast cells require the ATP produced upon ethanol formation for maintenance. If additional ATP is available, the yeast uses it for converting part of the sugars into new cells (cell growth). On the one hand, that growth is favorable for ethanol production, because a higher yeast concentration leads to faster ethanol production. On the other hand, we want to minimize sugar consumption for anything else than ethanol production, and growth decreases the yield of ethanol on sugars. The solution is to try to achieve a steady state in which the yeast is recycled and reaches a concentration level dictated by the available sugar amount (and hence ATP) and the requirement for maintenance (and hence ATP). Some yeast cells will lyse, and a small portion of the recycle flow is purged as the aforementioned "yeast bleed" to prevent accumulation of dead cells mass. The portion of lost cells is compensated by growth in the steady state that is achieved. The yeast recycle is possible because some tricks are used to prevent bacterial infections. Otherwise, fast-growing bacteria that might barely produce ethanol would take possession of the fermentor. The tricks involve centrifugation of yeast cells while minimizing sedimentation of the (smaller) bacterial cells and acidifying the yeast recycle stream to a pH not tolerated by harmful bacteria.

13.2.6 Energy Flows

Many material flows in the overall process need heating or cooling. All hot streams leaving the factory must have a temperature below 35°C. Heat integration is performed to maximize exchange of heat between streams that need cooling and streams that need heating. After optimization of the heat exchange network, it turns out that still 91,423 kW extra cooling duty and 7,281 kW heating duty are required.

13.2.7 Equipment

Some major equipment items are treated here.

13.2.7.1 Fermentors For designing the fermentors, the reaction kinetics for the growth of the yeast cells and for formation of ethanol and side products should be known, ideally in detail. So, information is required on the reaction rates as a function of the concentrations of yeast, sucrose, ethanol, ammonia, CO_2, acetic acid, etc. In fact, not only the extracellular concentrations but also all intracellular concentrations should be known, and quantitative relations describing all mechanisms inside and outside the cells as functions of these concentrations are required. Only a few years ago, large groups of researchers have started to measure the dynamic course of hundreds of intracellular concentrations and describe these using fundamental relations. This approach is too complicated for our design purpose, and even basic models that can approximate the course of a fermentation are not treated here.

Instead, based on experience, we simply assume an average volume-specific ethanol productivity of 8 kg ethanol per hour per m³ fermentation liquid. The mass balance then leads to 4529 m³ fermentation volume.

In the designed process, three large mixed continuous reactors are used in series. The first reactor consumes 60% of the sugar, and two smaller reactors each consume 20%. The intention of this configuration is a certain extent of plug flow, exposing yeast in each next reactor to a higher concentration of ethanol. Higher concentrations of ethanol inhibit the yeast, thus slowing down all reactions. This actual reaction course is not covered by our simplistic model that assumes constant reaction rates, but it leads to a reasonable approximation of the reactor sizes.

To distribute failure risk, two parallel cascades of 2265 m^3 each lead to the required fermentation volume of 4529 m^3. Consequently, the liquid volume of the fermentors per section is 1359, 453, and 453 m^3, respectively. Each fermentor is filled with liquid for only 90% (ao, because CO_2 bubbles are liberated), and therefore, the required vessels are even larger.

The residence time of liquid in, e.g., the 3rd fermentor is

$$\tau = \frac{\text{Volume}}{\text{Volume flow}} = \frac{\text{Volume}}{\text{Mass flow}/\text{density}}$$

$$= \frac{453 \text{ m}^3}{130 \text{ kg.s}^{-1}/2 \text{ cascades}/\sim 1000 \text{ kg}/\text{m}^3} \approx 7000 \text{ s} \approx 2 \text{ h}$$

The rising CO_2 bubbles can lead to a reasonable extent of mixing, but some additional stirring is used. For this, the exact geometry is important.

As compared to the complete microbial oxidation of sugars to CO_2 and H_2O, the formation of ethanol from sugars does not liberate much energy. However, heat production should not be neglected at this large scale. To maintain the fermentation at the desired 34°C, aqueous streams between the fermentors are cooled with cooling water using heat exchangers.

Since the aeration requirement is very small (a detail of the growth reaction), stirring is modest, and no cooling coils are required in the vessels, the required fermentors will be much less expensive than for aerobic (O_2-consuming) fermentations.

In general, cylindrical stainless steel vessels are used, with a height-to-diameter ratio of about 3:1.

13.2.7.2 Distillation   Among all equipment, distillation is the largest consumer of energy sources. The evaporation of ethanol and in particular water requires much energy, much more than for heating the liquid to its boiling point (see Table 13.5). This energy is to a large extent lost to the environment.

TABLE 13.5 Enthalpy of evaporation (h_{fg}), specific heat capacity, and enthalpy for heating from 30 to 80°C the boiling point of ethanol ($\Delta H_{heating} = c_p \Delta T$) for pure ethanol and water

	h_{fg} (kJ.mol^{-1})	h_{fg} (kJ.kg^{-1})	c_p (kJ.kg^{-1} K^{-1})	$\Delta H_{heating}$ (kJ.mol^{-1})	$\Delta H_{heating}$ (kJ.kg^{-1})
Ethanol	38.6	838	2.4	2.6	120
Water	40.6	2250	4.2	3.8	210

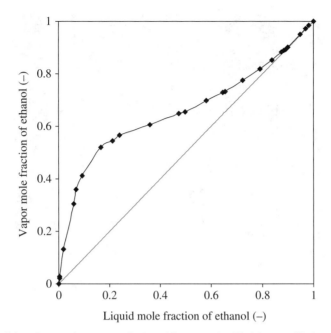

FIGURE 13.8   Measured contents of ethanol in water for liquid at equilibrium with vapor (data from Beebe et al. (1942)).

As mentioned before, distillation of ethanol–water mixtures at atmospheric pressure leads to an azeotropic mixture of 95.6 vol.% ethanol and 4.6 vol.% water. Figure 13.8 shows the vapor–liquid equilibrium, indicating that, from a certain ethanol content in the liquid onward, the vapor that is at equilibrium with the liquid has the same ethanol content as the liquid, so moving that vapor upward in a distillation column and cooling it to form a liquid do not lead anymore to enrichment.

In the simplest scheme of continuous distillation, the feed stream (containing 8.5 vol.% ethanol) is introduced somewhere near the middle of a distillation column, after preheating to ~100°C. The aforementioned azeotropic mixture leaves the column at the top, and water with nonvolatile side products leaves the column at the bottom. The vapor is largely condensed to liquid at the top and mainly recycled to the column to improve the degree of purification; by analogy, the liquid at the bottom is largely reboiled. The bottom temperature is ~115°C (so the pressure exceeds atmospheric pressure), and the top temperature is ~80°C.

In Brazilian practice, separate columns are used for the top and bottom sections, and additional equipment is used for recovering volatile side products and remaining heat.

13.2.8   Economy

The following economic data have been calculated using standardized methods, although it is not clear to which extent these are valid for the local situation in Brazil. The total investment for the project is 95,287,000 $, as derived from the purchase

FIGURE 13.9 Fuel prices in Brazil in 2006.

costs of numerous pieces of equipment. Main equipment items are, in order of cost contribution, the turbines for cogeneration, the sugar mills, the centrifuges (used after fermentation), and the boiler in the cogeneration section.

The annual cost to run the ethanol plant is 62,719,500 $, from which 85% is due to sugarcane costs. Much smaller percentages are due to costs of other raw materials, maintenance, local taxes, and labor. Utilities are not on the list because the required cooling water is reused almost completely and because energy is generated rather than consumed.

The two main products are ethanol, at 467.44 $.m^{-3}, and electricity, at 15.89 $.GJ^{-1}. Revenues from these products are 85,980,300 + 9,850,700 = 95,831,100 $ per year. Any coproducts have not been included in the economic analysis. The investment is earned back in 2.9 years, which is very acceptable for a factory that may be in operation for 20 years and which uses proven technology.

A more detailed analysis over the full 20 years indicates that the return on investment is 17%, which is clearly well above interest rates. In view of a huge expected demand for bioethanol and the availability of a lot of space for growing sugarcane, it is not surprising that about 100 bioethanol factories were planned to be built in Brazil at the time of this analysis.

The profitability of the process is sensitive to the costs of sugarcane and the selling price of ethanol, but not very sensitive to the selling price of electricity, the investment cost, or other costs to run the plant. Typically, sensitivity to taxes, subsidies, and mandates is also high. Figure 13.9 shows a situation where bioethanol is less expensive than fuels from fossil sources.

13.2.9 Ecological and Societal Impact

Like any human activity that is carried out at a very large scale, bioethanol production will lead to some degree of ecological problems. We can compare these with the

TABLE 13.6 Comparison of gasoline and bioethanol on the basis of LCA for driving 1 km[a]

Impact category	Gasoline impact	Bioethanol impact	Unit
Global warming potential	0.25	0.05	kg CO_2 equivalent
Abiotic depletion potential	1.7	0.3	10^{-3} kg antimony equivalent
Ozone layer depletion potential	3.1	1.5	10^{-8} kg CFC-11 equivalent
Photochemical oxidation potential	1.6	1.5	10^{-4} kg ethylene equivalent
Human and ecotoxicity potential	0.017	0.078	kg 1,4-dichlorobenzene equivalent
Acidification potential	7.5	11	10^{-4} kg SO_2 equivalent
Eutrophication potential	1.0	4.7	10^{-4} kg PO_4 equivalent

[a]Based on values obtained from Luo et al. (2009).

problems caused by gasoline, which bioethanol might partly substitute, using life cycle assessment (LCA). For comparison purposes, the functional unit used is the amount of bioethanol or gasoline required for driving 1 km. The results of a "cradle to grave" approach are shown in Table 13.6. Although bioethanol scores better on CO_2 emission than gasoline does, because the sugarcane fixes CO_2, there is still net CO_2 emission. Regarding the other six used impact categories, bioethanol and gasoline each score best three times. It depends on the importance attributed to each impact factor which fuel is to be preferred concerning LCA.

This LCA does not capture factors such as land use and competition between agriculture for food and fuels. Cutting tropic rain forest for bioethanol production is generally not accepted. However, much of the long-established agricultural area in Brazil is used for extensive cattle breeding and would be suitable for sugarcane. Also, the area used for soy far exceeds the sugarcane area. Thus, intensified sugarcane growth leads indirectly and unintentionally to increased pressure on rain forests.

Besides economy and ecology, other factors play a role when choosing bioethanol as fuel. It will lead to more jobs (in the agricultural sector) than gasoline does, and it decreases dependency on oil-producing countries. This type of reasons has dominated political discussions in the United States and led to subsidies for corn-based bioethanol production, although in many studies corn-based bioethanol scores ecologically lower than bioethanol based on sugarcane and even lower than gasoline.

13.3 SECOND-GENERATION BIOETHANOL PROCESSES

Alternative feedstocks that do not compete with food can be used for the production of ethanol (and other fuels and chemicals) by fermentation. Alternative feedstocks can be agricultural/forestry waste (often referred to as lignocellulosic or second-generation feedstock), sludge from wastewater treatment plants, algae biomass, and so on. In this section, we focus on lignocellulosic feedstocks.

13.3.1 Pretreatment

In order to make lignocellulosic feedstocks suitable for fermentation processes, a series of steps is required aiming at transforming the lignocellulosic feedstock (mainly containing cellulose, hemicellulose, and lignin) into a solution of fermentable sugars (i.e., pentoses and hexoses):

- Breakage of the bonds between cellulose, hemicellulose, and lignin, so that their structure is exposed to the hydrolysis catalysts
- Hydrolysis of hemicellulose and cellulose, which results in soluble fermentable pentose and hexose sugars
- Removal of the lignin fraction, a nonfermentable fraction that otherwise interferes with subsequent processing steps

Depending on the feedstock, an initial mechanical size reduction step (e.g., chipping, grinding, milling) might be required in order to facilitate handling and to increase the surface area for subsequent steps. Many pretreatment methods have been evaluated and implemented on a processing scale. Table 13.7 provides a summary of the most common methods and their functionality. In the following, these methods are shortly described.

13.3.1.1 Acid Hydrolysis   This method can be divided in two general approaches, namely, the use of diluted (or weak) acids at high temperature or the use of concentrated (strong) acids at low temperature. In the former, dilute (mostly sulfuric) acid is sprayed onto the raw material, and the mixture is held at high temperature (160–220°C) for up to a few minutes. Hydrolysis of the hemicellulose occurs, releasing soluble oligomers and monomers. As an alternative to inorganic acids, organic acids such as maleic acid and fumaric acid are being used.

TABLE 13.7   Most common pretreatment methods and their functionality

Method	Breakage of bonds	Hydrolysis of hemicellulose	Hydrolysis of cellulose	Hydrolysis/ removal of lignin
		Functionality		
Dilute/weak acid hydrolysis	X	X		
Acid hydrolysis	X	X	X	
Alkaline hydrolysis				X
Organosolv	X	X		X
Wet oxidation	X	X		X
Ozonolysis	X			X
Steam explosion	X	X		
Ammonia fiber explosion	X	X		X
Biological pretreatment		X	X	X

In the latter approach, concentrated (strong) acids such as H_2SO_4 and HCl have been widely used because of their hydrolytic effect on cellulose. However, given the corrosive nature of the reaction, this approach poses special requirements to the equipment. Further drawbacks of both approaches include the generation of degradation products that are inhibitors for subsequent steps and the requirement for neutralization afterward, resulting in waste generation.

13.3.1.2 *Alkaline Hydrolysis*
The major effect of alkaline pretreatment is the solubilization of lignin from the biomass. The reaction time can be long, but the process conditions are mild, resulting in limited degradation. Usually, lime (calcium hydroxide) or sodium hydroxide is used. This pretreatment is often used after acid hydrolysis; this combination, however, leads to the formation of poorly soluble salts, which need to be separated from the treated biomass.

13.3.1.3 *Organosolv*
With this method, the lignin is solubilized and the hemicellulose is hydrolyzed. Commonly used organic solvents include ethanol, methanol, acetone, and ethylene glycol. Temperatures used for the process can be as high as 200°C, but lower temperatures can be sufficient depending on whether a catalyst (organic or inorganic acid) is used. Solvent recovery is required for reducing costs and environmental impact but also because the solvent itself can be an inhibitor for the subsequent steps (both enzymatic hydrolysis and fermentation). This method results in a high-quality lignin fraction, which might facilitate further applications.

13.3.1.4 *Wet Oxidation*
Lignin removal can also be achieved by treatment with an oxidizing agent such as hydrogen peroxide, oxygen, or air. In combination with water and elevated temperature and pressure, the lignin polymer is then converted into carboxylic acids, among others. Since these acids are inhibitors for fermentation, they have to be neutralized or removed. In addition, a substantial part of the hemicellulose might be degraded and can no longer be used for sugar production.

13.3.1.5 *Ozonolysis*
This method for oxidation of lignin is performed at ambient temperature and pressure. The hemicellulose and cellulose are hardly decomposed.

13.3.1.6 *Steam Explosion*
This method uses injection of high-pressure saturated steam into a reactor filled with biomass. The temperature rises to 160–260°C. The pressure is then suddenly reduced, and the biomass undergoes an explosive decompression with hemicellulose degradation and lignin matrix disruption as a result. The hemicellulose degradation results in the formation of acetic acid, which in turn causes further hydrolysis of the hemicellulose fraction. Because of this, steam explosion is also referred to as "autohydrolysis." A variation of the method is catalyzed steam explosion, in which acids are added to increase the hydrolysis yield.

13.3.1.7 *Ammonia Fiber Explosion*
In a similar way as for steam explosion, the biomass is exposed to liquid ammonia at high temperature and pressure and then the pressure is released. This method reduces the lignin content and removes some

hemicellulose. The production of inhibitors is limited, although this method is not as efficient as acid hydrolysis or acid-catalyzed steam explosion.

13.3.1.8 Biological Pretreatment   This method employs microorganisms such as white-, brown-, and soft-rot fungi to degrade hemicellulose and lignin. Although the advantages of this method are that it has a low energy requirement and mild operation conditions, the rate of biological hydrolysis is usually too low for industrial implementation.

The best pretreatment method depends on the type of lignocellulosic feedstock. Current industrial activities for the production of ethanol show a preference toward acid-based pretreatment methods, probably due to the available information for a broad range of feedstocks. In acid-based methods, lignin is only removed after hydrolysis of the cellulose, which is a disadvantage for the hydrolysis step: lignin binds to the hydrolysis enzymes, reducing their activity and hampering enzyme recovery. In summary, the following aspects are important in the choice of pretreatment methods: efficiency, generation of inhibitors, neutralization requirement, need for solvent recovery, technical equipment demands (e.g., corrosion, operating temperature and pressure), use of chemicals and energy, separability of the resulting fractions, and waste generation.

13.3.2 Inhibitors and Detoxification

Next to the target fermentable sugars, during pretreatment, other products are formed to a varying extent depending on the process conditions used. Those products can have an inhibitory effect on the fermentation process, the level of toxicity depending, among others, on the concentration in the fermentation medium and on the cell physiological conditions. The main types of inhibitors are (see also Table 13.8):

TABLE 13.8   Main inhibitors produced during pretreatment

Fraction	Hydrolysis products	Degradation products	Inhibitory effect on fermentation
Cellulose	Hexoses (glucose)	HMF, levulinic acid	Affects sugar assimilation, cell growth, and respiration
Hemicellulose	Pentoses (xylose, arabinose)	Furfural	Affects sugar assimilation, cell growth, and respiration
	Hexoses (mannose, galactose, glucose)	HMF, levulinic acid	Affects sugar assimilation, cell growth, and respiration
	Acetic acid		Diffuses across cell membranes. Lowers cell pH
Lignin	Phenolic compounds (among others)		Affects the integrity of cell membranes. Affects cell growth and sugar assimilation

- Sugar degradation products: After hydrolysis and at high temperature and pressure, pentose and hexose sugars may dehydrate to furfural and 5-hydroxymethylfurfural (HMF), respectively. HMF, in turn, can be further degraded to levulinic acid.
- Acetic acid: This acid is derived from the acetyl groups in hemicellulose and is inherently formed during hemicellulose hydrolysis.
- Lignin products: A variety of compounds (e.g., aromatic, polyaromatic, phenolic, and aldehydic) may be released from the lignin fraction due to long exposure to high temperatures. At temperatures below 180°C, lignin degradation is negligible, provided that no strong acid or alkaline conditions prevail.

In general, the relative toxicity of these inhibitors is phenolics > furfural > HMF > acetic acid. However, these compounds also have synergistic effects. The yeast *S. cerevisiae*, e.g., grows in the presence of either furfural or HMF, but not in a mixture of both.

In order to deal with these inhibitors, the following approaches are being followed:

- Removal of the inhibitors before fermentation, also known as detoxification step. The most widely used methods include overliming (i.e., treatment with lime, typically at a pH of 9–11 and a temperature of 50–60°C), evaporation, steam stripping, and absorption.
- Optimization of the pretreatment method and process conditions for minimizing the generation of inhibitors while maximizing the production of fermentable sugars. Since such an optimum also depends on the feedstock, a comparison is usually based on the pretreatment severity (temperature, residence time, and pH) and the quality of the resulting hydrolysate (ratio of fermentable sugars to inhibitors).
- Strain engineering for higher tolerance to inhibitory compounds. The focus is currently mainly on acetic acid, as this compound is inherently formed during hemicellulose hydrolysis.

13.3.3   Enzymatic Hydrolysis

As discussed previously, many pretreatment methods partially hydrolyze the hemicellulose, but only a few result in cellulose hydrolysis. Hence, an additional step is required for obtaining the fermentable sugars. Enzymatic hydrolysis is preferred, since enzymes have high substrate specificity and require only mild operating conditions compared to chemical methods such as strong acid hydrolysis (e.g., a pH of 4.8 and a temperature of 45–50°C for cellulase).

The hydrolysis of cellulose requires cellulases, a group of enzymes that act synergistically in two steps: primary hydrolysis and secondary hydrolysis. In the primary hydrolysis, endoglucanases cleave the cellulose to form oligosaccharides, and subsequently, cellobiohydrolases (also known as exoglucanases) release the water-soluble dimer cellobiose. Primary hydrolysis occurs on the surface of the solid substrate and is the rate-limiting step. In the secondary hydrolysis, the cellobiose is further hydrolyzed by β-glucosidases to glucose. An important drawback of cellulase is that it suffers from end-product (cellobiose and glucose) inhibition. This results in reduction of the hydrolysis rate in the course of the reaction unless the product is removed as soon as it is being formed. As source for these enzymes, the same fungi used for biological pretreatment (e.g., *Trichoderma reesei*) are being used.

Hemicellulose can also be enzymatically hydrolyzed. Being a diverse group of heterogeneous polymers with various side groups, the complete hydrolysis of hemicellulose requires a more complex system of enzymes including xylanase, β-xylosidase, and several other complementary enzymes, such as acetylxylan esterase, α-arabinofuranosidase, α-galactosidase, and ferulic and/or p-coumaric acid esterase. While the number of enzymes required for hemicellulose hydrolysis is much greater than for cellulose hydrolysis, accessibility to the substrate is easier as xylan, the main constituent of hemicellulose, does not form tight crystalline structures (unlike cellulose).

The enzymatic hydrolysis of both cellulose and hemicellulose is confronted with a number of obstacles:

- *Price*: Although enzyme prices have decreased due to intensive research by, e.g., Novozymes, Genencor (acquired by DuPont), and DSM, enzyme loading should be minimized in order to reduce production costs. This, however, increases the time needed to complete hydrolysis. An alternative is to recycle the enzymes, since much of them remain active. Many methods, with or without enzyme immobilization, are currently being researched.
- *Product inhibition*: As mentioned previously, this results in lower performance and limits the use of high substrate concentrations. Alternative process configurations, in which the product is continuously removed, are being currently evaluated (see also Section 13.3.5).
- *Lignin interference*: Lignin shields the cellulose chains and adsorbs the enzymes, resulting in decreased efficiency and hampering enzyme recovery.

An alternative to the use of commercial enzymes is their production on the target lignocellulosic material. It has been reported that these enzyme preparations perform better than standard commercial enzyme preparations produced on substrates such as purified cellulose. This approach could beneficially be employed in biorefineries to produce enzymes on-site. Moreover, on-site enzyme production could reduce enzyme costs due to less need for purification and stabilization of enzyme preparations.

13.3.4 Fermentation of C₅ and C₆ Sugars into Ethanol

The hydrolysis of lignocellulosic feedstocks results in a mixture of hexose (C_6) and pentose (C_5) carbohydrates. Table 13.9 shows typical compositions for a few feed-stocks. Glucose and xylose are the most abundant sugars. There are many naturally occurring microorganisms that can ferment glucose into ethanol, *S. cerevisiae* being

TABLE 13.9 Major sugar composition (wt%) of common agricultural lignocellulosic feedstocks[a]

	Wheat straw	Bagasse	Sugar beet pulp
Hexoses (C_6)			
Glucose	32.6	39.0	24.1
Mannose	0.3	0.4	4.6
Galactose	0.8	0.5	0.9
Pentoses (C_5)			
Xylose	19.2	22.1	18.2
Arabinose	2.4	2.1	1.5
Other components	44.7	35.9	50.7

[a]Adapted from Van Maris et al. (2006).

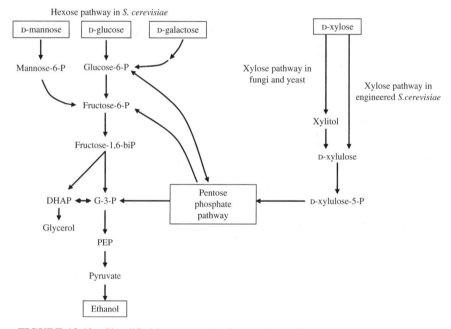

FIGURE 13.10 Simplified hexoses and xylose pathways for the production of ethanol.

the workhorse, as described in Section 13.2. The main features of *S. cerevisiae* for industrial ethanol production are (i) high ethanol productivity, (ii) high tolerance to ethanol, and (iii) tolerance to relatively low pH. The last two features are especially relevant for minimizing contamination by other microorganisms. This allows for the use of low-cost equipment and microorganism recycle, among others. Additionally, *S. cerevisiae* is able to ferment other hexoses (mannose and galactose) to ethanol (see Figure 13.10), and it shows tolerance to inhibitory compounds formed during pretreatment. Native strains, however, are unable to utilize xylose for growth or fermentation to ethanol.

Although *S. cerevisiae* cannot utilize xylose, the genes encoding the key enzymes in the pathway for xylose utilization are present in its genome; however, their expression levels are too low to allow for xylose utilization. In fact, only very slow growth on xylose has been observed when those genes are overexpressed. There are naturally occurring microorganisms able to ferment xylose into ethanol under oxygen-limited conditions (e.g., the yeast *Pichia stipitis*). However, the rate and yield of ethanol production from xylose in these strains are very low. Due to a redox imbalance in the cell, these strains also excrete xylitol as by-product, thereby reducing the yield of ethanol on substrate (see Figure 13.10).

Given all these aspects, a vast amount of research is currently being devoted to the engineering of microorganisms for hexose and pentose fermentation to ethanol. Rapid advances are being made, especially with *S. cerevisiae* as host microorganism. The engineering focuses on three aspects: (i) transport of the sugar across the cell membrane, (ii) coupling of the sugar metabolism to the main glycolytic pathway, and (iii) maintenance of a closed redox balance. For the latter, e.g., the introduction of a pathway for xylose utilization that does not lead to xylitol excretion (see Figure 13.10) has succeeded in providing a "proof of principle" under academic research and has now been taken up by industry for further development and implementation.

13.3.5   Process Configurations for Hydrolysis and Fermentation

Most process concepts for the production of bioethanol from lignocellulosic feedstocks start with pretreatment, followed by enzymatic hydrolysis of the cellulose part and a yeast-based fermentation of the resulting sugars. Lignin, the main by-product in the process, can be directly used as a solid fuel or as a source for higher added-value biorefinery products. This configuration is often referred to as sequential or separate hydrolysis and fermentation. In such a configuration, each process step can be optimized independently, resulting in different operating conditions (e.g., temperature, pH, and residence time). In addition, it is possible to have a separate C_5 and C_6 fermentation. The main advantage of this configuration is its flexibility. However, during hydrolysis, high levels of glucose accumulating in the reactor inhibit the activity of the enzyme.

An alternative that is receiving much attention is that of simultaneous saccharification and fermentation. In this configuration, cellulose hydrolysis and ethanol fermentation take place in the same processing step, resulting in less pieces of equipment. One of the main advantages of this configuration is that the glucose produced during hydrolysis is immediately consumed during fermentation, avoiding product inhibition on the enzymes and minimizing contamination by other microorganisms. A disadvantage is that the optimum temperature for enzymatic hydrolysis is typically higher than that of fermentation, and hence, a compromise in operating conditions is required. In order to overcome this, research is currently being done on ethanol fermentation at higher temperatures (Blanch, 2012).

This process configuration can be extended to simultaneous saccharification and cofermentation of pentose and hexose sugars. Another possibility is consolidated bioprocessing, where the fermenting microorganism produces the enzymes necessary to hydrolyze cellulose and hemicelluloses. Besides simplifying the process by performing everything in one step, synergism between enzyme and microbe has been reported, although the microorganisms are not efficient yet for tests at industrial scale.

13.3.6 Case Study: Production of Ethanol from Bagasse

The transition toward lignocellulosic feedstocks will have an impact on the process scheme; the mass, energy, and waste streams in the process; and the equipment used. Consequently, it will also have an impact on the economic performance of the process. In a similar way as described Section 13.2, a conceptual study for a plant located in Brazil was performed (Efe et al., 2007). In this case, however, the plant produces both sugar and ethanol. Nevertheless, the process steps required for ethanol production are similar to those already described. Table 13.10 shows a summary of the input and output flows for the first-generation scenario of the combined sugar and ethanol plant. Just as for the plant producing ethanol only (Section 13.2), the process requires large amounts of sugarcane, water, and air.

In this case, 9 kg sugarcane results in 1 kg of product (both sugar and ethanol). In other words, about 13 kg sugarcane is required for producing 1 kg sugar and 0.4 kg ethanol. In the first-generation scenario, the sugarcane bagasse is used for cogeneration (in the same way as shown in Figure 13.7). In the second-generation scenario, all bagasse is pretreated so that it can also be used as feedstock for the fermentation. A simplified block scheme is shown in Figure 13.11.

TABLE 13.10 Input and output flows for a first-generation sugar and ethanol plant[a]

Input flow	$kg.s^{-1}$	Output flow	$kg.s^{-1}$
Sugarcane	332.58833	Ethanol	11.03899
Imbibition water	93.12473	Sugar	24.63819
(Recycle) water	174.02778	Carbon dioxide	11.31399
Air	407.43717	(Waste) water	308.87555
Lime	0.30309	Exhaust gas	488.09758
Flocculant	0.00069	Ash	1.240000
Sulfuric acid	0.14199	Filter cake	15.27793
Sulfur	0.10809	Vinasse	16.99999
		Fusel oil	113.29699
		Yeast	3.51859
		Exhaust air	13.43279
Total in	1007.73187	Total out	1007.73059

[a]The plant capacity is determined by the feedstock supply (5 million tonnes per year of sugarcane). The plant operates 174 days per year (4176 h per year).

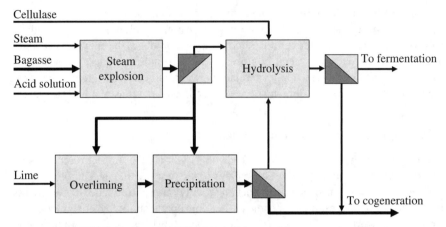

FIGURE 13.11   Simplified block scheme of bagasse pretreatment.

The main features of this pretreatment scheme are:

- A size reduction step is not necessary in this case.
- For the breakage of bonds and hydrolysis of hemicellulose, acid-catalyzed steam explosion is used. The fibers are first heated up to 100°C in a steaming vessel and then mixed with hot (100°C) acid solution (1% H_2SO_4 (w/w) dry bagasse) and pumped into the steam explosion vessel using a high-pressure (14 bar) pump.
- For detoxification, flash cooling and overliming are applied. Since the steam explosion is carried out at high temperature, no extra heat input is required for steam stripping, and just by flashing the effluent of the steam explosion, the volatile toxic materials (acetic acid and sugar degradation products) can be stripped of the medium. The liquid portion is removed from the fibers and cooled down to the overliming temperature (50°C). Overliming is performed at a pH of 10; afterward, the pH is brought down to 4.5 by using a bypass from the liquid fraction after steam explosion. This pH adjustment results in precipitates that are removed by clarifiers. The remaining toxic materials, which are phenolic lignin degradation products, can be absorbed by activated charcoal if the concentrations are above the threshold inhibition concentrations (not shown in Figure 13.11).
- The detoxified liquid fraction contains the pentose (C_5) sugars. This stream is used to dilute the fibers after steam explosion to about 20 wt% solids. These fibers contain the cellulose and the lignin. Commercial cellulase enzymes are added to break the cellulose in hexose (C_6) sugars. Cellulose hydrolysis takes place at 65°C with a large residence time (36 h), after which 90% of the cellulose is hydrolyzed. The enzymes are not recovered.
- The lignin residue is separated from the mixed sugar solution and sent to cogeneration to be combusted for the generation of energy.
- The liquid fraction (or mixed sugar solution) is sent to fermentation for ethanol production.

TABLE 13.11 Input and output flows for a second-generation sugar and ethanol plant

Input flow	kg.s^{-1}	Output flow	kg.s^{-1}
Sugarcane	332.58833	Ethanol	26.39900
Imbibition water	93.12473	Sugar	24.63849
(Recycle) water	500.72025	Carbon dioxide	26.92400
Air	236.91997	(Waste)water	546.81902
Lime	0.92509	Exhaust gas	281.63719
Flocculant	0.00069	Ash, gypsum, soil	6.36579
Sulfuric acid	0.93759	Fusel oil	239.54099
Sulfur	0.10809	Exhaust air	13.43740
Cellulase	0.36579		
Natural gas	0.05189		
Total in	1165.74240	Total out	1165.76190

For the fermentation and given the rapid developments that are being made, recombinant yeast able to ferment C_5 and C_6 sugars has been considered. For the calculations, a future scenario is assumed in which such microorganism has a performance similar to current yeast. After the fermentation, the downstream process has the same structure as the first-generation scenario. Table 13.11 shows a summary of the input and output flows for the second-generation scenario.

From this table, it can be seen that, for the same sugarcane input as in the first-generation scenario, more ethanol can be produced. In fact, roughly 13 kg sugarcane is now required for producing 1 kg sugar and 1 kg ethanol. Other important differences are:

- *Increase in input flows associated with the bagasse pretreatment*: More sulfuric acid for the steam explosion, more lime for detoxification, and commercial cellulase for hydrolysis.
- *Less waste streams*: As an alternative to bagasse, other organic waste streams are sent to cogeneration, namely, filter cake, vinasse, and yeast. The solids from the furnace (ash, gypsum, and dirt) are sent to landfill.
- *Less steam and electricity generation*: Since the bagasse is now directed to fermentation, the excess electricity demand is met by purchasing it from the grid. For the same reason, less air but more natural gas is now required for cogeneration.

The main economic figures for both scenarios are shown in Table 13.12. Although somewhat dated, these values are still valuable for comparison purposes. Despite the increase in investment costs due to the additional pretreatment line and an increase in energy consumption, the plant revenues in the second-generation scenario are 60% higher, resulting in a significant reduction in ethanol production costs.

TABLE 13.12   Main economic figures for ethanol production at a location
in Brazil (in 2005 US$)

	First generation	Second generation
Total capital investment ($)	101,907,323	132,597,219
Total operating costs ($ per year)	76,464,948	106,393,893
Revenues ($ per year)	145,462,674	229,886,830
Production cost of ethanol ($.$t^{-1}$)	221	168
Production cost of sugar ($.$t^{-1}$)	107	107

13.4   BUTANOL

Already in 1861, Pasteur observed 1-butanol production by anaerobic bacterial
fermentation. In the beginning of the twentieth century, the so-called acetone–
butanol–ethanol (ABE) fermentation using a *Clostridium* species was developed.
The products are produced in a molar ratio of about 3:6:1. Potato starch was the first
carbohydrate feedstock used on an industrial scale. The process was the main source
of 1-butanol and acetone, which were used as solvents and for the production of
other chemicals.

In the early 1960s, petrochemical production methods led to the decline of the ABE
fermentation industry, although production continued much longer in South Africa,
the former Soviet Union, and China. Nowadays, there is a renewed interest in
ABE fermentation for reasons of sustainability and economic opportunities due to
increasing oil prices.

The main incentive for focusing on butanol rather than ethanol is that on a mass
basis butanol has a 31% higher combustion value than ethanol, which is very important for an automotive fuel. Also, the lower volatility and higher hydrophobicity of
butanol allow blending with conventional gasoline in any proportion, whereas the
use of ethanol can lead to corrosion and may require adaptation of transportation lines
and engines.

ABE fermentation is performed by a large variety of *Clostridia* strains. When only
butanol is desired, the ideal stoichiometry starting from hexose sugars is

$$C_6H_{12}O_6 \rightarrow C_4H_{10}O + 2CO_2 + H_2O \qquad \text{(RX. 13.4)}$$

This overall reaction can be coupled to the formation of 2 ATP, thus enabling
cell growth and maintenance. Metabolic engineering of *Clostridia* and many other
microorganisms is used to adapt the original butanol fermentation in order to:

- Maximize butanol yield on carbohydrate (by minimizing acetone and ethanol
  formation and minimizing the need for cell disposal and regrowth)
- Ferment lignocellulosic sugars into butanol
- Increase the rate of butanol production by the used microorganism
- Increase the tolerance of the used microorganism to butanol

The last point remains the most challenging one. Wild-type organisms may tolerate butanol up to aqueous concentrations of only ~15 g.L^{-1}, while engineered organisms are still limited to ~30 g.L^{-1}. In contrast, a tolerance of 170 g.L^{-1} has been achieved in some ethanol fermentations. Ethanol is less hydrophobic and has less tendency to interfere with the lipid bilayer of the cell membrane.

The low final butanol concentration leads to relatively short batch fermentations and relatively long downtimes after each batch and thus to inefficient fermentor use and high investment cost. Moreover, for both batch and continuous fermentation, the recovery of butanol from dilute aqueous solutions is energy intensive, which is expensive and also energetically unfavorable if the butanol is to be used as fuel. The default recovery method is distillation.

Because of these issues, 1-butanol does not yet seem to be competitive with ethanol as biofuel, despite its much better properties. Nevertheless, significant commercial fermentative 1-butanol production has started again in China and is planned at several other locations. This butanol seems to be used as chemical rather than as fuel. Similarly, isobutanol production using recombinant bacteria is reaching large-scale commercialization.

13.5 DIESEL-LIKE PRODUCTS

Current biofuels, namely, first-generation ethanol and biodiesel, supply 2% of the global transport energy demand. This number is expected to increase to 27% by 2050 (EIA, 2011), in particular due to the need for replacement of liquid fuels for planes, marine vessels, and other heavy transport vehicles that are not suited for other renewable energy sources such as electricity.

Biodiesel is a mixture of fatty acid alkyl esters (FAAE) obtained from free fatty acids and triglycerides (see also Chapter 18). Currently, biodiesel is being produced from vegetable and animal oils or fats and cooking oil waste, and it is mostly used in conventional diesel engines in blends of up to 20 vol.%. Alternative feedstocks and routes are being evaluated in the search for products that resemble current petroleum fuels and, hence, can make use of existing infrastructure. These are also referred to as advanced or drop-in biofuels. In this section, we look at the routes involving industrial fermentation. At the time of writing, most of these routes were still in research or pilot-scale stage.

13.5.1 Routes Using Oleaginous Microorganisms

All microorganisms have the ability to synthesize lipids as a fundamental constituent of the cell membranes. However, a few genera of bacteria, yeasts, molds, and algae— referred to as oleaginous microorganisms—are also able to accumulate lipids to more than 20% of their dry cell mass. In such microorganisms, lipid accumulation starts when a carbon source is present in excess and an element in the growth medium becomes limiting. The microorganisms are not able to grow and hence convert the excess carbon into lipids, storing them as energy reserve in a vacuole inside the cell.

In most oleaginous microorganisms, the stored lipids consist of 80–90% triacylglycerols (TAG) with a fatty acid composition similar to that of vegetable oils. It has been demonstrated that such microbial oils, also called single-cell oils, can be used as feedstock for biodiesel production. Under conditions of nitrogen limitation, accumulation has been reported to increase up to 70–80% of the dry cell mass (Subramaniam et al., 2010).

After fermentation, the cells need to be harvested from the medium. Due to the lipid content, the density difference with the medium is low, and separation by, e.g., centrifugation is not efficient. Cross-flow filtration has been used as an alternative, but fouling of the filtration membrane by the cells and other medium components results in low flux, and thus, large membrane areas are required to achieve satisfactory throughputs.

Following cell harvest, the oil has to be removed from inside the cells. This is usually done by a combination of mechanical cell disruption and solvent extraction or by direct contact of the cells with a solvent, which results in membrane permeabilization and cell rupture. In either case, the extracted oil is then further refined and transesterified into biodiesel.

Clearly, the intracellular production has disadvantages with respect to reaction and product recovery:

- The cell density becomes considerable lower, affecting mixing in the reactor and the use of separation equipment like centrifuges.
- The need for cell disruption for product recovery limits the applicability of reactor intensification techniques such as cell recycle.

Additional disadvantages of this route include the low genetic accessibility of the microorganisms, hampering improvements in the achievable product concentrations and the type of product obtained, and the need for additional conversion steps for direct use as biofuel. Because of this, biodiesel production from oleaginous microorganisms has not yet reached industrial implementation. Nevertheless, the lipids produced by these microorganisms are still receiving attention for higher-value niches in the food and pharmaceutical sectors.

13.5.2   Routes Using Engineered Microorganisms

The metabolism of well-studied industrial microorganisms (mostly *Escherichia coli* and *S. cerevisiae*) and photosynthetic organisms (e.g., cyanobacteria) is being directed toward the production of biodiesel and diesel and jet fuel replacements. Much of this research is being carried out in biotechnology companies (e.g., REG Life Sciences, Amyris, Joule Unlimited). The emphasis has been on the production of FAAE, isoprenoid compounds such as farnesene ($C_{15}H_{24}$), and fatty acid-derived alkanes such as pentadecane ($C_{15}H_{32}$). FAAE can be used in the same way as biodiesel. Farnesene is highly unsaturated, resulting in a low cetane number and low oxidative stability. Therefore, it requires an additional hydrogenation step for fuel use. This chemical hydrogenation yields farnesane, which has a cetane number of 58 and good cold-flow properties and can be blended up to 35 vol.%. In contrast, pentadecane has a cetane number of 95 and may be used directly as fuel. Both farnesane and pentadecane have been reported to be produced extracellularly. The main potential advantages of these routes are the extracellular production, making cell retention possible and facilitating product recovery in a separate lipid phase, and the possibility of tailoring the type of product obtained.

13.5.2.1 Fatty Acid Ethyl Esters (FAEE): in vivo Transesterification   This route aims at the complete synthesis of fatty acid ethyl esters (FAEE) from renewable carbon sources. This has been achieved by modifying *E. coli* in two steps: firstly, by establishing a pathway for ethanol synthesis and, secondly, by including an enzyme for the transesterification of ethanol and the fatty acid moiety of acyl-coenzyme A (acyl-CoA) to FAEE. The latter is especially important, since it saves the costly chemical transesterification step of first-generation biodiesel production.

Results to date have shown intracellular FAEE accumulation of up to 26% of dry cell mass (Kalscheuer et al., 2006). However, substantial FAEE biosynthesis was only achieved when fatty acids were also present in the feedstock. Current research is looking for alternative host microorganisms, such as oleaginous microorganisms, so that the flux of fatty acids can be directed from TAG toward FAEE biosynthesis.

13.5.2.2 Production of Isoprenoids   Isoprenoids are a diverse class of chemicals derived from isopentenyl pyrophosphate (IPP) and comprise one or more five-carbon isoprene units. Isoprenoids are involved in many cellular processes including respiration, cell membrane structure, signaling, photosynthesis, cell defense, and vitamin production. Plant isoprenoids, in particular, have long been used as flavor and fragrance agents.

In nature, two independent biosynthetic pathways, the mevalonate (MEV) pathway and the methylerythritol phosphate (MEP) pathway, are responsible for the production of the key intermediate, IPP. Although both pathways exist in plants, the MEV pathway is responsible for all the isoprenoid production in archaea (prokaryotic single-cell microorganisms), some bacteria, and most eukaryotes (including the yeast *S. cerevisiae*), while the MEP pathway is present in most bacteria (including *E. coli*) and green algae. The biosynthesis of farnesene, however, involves the enzymatic conversion of one of the pathway intermediates, farnesyl pyrophosphate (FPP). In *S. cerevisiae*, this has been achieved by incorporating the enzyme from a plant, *Artemisia annua*, into the pathway and by overexpressing the enzymes from the native MEV pathway. Farnesene is secreted to the production medium and after recovery is converted by chemical hydrogenation to farnesane. This approach has been demonstrated at pilot and production scales (Chandran et al., 2011). It is also being used for the production of flavors and fragrances.

13.5.2.3 Production of Alkanes and Alkenes   This route focuses on the biosynthesis of long-chain hydrocarbons, such as pentadecane, derived from a fatty acid pathway. This has been achieved by incorporating an alkane biosynthesis pathway from cyanobacteria in *E. coli* (Schirmer et al., 2010). The pathway converts the fatty

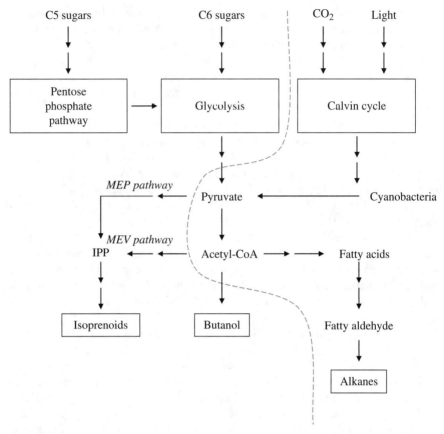

FIGURE 13.12   Simplified pathways for the production of liquid biofuels other than ethanol. The pathways of *Cyanobacteria* are at the right-hand side of the dashed line.

aldehyde—an intermediate of fatty acid metabolism—to an alkane or alkene (olefin). So far, this approach has led to the secretion of C_{13}–C_{17} mixtures of hydrocarbons of ca. 0.3 g L^{-1}. The process is currently under optimization, and pilot-plant fermentations (1 m^3 scale) have already been performed (Domínguez de María, 2011).

Figure 13.12 shows simplified pathways for the production of isoprenoids, butanol, and long-chain alkanes.

13.6   STOICHIOMETRIC AND THERMODYNAMIC COMPARISON OF FERMENTATIVE BIOFUELS

Worldwide, there are many studies for the development of novel biofuels by fermentation as an alternative to the established ethanol. Examples include butanol, branched alcohols, alkanes, alkenes, esters, and ketones. Evaluating the potential of certain fermentative production routes of biofuels can be a daunting task, especially when metabolic routes toward these fuels are hypothetical or even unknown. Still, researchers and people funding and judging research have to make proper choices early on.

This section is meant to partly simplify such choices. Using basic stoichiometry and thermodynamics, one can determine the potential of fermentative processes for hypothetical biofuels and compare these to ethanol. Kinetic limitations are of a less fundamental nature and will not be taken into account in the following discussion.

Assuming glucose or an equivalent carbohydrate as the feedstock, the overall fermentation reaction considered is

$$C_6H_{12}O_6 \rightarrow xC_aH_bO_c + yCO_2 + zH_2O \qquad \text{(RX.13.5)}$$

Thus, only anaerobic fermentative production of biofuels is considered. Aerobic fermentation is excluded from this evaluation as the reaction with O_2 leads to partial combustion of the available glucose and thus does not lead to the highest possible biofuel yields. In addition, anaerobic fermentations may be carried out in simpler equipment than aerobic fermentations, since no aeration is required. However, under anaerobic conditions, the microorganisms may not generate sufficient ATP to grow. Therefore, we will assume that (previously grown) cells are retained in the fermentor. We will also assume that the small amount of ATP produced by the reaction of the aforementioned equation (RX. 13.5) is consumed in cell maintenance reactions. If thermodynamics show that no ATP can be produced by the reaction, the reaction product will be excluded as potential biofuel. The cell is considered as a black box, so this analysis also includes products for which the metabolic pathway is presently unknown.

The product may accumulate extracellularly up to a concentration that is toxic to the cell. At that point, maintenance processes require ATP beyond what is produced. Because this constraint has a kinetic nature and can potentially be solved by using product-tolerant strains, it will be neglected in this analysis. Instead, it is assumed that the achievable product concentration is determined by the glucose feed concentration and the reaction stoichiometry. The fermentation is assumed to occur at thermodynamic standard conditions (25°C and 1 bar). For thermophilic fermentations, however, this may be a poor assumption.

Using the three elemental balances (for C, H, and O) derived from the equation (RX. 13.5), the stoichiometric coefficients x, y, and z can be calculated using the product values for a, b, and c. For certain products (mainly carboxylic acids), negative values of the stoichiometric coefficient result for CO_2, implying that CO_2 is a cosubstrate rather than a coproduct. For the production of H_2, water is a cosubstrate.

The maximum theoretical mass yield of the biofuel product on glucose, $Y_{\text{prod/glc}}^{\text{max}}$, can be calculated using the molecular weights (MW) according to Equation (13.1):

$$Y_{\text{prod/glc}}^{\text{max}} = \frac{x \times MW_{\text{product}}}{MW_{\text{glucose}}} \left(\text{kg.kg}^{-1} \right) \qquad \text{(Eq. 13.1)}$$

TABLE 13.13 Stoichiometric and thermodynamic data for some biofuels, with glucose as reference

| | Formula | Density ($kg.m^{-3}$) | $\Delta_r G^0$ ($kJ.mol^{-1}$) | $-\Delta_c H^0$ ($kJ.mol^{-1}$) | $Y^{max}_{prod/glc}$ ($kg.kg^{-1}$) | Energy content | | $-\Delta_c H^0 \times Y^{max}_{prod/glc}$ ($MJ.kg^{-1}_{glucose}$) |
						$MJ.kg^{-1}$	$GJ.m^{-3}$	
Hydrogen	H_2	0.09	−35	286	0.13	141	0.01	19.0
Methane	CH_4	0.72	−427	891	0.27	55.5	0.04	14.8
Ethanol	C_2H_6O	790	−230	1367	0.51	29.7	23.5	15.2
1-Butanol	$C_4H_{10}O$	810	−280	2676	0.41	36.1	29.3	14.8
Glucose	$C_6H_{12}O_6$		0	1556	1.00	15.6		15.6
Butyl butyrate	$C_8H_{16}O_2$	870	−285	4982	0.44	34.5	30.0	15.1
Farnesene	$C_{15}H_{24}$	840	−386	9428	0.29	46.1	41.3	14.7

Results are shown in Table 13.13. The achieved yield values for well-known fermentation products are significantly lower than this theoretical yield, but the trends shown in Table 13.13 are correct.

For the reaction to occur, the Gibbs energy must be negative. It is relatively easy to calculate standard Gibbs energies of reaction ($\Delta_r G^0$) taking the pure component standard states (crystalline glucose, gaseous carbon dioxide, liquid water, and liquid ethanol) (Alberty, 1998). Although the phases and concentrations are very different at fermentation conditions, this gives a reasonable estimate of the possibility of the reaction. For hydrogen, Table 13.13 shows only a very slightly negative value of $\Delta_r G^0$, as compared to the other possible products. This implies that the ideal stoichiometry, leading to the indicated value of $Y^{max}_{prod/glc}$, is not achievable with any metabolic pathway in any microorganism at ambient conditions. Hydrogen production by fermentation is pursued, though, but the yields at which researchers aim are much lower than the one in the table, because with realistic pathways part of glucose is fermented to carboxylic acids to obtain an overall $\Delta_r G^0$ with a sufficiently negative value.

The energy content of a biofuel is another important number. It is taken to be equal to the standard enthalpy of combustion, $\Delta_c H^0$. The energy is assumed to be released by combustion with oxygen at normal atmospheric pressure, with gaseous CO_2 and liquid H_2O as products at room temperature (i.e., the higher heating value). The standard enthalpy of combustion, in turn, can be calculated from the standard enthalpies of formation, $\Delta_f H^0$, of the reactants and products. For a compound containing only carbon, hydrogen, and oxygen, the general combustion reaction is

$$C_a H_b O_c + \left(a + \frac{1}{4}b - \frac{1}{2}c \right) O_2 \rightarrow a CO_2(g) + \frac{1}{2}b H_2 O(l) \qquad \text{(RX. 13.6)}$$

The standard enthalpy of combustion $\Delta_c H^0$ is given by (Lide 2011)

$$\Delta_c H^0 = -a\Delta_f H^0(CO_2, g) - \frac{1}{2}b\Delta_f H^0(H_2O, l) + \Delta_f H^0(C_a H_b O_c)$$

$$= 393.51a + 142.915b + \Delta_f H^0(C_a H_b O_c) \ \left(kJ.mol^{-1} \right) \qquad \text{(Eq. 13.2)}$$

The energy content is usually expressed per unit of mass or per unit of volume.

Some produced biofuels have a much higher energy content than others, but the maximum energy yield, which is the product of the product's energy content and achievable product yield ($-\Delta_c H^0 \times Y^{max}_{prod/glc}$), does not vary much, as shown in Table 13.13. The reason is that the combustion energy of glucose is transferred virtually completely to the biofuel, since the coproducts of the fermentation (carbon dioxide and water) have no combustion value. Differences between values of $-\Delta_c H^0 \times Y^{max}_{prod/glc}$ of the biofuels originate from a different production of entropy in the different fermentation cases.

The maximum captured combustion energy is very similar for the fermentative biofuels considered, but the energy required for recovering the biofuels from the fermentation mixture is very different. Several cases can be distinguished:

- The biofuel is a gas, e.g., methane. This gas is present in the off-gas of the fermentor mixed with the produced CO_2. Theoretically, the energy to demix these gases is the negative of the mixing energy, but in order to achieve practical rates, the operation cannot be done close to equilibrium, and at least ten times the mixing energy is required.
- The biofuel is a water-miscible liquid such as ethanol. This liquid is dissolved in the fermentation broth. A large amount of energy may be required for distillation.
- The biofuel is a water-immiscible liquid such as pentadecane. If no stable emulsion is formed, this liquid can be recovered without much energy requirement.
- The biofuel is a solid. Processing of solids is significantly more complicated than processing of fluids. Their use as biofuel is not advocated.

The outcome of this analysis is that water-immiscible liquids are the most attractive ones considering product recovery. Diesel-like products are good examples, since the Gibbs energies of reaction for the formation of these products from glucose are favorable and, although yields are low, the energy density of these compounds is high. This explains the current attempts to produce diesel-like products by fermentation (see Section 13.5).

13.7   OUTLOOK

The (bio)fuel market is characterized by tight economic margins that require efficient, low-cost processes. The routes described in this chapter concern pure cultures that make use of conventional fermentation equipment and require sterilization of the nutrients and inflow gases. A successful process would have, most likely, the following characteristics:

- *Use of alternative feedstocks*. The same developments that are being made in the production of lignocellulosic ethanol (see Section 13.3) can be applied in the microbial production of other biofuels.
- *Sepsis and microorganism robustness*. Most pure cultures require sterilization of the equipment, nutrients, and inflow gases in order to avoid contaminations. In the ethanol production process, contaminations are naturally controlled since most microorganisms cannot tolerate the ethanol levels in the fermentation medium. As a result, low-cost (open) fermentors including cell recycle loops can be used. In the production of diesel and jet fuel type of molecules, however, suppression of growth of competing microorganisms by the product itself is not expected. Hence, inherently robust microorganisms able to tolerate unfavorable conditions for competing microorganisms are being used. Examples include *S. cerevisiae*, which tolerates low pH, and cyanobacteria, which tolerate high salt concentrations. Another approach that is receiving much attention is the use of mixed culture techniques. The production of medium-chain fatty acids has already been accomplished (Steinbusch et al., 2011).

- *Anaerobic processes.* Performing a process under anaerobic conditions has the potential advantage of reducing the investment and operating costs by increasing the yield of product on substrate and eliminating the input of (compressed) air. Additionally, the produced fermentation gases can be used for providing sufficient mixing in the reactor. However, anaerobic product formation is not inherent to the routes for diesel-like products described in Section 13.5, so metabolic engineering will probably be required. Research is currently being conducted in this area (Weusthuis et al., 2011).

- *High cell densities and cell reuse.* Since the substrate is one of the major cost drivers in industrial fermentation, it should be directed optimally to product formation. This can be achieved by separate cell growth and product formation regimes and by extracellular production. In this way, high cell densities can be obtained first (in the cell growth regime) and maintained by cell retention or recycle (in the product formation regime). This approach has been proven successful in ethanol production. Microorganism robustness is an important element in this approach.

- *Simple product recovery.* One advantage that diesel-like molecules have over short-chain alcohols such as ethanol and butanol is their very low solubility in water, which could result in simple product recovery by, e.g., gravity separation. The cells and other medium components, however, can act as surfactants, hindering the coalescence of the product droplets and forming stable emulsions. These emulsions may require intensive centrifugation and/or de-emulsification techniques such as the use of additives and pH and temperature shifts. These will obviously increase the investment and operating costs of the process. Research is currently being performed on coalescence improvement by equipment and medium design.

- *Tailored products.* As mentioned earlier, products that do not require further conversion steps and that can use the existing infrastructure, i.e., drop-in biofuels such as long-chain alkanes, are preferred.

CHAPTER SUMMARY AND STUDY GUIDE

This chapter focuses on the production of liquid biofuels by means of industrial fermentation. The most successful example, ethanol, has been presented in detail. First-generation ethanol processes have reached worldwide implementation mostly based on feedstocks such as sugarcane and corn. Second-generation processes, which consume nonfood sugar sources such as lignocellulosic material, require a series of chemical, physical, and enzymatic steps for obtaining fermentable sugars. An overview of the current developments in this area is provided. Next, developments in other liquid biofuels such as butanol and diesel-like compound are also described. Finally, a comparison of biofuel routes by industrial fermentation is provided on the basis of stoichiometry and thermodynamic data.

KEY CONCEPTS

Anaerobic fermentation
Yield on substrate
Enzymatic hydrolysis
Metabolic pathways
Cell growth and maintenance
ATP production
Biofuel energy content
Fermentable sugars
Product inhibition
Engineered or recombinant microorganisms

SHORT-ANSWER QUESTIONS

13.1 What crops are typically used as a feedstock for ethanol fermentation?

13.2 What are bagasse and vinasse?

13.3 What is the advantage of producing a liquid biofuel that is immiscible with water?

13.4 Calculate the ethanol production per year from the data given in Table 13.3 or 13.5 on the basis of 167 days of operation per year (24 h per day).

13.5 What is a typical energy content of biofuel produced from a 1 kg glucose by fermentation?

PROBLEMS

13.1 Combine Figures 13.2–13.7 in one drawing.

13.2 Does the material balance of Table 13.4 close?

13.3 Calculate the maximum theoretical yield of ethanol on sucrose in $kg.kg^{-1}$ on the basis of the reaction equations in 13.2.5 (molecular weights ($g.mol^{-1}$): sucrose, 342; ethanol, 46; CO_2, 44), and compare to the actual yield according to Table 13.4.

13.4 The material balance of Table 13.3 does not exactly close, seemingly a small calculation error. But how many kg per day has an unknown source?

13.5 Check the fermentor volume calculation in Section 13.2.7 using Table 13.4.

13.6 Derive the 85,980,300 $ per year revenues of the first-generation bioethanol process from the data given in 13.2.8 and the answer to short question 13.4.

13.7   For pentadecane ($C_{15}H_{32}$, $\Delta_f H^0 = -428.8$ kJ.mol^{-1}), calculate $\Delta_c H^0$, $Y^{max}_{prod/glc}$, and their product. Compare to the values for other biofuels in Table 13.13. What can you conclude?

PROJECTS

P13.1   Make a block scheme for the production of pentadecane using a lignocellulosic feedstock. Estimate the size of the main input and output streams.

REFERENCES

Alberty RA Calculation of standard transformed Gibbs energies and standard transformed enthalpies of biochemical reactants. Archives of Biochemistry and Biophysics 1998;353 (1):116–130.

Beebe H, Coulter KE, Lindsay RA, Baker EM. Equilibria in ethanol–water system at pressures less than atmospheric. Industrial and Engineering Chemistry 1942;34:1501–1504.

Blanch HW. Bioprocessing for biofuels. Current Opinion in Biotechnology 2012;23:390–395.

Chandran SS, Kealey JT, Reeves CD. Microbial production of isoprenoids. Process Biochemistry 2011;46(9):1703–1710.

Domínguez de María P. Recent developments in the biotechnological production of hydrocarbons: paving the way for bio-based platform chemicals. Chemsuschem 2011;4(3):327–329.

Efe Ç, Straathof AJJ, van der Wielen LAM. *Technical and economical feasibility of production of ethanol from sugar cane and sugar cane bagasse*. Delft: Delft University of Technology; 2007. Available at http://repository.tudelft.nl/view/ir/uuid:5f3b7381-0da3-4d26-b334-9b4856ecacda/. Accessed May 30, 2014.

Kalscheuer R, Stölting T, Steinbüchel A. Microdiesel: *Escherichia coli* engineered for fuel production. Microbiology 2006;152:2529–2536.

Lange HC, Heijnen JJ. Statistical reconciliation of the elemental and molecular biomass composition of *Saccharomyces cerevisiae*. Biotechnology and Bioengineering 2001;75(3): 334–344.

Lide DR. *CRC Handbook of Chemistry and Physics*, CRC Press; 2011.

Luo L, van der Voet E, Huppes G. Life cycle assessment and life cycle costing of bioethanol from sugarcane in Brazil. Renewable & Sustainable Energy Reviews 2009;13(6–7): 1613–1619.

Schirmer A, Rude MA, Li X, Popova E, del Cardayre SB. Microbial biosynthesis of alkanes. Science 2010;329(5991):559–562.

Steinbusch KJJ, Hamelers HVM, Plugge CM, Buisman CJN. Biological formation of caproate and caprylate from acetate: fuel and chemical production from low grade biomass. Energy & Environmental Science 2011;4(1):216–224.

Subramaniam R, Dufreche S, Zappi M, Bajpai R. Microbial lipids from renewable resources: production and characterization. Journal of Industrial Microbiology and Biotechnology 2010;37:1271–1287.

US Energy Information Administration (EIA). Annual energy review 2010, Washington, DC: US EIA; 2011.

Van Maris AJA., Abbott DA, Bellissimi E, van den Brink J, Kuyper M, Luttik MA, Wisselink HW, Scheffers WA, van Dijken JP, Pronk JT. Alcoholic fermentation of carbon sources in biomass hydrolysates by *Saccharomyces cerevisiae*: current status. Antonie Van Leeuwenhoek International Journal of General and Molecular Microbiology 2006;90(4):391–418.

Weusthuis RA, Lamot I, van der Oost J, Sanders JPM. Microbial production of bulk chemicals: development of anaerobic processes. Trends in Biotechnology 2011;29(4):153–158.

14

BIOCHEMICAL CONVERSION: ANAEROBIC DIGESTION

Robbert Kleerebezem

Department of Biotechnology, Environmental Biotechnology Group, Faculty of Applied Sciences, Delft University of Technology, Delft, the Netherlands

ACRONYMS

CHO	carbohydrates
CHP	combined heat and power plant
LIP	lipids
MSW	municipal solid waste
NVMG	normal volume of 1 mole of gas
OFMSW	organic fraction of municipal solid waste
P-COD	particulate chemical oxygen demand
PR	protein
VFA	volatile fatty acids

SYMBOLS

COD	chemical oxygen demand	$[g_{O_2} \cdot kg^{-1}]$
E	energy content of gas	$[MJ \cdot m_n^{-3}]$
f_I	fraction of inert substrate	$[-]$
ΔG^0	standard Gibbs free energy	$[kJ \cdot mol^{-1}]$

Biomass as a Sustainable Energy Source for the Future: Fundamentals of Conversion Processes,
First Edition. Edited by Wiebren de Jong and J. Ruud van Ommen.
© 2015 American Institute of Chemical Engineers, Inc. Published 2015 by John Wiley & Sons, Inc.

ΔG^{01}	standard Gibbs free energy, corrected for pH = 7	[kJ per electron] or [kJ per reaction]
K_a	equilibrium constant for acid–base equilibrium	[−]
K_h	equilibrium constant for gas–liquid partitioning	[−]
k_h	hydrolysis rate coefficient	[s^{-1}]
$k_l a$	volumetric gas–liquid mass transfer coefficient	[s^{-1}]
K_S	half-saturation coefficient in the Monod equation	[g$_{\text{S-COD}} \cdot$ L^{-1}]
ODM	organic dry matter content	[g$_{\text{ODM}} \cdot$ kg^{-1}]
P	particulate substrate concentration	[g$_{\text{P-COD}} \cdot$ L^{-1}]
P	energy production	[MW]
q_S	actual biomass-specific substrate uptake rate	[g$_{\text{S-COD}} \cdot$ g$_{\text{X-COD}}^{-1} \cdot$ h^{-1}]
q_S^{max}	maximum biomass-specific substrate uptake rate	[g$_{\text{S-COD}} \cdot$ g$_{\text{X-COD}}^{-1} \cdot$ h^{-1}]
r_h	rate of hydrolysis	[g$_{\text{P-COD}} \cdot$ L$^{-1} \cdot$ h^{-1}]
r_{met}	rate of methanogenesis	[g$_{\text{S-COD}} \cdot$ g$_{\text{X-COD}}^{-1} \cdot$ h^{-1}]
R	recirculation ratio	[−]
S	soluble substrate concentration	[g$_{\text{S-COD}} \cdot$ L^{-1}]
TDM	total dry matter	[g$_{\text{TDM}} \cdot$ kg^{-1}]
X	concentration of methanogenic biomass responsible for converting VFA to biogas	[g$_{\text{X-COD}} \cdot$ L^{-1}]
Y	stoichiometric factor	[typically g$_{\text{COD}} \cdot$ g$_{\text{COD}}^{-1}$]
γ	oxidation state of the substrate	[mol$_e \cdot$ mol$_C^{-1}$]
η	fraction of a given compound in substrate mixture	[−]
λ_{CH_4}	stoichiometric coefficients of biomass degradation to CH$_4$	[−]
λ_{CO_2}	stoichiometric coefficients of biomass degradation to CO$_2$	[−]

14.1 INTRODUCTION

14.1.1 What Is Anaerobic Digestion?

Anaerobic digestion is a biotechnological process by which a complex organic feedstock is first converted into a range of simpler water-soluble organic compounds that are subsequently converted into methane-containing biogas. The process is anaerobic (literally without air), meaning here that it occurs in absence of oxygen or any other external electron acceptor like nitrate or sulfate. In nature, anaerobic digestion occurs in environments rich in organic carbon and limited by input of electron acceptors and/or energy sources such as light. Example natural environments include wetlands, rice paddies, and the rumen of animals and insects.

One of the main conceptual advantages of anaerobic digestion as bioenergy-producing bioprocess is the gaseous end product that evidently implies directly *in situ* product separation. The biogas generated typically consists of 50–70 vol.% methane and 30–50 vol.% carbon dioxide, as well as minor amounts of water, molecular hydrogen, and hydrogen sulfide. Depending on the actual biogas composition, it can both be applied directly for heat and electricity generation in a combined heat and power plant (CHP) or upgraded to natural gas quality (green gas) by removal of impurities from the biogas in order to obtain a methane concentration of at least 90 vol.%. The anaerobic digestion process typically is applied for treatment of heterogeneous organic residues with a relatively high water content (>70 vol.%) and a high fraction biodegradable matter such as the organic fraction of municipal solid waste (OFMSW), manure, and numerous agro-industrial residues. Technologies applied for anaerobic wastewater treatment are strongly different from those implemented for treatment of streams with a high solid content and are therefore excluded from this chapter. For an introduction into anaerobic wastewater treatment, see, for instance, the review by van Lier et al. (2008).

14.1.2 Anaerobic Digestion as Bioenergy Process

In many cases, bioenergy production is not the only driver of the anaerobic digestion process. This is partly due to the low price of methane compared to the high costs for handling and processing organic residues. To date, cost-effective production of methane-containing biogas is only possible in highly specific situations or when long-term subsidies are available like in Germany and Sweden.

The main driver of the anaerobic digestion process is waste management through reduction of the amount of waste and waste stabilization. An example application where sludge volume reduction is an important argument for application of anaerobic digestion is the treatment of sludge generated in sewage treatment plants. The nonbiodegradable fraction of sewage sludge typically is incinerated at centralized incineration plants (in the Netherlands), but the sludge volume reduction achieved by anaerobic digestion has an important impact on the transportation and processing costs. In this respect, also the enhanced dewaterability due to anaerobic digestion of sludge contributes to a reduction of transportation costs and an increase of the heating value upon incineration of the sludge.

Another argument for implementation of the anaerobic digestion process is *stabilization* of the organic material. Stabilization consists of the mineralization of biodegradable organic material, which is required for reuse of digested material on agricultural land. Anaerobic digestion of the OFMSW is often combined with aerobic composting of the material left over after anaerobic digestion.

14.2 BIOCHEMICAL FUNDAMENTALS

14.2.1 Degradation Pathway

In all nonphototrophic microbial systems microorganisms rely on the catalysis of redox reactions for energy generation and growth. In the absence of electron acceptors, microorganisms depend on fermentation reactions in which the organic substrate

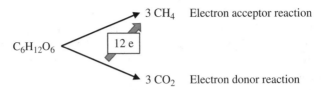

FIGURE 14.1 Electron transfer during glucose fermentation to a mixture of methane and carbon dioxide.

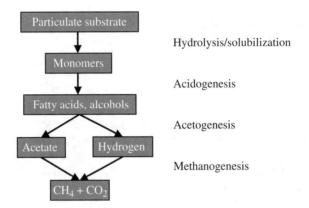

FIGURE 14.2 Simplified scheme of the sequence of conversion reactions occurring during the anaerobic digestion process.

is both electron donor and acceptor of the chemical redox reaction. As an example, the anaerobic degradation of glucose to a mixture of methane (CH_4) and carbon dioxide (CO_2) in terms of an electron donor and acceptor reaction is shown in Figure 14.1. From the six carbon atoms in glucose, three are oxidized to carbon dioxide, and the electrons are donated to the other three carbon atoms that are reduced to methane.

As opposed to most biochemical degradation schemes, anaerobic digestion requires the activity of a range of specialized microorganisms to catalyze the overall fermentation process. Complex organic matter is degraded in a sequence of degradation reactions to methane-containing biogas according to the scheme shown in Figure 14.2. Polymeric (particulate) organic material is first hydrolyzed and solubilized. This reaction typically is catalyzed by extracellular enzymes produced by acidogenic bacteria. Polymeric substrates may be proteins, polymeric carbohydrates, or lipids. Hydrolysis products of these substrates are amino acids, sugar monomers, and long-chain fatty acids and glycerol, respectively. The monomers are subsequently fermented in the acidogenesis process, and volatile fatty acids (VFA) (acetic, propionic, and butyric acid), carbon dioxide, hydrogen, and/or alcohols or other short-chain organic acids are produced. Amino acid degradation is accompanied by the release of ammonium in solution. The intermediately formed VFA and other small organic molecules are converted to the precursors for methane production by acetogenic bacteria: acetates and molecular hydrogen.

Acetates and molecular hydrogen (with carbon dioxide) are the main substrates for methanogenic archaea that are responsible for methane production. Archaea are a distinct group in the evolutionary tree of life and are positioned between prokaryotes (bacteria) and eukaryotes (plants and animals). The range of substrates that can be used by methanogenic archaea is limited to carbon compounds with no more than two carbon atoms, explaining the sequence of reactions required to establish the anaerobic digestion process. It should be noted that actual removal of organic carbon from the solid/liquid system is only achieved in the final step of the anaerobic digestion process. The other steps in the process have no other function than to generate the precursors for methane production.

14.2.2 Chemical Substrate Characterization

The overall stoichiometry of the anaerobic digestion can be estimated from the elemental composition of the biodegradable organic substrate. Assuming an elemental composition of the substrate of $C_cH_hO_oN_n$, stoichiometric coefficients can be estimated for the end products of the anaerobic digestion process: methane (CH_4), carbon dioxide (CO_2), water (H_2O), and ammonium bicarbonate (NH_4HCO_3). A solution is obtained by setting up elemental balances for C, H, O, and N:

$$C_cH_hO_oN_n \rightarrow \left(\frac{c}{2} + \frac{h}{8} - \frac{3n}{8} - \frac{o}{4}\right)CH_4 + \left(\frac{c}{2} - \frac{h}{8} - \frac{5n}{8} + \frac{o}{4}\right)CO_2$$

$$+ n\,NH_4HCO_3 + \left(\frac{h}{4} - c - \frac{7n}{4} + \frac{o}{2}\right)H_2O \qquad \text{(RX. 14.1)}$$

Evidently, more reduced substrates result in a higher CH_4:CO_2 ratio in the biogas produced. A higher nitrogen content in the substrate also results in a higher production of ammonium bicarbonate and, consequently, a (slightly) lower carbon dioxide content in the biogas. More importantly, ammonium bicarbonate production results in the production of pH buffer capacity. It should be noted that not only ammonium production from organic carbon results in bicarbonate production but also other forms of alkalinity result in the formation of bicarbonate, which is essential to maintain an approximately neutral pH value (pH ~7) as strictly required for a stable anaerobic digestion process.

Alkalinity can be defined as the stoichiometric sum of the bases in solution. It is typically expressed in base equivalents per unit volume (eq.L^{-1}). In the anaerobic digestion process, bicarbonate–carbonate alkalinity is the most important form of alkalinity, and the bicarbonate–carbon dioxide chemical equilibrium typically determines the pH of the system. Neutralized organic acids such as common intermediates in the anaerobic digestion process, e.g., acetate, propionate, and butyrate, also contribute to the substrate alkalinity. This is due to the fact that upon degradation of a neutralized organic acid, the cation responsible for neutralization receives a bicarbonate molecule as counterion, herewith contributing to the pH buffer capacity of the system. Also, other forms of alkalinity may play an important role in the anaerobic digestion process, e.g., phosphates or precipitates of cations like calcium or magnesium that upon solubilization at decreasing pH values will contribute to the pH buffer capacity of the system. Figure 14.3 (left) shows an overview of the main pH-determining reactions occurring in the anaerobic digestion process.

Figure 14.3 (right) shows a graph of the pH and biogas composition upon degradation of 50 mM glucose to biogas as a function of the alkalinity in the system. The graph shows that at increasing alkalinity the pH increases. This is due to two reasons: first, the alkalinity results in elevated bicarbonate concentrations in the system, and second, due to bicarbonate formation, the carbon dioxide partial pressure in the biogas decreases. In the bicarbonate–carbon dioxide pH buffering system, bicarbonate represents the base and carbon dioxide is the acid, and more base and less acid evidently mean an increasing pH. It is furthermore evident that at high alkalinity values ($> \sim 50$ mM), the pH in the anaerobic digestion process is strongly buffered, and pH instabilities due to instabilities in biological acid or base production will not result in strong variations in the pH. A limitation of strong pH buffering is that problems with the process performance can hardly be observed from changes in the pH at high alkalinity concentrations.

To predict the biogas composition from substrate characteristics, the elemental composition needs to be determined. However, this is not a standard measurement for heterogeneous substrates typically used for the anaerobic digestion process. The elemental substrate composition can be estimated from three standard measurements

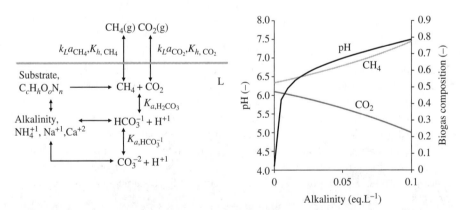

FIGURE 14.3   Schematic overview of the relation between substrate composition, alkalinity, biogas composition, and pH in the anaerobic digestion process. $k_L a$ is the volumetric gas–liquid mass transfer coefficient, whereas K_h and K_a are thermodynamic equilibrium constants for gas–liquid partitioning and acid–base equilibrium, respectively. The graph on the right shows the pH and biogas composition upon degradation of 50 mM glucose to methane and inorganic carbon as a function of the alkalinity concentration in the system.

that are routinely conducted on organic substrates (Kleerebezem and Van Loos-drecht, 2006):

- Chemical oxygen demand (COD, $g_{O_2} \cdot kg^{-1}$), corresponding to the amount of oxygen equivalents required to oxidize all organic carbon to carbon dioxide.
- Organic dry matter content (ODM, $g_{ODM} \cdot kg^{-1}$), typically corresponding to 50 – 90 wt% of the total dry matter (TDM, $g_{TDM} \cdot kg^{-1}$). The difference between the TDM and the ODM values is the inorganic ash concentration.
- Organic nitrogen content (N_{org}, $g_N \cdot kg^{-1}$), typically estimated from the total concentration of reduced nitrogen compounds (so-called *Kjeldahl* nitrogen) minus the concentration of ammonium (NH_4^{+1}).

Three basic equations can be established that allow for identification of the elemental composition of the organic feedstock. First of all, the COD value of organic matter can be identified using the combustion equation, which in general terms can be expressed as

$$C_c H_h O_o N_n + x O_2 \rightarrow y H_2 O + z CO_2 + w NH_3 \qquad \text{(RX. 14.2)}$$

By solving the elemental balances for C, H, O, and N, we can identify the stoichiometric coefficients x, y, z, and w. The stoichiometric coefficient $x = (4 + h - 2o - 3n)/4$ represents the number of oxygen molecules required to combust one mole of organic carbon ($c = 1$). Using the molar weights or the various compounds (e.g., 32 g.mol^{-1} for O_2) consequently, the COD/ODM ratio ($g_{O_2} \cdot g_{ODM}^{-1}$) can be defined as

$$\frac{\text{COD}}{\text{ODM}} = \frac{32(4 + h - 2o - 3n)/4}{(12 + h + 16o + 14n)} \qquad \text{(Eq. 14.1)}$$

From the organic nitrogen measurement, an equation can be derived for the stoichiometric coefficient n:

$$n = \frac{N_{org}(12 + h + 16o + 14n)}{14 \text{ODM}} \qquad \text{(Eq. 14.2)}$$

The third equation is based on the fact that carbon is always tetravalent, and the total number of its bonds will be 4. Assuming C–H, C = O, and C–NH$_2$ as carbon bonds, the following equation can be derived:

$$4 = h + 2o - n \qquad \text{(Eq. 14.3)}$$

These three equations can be solved to obtain the elemental composition of the organic carbon compound (with $c = 1$):

$$h = \frac{308 \, \text{COD} + 704 \, N_{org}}{49 \, \text{COD} - 64 \, N_{org} + 112 \, \text{ODM}} \qquad \text{(Eq. 14.4)}$$

$$o = \frac{224\,\text{ODM} - 56\,\text{COD} - 304\,N_{\text{org}}}{49\,\text{COD} - 64\,N_{\text{org}} + 112\,\text{ODM}} \qquad \text{(Eq. 14.5)}$$

$$n = \frac{352\,N_{\text{org}}}{49\,\text{COD} - 64\,N_{\text{org}} + 112\,\text{ODM}} \qquad \text{(Eq. 14.6)}$$

From the stoichiometric coefficients of organic feedstock degradation to methane (λ_{CH_4}) and carbon dioxide (λ_{CO_2}) in the overall stoichiometric equation for anaerobic digestion, the biogas composition can be estimated:

$$\%\text{CH}_4 = \frac{\lambda_{\text{CH}_4}}{\lambda_{\text{CH}_4} + \lambda_{\text{CO}_2}} \cdot 100\% \quad \%\text{CO}_2 = \frac{\lambda_{\text{CO}_2}}{\lambda_{\text{CH}_4} + \lambda_{\text{CO}_2}} \cdot 100\% \qquad \text{(Eq. 14.7)}$$

The stoichiometric equation furthermore allows for calculation of the biogas production upon full organic carbon degradation, using the definition of the normal volume of 1 mole of gas (NVMG ~22.4 $l\cdot mol^{-1}$):

$$\text{Biogas}\left(L_n.\text{kg}^{-1}\right) = \frac{\text{ODM}}{12 + h + 16o + 14n}(1 - n)\text{NVMG} \qquad \text{(Eq. 14.8)}$$

Besides the derivation of the elemental composition of the organic substrate, these basic three measurements also enable identification of the composition of the substrate in terms of the principal building blocks, i.e., proteins, carbohydrates, and lipids, as outlined in Table 14.1.

From the elemental composition derived earlier, the carbon mole fraction (η) of CHO, PR, and LIP in the organic substrate can readily be calculated from the elemental compositions shown in Table 14.1. The derivation of the equation for the PR fraction is straightforward since PR is the only nitrogen-containing compound assumed and VFA is the only charged compound considered. Therefore, the fraction of proteins (η_{PR}) in the waste follows directly from the N-content of

TABLE 14.1 Principal compounds considered as general organic feedstock constituents

Compounds[a]	Abbr.	Chemical composition	Oxidation state[b] (γ) $\left(\text{mol}_e\cdot\text{mol}_C^{-1}\right)$	N-content $\left(\text{mol}_N\cdot\text{mol}_C^{-1}\right)$
Proteins	PR	$C_1H_{2.52}O_{0.87}N_{0.26}$	4.0	0.26
Carbohydrates	CHO	$C_1H_2O_1$	4.0	0.0
Lipids	LIP	$C_1H_{2.85}O_{0.575}$	5.7	0.0

The elemental composition of proteins is based on the generalized amino acid composition of proteins as proposed by Batstone et al. (2002). Carbohydrates are based on the composition of glucose $C_6H_{12}O_6$.
[a] The model lipid chosen is composed of one glycol molecule and three n-palmitic acid side chains, resulting in $C_{51}H_{98}O_6$ as elemental composition.
[b] The oxidation state of the substrate is defined as the number of electrons liberated per C-mol substrate upon full oxidation to carbon dioxide, equivalent to the COD definition described earlier (Heijnen and Kleerebezem, 2010).

the organic substrate (n) and the N-content of the standard amino acid composition of the protein chosen (N_{PR}):

$$\eta_{PR} = \frac{n}{N_{PR}} \qquad \text{(Eq. 14.9)}$$

The fraction of LIP (η_{LIP}) can be calculated from the different oxidation states of lipids (γ_{LIP}) compared to the other organic substrates defined:

$$\eta_{LIP} = \frac{h - 2o - 3n}{\gamma_{LIP} - 4} \qquad \text{(Eq. 14.10)}$$

The fraction of carbohydrates subsequently follows from the total balance:

$$\eta_{CHO} = 1 - \eta_{PR} - \eta_{LIP} \qquad \text{(Eq. 14.11)}$$

One should realize that assumptions have been made while deriving these equations and results should therefore be handled with care. The calculated fraction of lipids, e.g., may be overestimated in the presence of other reduced organic compounds like humic substances or alcohols. The estimation of the fraction of lipid compounds versus carbohydrates has furthermore shown to be strongly susceptible to measurement errors in the COD/TOC ratio (Heijnen and Kleerebezem, 2010).

Figure 14.4 shows an example that demonstrates how a few simple waste measurements allow for feedstock characterization and prediction of the biogas composition.

14.2.3 Factors Affecting the Process

14.2.3.1 Electron Acceptors The presence of electron acceptors in the anaerobic digestion process may give rise to the generation of side products in the process. Due to its poor solubility, dissolved oxygen ($\sim 8 \; mg_{O_2} \cdot L^{-1}$) will hardly affect the anaerobic digestion process since the electron equivalent amount of organic carbon is many orders of magnitude higher than the maximum oxygen solubility. Oxidized forms of nitrogen (nitrite or nitrate) are only encountered in environments that have been exposed to aerobic conditions. This normally means that most organic carbon is degraded as well and therefore a mixture of organic carbon and strong inorganic electron acceptors is unlikely to be present in a single substrate.

Sulfate is often present in significant concentrations in substrates for anaerobic digesters. Since sulfate is preferred as electron acceptor for organic carbon oxidation in the anaerobic digestion process, it will normally be reduced to hydrogen sulfide, which at approximately neutral pH will end up in the biogas. The hydrogen sulfide concentration in the biogas can be estimated by including sulfate reduction in the stoichiometric equations derived earlier (RX. 14.1). Typical hydrogen sulfide concentrations in biogas range from 100 ppm to 1 or 2 vol.%. A more detailed description of the sulfate reduction process during anaerobic digestion can be found elsewhere

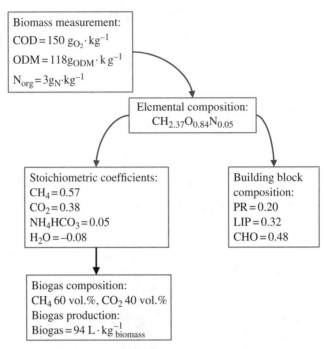

FIGURE 14.4 Calculation example demonstrating the characterization of a wet organic substrate in terms of its elemental composition, building block composition, and end product formation stoichiometry, enabling the estimation of biogas production and composition. Full biodegradation is assumed, but a correction for the actual biodegradable fraction can readily be implemented.

(van Lier et al., 2008). In the section on *biogas upgrading and utilization*, it is shown that hydrogen sulfide removal from biogas is required to avoid sulfuric acid production and corrosion in a biogas burner.

14.2.3.2 Temperature Three major temperature operating ranges can be distinguished in anaerobic digestion. These are psychrophilic (4–15°C), mesophilic (20–40°C), and thermophilic (45–65°C) temperatures. While reactors can operate between these ranges effectively, optimal temperatures for mesophilic and thermophilic organisms are approximately 35 and 55°C, respectively.

The influence of the temperature on biochemical systems can be summarized as follows:

- Increase in reaction rates with increasing temperature as predicted by the Arrhenius equation
- Rapid decrease in reaction rate with increasing temperature above the optimum temperature (>40°C for mesophilic and >65°C for thermophilic systems)
- Increase in biochemical energy requirements for maintenance purposes with increasing temperature, resulting in lower biomass yield on substrate

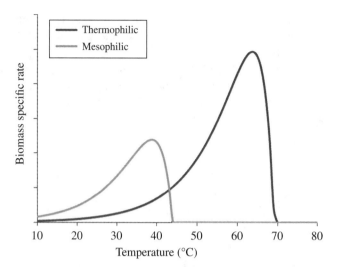

FIGURE 14.5 Biomass-specific substrate conversion rates (q-rates) as a function of temperature. Two types of microorganisms with a different optimum temperature can be identified: mesophilic (38°C) and thermophilic (65°C).

The influence of the temperature on biomass-specific conversion rates (q-rates, $g_{substrate} \cdot g_{biomass}^{-1} \cdot h^{-1}$) is shown in Figure 14.5. The figure shows that below the optimum temperature of both mesophilic and thermophilic microorganisms, the q-rates increase with increasing temperature. At temperatures exceeding the optimum temperature, the q-rates rapidly decrease. Figure 14.5 also shows that in the mesophilic temperature range, the thermophilic microorganisms have lower q-rates than mesophilic microorganisms, explaining the dominance of mesophilic microorganisms at moderate temperatures. Furthermore, it is evident that the maximum q-rates as established at the optimum temperature are significantly higher for thermophilic than for mesophilic microorganisms.

Compared to aerobic (combustion) processes, the effect of microbial metabolic activity on the temperature of the process is limited because the enthalpy effects of anaerobic digestion are small and normally the temperature increase due to biological activity is negligible. Furthermore, biogas produced during anaerobic digestion is saturated with water resulting in a decrease of the temperature of the bioreactor. Overall, this means that for most substrates the bioreactor (or the substrate) needs to be heated in order to maintain the mesophilic or thermophilic temperatures required.

In practice, anaerobic digesters are operated at both mesophilic and thermophilic temperatures. Retention times typically are shorter in thermophilic digesters (10–20 days) than in mesophilic digesters (20–40 days) due to the higher biomass-specific conversion rates, reducing the bioreactor volume required. For hygienization of the digestate of anaerobic digesters, as required for reuse of the material in agriculture, the feedstock needs to be heated to 70°C and kept at that temperature for a short period of time (10–30 min). Depending on country-dependent legislation, hygienization is required to minimize the number of pathogenic bacteria that may originate from the feedstock (like sewage sludge and pig manure) and thrive at around 37°C.

14.2.3.3 Potential Process Inhibition Inhibition of the anaerobic digestion process may occur due to accumulation of inhibitors during the process. Typically, the acetate-degrading methanogens (see Figure 14.2) are most strongly inhibited in the process, due to the low Gibbs free energy yield of the process they catalyze. Unfavorable process conditions such as the accumulation of inhibitors require expenditure of metabolic energy by the methanogens. Herewith, less metabolic energy is available for growth and maintaining cellular homeostasis resulting in inhibition.

The toxicity of many compounds in anaerobic digestion is strictly related to the operational pH. This is due to the fact that undissociated weak acids and bases can penetrate the cell membrane of microorganisms and dissociate after entering the cytoplasm, resulting in interference with the membrane potential, which is essential for cellular homeostasis. A graph demonstrating the undissociated fractions of common inhibitors in the anaerobic digestion process as a function of the pH is shown in Figure 14.6.

Acetic acid is the main precursor for methane production, and elevated acetic acid (and propionic and butyric acid, in general named VFA) concentrations can be obtained with readily degradable substrates when the rates of the first steps in the degradation scheme shown in Figure 14.2 exceed the methanogenic capacity of the biomass. In that case, VFA will accumulate and the pH will decrease. Both the increase in VFA concentration and the lower pH will increase the concentration of undissociated acetic acid, amplifying the toxicity effect of VFA in the system. The toxicity of hydrogen sulfide is related to the ratio of sulfate to organic carbon in the substrate. Elevated ratios of sulfate to organic carbon result in a high partial pressure of hydrogen sulfide in the biogas, and at low pH values, significant inhibition can be the result. Ammonia inhibition occurs during anaerobic digestion of PR-rich substrates (like manure) at high pH values.

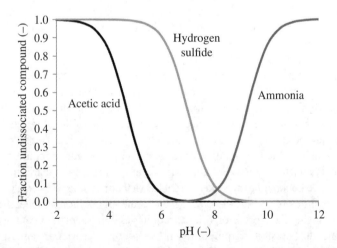

FIGURE 14.6 pH dependency of the concentration of undissociated weak acids and bases as encountered in the anaerobic digestion process.

14.3 THERMODYNAMIC FUNDAMENTALS

The microbial conversion of organic substrates to methane-containing biogas is based on a firm thermodynamic foundation. In a thermodynamically closed system, redox reactions are driven in the direction of the thermodynamic equilibrium state. This can most easily be explained by identifying the half reactions that describe organic carbon oxidation to the most oxidized state of carbon (carbon dioxide) and comparing the Gibbs free energy change per electron of the different reactions. Four example half reactions are shown below:

$$\text{Glucose}: C_6H_{12}O_6 + 6H_2O \rightarrow 6CO_2 + 24H^{+1} + 24e^{-1}$$
$$\Delta G^{01} = -40.9 \text{ kJ per electron} \qquad \text{(RX. 14.3)}$$

$$\text{Lactate}: C_3H_5O_3^{-1} + 3H_2O \rightarrow 6CO_2 + 11H^{+1} + 12e^{-1}$$
$$\Delta G^{01} = -32.8 \text{ kJ per electron} \qquad \text{(RX. 14.4)}$$

$$\text{Acetate}: C_2H_3O_2^{-1} + 2H_2O \rightarrow 2CO_2 + 7H^{+1} + 8e^{-1}$$
$$\Delta G^{01} = -28.1 \text{ kJ per electron} \qquad \text{(RX. 14.5)}$$

$$\text{Methane}: CH_4 + 2H_2O \rightarrow CO_2 + 8H^{+1} + 8e^{-1}$$
$$\Delta G^{01} = -23.6 \text{ kJ per electron} \qquad \text{(RX. 14.6)}$$

Organic carbon is a strong electron donor, suggesting that upon oxidation of organic compounds, Gibbs energy is generated as reflected in the negative ΔG^{01} values shown. In the absence of an external electron acceptor, the oxidation of organic carbon needs to be combined with the reduction of carbon dioxide to another organic compound to close the electron balance. Evidently, the reduction of carbon dioxide to organic carbon is thermodynamically unfavorable (endergonic). The overall redox reaction is thermodynamically only favorable if the Gibbs energy content per electron decreases upon conversion of one organic compound into another. Only thermodynamically favorable (exergonic) redox reactions can sustain growth and maintenance of a microbial ecosystem. Therefore, the four compounds shown in RX. 14.3–14.6 will be degraded through the sequence shown in Figure 14.7.

In more generalized terms, it can be stated that in a thermodynamically closed microbial system (no input of energy source and electron acceptor), organic carbon will be converted to the compound with the lowest energy content per electron. It is therefore no surprise that when we compare the energy content per electron of all organic molecules, the lowest value is obtained for gaseous methane (see Figure 14.8). There is (of course!) one exception to this firm rule, and that is elemental carbon (graphite), which has a Gibbs energy content per electron even lower than methane, suggesting that (microbial) conversion of methane to graphite is exergonic, but to date, no microorganism has been identified that is capable of catalyzing this reaction.

The thermodynamic basis of the anaerobic digestion process makes it particularly suitable for bioenergy production from heterogeneous substrates. Independent of the origin and composition of the substrate, the end product of the microbial degradation

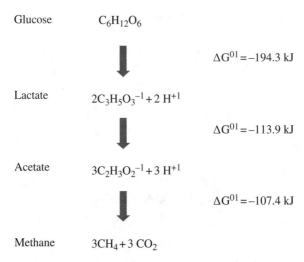

Glucose $\quad C_6H_{12}O_6$

$\Delta G^{01} = -194.3$ kJ

Lactate $\quad 2C_3H_5O_3^{-1} + 2\,H^{+1}$

$\Delta G^{01} = -113.9$ kJ

Acetate $\quad 3C_2H_3O_2^{-1} + 3\,H^{+1}$

$\Delta G^{01} = -107.4$ kJ

Methane $\quad 3CH_4 + 3\,CO_2$

FIGURE 14.7 Thermodynamic properties of sequential substrate degradation in anaerobic digestion. The values of ΔG^{01} are for the reactions as shown.

process will always be methane-containing biogas. It should be noted that by definition all other biotechnological conversion processes lack this principal advantage. Fermentation for the production of ethanol, e.g. (see Chapter 13), can only be achieved because high ethanol concentrations are strongly inhibitory to methanogenic archaea, and therefore, biogas production can effectively be avoided in the bioethanol process.

Another implication of the decreasing Gibbs energy content per electron during anaerobic digestion is that upon conversion to biogas, the energy content is lowered. Even though this is theoretically true, the extent at which the energy yield decreases by conversion is limited because energy generation upon combustion of methane is dominated by the energy content of the electron acceptor during combustion, molecular oxygen:

$$O_2 + 4H^{+1} + 4e^{-1} \rightarrow 2H_2O$$
$$\Delta G^{01} = -78.7 \text{ kJ per electron} \qquad \text{(RX. 14.7)}$$

14.4 PROCESS ENGINEERING

14.4.1 Stoichiometry and Kinetics

To enable rational bioreactor engineering, it is crucial to describe the anaerobic digestion process in terms of reaction stoichiometry and kinetics. In the past decades, a range of mathematic models have been developed that aim to describe the anaerobic digestion process. Most of these models follow the scheme shown in Figure 14.2 and describe the processes catalyzed by the different types of microorganisms. Typically,

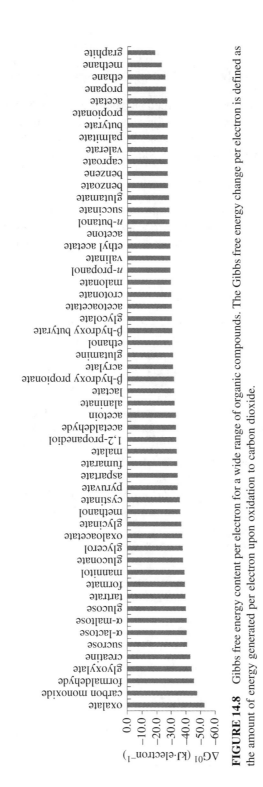

FIGURE 14.8 Gibbs free energy content per electron for a wide range of organic compounds. The Gibbs free energy change per electron is defined as the amount of energy generated per electron upon oxidation to carbon dioxide.

these models are very complex with six to eight groups of microorganisms that are characterized with at least four or five kinetic and stoichiometric parameters for each group of microorganisms.

Here, we describe a strongly simplified version of the anaerobic digestion model that focuses on the most common rate-limiting steps in the process. For anaerobic digestion of particulate substrates like those considered here, there are two potentially rate-limiting steps in the overall process and one boundary condition that need to be fulfilled.

14.4.1.1 *Rate-Limiting Step 1*

Hydrolysis of complex particulate organic matter (first step in Figure 14.2), typically described as a first-order process in the concentration of the biodegradable particulate substrate (P, $g_{P\text{-COD}} \cdot L^{-1}$) under formation of a soluble substrate (S, $g_{S\text{-COD}} \cdot L^{-1}$):

$$r_h = k_h P \qquad \text{(Eq. 14.12)}$$

This process step is generally catalyzed by extracellular enzymes that are excreted by bacteria growing on the soluble substrate. Hydrolysis is lumped with the (non-rate-limiting) second step in Figure 14.2, the microbial growth reaction (acidogenesis) that converts monomeric carbohydrates, amino acids, and fatty acids to mainly VFA.

14.4.1.2 *Rate-Limiting Step 2*

Methanogenesis of VFA, which includes the acetogenesis and methanogenesis steps (last two steps in Figure 14.2). This process is described as a typical microbial growth process, in which the microbial substrate conversion is a first-order process in the concentration of biomass (X, $g_{X\text{-COD}} \cdot L^{-1}$). It also depends on the actual biomass-specific uptake rate (q_S, $g_{S\text{-COD}} \cdot g_{X\text{-COD}}^{-1} \cdot h^{-1}$), which in turn is a function of the maximum biomass-specific substrate uptake rate (q_S^{max}, $g_{S\text{-COD}} \cdot g_{X\text{-COD}}^{-1} \cdot h^{-1}$) and the substrate concentration (S, $g_{S\text{-COD}} \cdot L^{-1}$) in a so-called Monod term, $S/(K_S + S)$.

The rate of this step can be expressed as

$$r_{met} = q_S = q_S^{max} \frac{S}{K_S + S} X \qquad \text{(Eq. 14.13)}$$

14.4.1.3 *Boundary Condition*

A certain fraction of the particulate substrate is not susceptible to biodegradation and is therefore referred to as the fraction of inert substrate (f_I). The initial concentration of biodegradable substrate (P_0, $g_{P\text{-COD}} \cdot L^{-1}$) can be calculated from the total particulate substrate concentration (P_{T0}, $g_{P\text{-COD}} \cdot L^{-1}$) using $P_0 = (1 - f_I) P_{T0}$.

The resulting process stoichiometry and kinetics are shown in Table 14.2. In Section 14.4.2, this process model is implemented in a bioreactor model to clarify the impact of the two rate-limiting steps on the overall design of an anaerobic digester.

TABLE 14.2   COD-based stoichiometry and kinetics of the two-step anaerobic digestion process (hydrolysis and methanogenesis)

Step	P	S	X	CH$_4$	Rate equation
Hydrolysis	−1	1	0	0	$k_h P$
Methanogenesis	0	−1	Y_X	Y_{CH_4}	$q_S^{max} \frac{S}{K_S + S} X$

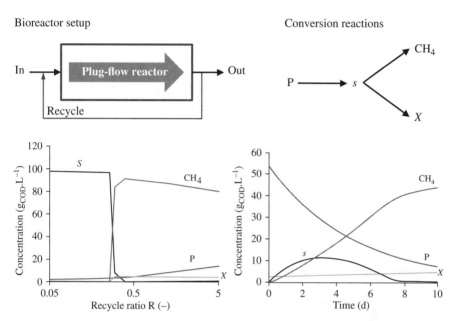

FIGURE 14.9   Impact of recirculation on a plug flow anaerobic digester. Pictures show the bioreactor with variable recirculation flow considered and the conversion reactions assumed. The left graph shows the effluent concentrations of particulate substrate (P), soluble intermediate (S), and the end products methane (CH$_4$) and methanogenic biomass (X) as a function of the recirculation ratio R. The right graph shows the concentration profile over the reactor at a recirculation ratio of 1.

14.4.2   Bioreactor Design

14.4.2.1   Reactors   The two-step process proposed to describe the anaerobic digestion process is implemented in a bioreactor model. The two main bioreactor configurations are the plug flow reactor (PFR) and the continuous stirred tank reactor (CSTR). In a PFR, mixing is assumed to be limited to the radial direction, resulting in a concentration gradient of the substrate, product, and biomass along the length of the reactor (see Figure 14.9). In a CSTR, concentrations are homogeneously distributed over the reactor volume, and the concentrations inside the reactor equal the effluent concentrations. Effluent recirculation from a PFR results in a hybrid between a PFR and a CSTR.

Derivation of the mass balances and their solution are outside the scope of this chapter, but a qualitative idea of the impact of the reactor design can be obtained from

Figure 14.9. The left graph in Figure 14.9 depicts the effluent composition of a PFR with recirculation as a function of the recirculation ratio (R) imposed. R is defined as the ratio of the recirculation flow rate to the influent flow rate. The influent of the reactor is assumed to be composed of 100 $g_{P\text{-}COD} \cdot L^{-1}$, and the retention time in the system is set to 20 days.

From Figure 14.9, it can readily be seen that at low values of R ($R < 0.3$), only hydrolysis occurs, and P is converted to S. This is due to the fact that a PFR with no recirculation cannot sustain a microbial growth process because all microorganisms in the system are washed out if the influent does not contain the required microorganism. For batch processes, this is probably more easy to imagine: if in a batch process no microorganisms are supplied (no recirculation), no microbial growth will occur, so no methane-containing biogas will be produced.

The extent of particulate substrate conversion is higher in a PFR than in a CSTR. This is due to the particulate substrate (P) gradient in a PFR, resulting in a higher average substrate concentration and thus a higher average hydrolysis rate compared to that in a CSTR. Consequently, the extent of particulate substrate (P) conversion in a PFR decreases with increasing recirculation.

Evidently, the theoretically optimal reactor configuration is a PFR reactor with a recirculation rate that is adequate to maintain the microorganisms that are catalyzing the second step in the anaerobic digestion model proposed. In the specific case described here, this implies a recirculation ratio of approximately 0.5.

14.4.2.2 Wet versus Dry Digestion Commercial forms of anaerobic digestion are available for treatment of both substrates with a high solid content (30–40 wt% TDM, dry anaerobic digestion) and substrates with a lower solid content (<30 wt%, wet anaerobic digestion). Wet anaerobic digestion is also implemented for dry substrates by recycling of the water after filtering of the digestate. An evident advantage of wet anaerobic digestion is easier handling (i.e., pumping and mixing) of the slurry, but a major disadvantage is the requirement for dewatering after anaerobic digestion and treatment of the liquid residue.

Dry anaerobic digestion typically is conducted in plug flow bioreactors with recirculation. The orientation of the reactor can be vertical like in the Dranco process or horizontal like in the Kompogas process. Schematics of both types of dry anaerobic digesters are shown in Figure 14.10. Both the Dranco and Kompogas processes are operated at thermophilic temperatures (~60°C) and at a retention time of approximately 20–25 days, resulting in a volumetric loading rate of approximately 10 kg ODM.m^{-3}.d^{-1}. Reported biogas yields in the process are 100–200 m^3 per tonne of waste processed.

Wet anaerobic digestion typically is conducted in CSTRs. Sometimes, a limited degree of plug flow is established by placement of two bioreactors in series. Commercial forms of the wet anaerobic digestion process are the Citec process and the Biogen Greenfinch process. Both processes are operated at mesophilic temperatures. Reported retention times range from 20 to 100 days. Evidently, the organic loading rates at comparable retention times are two to three times lower for wet anaerobic digestion processes compared to dry anaerobic digesters due to the higher water

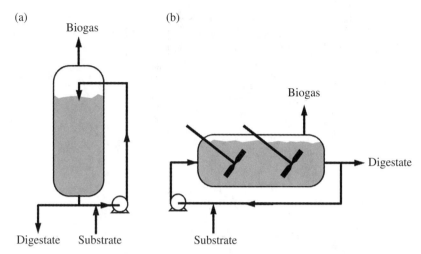

FIGURE 14.10 Vertical Dranco dry anaerobic digestion bioreactor (a) and horizontal Kompogas dry anaerobic digestion bioreactor (b).

content, and consequently, these bioreactors are two to three times larger than the reactors used in dry anaerobic digestion for treatment of the same amount of substrate.

14.4.2.3 Substrate Properties A wide range of substrates are currently used in the anaerobic digestion process, and they exhibit a big variation in biogas potential. Table 14.3 provides indicative values for some feedstocks for the anaerobic digestion process. The different substrates vary strongly in terms of biodegradability, hydrolysis kinetics, and biogas potential. Pig manure, e.g., originates from a highly efficient anaerobic digester—the intestinal tract of the pig—and therefore, its biodegradability is limited. Furthermore, due to the high nitrogen content of pig manure, ammonia inhibition is potentially of importance.

A similar line of reasoning can be followed for sludge obtained from a sewage treatment plant. Sewage sludge often consists of particulate organic matter in sewage secondary sludge: biomass that is grown on water-soluble organic compounds during secondary treatment. Due to the long residence times in the sewer as well as in the sewage treatment plant, most of the readily degradable organic carbon is converted to new biomass (poorly degradable organic carbon) and mineralized, and only poorly degradable organic carbon remains. Evidently, this results in a limited biodegradability of the remaining organic carbon that is used as substrate for the anaerobic digestion process. Removal of particulate organic matter from sewage sludge prior to biological secondary treatment (primary sludge) yields a stream containing more biodegradable particulate organic matter and thus offers a higher biogas potential compared to secondary sludge.

The situation is different and typically more favorable for substrates that have not been exposed to microbial degradation prior to the anaerobic digestion process, like the OFMSW, and residues from bioenergy crops like maize silage. These substrates

TABLE 14.3 Properties of typical feedstocks for the anaerobic digestion process (Batstone et al., 2002; Nasir et al., 2012a; Nasir et al., 2012b)

Feedstock	TDM $(g.kg^{-1})$	ODM $(g.kg^{-1})$	N $(g.kg^{-1})$	COD $(g.kg^{-1})$	Biodeg (%)	Composition	$k_h^{\,b}$ (d^{-1})	Biogas $(L.kg^{-1})$	$CH_4^{\,c}$ (%)
Pig manure	200	160	7.0	176	20	$CH_{2.27}O_{0.91}N_{0.094}$	0.05	23	57
MSW	300	240	1.8	264	25	$CH_{2.08}O_{0.97}N_{0.016}$	0.20	48	52
OFMSW	300	270	3.0	297	50	$CH_{2.10}O_{0.96}N_{0.024}$	0.20	106	53
Slaughterhouse	200	180	6.0	260	80	$CH_{2.59}O_{0.74}N_{0.065}$	0.10	118	66
Maize silage	400	360	1.2	396	80	$CH_{2.05}O_{0.97}N_{0.007}$	0.20	231	51
Grass	350	315	1.1	347	60	$CH_{2.05}O_{0.97}N_{0.007}$	0.15	151	52
Sewage sludge[a]	50	45	1.2	50	35	$CH_{2.18}O_{0.94}N_{0.057}$	0.07	12	55
Food waste	300	270	2.5	351	75	$CH_{2.32}O_{0.85}N_{0.019}$	0.25	169	58

The reported values have been compiled from a wide range of literature sources and should be considered as indicative, since reported values may vary by a factor of two at least.

[a] A mixture of primary and secondary sludge is assumed.

[b] Indicative hydrolysis rate constants reported are for mesophilic digestion; typically, thermophilic rate constants are a factor of two higher.

[c] It is assumed that the biodegradable and nonbiodegradable fractions of organic carbon have the same elemental composition.

can be degraded to a higher extent and have a much higher biogas potentials of 100 – 200 m^3 per tonne (wet) substrate.

Codigestion concerns the anaerobic digestion of mixtures of feedstocks. Examples of full-scale trials described in the literature include the digestion of sewage sludge with OFMSW (Italy and Slovenia), pig manure and energy crops (Switzerland), and algae biomass and wastepaper. Codigestion provides a number of potential advantages over digestion of a single substrate:

- Mixing of substrates may improve the properties of the feedstock through optimization of the nutrient concentration (e.g., using manure as nutrient supply), water content, or alkalinity concentration.
- When more readily degradable substrates are added to, e.g., manure digesters, the biogas production is increased, minimizing the need for mechanical mixing and associated electricity consumption.
- Codigestion of a mixture of locally produced substrates minimizes transport distances and allows for larger centralized anaerobic digesters, which can be constructed and operated more efficiently (economies of scale).

Codigestion also imposes limitations on the process. Whereas anaerobic digestion of a specific feedstock enables reuse of the digestate as fertilizer in agriculture, the use of digestate of codigestion cannot in all cases be used in agriculture due to the presence of specific compounds in the digestate of specific substrates (e.g., metals in manure). To which extent digestate can be reused in agriculture is strongly dependent on the local legislation and the nutrient requirements of local agriculture. Some degree of hygienization will be required as typically is established by heating the feedstock to 70°C. In terms of process operation, it has been demonstrated that the performance of codigestion basically is comparable to the sum of the individual substrates, provided that specific limitations (e.g., N) are overcome.

14.4.3 Process Integration

Successful implementation of the anaerobic digestion process for bioenergy production largely depends on the development of an efficient processing scheme. In practice, the anaerobic digestion process is the central process in a treatment scheme involving a large range of technologies. Additional processing steps required for implementation of the anaerobic digestion process include:

- Upgrading of the biogas, through removal of hydrogen sulfide (H$_2$S) for effective combustion of biogas in a CHP plant and removal of carbon dioxide (CO$_2$) to facilitate distribution in the natural gas network.
- Feedstocks like municipal solid waste (MSW) require several pretreatment steps for separation of specific fractions (like sand, plastics, metals) from the waste prior to anaerobic digestion. A typical mass flow chart including mass balances for MSW treatment is shown in Figure 14.11.

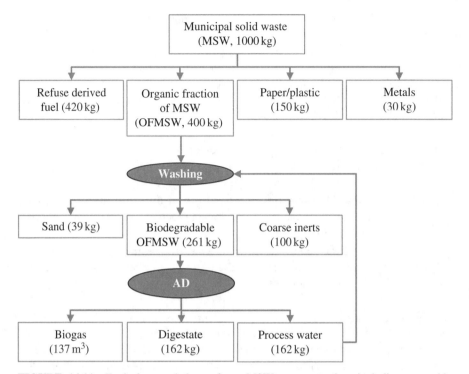

FIGURE 14.11 Typical mass balance for a MSW treatment plant including anaerobic digestion (AD) of the biodegradable organic fraction. This example is based on the Vagron plant for treatment of 230,000 tonnes MSW per year, located in Groningen.

- Depending on the quality and potential local application of the digestate, it will require several postprocessing steps. Usually, these processing steps involve phase separation resulting in a nutrient-rich liquid fraction and a solid fraction that after drying is suitable for combustion in a power plant. Nutrient removal or recovery strategies are available for treatment of the liquid fraction of the digestate.
- Utilization of the biogas for electricity and heat production requires the construction of a CHP plant.

In the following paragraphs, some integration aspects of the anaerobic digestion process will be discussed.

14.4.3.1 Biogas Upgrading and Utilization Biogas is composed of methane (CH_4) and carbon dioxide (CO_2) with smaller amounts of hydrogen sulfide (H_2S), ammonia (NH_3), and dinitrogen gas (N_2). Usually, biogas is saturated with water vapor. In principle, biogas can be used for all applications designed for natural gas. Not all gas applications require the same gas standards.

For utilization of biogas in a CHP, hydrogen sulfide needs to be removed in order to avoid corrosion of the boiler due to sulfuric acid (H_2SO_4) production by oxidation of hydrogen sulfide. Hydrogen sulfide removal from biogas can be accomplished using different methods:

- Absorption of H_2S in an alkaline solution and subsequent microbial oxidation of sulfide with molecular oxygen to (solid) elemental sulfur (S^0) that can be recovered. This process is called the *Thiopaques* process and is commercialized by Paques, the Netherlands.
- Reactive absorption of H_2S in a chelated or acidic ferric iron (Fe^{+3}) solution under formation of elemental sulfur (S^0) and ferrous iron (Fe^{+2}). Sulfur can readily be separated from the solution, and ferric iron can be regenerated by microbial or chemical oxidation with oxygen.
- Ferrous iron dosage to the digester in order to precipitate sulfide formed as iron sulfide (FeS).

For effective use of biogas as transportation fuel or for introduction into the natural gas network, biogas has to be enriched in methane. This is primarily achieved by carbon dioxide removal, which increases the heating value of the gas to enable longer driving distances with a fixed gas storage volume. Four different techniques for CO_2 removal from biogas are commercially applied:

- Absorption with water
- Pressure swing adsorption (PSA)
- Absorption with Selexol^{TM}
- Chemical absorption with amines

Absorption of CO_2 with water or water scrubbing is the most commonly used technique. The technique is based on the higher solubility of CO_2 in water compared to CH_4.

Biogas can be utilized directly or after upgrading to natural gas quality for electricity and heat production in a CHP plant. Guidelines for conducting calculations on the conversion of chemical redox energy available in methane combustion to electrical energy production are shown in Figure 14.12.

Integration of material and energy balances in anaerobic digestion facilities is another challenge for engineers involved in the implementation of anaerobic digestion. Figure 14.13 shows an example scheme for an energy self-sufficient, integrated waste processing plant based on anaerobic digestion, which uses the heat produced in the CHP for upgrading (drying) of the solid and liquid fractions of the digestate.

14.5 OUTLOOK AND DISCUSSION

One of the fundamental strong points of the anaerobic digestion process for sustainable energy production is its firm foundation on thermodynamic grounds. Methane is the organic compound with the lowest energy content per electron, and therefore, all

Energy content of biogas

$$CH_4 + 2O_2 \rightarrow CO_2 + 2H_2O$$

Normal volume of 1 kmol gas = 22.4 m^3:

CH_4 content of biogas = 60 vol.%:

Efficiency of electricity generation = 40 vol.%:

1 kWh corresponds to 3.6 MJ:

$\Delta G^0 = -818\,kJ.mol_{CH4}^{-1}$

$E = 36.5\,MJ.m_{n,CH4}^3$

$E = 21.9\,MJ.m_{n,biogas}^3$

$E = 8.8\,MJ_{electricity}.m_{n,biogas}^3$

$E = 2.1\,kWh_{electricity}.m_{n,biogas}^3$

Energy production of anaerobic digestion

Substrate: 100,000 t.y^{-1}; 150 m$_{n,biogas}^3$.t^{-1}

31.5·10^6 per year:

60 vol.% CH_4; 22.4 m$_n^3$.kmol^{-1}:

818 MJ.kmol$_{CH4}^{-1}$:

Efficiency of electricity generation = 40%:

$\varphi_V = 0.48\,m_{n,biogas}^3.s^{-1}$

$\varphi_n = 0.012\,kmol_{CH4}.s^{-1}$

$P = 9.7\,MJ.s^{-1}$ or MW total energy

$P = 3.9\,MJ.s^{-1}$ or MW electricity

FIGURE 14.12   Energy calculations for methane-containing biogas.

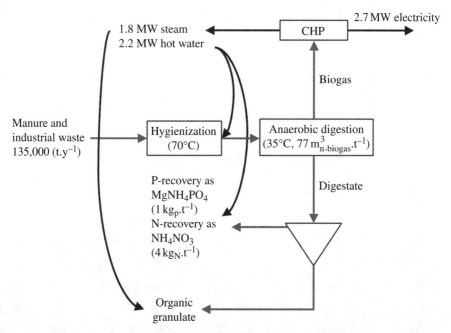

FIGURE 14.13   Example of an energetically autonomic anaerobic digestion scheme for treatment of 135,000 tonnes pig manure and industrial waste per year. Besides electricity, the CHP generates high- and low-value heat (steam and hot water), which are used for substrate hygienization, concentration of the nutrients recovered, and drying of the solid residue facilitating effective combustion.

microbial organic carbon conversions in the absence of strong electron acceptors or energy sources will eventually lead to methane production. Evolution has taken care of an enormous microbial diversity that is capable of catalyzing most organic carbon conversions as long as energy is available for the reaction. This thermodynamic foundation makes anaerobic digestion very suitable for treatment of heterogeneous mixtures of substrates since all biodegradable organic compounds are converted to methane and no specific measures need to be taken to avoid side-product formation.

A second main selling point of anaerobic digestion is the production of a gaseous end product, minimizing the efforts required for downstream processing. Even though some degree of upgrading of the biogas will be required for either combustion in a CHP plant or for distribution into the natural gas network, the phase transition during anaerobic digestion from solid and water-soluble organic carbon to gaseous organic carbon is a major advantage compared to, e.g., bioethanol production. The limited energy requirements for anaerobic digestion compared to bioethanol and biodiesel production are also the main reasons why anaerobic digestion has a two to three times higher net energy yield (in MJ per ha) for energy crops compared to bioethanol and biodiesel production.

Despite these intrinsic advantages of anaerobic digestion, the gaseous end product also represents the main limitation of the process. Current infrastructure for transportation fuels relies on liquid fuels, and European legislation aims for replacement of a significant fraction of these fuels with biomass-based fuels. This indirect subsidy for bioethanol and biodiesel has increased the price of these liquid fuels and has stimulated the production of bioethanol and biodiesel to a much higher extent than biogas production. One could easily argue that instead of replacement of a significant fraction of liquid transportation fuels with bioethanol or biodiesel, part of the infrastructure could be replaced by biogas or natural gas-powered transportation, but this argument is only scarcely heard (Tilche and Galatola, 2008). Consequently, anaerobic digestion is currently primarily used for bioenergy generation from low-value feedstocks like OFMSW, manure, and agro-industrial residues. These processes can currently only be developed cost-efficiently due to subsidies for sustainable energy. This has even become more true in recent years in which the natural gas price has decreased significantly to below US$ 0.5 per m^3 due to the discovery of novel natural gas reserves (shale gas).

In general, the anaerobic digestion process *itself* is not limiting implementation. Even though it is a relatively slow process, requiring retention times of 10–30 days and consequently large bioreactors, it is a very robust process with a well-defined end product. The bottleneck for industrial implementation of anaerobic digestion is related to the complex infrastructure required. Effective upgrading of the residues that remain after anaerobic digestion and complex thermal energy integration upon biogas combustion are prerequisites for successful implementation. Other complications include the need for nutrient recovery or removal from the digestate and definition of a destination of the solid fraction of the digestate. A final concern is the large scale required for effective implementation of anaerobic digestion (typically more than 100,000 tonnes per year), combined with the infrastructure required for feedstock supply and reuse of the digestate.

Overall, the anaerobic digestion process may contribute significantly to the sustainable energy supply of the future. However, at the current price levels for natural gas, bioenergy production as driver for development of the anaerobic digestion process will only be of limited value compared to waste stabilization and reuse-related objectives of the process.

CHAPTER SUMMARY AND STUDY GUIDE

This chapter describes the anaerobic digestion process for production of methane-containing biogas from various feedstocks. The biochemical and thermodynamic bases of the process are described, and some basic tools for feedstock characterization and estimation of the biogas potential of a substrate are described. Fundamental bioprocess engineering considerations that enable bioreactor design are presented. Integration of the process in a treatment scheme including nutrient recovery, heat integration, and electricity production is described.

KEY CONCEPTS

Anaerobic digestion
Methane-containing biogas
Natural gas
Degradation pathway
Cell growth and maintenance
Process thermodynamics
Carbohydrates, proteins, and lipids
Nutrient recovery
Process integration

SHORT-ANSWER QUESTIONS

14.1   Reproduce the elemental composition of the feedstocks shown in Table 14.3 from the basic ODM, COD, and organic N measurements.

14.2   Estimate the feedstock composition in terms of the fraction lipids, proteins, and carbohydrates.

14.3   Reproduce the potential biogas production (L.kg^{-1}) and the biogas composition (vol.% CH$_4$) of the feedstocks shown in Table 14.3.

14.4   Why do you think the biodegradable fraction of pig manure is low compared to that of the other feedstocks?

14.5   Why is the end product of anaerobic digestion methane-containing biogas and not, e.g., ethanol?

PROBLEMS

14.1 The hydrolysis rate coefficient of the biodegradable fraction of a feedstock is $0.05\,h^{-1}$. What retention time is required to obtain 80% degradation of the biodegradable fraction of the substrate in a continuous stirred tank reactor (CSTR)? What happens to the retention time required if we place two CSTRs in series or in parallel?

14.2 How much MW electricity can an anaerobic digester treating 100,000 tonnes of OFMSW per year produce? In the Netherlands, approximately 1400 kt of OFMSW is collected per year. If all this OFMSW would be treated by anaerobic digestion, estimate the potential contribution to the natural gas consumption in the Netherlands $(50 \cdot 10^9\,m^3.y^{-1})$. Discuss the results obtained.

PROJECT

A food processing company generates 60,000 tonnes waste per year and is considering anaerobic digestion as a potential method for waste valorization. Measurements on the waste provide the following data: TDM = 300 g.kg^{-1}, ODM = 280 g.kg^{-1}, N = 3 g. kg^{-1}, and COD = 400 g.kg^{-1}. From the feedstock characteristics, estimate:

1. The elemental composition of the organic fraction of the biomass
2. The fractions of lipids, proteins, and carbohydrates in the feedstock

    To assess (i) the biodegradable fraction of the feedstock and (ii) the first-order rate coefficient for hydrolysis of the biodegradable fraction of the feedstock, a batch biodegradability survey is conducted. The biodegradability survey is conducted by incubating 25 g waste in a one liter bottle with methanogenic biomass. The net methane production is monitored as a function of time, and the following data are obtained:

Time (d)	CH$_4$ (L)
0	0.00
5	0.78
10	1.32
15	1.71
20	1.98
30	2.30
40	2.47
50	2.55

Due to the supply of methanogenic biomass in the test, the methane production directly reflects the amount of substrate degraded. Full conversion of biodegradable substrate into methane can be assumed, and therefore, side reactions like partial conversion of substrate into biomass or intermediate accumulation of VFA are neglected. In a COD-based mass balance, this means that degraded COD material is converted into methane COD.

3. From the data obtained in the biodegradability survey, estimate the inert fraction of the organic substrate (f_I) and the first-order rate coefficient for hydrolysis of the biodegradable fraction of the organic substrate (k_h). Discuss the results obtained. By identifying the biodegradable fraction of the substrate and the hydrolysis rate coefficient, we can design a continuous stirred tank bioreactor. The extent of degradation of the substrate we want to achieve is the main variable in the design of the process, and we therefore aim our design at achieving 60, 70, 80, 90, or 95% degradation of the biodegradable fraction of the substrate to biogas.

4. Estimate the retention times and reactor volumes required to obtain the required degrees of degradation.

5. Estimate the potential biogas production ($m^3.h^{-1}$) and the biogas composition (vol.% CH_4 and vol.% CO_2).

6. Estimate the electricity production (MW) that can be obtained from the biogas in a CHP plant.

7. Discuss the results obtained in terms of the volume required versus the electricity production.

8. Discuss the additional technology and measures that are needed for successful implementation of the process. Think about pretreatment methods or technologies for nutrient recovery, heat integration, etc.

REFERENCES

Batstone DJ, Keller J, Angelidaki I, Kalyuzhnyi SV, Pavlostathis SG, Rozzi A, Sanders WTM, Siegrist H, Vavilin VA. *Anaerobic Digestion Model No. 1 (ADM1)*. London: IWA publishing; 2002.

Heijnen JJ, Kleerebezem R. Bioenergetics of microbial growth. In: Flickinger MC, editor. *Encyclopedia of Industrial Microbiology: Bioprocess, Bioseparation and Cell Technology*. Volume 1, New York: John Wiley & Sons, Inc.; 2010. p 594–617.

Kleerebezem R, Van Loosdrecht MCM. Waste characterization for implementation in ADM1. Water Sci Technol 2006;54(4):167–174.

Nasir IM, Ghazi TIM, Omar R. Anaerobic digestion technology in livestock manure treatment for biogas production: a review. Eng Life Sci 2012a;12(3):258–269.

Nasir IM, Ghazi TIM, Omar R. Production of biogas from solid organic wastes through anaerobic digestion: a review. Appl Microbiol Biotechnol 2012b;95(2):321–329.

Tilche A, Galatola M. The potential of bio-methane as bio-fuel/bio-energy for reducing greenhouse gas emissions: a qualitative assessment for Europe in a life cycle perspective. Water Sci Technol 2008;57(11):1683–1692.

van Lier JB, Mahmoud N, Zeeman G. Anaerobic wastewater treatment. In: Henze M, Van Loosdrecht MCM, Ekama GA, Brdjanovic D, editors. *Biological Wastewater Treatment Principles, Modelling and Design*. London: IWA Publishing; 2008. p 401–442.

15

BIOREFINERIES: INTEGRATION OF DIFFERENT TECHNOLOGIES

WIEBREN DE JONG

Department of Process and Energy, Energy Technology Section, Faculty of Mechanical, Maritime and Materials Engineering, Delft University of Technology, Delft, the Netherlands

ACRONYMS

BTX	benzene, toluene, xylene(s)
CHP	combined heat and power
DME	dimethyl ether
FT	Fischer–Tropsch
LPG	liquefied petroleum gas(es)
PE	potential energy
TAI	total of added investment
TFC	total fixed capital
TDI	total direct investment
TPI	total process investment

SYMBOLS

BCI	biorefinery complexity index	[–]
C	costs (sum of production expenses and depreciation)	[currency]
CF	cash flow	[currency]
D	depreciation	[currency]

Biomass as a Sustainable Energy Source for the Future: Fundamentals of Conversion Processes,
First Edition. Edited by Wiebren de Jong and J. Ruud van Ommen.
© 2015 American Institute of Chemical Engineers, Inc. Published 2015 by John Wiley & Sons, Inc.

FV	feature value	[–]
i	interest rate (also called discount rate)	[–]
I	total investment costs	[currency]
I_d	demolition value of the investment	[currency]
I_{wc}	working capital	[currency]
I_0	investment in production facility and supply structure	[currency]
n	number of years	[year]
NPV	net present value	[currency]
P	profit	[currency]
POT	payout time (alternatively called payback period)	[year]
ROI	return on investment	[%]
S	total sum of income	[currency]

Subscripts

$DCFRR$	discounted cash flow rate of return
j	year

15.1 WHAT IS A BIOREFINERY AND WHAT IS THE DIFFERENCE WITH AN OIL REFINERY?

Currently, modern societies still are depending on fossil fuels to satisfy their energy and material needs with a prominent role for oil as feedstock in petrochemical industrial complexes. Oil prices have been fluctuating and its availability becomes increasingly challenging, giving progressively rise to pressure on prices and the environment. Whatever path short-term oil price development will follow, the dropping amount of available and affordable fossil energy sources, as well as conventional material sources and the need for sustainable, climate-neutral production technologies, have led to a renewed interest in the utilization of biomass. Biomass is indeed an abundant, renewable, and natural source, which is more evenly spread over the world in both terrestrial and aquatic environments compared to conventional fossil resources. Its annual global production is of the order of 170–200 billion tonnes (see also Chapter 1).

The drivers for a biomass-based economy are clean and sustainable environmental development, economic growth, and green politics. In December 2008, the European Parliament approved a directive that set mandatory targets on the EU-25 Member States in order to increase energy efficiency by 20% compared to the present levels, cut greenhouse gas emissions by 20% compared to the 1990 levels, and reach a 20% renewable energy share of final consumption, within the year 2020 (EC, 2008).

Already in 1949, Glesinger described different concepts to better integrate loosely operating wood processing factories to create synergies of material production (wood logs, pulp, and paper), chemicals (e.g., from lignin), energy (gasification of residues),

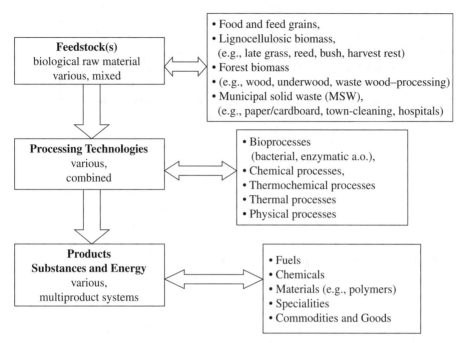

FIGURE 15.1 Biorefinery concept. (Source: Reproduced with permission from Kamm and Kamm (2004). © Springer Science + Business Media, Figure 1.)

and even food and feed sugars (Glesinger, 1949). Glesinger gave many examples of (pre-)WWII systems and the German dependency on wood as a major raw material and energy source. This is a typical example of the biorefinery approach. A biorefinery is aimed at fractionating and further processing of biomass to generate versatile (organic) products. It is in fact the analogue to the petrochemical refinery. The main constituents of biomass, though, differ from crude oil, the feedstock of a petrochemical refinery. Biomass has a complex and to some extent variable composition. Its constituents are mainly oxygenated species as presented in Chapter 2: carbohydrates (cellulose, hemicellulose) and lignin. Often, minor amounts (generally <5 wt% of the total as received (ar) mass) of other compounds are present in biomass, e.g., proteins, lipids/oils, terpenes, vitamins, dyes, flavors, and minerals/salt (Kamm et al., 2006). A very generic picture of the biorefinery concept is shown in Figure 15.1. A shift toward bio-based industries is already starting and means nothing less than a worldwide technology revolution. The term biorefinery is yet comparatively new and has come up only toward the end of the twentieth century. However, such units in certain forms already have been among us since long times. The term "biorefinery" refers to both a factory and a concept. The definitions given nowadays are dealt with in this part of the book.

The American institute NREL (United States) defines the biorefinery as follows (tinyurl.com/klwrfa7): "A biorefinery is a facility that integrates biomass conversion processes and equipment to produce fuels, power, and chemicals from biomass."

The US Department of Energy summarizes its unique features as: "biorefinery is an overall concept of a promising plant where biomass feedstocks are converted and extracted into a spectrum of valuable products" (tinyurl.com/kzhn5ej). The National Non-Food Crops Centre in the United Kingdom uses a broad definition: "A biorefinery is a manufacturing site involved in the refining of biomass material to yield purified materials and molecules. This conversion can be achieved using biological or thermochemical processing or a mixture of both. Downstream manufacturing sites processing materials/molecules from biomass are often termed secondary biorefineries" (NNFCC, 2007). In the definitions, often two aspects are not considered to their full extent: food and feed coproduction and the self-supply of heat and electricity of the refinery. IEA has dedicated a separate task (nr. 42) since early 2007 on biorefineries and has defined this concept of biomass processing clearly, emphasizing the driving force from a broader context of sustainable development as follows (Jungmeier et al., 2013): "Biorefining is the sustainable processing of biomass into a spectrum of marketable bio-based products (food/feed ingredients, chemicals, materials) and bioenergy (biofuels, power and/or heat)." In biorefinery development, different stages can be discriminated, which can be ranked into three generations as has been worked out by Kamm and Kamm (2004):

Generation I biorefineries are characterized by the lowest flexibility because feedstock type and products as well as resulting by-products are fixed. An example is the dry-milling ethanol plant using grain as feedstock.

Generation II biorefineries are more flexible in the end products, for instance, a wet-milling ethanol production plant can produce different products depending on demand, which may include ethanol, starch, high-concentration fructose syrups, oils, and animal feed.

Generation III biorefineries are the most flexible, as these can process a multitude of biomass feedstocks into variable end products. Given the situation that often a single type of biomass feedstock's availability within a reasonable distance might not be sufficient to operate a plant at full capacity during the entire year, this flexibility will become more and more necessary (Zinoviev et al., 2010).

A typical oil-based refinery is depicted in Figure 15.2 (Moulijn et al., 2001). A wide range of products is generated from such a complex in all aggregation states: from gaseous fuels (e.g., refinery fuel gas and intermediate synthesis gas/hydrogen), via liquid products (liquefied petroleum gas (LPG), liquid transportation fuels, solvents, lubrication oil, and greases), to solid coke.

Many processes have been developed for conversion of oil to intermediate and end products. Some add hydrogen (hydrotreating, hydrocracking), and some remove carbon (flexicoker) to improve the H/C ratio of the desired fuel product. Catalysts play a key role in many processes. An example is the well-developed fluid catalytic cracking unit in which heavy oil is cracked to produce gasoline using a catalyst in a riser (vertical transport) reactor. The mostly linear alkanes produced need to be catalytically

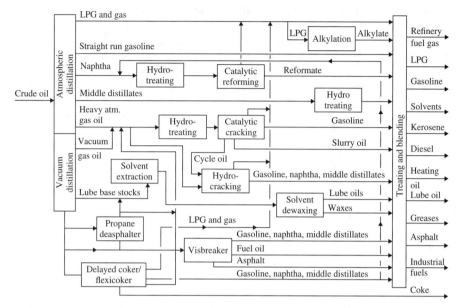

FIGURE 15.2 A modern oil refinery; conversion technologies and products. (Source: Reproduced with permission from Moulijn et al. (2001). © John Wiley & Sons Ltd.)

reformed to their isomers, which is needed for a properly defined gasoline that shows improved engine knock behavior; this is another example of a catalytic reactor in the oil refinery.

Downstream of the depicted oil refinery, a myriad of processes have been developed in petrochemistry, generating products that are used widely in our daily lives (see, e.g., the website of the Association of Petrochemicals in Europe, tinyurl.com/o4jemkt). One should keep in mind, though, that this product manufacturing sector is much smaller in capacity than fuel production (e.g., gasoline). The key petrochemical base chemicals are relatively simple molecules, namely, lower olefins (ethylene, propylene, and butadiene), aromatics (benzene, toluene, and xylene, taken together as BTX), and ammonia (NH_3) and methanol (CH_3OH) (Moulijn et al., 2001).

The main differences between a biorefinery and a petrochemical refinery are summarized in Table 15.1.

There are two fundamentally different approaches that can be followed in the trail to mass implementation of biorefineries worldwide. First, biorefineries can be designed and set up in such a way that production of core molecules of the current petrochemistry is targeted. In this approach, the connectivity to existing infrastructures—at least of product distribution—is ensured. The second approach is to identify new platform chemicals that do not (yet) show a large market availability today and make brand new products that are more closely resembling major biomass component structures. In this approach, new products, processes, and systems (also for product distribution) must be developed.

TABLE 15.1 Differences between oil refineries and biorefineries

	Oil refinery	Biorefinery
Feedstock sourcing	Oil wells, sedimentary rocks or shale	Agriculture, also organic waste collection
Feedstock type	Alkanes, cycloalkanes, (poly) aromatic hydrocarbons; smaller amounts of S, N organic species	Mainly carbohydrates, but also lignin, terpenes, vegetable oils and fats, and other living matter-derived molecules
Functionality of feedstock	Largely oxygen-free, lacking functional groups	Abundant in oxygen, overfunctionalized (e.g., OH groups)
Processing	Primarily via distillation and subsequently (catalytic) upgrading to end products	Primarily via widely differing processes, ranging from (dry) thermochemical to aqueous (bio)chemical
Main (current) products	Diesel, gasoline, aviation fuel, olefins, aromatics, lubricants, tar, monomers, polymers, solvents	Ethanol, biodiesel, diverse, mainly oxygenated organic chemicals
Scale	Mostly huge	Small to medium, adapted often to local biomass economic availability
Status	Mature	Upcoming, some concepts up and running

15.2 TYPES OF BIOREFINERIES

Several attempts have been made to name and classify biorefinery schemes (see, e.g., Kamm et al., 2006). Recently, IEA Task42 "Biorefineries" (tinyurl.com/pj2x3or) constructed a very comprehensive and useful classification diagram. It is based on the following four main features:

1. Feedstocks
   - Grasses
   - Lignocellulosic crops
   - Lignocellulosic residues
   - Marine biomass
   - Oil crops
   - Oil-based residues
   - Organic residues and others (including manure)
   - Starch crops
   - Sugar crops
2. Processing methods
   - Biochemical
   - Chemical

- Physical
- Thermochemical

3. Platforms (also combinations)
   - Biogas
   - C5 and/or C6 sugars
   - Green pressate (also called "juice")
   - Hydrogen
   - Lignin (solid)
   - Oil (liquid)
   - Organic juice
   - Power and/or heat
   - Proteins
   - Pulp
   - Pyrolytic liquid ("bio-oil")
   - Syngas

4. Products
   - Material products
   - Energy products

Figure 15.3 shows examples of processes, with each of these features depicted in the flow diagrams with its own box symbol.

Based on this classification method, recently, attempts have been made to identify the complexity of different concept schemes, similar to the Nelson Complexity Index for oil refineries (Jungmeier et al., 2012). It has been proposed to represent the biorefinery complexity index (*BCI*) as

$$BCI = n_{products}FV_{products} + n_{platforms}FV_{platforms} + n_{feedstocks}FV_{feedstocks}$$
$$+ n_{processes}FV_{processes} \qquad \text{(Eq. 15.1)}$$

In this equation, n is the number and FV is the feature value, a parameter indicating the state of the art of products, platforms, feedstock, and processes. Year 2010 is taken as 1 (already existing commercially), 2015 = 1.5, 2020 = 2, …, 2050 = 5.

Example 15.1   *BCI* characterization of a vegetable oil-based biodiesel process

A biorefinery is based on the supply of rapeseeds (an oil crop). These seeds are pressed, resulting in an oil stream. The oil is transesterified with methanol to produce biodiesel and crude glycerol. This last product is distilled to reach the required purity for this by-product. The press cake is used as animal feed. What is the BCI?

Solution
All technologies and products already exist commercially on the market, so their *FV* are all 1. We can identify three products (one energy product and two material

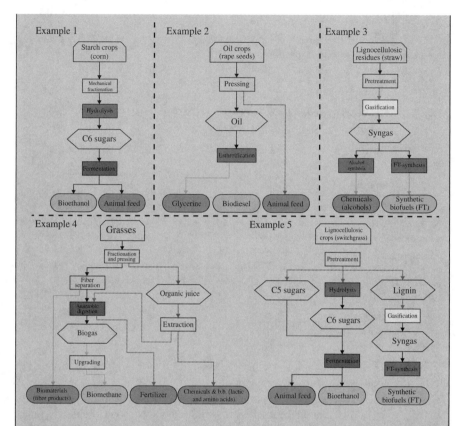

FIGURE 15.3 Classification schemes for different biorefinery concepts for determination of the BCI. (Source: Reproduced with permission from Cherubini et al. (2009). © Society of Chemical Industry and © John Wiley & Sons Ltd.)

products), one platform (oil), one bio-based feedstock (rapeseeds), and two processes (one chemical process, esterification, and one physical process, pressing). Thus, $BCI = 3 + 1 + 1 + 2 = 7$ (see Figure 15.3 and Example 15.2).

15.2.1 Thermochemical Biorefinery Concepts

In Chapters 10–12, thermochemical conversion technologies leading to a product spectrum of fuel components still containing chemical energy have already been described. For gasification (Chapter 10), this is a product gas that is rich in CO and H_2. Pyrolysis (Chapter 11) results in a wide range of organic products, and torrefaction (Chapter 12), a form of mild pyrolysis, produces a carbon-enriched char. In this chapter, we describe some possible ways to integrate such techniques for the production of a multitude of potential energy (PE) carriers and chemicals/materials.

Figure 15.4 (Boerrigter et al., 2004) presents a good example of an extended, highly integrated thermochemical biorefinery based on torrefaction/gasification.

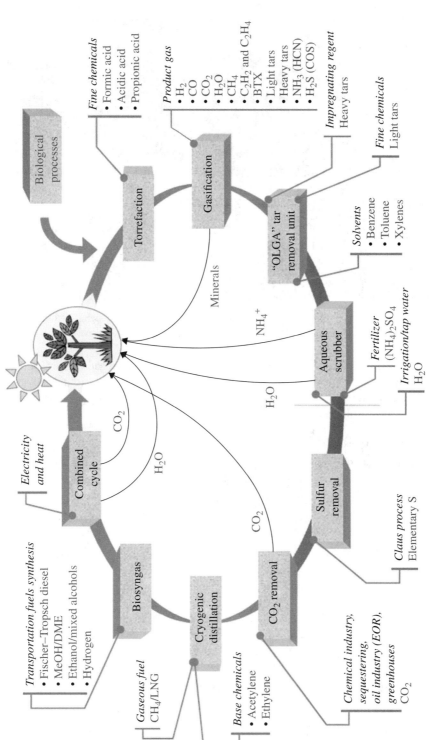

FIGURE 15.4 Thermochemical biorefinery concept. Data from Boerrigter et al. (2004).

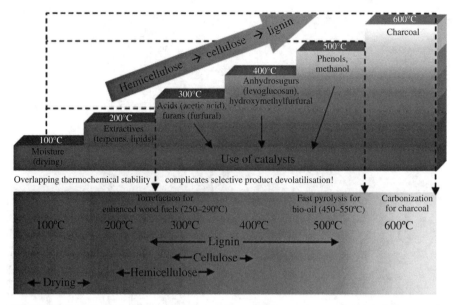

FIGURE 15.5 Stepwise pyrolysis for generation of different classes of chemical compounds. (Source: Reproduced with permission from De Wild et al. (2011). © Future Science Ltd.)

It is aimed at closing different elemental cycles. Thus, based on a wide variety of biomass input, even very wet biomass—as torrefaction can be performed under hydrothermal conditions (see Chapter 12)—the following classes of energy carriers and compounds can be generated:

- Gaseous fuels
- Transportation fuels
- Base chemicals, both organic and inorganic
- Fertilizers
- Fine chemicals and wood impregnation organics
- Clean water
- Heat and power

Another concept has been published by De Wild et al. (2011) (see Figure 15.5), which illustrates that pyrolysis as thermochemical conversion technology can be used in a stage-wise manner by increasing the temperature so as to produce different classes of compounds until a charcoal is produced, which can have multifunctional use. Catalysts are needed to selectively produce targeted products.

15.2.2 (Bio)chemical Biorefinery Concepts

Biorefineries based on biochemical conversion technologies make use of lower temperature processes, and processing takes place in the aqueous phase. Pretreatment is needed to open up the tightly bound structures contained in biomass. These are mainly entanglements of lignin with the carbohydrate content that is to be released and further processed, enabling sugar conversions (fermentation [see Chapter 13] and chemical conversion strategies [see Chapter 18]). Pretreatment is a key step to ensure sufficient conversion in such processes, and it is aimed at the realization of the following effects (Aresta et al., 2012; Mosier et al., 2005; Sun and Cheng, 2002):

- Increase of the accessible surface area
- Decrystallization of cellulose
- Partial depolymerization of cellulose
- Dissolution of hemicellulose and/or lignin
- Modification of lignin structure

Basically, in biorefineries, one can divide pretreatment based on chemical conversion into two distinct groups (Lersch, in (Aresta et al., 2012)):

1. Hydrolysis processes that are accompanied with cellulose dissolution ("sugarification")
2. Pulping processes in which lignin is extracted from the biomass

An overview of such pretreatment techniques is given in Chapter 13 (see Table 13.7).

15.2.3 Hybrid Concepts

Figure 15.6 shows a block diagram of a biorefinery based on a sugar platform (obtained via (bio)chemical processing) and a syngas platform (generated via thermochemical gasification). The fractionation step is not needed in all possible hybrid schemes; it is needed, however, when biochemical conversion is intended to be the first step, as molecular structure opening is then required.

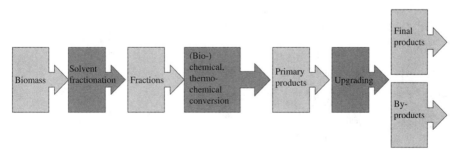

FIGURE 15.6   Generic schematic sequence of hybrid techniques. Based on De Wild (2011).

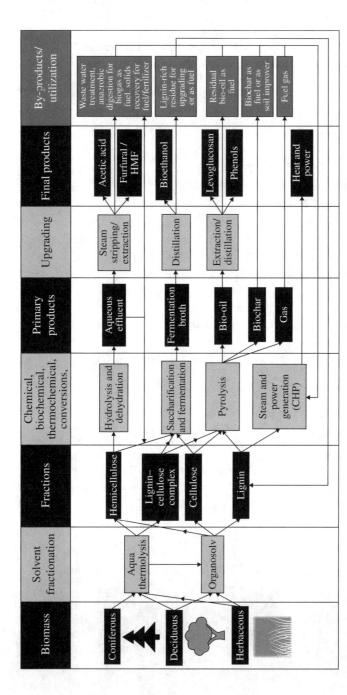

FIGURE 15.7 Worked-out example of a biorefinery based on hybrid technology. (Source: Adapted from De Wild (2011). © P. de Wild.)

Lignin conversion by biochemical techniques is not straightforward—remember that one of its functions in plants is to protect them from microbial attack—and is still under investigation. Thus, in most (bio)chemical processes developed so far, the carbohydrate part of biomass is converted using (enzymatic) hydrolysis and subsequent or simultaneous sugar fermentation. Thermochemical conversion of the lignin part is usually carried out in industrial practice by combustion to generate steam as the simplest technology; this combination of techniques is already current practice. A more advanced development is to gasify (part of) the lignin to generate syngas or hydrogen that can be used in the plant or can be sold. Hydrothermal conversion is even more promising as the residue streams contain high amounts of water. An example of such a detailed hybrid plant based on the above generic scheme is given by De Wild (2011) (Figure 15.7).

Another approach in hybrid biorefining is that a thermochemical conversion process is placed upfront (bio)chemical conversion. In this case, gasification and gas cleaning are used to produce a product gas or synthesis gas, which in a subsequent step is biochemically converted to, e.g., ethanol. This configuration makes use of the advantage of thermochemical conversion to convert the largest part of biomass (including lignin), while ensuring selective biofuel production (in particular ethanol) by fermentation, a biochemical process. A good overview of such a biorefinery constellation is given by Mohammadi et al. (2011).

There are microbes that ferment syngas under anaerobic conditions so as to perform effectively a water–gas shift reaction aimed at H_2 production (see, e.g., Jung et al., 1999; Merida et al., 2004; Younesi et al., 2008) and conversion of syngas into a mixture of acetic acid and ethanol (see, e.g., Allen et al., 2010; Lorowitz and Bryant, 1984; Sakai et al., 2004; Tanner et al., 1993); yet other microbes produce mixtures of ethanol, butanol, acetic acid, and butyric acid from syngas components (see, e.g., Grethlein et al., 1990; Heiskanen et al., 2007; Liou et al., 2005). Methane can also be formed from syngas using microbes (see, e.g., Klasson et al., 1991). Industrial application is currently developed by the company Coskata based on multifuel input (tinyurl.com/les4tw4).

15.2.4   Examples of Current Biorefineries

Table 15.2 gives examples of biorefineries that are currently running (not exhaustive; the reader is encouraged to look up and evaluate novel developments; see also, e.g., tinyurl.com/nl7qt3n).

15.3   ECONOMIC CONSIDERATIONS EVALUATING BIOREFINERY CONCEPTS: BASIC METHODS FOR ASSESSING INVESTMENTS AND COST PRICES

One can imagine and design a myriad of different biorefineries, for sure, for generation III types given their multioutput multi-input nature. How to evaluate these from an economical point of view? There are extensive ways of making economic

TABLE 15.2 Some (semi)commercially operating biorefinery examples (sizes mostly larger than 10 kt·year⁻¹ input only)

Company, place/country	Feedstock	Products	Capacity (kt·year⁻¹), input	Short description, reference(s)
Abengoa, Hugoton, Kansas, United States	Corn stover, wheat straw, switchgrass	Ethanol, CHP	320	Steam explosion pretreatment coupled with biomass fractionation, C5/C6 fermentation, distillation for ethanol recovery. Heat and power is provided by means of biomass gasification. Cogeneration of 18 MW gross electrical power. Construction phase (tinyurl.com/n2bhpmx)
Arkema, Marseille, France	Castor oil	Aminoundecanoic acid, nylon-11 monomer Glycerol Heptanaldehyde Heptanol Heptanoic acid "Esterol" (mix of esters)	n.i.	Transesterification; thermal cracking of the methylester to methylundeylenate and heptanaldehyde; hydrolysis, reaction of the organic acid with HBr; and finally reaction with NH_3. Commercial operation (Aresta et al., 2012)
Beta Renewables, Crescentino, Piedmont, Italy	Lignocellulosics	Ethanol, CHP	270	PROESA™ process; hydrothermal pretreatment, enzymatic hydrolysis, and fermentation (tinyurl.com/ltuzewk)
BioMCN, Delfzijl, The Netherlands	Glycerin from 1st gen. biodiesel	Methanol	200	Purification, reforming methanol synthesis (tinyurl.com/l3pfuy5)
Borregaard Industries Ltd., Sarpsborg, Norway	Lignocellulosics, sulfite spent liquor (SSL, 33% dry content) from spruce wood pulping	Ethanol, lignosulfonate	400	Pulp for the paper mill is produced by cooking spruce chips with acidic calcium bisulfite cooking liquor. Hemicellulose is hydrolyzed to various sugars during the cooking process. After concentration of the SSL, the sugars are fermented and ethanol is distilled off in several steps. Part of the 96% ethanol

Company, Location	Feedstock	Product	Capacity	Description
Chemrec, Sweden	Black liquor	Methanol, dimethyl ether (DME)	~5	is dehydrated to produce absolute ethanol (tinyurl.com/myclw39) In the pulp and paper industry, next to the main products, black liquor is produced. This is gasified to syngas, which in this process is converted into methanol and DME. Process demonstration unit (tinyurl.com/jvvr7hw). Capacity estimation based on Knoef, Chapter 7 (Knoef, 2012)
Inbicon, Kalundborg, Denmark	Straw	Ethanol, C5 sugar molasses, lignin pellets	30	Straw pretreatment by pressurized cooking, enzymatic hydrolysis, pressing out C5 molasses, sugar fermentation, and ethanol distillation, lignin drying, and pelletizing (tinyurl.com/m4u2u93)
Lanzatech, Soperton, Georgia, United States	Lignocellulosic residues	Ethanol	~15[a]	Fermentation of syngas obtained from biomass gasification (tinyurl.com/nqf7s5s)
NatureWorks LLC	Corn syrup, other C6 sugar	Polylactide	150	Biochemical conversion (tinyurl.com/6e7nks)
Neste Oil, Porvoo, Finland, 2 units; Rotterdam, Netherlands; Singapore	Vegetable oils	Biodiesel, gas, biogasoline	190, 800, 800[b]	Hydrogenation of vegetable oil (tinyurl.com/lj7nws2)
POET-DSM, Emmetsburg, Iowa, United States[a]	Corn crop residues	Ethanol and biogas	75[b]	Pretreatment, enzymatic hydrolysis, fermentation (tinyurl.com/nl7qt3n)

n.i., not indicated.
[a] Under construction.
[b] Output based.

evaluations (see, e.g., the books of Peters and Timmerhaus (1991) and Towler and Sinnott (2012)); however, the aim of this text is to provide the reader just with simple first screening tools.

A company deciding to invest in novel technologies as, e.g., biorefineries aims at making profit (P). The net profit in a certain year j is defined as

$$P_j = (S_j - C_j)(1 - t_j)$$ (Eq. 15.2)

with S being the total sum of income generated and C the costs made; t is the relative tax level. Profit before tax raising ("pretax") is considered when $t_j = 0$. The costs (C) consist of expenses (E) and depreciation on investment in the plant (D):

$$C_j = E_j + D_j$$ (Eq. 15.3)

Now, an important term in economics is cash flow (CF), and it is the total income minus expenses made. In year j, it is

$$CF_j = S_j - E_j$$ (Eq. 15.4)

Thus, the relation between profit and CF becomes

$$CF_j = \frac{P_j}{(1 - t_j)} + D_j$$ (Eq. 15.5)

By making the decision to invest a money value equal to I, the company targets at generating CF. This investment, I, consists of two distinctive parts, I_0 and I_{wc}, so that

$$I = I_0 + I_{wc}$$ (Eq. 15.6)

I_0 consists of investment regarding the production facility and the accompanying supply structure. This investment part's value decreases in time by wear and technical aging. At the end of the investment's lifetime, it only has a low value left, the demolition value I_d. Total depreciation in n years on the investment, D, then is defined as

$$D = \sum_{j=1}^{n} D_j = I_0 - I_d$$ (Eq. 15.7)

Often, I_d is neglected in economic evaluations. I_{wc} is the working capital accompanying the targeted investment. The difference between I_0 and I_{wc} is that the working capital keeps its value; one can think of (durable) supply stocks of raw materials, (semi)ready products, cash money needed for the investment, etc. In the beginning (end of year 0), the company must invest in both I_0 and I_{wc}. At the end of the lifetime, I_{wc} is (to be) released again together with I_d.

As depreciation has an immediate impact on the profit (question: why?), governments have framed fiscal regulations that do not allow, e.g., immediate, complete depreciation at the start. A default way of depreciating for an economic lifetime of n years, though more forms exist (see, e.g., (Peters and Timmerhaus, 1991)), is linear depreciation:

$$D_j = \frac{(I_0 - I_d)}{n} \qquad \text{(Eq. 15.8)}$$

One of the indicators of profitability is the payout time (POT), alternatively called payback period (PBP), and it is defined as

$$POT = \frac{I_0}{CF_j} \qquad \text{(Eq. 15.9)}$$

Here, it is assumed that the yearly CF is constant; if this is not the case, then an average value over the years can be assumed. The indicator is a comparatively simple one and more often applied for small-scale investments.

An alternative, graphical way is to sum all the yearly CFs until reaching 0 (this is called the CF curve method).

Another indicator for the profitability of a project is the return on investment (ROI), which is

$$ROI = \frac{P_j}{I} \times 100\% \qquad \text{(Eq. 15.10)}$$

The ROI is usually considered "after tax" with a tax rate, t, taken into account in the profit. Regarding the investment, also the contribution of the working capital is considered.

Example 15.2   *ROI* and *POT* of a bioinvestment

An engineer wants to set up a biorefinery project that requires an investment of 20 M€ (on-site, working capital neglected). The economic lifetime of the installation is 10 years. An after-tax CF with a positive value of 2.5 M€·year^{-1} is ensured.

a. Calculate the depreciation per year using the linear method.
b. What is the (pretax) ROI (taxation is neglected)?
c. Draw a diagram in which you show the cumulative CF and determine the POT value.

Solution
a. Depreciation is linear; thus, $D_j = $ 20 M€/(10 year) = 2 M€·year^{-1}.
b. $ROI = $ (profit before taxes/investment) $\times 100\% = ((2.5 - 2)/20) \times 100\% = 2.5\%$.

> c. This is a straight line starting at year 0 at a value of -20 M€ and crossing the time axis at year 8 (= *POT*).
>
> *Question*: What changes when a tax rate, t, of 0.4 is employed?

The aforementioned indicators do not appreciate the fact that the value of money is time dependent. The most important factor is that one can profit from interest obtained yearly.

Now, two basic approaches exist to evaluate profitability accounting for the time value of money. These are the discounted cash flow rate of return (*DCFRR*) and the net present value (*NPV*).

The discounted *CF* in year j is

$$DCF_j = \frac{CF_j}{(1+i)^j} \tag{Eq. 15.11}$$

Thus, a certain *CF* generated in the early years of a project is worth more than the *CF* generated in later years as expressed in the division by the factor $(1+i)^j$. The *NPV* given a relative interest rate i ($NPV_{i\%}$) is directly related to the DCF_j values via the sum from year 0 to year n:

$$NPV_{i\%} = \sum_{j=0}^{n} \frac{CF_j}{(1+i)^n} \quad \text{with} \quad CF_0 = -I_0 \tag{Eq. 15.12}$$

The advantage of calculating the *NPV* is that it is independent of the way of paying off the financed I_0 and the way of depreciation when one evaluates a project or investment; on the "higher" company level, the way of depreciation, though, plays a role as tax paid over profit plays a significant role.

Now, if the situation is such that each year the same CF_j is generated and if $I_d = 0$, then Equation (15.12) simplifies to

$$NPV_{i\%} = \sum_{j=0}^{n} \frac{CF_j}{(1+i)^n} = -I_0 + CF \frac{(1+i)^n - 1}{i(1+i)^n} \tag{Eq. 15.13}$$

The *DCFRR* method is based on the principle that the value *NPV* becomes zero at a certain value of the internal interest rate. This method's background is that when a project has a *DCFRR* of x% than the invested capital in it, the project's lifetime yearly generates a profit equal to x% of the part left of the original investment in that year next to a complete payoff of the original investment:

$$NPV_{DCFRR} = 0 = \sum_{j=0}^{n} \frac{CF_j}{(1+DCFRR)^n} \quad \text{with} \quad CF_0 = -I_0 \tag{Eq. 15.14}$$

Example 15.3 *NPV and sensitivity analysis to major investment parameters*

Taking the same data as presented in Example 15.2, assuming in addition an interest rate of 6%, calculate both the *NPV* and the sensitivities of changes of the initial investment (I_0), *CF*, expected lifetime of the installation (n), and interest rate (i).

Solution
As the *CF* pattern is constant, Equation (15.11) holds

$$NPV_{i\%} = -I_0 + CF\frac{(1+i)^n - 1}{i(1+i)^n} = -20 \text{ M€} + 2.5 \text{ M€}\frac{(1+0.6)^{10} - 1}{0.06(1+0.06)^{10}} = -1.6 \text{ M€}$$

The sensitivity of $NPV_{i\%}$ with respect to I_0 can be found by simple differentiation:

$$\frac{\partial NPV_{i\%}}{\partial I_0} = -1$$

Regarding the sensitivity to *CF*,

$$\frac{\partial NPV_{i\%}}{\partial CF} = \frac{(1+i)^n - 1}{i(1+i)^n} = \frac{(1+0.06)^{10} - 1}{0.06(1+0.06)^{10}} = +7.4$$

With respect to the expected lifetime, and therefore depreciation time considered (n), the sensitivity is

$$\frac{\partial NPV_{i\%}}{\partial n} = CF\frac{\ln(1+i)}{i(1+i)^n} = 2.5 \text{ M€}\frac{\ln(1+0.06)}{0.06(1+0.06)^{10}} = 1.4 \text{ M€}$$

Finally, with respect to the interest rate, the sensitivity is

$$\frac{\partial NPV_{i\%}}{\partial i} = CF\frac{i(n+1)-(1+i)^{n+1}+1}{i^2(1+i)^{n+1}} = 2.5 \text{ M€}\frac{0.06(10+1)-(1+0.06)^{10+1}+1}{0.06^2(1+0.06)^{10+1}}$$

$$= -87.1 \text{ M€}$$

This example shows that the time value of money matters; whereas a *POT* analysis shows payback time of 8 years, the *NPV* is—given the interest rate of 6%—still negative!

A way to estimate the investment costs of process equipment is illustrated in the calculation sheet presented in Table 15.3 with an explanation of terms in Table 15.4.

Once the investment cost is known for a certain process equipment ($I_{\text{equipment}}$) with a capacity A (e.g., expressed in terms of mass flow, volumetric flow, or power unit),

TABLE 15.3 Example calculation sheet for investment calculation

Location	Country	Start-up year	Year	Currency	US $year
Process description					
Design capacity (kt.year⁻¹) → $kt.year^{-1}$				Hours per year	8000
Depreciation of direct investment	Years			Method	Linear
Depreciation of indirect investment	Years			Method	Linear
Process equipment	*Number*		*FOB price*	*Hand factor* (Couper, 2003)	*Module investment*
	Nr.				
Reactors	R-x			4[a]	
Fractionation columns	C-x			4	
Vertical vessels	V-x			4[a]	
Horizontal vessels	D-x			4[a]	
Heat exchangers	E-x			3.5	
Pumps	P-x			4	
Compressors	K-x			2.5	
Furnaces	F-x			2[b]	
Other equipment				2.5	
Unforeseen	25	% of on-site investment	—	—	—
Total direct investment (TDI) (on-site investment)	—	—	—	—	
Instrumentation	15–20[b]	% of on-site	Investment		
Tankage, storage, and handling	5–25[b]	% of on-site	Investment		
Utilities	15–25[b]	% of on-site	Investment		
Off-sites	10–25[b]	% of on-site	Investment		
Environmental supplies	10–20[b]	% of on-site	Investment		

Total added investment (TAI)

Total investment (TDI+TAI)
"One-time" investment

Total process investment (TPI)
Correction factors — Time factor[c] — Site/location factor

TPI corrected
Working capital (I_{wc}) — % of yearly — Production costs
Total fixed capital, TFC (=TPI$_{corr}$ + I_{wc})

[a] When pressurized vessels are applied.
[b] Values depend on whether a process installation is built on "grass-roots" (highest values) or newly built (intermediate values) in an industrial area or is just an extension of an existing process installation (lowest values).
[c] Here, the ratio of CEPCI values can be used (see Equation (15.16)).

TABLE 15.4   Explanation of some terms in Table 15.3

Hours per year	Usually set at 8000 h
Depreciation	Process installations 10 years; accompanying infrastructure 20 years
Process equipment	Identify major equipment and their nature; look for free-on-board (FOB) prices, multiply with the Hand factor to account for a unit that can be integrated in a modular way
TDI	The primary process investment, also called "within/inside battery limits" or on-site
Instrumentation	Not only measurement equipment but also control equipment, like valves and supplies for safety (alarms)
Tankage, storage, and handling	For example, storage of raw material stocks and ready products; supplies to move such stocks
Utilities	All that is needed to supply process equipment with power, fuel, process and cooling water, steam, pressurized air, inertizing gases
Off-sites	Facilities like a control room, a lab, administration building, canteen, fire security department, safety department, medical service, storage building for spare parts, garage, waiting room, roads, parking places, lighting, etc.
Environmental supplies	For example, chimneys, measures to reduce noise, remove contaminations (unless these are major reactors)
One-time investments	Costs that are immediately depreciated, e.g., certain license costs, start-up costs

one can use a rule of thumb for scale up concerning the costs, according to the Williams' 0.6 scale factor rule:

$$I_{\text{equipment\_capacity}A} = I_{\text{equipment\_capacity}B} \times \left(\frac{\text{capacity}A}{\text{capacity}B}\right)^{0.6} \qquad \text{(Eq. 15.15)}$$

Please note that each specific large piece of equipment may have varying exponents; more info can be found in, e.g., Peters and Timmerhaus (1991). When values of investments are presented in a certain year 0, then one can calculate the amount of money involved in a more recent year 1 according to

$$\frac{I_{\text{equipment\_1}}}{I_{\text{equipment\_0}}} = \left(\frac{CEPCI_1}{CEPCI_0}\right) \qquad \text{(Eq. 15.16)}$$

where $CEPCI_j$ is the Chemical Engineering Plant Cost Index in year j. The value of $CEPCI_j$ is reported in the Chemical Engineering magazine (tinyurl.com/lhppp6d).

The price of a product can be calculated as detailed in Table 15.5 and Table 15.6.

Some studies have been performed regarding technoeconomic analyses of biorefineries. Wright and Brown (2007) compared first-generation ethanol production with (ligno)cellulosic ethanol production as well as with Fischer–Tropsch (FT) diesel production and methanol and hydrogen generation. The basis of comparison was a 150 million gallon (gasoline equivalent) plant (reference year for all calculations: 2005).

TABLE 15.5   Calculation sheet for cost + return price of a chemical product

Production costs	US \$.(unit)$^{-1}$	Unit.t^{-1}	US \$.t^{-1}	US \$.year^{-1}
Raw materials				
Added materials				
Utilities				
Miscellaneous				
Yield of by-products (negative value!)				
Total variable costs				
Operation				
Maintenance				
Lab				
Staff/support				
Rent of land				
Insurances				
Total fixed costs				
Total expenses production (= variable + fixed costs)				
Depreciation				
Costs for sales, research				
Total production costs				
Profit at x% *ROI* (before taxation)				
Cost + return price				

TABLE 15.6   Explanation of cost items in Table 15.5

Production costs	"Unit" can be mass, volume, or energy, e.g., t, kg, m^3, and kWh. The third-column value is the multiplication of the first two; the fourth column is the mass flow per year times the third-column value
Raw materials	Follows from material balance; prices can, e.g., be retrieved from the "Chemical Marketing Reporter." Use consistent sources for prices
Added materials	These comprise of materials that are used but are not included in the end product, e.g., catalysts
Utilities	An internal calculation price consisting of a fixed and variable part but without depreciation and profit
Miscellaneous	For example, environmental taxes
Yield of by-products	Use a conservatively estimated market price
Operation	Salaries of operators and their chiefs; select number of employees and number of shifts
Maintenance	Materials and labor, can be estimated as 5% of TDI
Lab	Estimated as 15% of operation costs
Staff/support	Salaries for local management, administration personnel, process engineers, safety manager, etc. Estimated as 50% of operation costs
Rent of land	Estimated as 1% of the TPI
Insurances	Estimated as 1% of TFC
Depreciation	Guideline; 10% of TDI plus 5% of total added costs and TAI
Sales and research	Estimated at 10% of the total production costs
ROI	Rate of return ("imposed" profit margin before taxation); here, defined as x % of the sum of on-site investments (I_0) and working capital (I_{wc})

TABLE 15.7   Capital costs for 150 million gallons per year plants (gasoline equivalent, 2005 US$)

Fuel	Total capital costs (M US$)	Capital cost per unit production (pbpd)[a]	Operating cost (US$ per gallon)[b]
Grain ethanol (1st gen.)	111	13,000	1.22
Cellulosic ethanol	756	76,000	1.76
Methanol via gasification	606	66,000	1.28
Hydrogen via gasification	543	59,000	1.05
FT diesel	854	86,000	1.80

[a] Per barrel per day gasoline equivalent.
[b] Gallons gasoline equivalent.

Cellulosic feedstock costs were assumed to be 50 US$·t^{-1} and of the price of corn was assumed to be 2.12 US$ per bushel (1 US bushel = 35.1 l). A summary of their main results is presented in Table 15.7. It can be seen that the capital costs are predicted to be much higher than for first-generation technology. The smaller number of unit operations causes thermochemical hydrogen production to show lower capital costs than the other gasification-based options (methanol, FT diesel). Hydrogen, though still has a long way to go for fuel implementation as storage and distribution systems, must be developed, whereas the liquid fuels can be implemented in current infrastructure fordistribution. Methanol, however, is not yet widely distributed as fuel, which is related to its associated relative toxicity. Cellulosic ethanol and FT diesel come out at approximately similar values. Future enhancement in biochemical and thermochemical platform technologies has a real potential for the reduction of biofuel cost prices.

15.4   OUTLOOK TO THE FUTURE OF BIOREFINERIES

The future sustainable society will expectedly be based on biomass as one of the key renewable sources for food, feed, materials, chemicals, transportation fuels, and (combined) heat and power (CHP). A framework for this is offered by biorefineries. The successful further development of such biorefineries is an exciting area of multidisciplinary science integration. Knowledge of plant biology, chemistry, physics, geoscience, economics, logistics and infrastructure, and engineering needs to be integrated and applied in this field. New synergies between such widely differing science fields thereby need to be elaborated and established (Aresta et al., 2012). Next to such a development, one needs to consider or rather envision that the context will change, such as novel technologies for transportation, new ways of logistics, changed policies, economy, and social science boundary conditions. In order to open up the full potential of biorefineries, further system and technology development is needed; research, development, and demonstration programs thereby can link industry, research institutes, universities, governmental bodies, and nongovernmental organizations, while market introduction strategies are required to be developed in parallel.

CHAPTER SUMMARY AND STUDY GUIDE

In this chapter, the biorefinery is introduced as an integrated concept for processing of multisource biomass into a spectrum of different products for supply of energy forms, chemicals, and materials. The basic types of biorefineries are based on thermochemical, biochemical, and hybrid concepts. A framework for economic evaluation of different processes is presented.

KEY CONCEPTS

Definition of a biorefinery
Identification of types of biorefineries, different platforms (sugar, thermochemical via syngas as examples)
Selection of processing routes for production of compounds from multiple biosources
Payout Time
Time value of money: net present value, discounted cash flow rate of return principles to evaluate the economic impact of a project and to compare alternatives
Evaluation of investment costs of process installations
Calculation of the cost + return price

SHORT-ANSWER QUESTIONS

15.1 Mention the main differences between an oil refinery and a biorefinery.

15.2 Sometimes, one sees the term "oleochemical-based biorefinery." What is meant by this term and can you give a few examples?

15.3 One sometimes uses the production of base chemicals as a useful indicator for the growth of the petrochemical industry. What would you use as indicator for growth in the newly established bio-based economy?

15.4 Does crude oil need to be pretreated before entering the atmospheric distillation unit? If so, which pretreatment is required?

15.5 Figure 15.2 shows a schematic of a modern oil refinery. Identify routes based on biomass for producing all the products/product classes in this schematic.

15.6 A catalytic cracking unit in an oil refinery produces the so-called amylenes that can be dehydrogenated (abstraction of hydrogen). Which product is formed? Is there a simple biomass-derived process possible to produce the same chemical?

15.7 Write out the reaction equations for syngas fermentation to acetic acid with either CO or CO_2 as reacting species.

15.8   An EIA report (tinyurl.com/bpnnuwh) refers to four classes of liquid biofuels as "biomass-based biodiesel," "conventional biofuels," "cellulosic biofuels," and "noncellulosic advanced biofuels." Can you mention at least two examples of fuels per class?

15.9   In a process, ethanol is produced from corn using hydrolysis and subsequent fermentation of the starch-derived sugar fraction. The residues are used to feed cattle.

   a. Which type of biorefinery is this?
   b. Characterize the complexity using the *BCI*.

15.10 When biomass is pyrolyzed and the bio-oil/char mixture that is produced is fed with heavy oil residues into a fluid catalytic cracker in an oil refinery, is this a biorefinery?

15.11 Could an anaerobic digester also be called a biorefinery? If not, under which conditions could it be set up as such? When manure is used as feedstock only and biogas is produced together with digestate, what is then the *BCI* value?

15.12 Name examples of biorefineries that were operated already more than 100 years ago.

15.13 Name at least five different pretreatment techniques for biomass by which fractionation of the main biomass contained polymers can be realized.

15.14 In Figure 15.4, identify what you think are major technological challenges and discuss these in a group. Which cycles are aimed to be closed?

15.15 Why would lignin valorization to high added value products in newly established modern biorefineries be rather a necessity than just an option?

15.16 Gonzalez et al. (2012) report on the cost of different types of equipment in a biorefinery concept for cellulosic ethanol production using gasification. What is the explanation for investment scale factors smaller than one? What is the explanation for scale factors larger than one? In case of a huge scale factor, what would you propose?

PROBLEMS

15.1   Set up two process schemes for the production of ethanol from wood, one based on a thermochemical platform and the other based on a biochemical platform. What are advantages and disadvantages of both schemes?

15.2   In a biorefinery concept, biomass is pretreated using organosolv so that a cellulose stream is produced that does not dissolve in the organic solvent and is used for paper production; the lignin and hemicellulose parts, though, dissolve and are further separated; the lignin is burned for steam production and the hemicellulose is hydrolyzed to sugars with production of furanic compounds for blending into transportation fuel. Which type of biorefinery is this?

15.3   In the petrochemical industry, the production of olefins, in particular ethylene, is important. This is the raw material for polyethylene production. Now, based on biomass, propose a thermochemical and a biochemical biorefinery concept in which ethylene can be (co)produced. Can you say something about the economics of these processes?

15.4   A consortium of parties decides to work out a biorefinery concept based on rapeseed input only; it produces a biodiesel by transesterification with methanol, crude glycerol stream, and a press cake that can be used as cattle feed. The input is 675 t·d^{-1} of rapeseed (consider it dry), and the press cake production is 375 t·d^{-1}. Neglect catalyst input (KOH) needed. For simplification, consider rapeseed oil as the triglyceride ester of oleic acid (formula: $CH_3(CH_2)_7CH=CH(CH_2)_7COOH$), which is the most important oil constituent, and glycerol.

a. How much methanol (t.d^{-1}) would be needed at stoichiometric, complete conversion? Can this be produced with minimal fossil footprint?
b. How much crude glycerol is produced?
c. Name some applications of glycerol.
d. What might limit the production of biodiesel to less than the calculated amount?
e. Which type of biorefinery is this?

15.5   A company producing cola wants to increase the production of "green bottles" made of PET (polyethylene terephthalate). Therefore, they plan to build a plant to produce ethylene glycol (one of the two monomers of PET) from sugarcane residues (bagasse). The capacity will be 500 kt.year^{-1}.

a. Make a block process diagram of how this biorefinery could look like.
b. Give the main reactions taking place in the process leading to ethylene glycol.
c. How much bagasse (kt.year^{-1}) would be needed in this process when bagasse is assumed to consist of 38 wt% cellulose (db), 27 wt% hemicellulose, 20 wt% lignin, 3 wt% proteins, 9 wt% extractives (water soluble), and 3 wt% ash. Consider ethanol to be an intermediate product, which is only made from the cellulose part via enzymatic hydrolysis and subsequent sugar fermentation.
d. Which products can be made from the noncellulosic part of bagasse?

15.6   Check the derivation of Equation (15.11) for the calculation of $NPV_{i\%}$ taking into account the value of the particular sum called the geometric progression (also called geometric sequence).

15.7   Regarding Example 15.3, investigate the impact of the interest rate on NPV and the sensitivity values. What can you conclude from this exercise?

15.8   A 200 ML/year producing corn-based ethanol production facility is reported to have cost 80 million dollar in 2008. What would the investment be now?

15.9   Carefully read the article of Haas et al. (2006), describing an economic evaluation of a biodiesel production facility based on soybean oil. Use the methodologies for estimating the investment presented in this chapter for the case presented in the article. Can you comment on the differences?

15.10   A biorefinery process called "Biofine" has been presented in the recent past (Kamm and Kamm, 2004). It is a biomass-based process route making use of acid hydrolysis and dehydration subprocesses and esterification with ethanol to ethyl levulinate (EL) (an ester of levulinic acid and ethanol). By-products considered are power and formic acid (FA). The production of EL is 133 kt. year^{-1}. The capital cost is 150 million US$ (consider linear depreciation in 10 years). Table 15.8 gives an overview of the prices of the raw materials and by-products. In addition, the water supply costs are US$ 500,000/year. Regarding labor, there are 17 operators per shift working at a salary of US$ 20/h and two supervisors per shift working at a salary of US$ 24/h. Assume an ROI of 15%. For other costs, take the guidelines given in this chapter (Table 15.6).

a. Calculate the cost and return price in US $ per tonne EL produced.
b. What is the price in US $ per GJ HHV? (hint: calculate the heat of combustion of EL).
c. Is it possible to produce the required ethanol in the process itself?

TABLE 15.8   Overview of costs, yields of by-products, and material amounts for the "Biofine" process

Raw material/utility/by-product	Amount	Price in US$
Feedstock	350 kt·year^{-1}	40·t^{-1}
Sulfuric acid	3.5 kt·year^{-1}	100·t^{-1}
Caustic soda	0.5 kt·year^{-1}	120·t^{-1}
Ethanol	35 kt·year^{-1}	350·t^{-1}
Hydrogen	0.12 kt·year^{-1}	1500·t^{-1}
Ash disposal	17.5 kt·year^{-1}	35·t^{-1}
Power exported	3.1 MW	60 MWh^{-1}
Formic acid sold	38.5 kt·year^{-1}	110·t^{-1}

PROJECTS

P15.1 Visit an existing biorefinery, like a sugar mill or a potato factory. What type of biorefinery is it? Try to set up mass and energy balances for this plant.

P15.2 You can make your own biodiesel and glycerol from a biomass-derived source. Find a synthesis description in literature. Select a vegetable oil and perform in your lab the transesterification reaction producing this oil and glycerol by-product. Now study the impact of air and/or light admission to a test tube on the quality of the biodiesel obtained. Which physical characteristics of the biodiesel can you use to perform this study as a function of time? Which would be the best/most simple?

P15.3 Identify, by performing a literature survey, potential processes that might be applied for processing lignin in a biorefinery completely based on biochemical processes.

P15.4 This project deals with biorefining based on a cultivated crop in Kenya. The main issue in producing biofuel from dedicated crops is that their relatively high energetic value is generally directly linked to extensive requirements with respect to growing conditions such as water and fertility of soil and the intensive agricultural needs. This means that crops providing fuel(s) compete with food crops for nurturing a society. Furthermore, locations that depend on large amounts of fuel are predominantly situated in densely populated areas with limited means of agricultural activities. Relocation to cheaper, less populated regions would induce an imbalance regarding these locations that are prone to inequality in food yields to start with.

Jatropha curcas is a plant that may positively address the aforementioned issues. This plant has originated in Central America and is now found throughout the tropics, including Africa and Asia. The plant generates seeds that are poisonous to humans and is a robust crop species, capable of growing on infertile, gravelly, sandy, or even saline soils. Therefore, it does not require arable land, and since it thrives in warmer climates and is even tolerant to severe heat, it can be cultivated in areas that are not suitable for food-based agriculture. Although it needs water to produce seeds, it can withstand drought for as long as 2 years to regrow when irrigated. Its robust characteristics reduce the labor effort otherwise involved in growing vegetation and add to its benefits of yielding from otherwise arid patches of land.

Estimates of yields of *Jatropha* seeds vary from 1.5 to 3.5 tonnes/ha, which translates into roughly 540–1260 l of *Jatropha* oil per ha (Dar, 2007; Ofori-Boateng and Teon, 2011), with potential for improvement. With the large amounts of oil needed, it is hard to stake claims on its production not diverting resources away from crops that are used for food

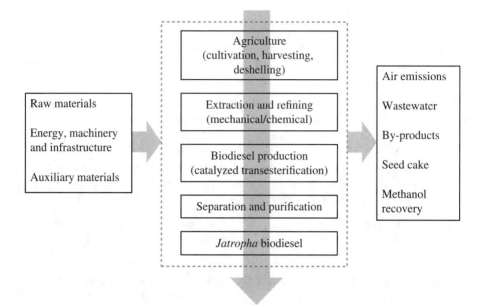

FIGURE 15.8   *Jatropha* biodiesel generation, a generic scheme of processing.

production, since substantial soil capacity is required. However, as older plants are able to sequester carbon, absorbing around 7.9 kg of CO_2 annually, *Jatropha* harvesting can serve purposes other than filling the looming fuel gap. A study by Yale University has confirmed that renewable jet fuel made from *Jatropha* oil can reduce greenhouse gas emissions by as much as 91% (Bailis and Baka, 2010).

Oil extracted from the *Jatropha* has characteristics similar to fossil-based diesel fuel. Due to this similarity, it can be used in most diesel engines, and others can run on it with slight modifications (Nahar and Ozores-Hampton, 2011). The oil can be used directly in fuel combustion engines or may be subjected to transesterification in order to produce biodiesel. In this example, we will be taking this extra step to produce biodiesel.

The *Jatropha* oil is obtained by simply mechanically extracting it from the seeds by using a screw press. In order to produce biodiesel, an esterification process is used to produce the biofuel, which then only needs to be purified. This process is shown in Figure 15.8, which also shows the extra ingredients (left side) and by-products (right side) of this process.

Various processes can yield different rates of extraction; we choose high-efficiency mechanical extraction, which is 91% efficient and uses 0.4 kWh·kg^{-1} (Nahar and Ozores-Hampton, 2011). The seeds contain on an *ar* mass basis between 27 and 40% oil, with an average of 34% (Achten et al., 2007). The residue of the seed is substantial and can be used in various industries including in our own process.

Assumptions made for the Project

For this project aiming at making an economic analysis of the conversion of *Jatropha* seeds into biodiesel, the following input parameters are given:

- The land area is 300 ha.

- Seeds will be harvested once a year.

- Fertilizer addition is required once in the beginning of the projects at 32.4 kg. ha^{-1}; costs are 240 US$.t^{-1}.

- Local power price is estimated at 0.18 US$·(kWh)$^{-1}$.

- Lifetime for scenario A is 50 years, and that for scenario B 30 years.

- Oil content is 40 wt% for scenario A and 27 wt% for scenario B.

- Extraction efficiency by mechanical pressing is 91% in both scenarios.

- Oil-to-biodiesel conversion rate is 87%.

- Potassium hydroxide (30.8 US$·kg^{-1}) is needed at a rate of 0.15 mol·l^{-1} oil and methanol (389 US$·t^{-1}) at 5.1 mol·l^{-1} oil.

- Biodiesel price is assumed at 168 US$ per barrel.

- Tax on profit (t) is 37.5%.

- Interest rate (i) is 12%.

- Taking into account off-line periods for maintenance, repairs, and holidays, the plant will be in operation for 230 days/year.

- Linear depreciation of investments.

- Assume 0.4 labor forces per ha.

- Make your own assumptions based on literature search for labor wages and land cost given the Kenyan context.

- Assume no irrigation costs due to sufficient rainfall.

The calculations are to be made for two scenarios, in which only the yield from the crops differs, in order to be able to compare best- and worst-case scenarios. Scenario A will represent the best-case scenario (with agricultural practice improvements) in which the yield is 12.5 t.ha^{-1}, whereas scenario B will represent the worst-case scenario where the yield is only 5.25 t.ha^{-1}.

In order to quantify the inputs that are required to produce a viable source of fuel from the *Jatropha* crop, economic calculations will be made, based on the parameters explained in Section 15.3. For both scenarios, calculate the fixed capital investment, total capital investment, profit overview per year, *CF* overview per year, *POT*, *ROI*, and finally the *NPV*. What are your conclusions about the project scenarios?

INTERNET REFERENCES

tinyurl.com/bpnnuwh
www.eia.gov/forecasts/aeo/pdf/0383(2012).pdf

tinyurl.com/jvvr7hw
www.chemrec.se

tinyurl.com/klwrfa7
http://www.nrel.gov/biomass/biorefinery.html

tinyurl.com/kzhn5ej
http://www1.eere.energy.gov/bioenergy/

tinyurl.com/les4tw4
www.coscata.com

tinyurl.com/lhppp6d
www.che.com

tinyurl.com/lj7nws2
www.nesteoil.com

tinyurl.com/ltuzewk
www.betarenewables.com/PROESA-technology.html

tinyurl.com/l3pfuy5
www.biomcn.eu

tinyurl.com/myclw39
www.borregaard.com

tinyurl.com/nl7qt3n
http://demoplants.bioenergy2020.eu/projects/mapindex

tinyurl.com/nqf7s5s
www.lanzatech.com

tinyurl.com/n2bhpmx
www.abengoabioenergy.com/web/en/2g_hugoton_project

tinyurl.com/o4jemkt
www.petrochemistry.net

tinyurl.com/pj2x3or
www.iea-bioenergy.task42-biorefineries.com

tinyurl.com/6e7nks
www.natureworksllc.com

REFERENCES

Achten WMJ, Mathijs E, Verchot L, Singh VP, Aerts R, Muys B. *Jatropha* biodiesel fueling
    sustainability. Biofuels Bioprod Biorefin J 2007;1(4):283–291.

Allen TD, Caldwell ME, Lawson PA, Huhnke RL, Tanner RS. Alkalibaculum bacchi gen. nov., sp. nov., a CO-oxidizing, ethanol-producing acetogen isolated from livestock-impacted soil. Int J Syst Evol Microbiol 2010;60:2483–2489.

Aresta M, Dibenedetto A, Dumeignil F. *Biorefinery: From Biomass to Chemicals and Fuels.* Berlin/Boston: De Gruyter; 2012.

Bailis RE, Baka JE. Greenhouse gas emissions and land use change from *Jatropha curcas*-based jet fuel in Brazil. Environ Sci Technol 2010;44:8684–8691.

Boerrigter H, Deurwaarder EP, et al. Contributions ECN biomass to the 2nd world conference and technology exhibition on biomass for energy, industry and climate protection. Petten (The Netherlands): ECN; 2004. p 67–72.

Cherubini F, Jungmeier G, Wellisch M, Willke T, Skiadas I, Ree RV, De Jong E. Toward a common classification approach for biorefinery systems. Biofuels Bioprod Biorefin J 2009;3:534–546.

Couper JR. *Process Engineering Economics.* New York: CRC Press; 2003.

Dar WD. Research needed to cut risks to biofuel farmers. *Science and Development Network.* London, UK; December 6, 2007. Available at www.SciDevNet. Accessed May 30, 2014.

De Wild PJ. Biomass pyrolysis for chemicals [PhD thesis]. Groningen (The Netherlands): Groningen State University; 2011.

De Wild PJ, Reith H, Heeres HJ. Biomass pyrolysis for chemicals. Biofuels 2011;2(2):185–208.

European Commission. Directive of the European parliament and of the council on the promotion of the use of energy from renewable sources, COM (2008) *30 Final.* Brussels: European Commission; 2008. p 61.

Glesinger E. *The Coming Age of Wood.* New York: Simon and Schuster, Inc; 1949.

Gonzalez R, Daystar J, Jett M, Treasure T, Jameel H, Venditti R, Phillips R. Economics of cellulosic ethanol production in a thermochemical pathway for softwood, hardwood, corn stover and switchgrass. Fuel Process Technol 2012;94(1):113–122.

Grethlein AJ, Worden RM, Jain MK, Datta R. Continuous production of mixed alcohols and acids from carbon monoxide. Appl Biochem Biotechnol 1990;24:875–884.

Haas MJ, McAloon AJ, Yee WC, Foglia TA. A process model to estimate biodiesel production costs. Bioresour Technol 2006;97:671–678.

Heiskanen H, Virkajärvi I, Viikarib L. The effect of syngas composition on the growth and product formation of *Butyribacterium methylotrophicum.* Enzyme Microb Technol 2007;41:362–367.

Jung GY, Kim JR, Jung HK, Park J Y, Park S. A new chemoheterotrophic bacterium catalyzing water-gas shift reaction. Biotechnol Lett 1999;21:869–873.

Jungmeier G, Hingsamer M, et al. *Biofuel-driven Biorefineries—a selection of the most promising biorefinery concepts to produce large volumes of road transportation biofuels by 2025.* Wageningen (the Netherlands): IEA Bioenergy—Task 42 Biorefinery; 2013. p 34.

Jungmeier G, Jorgensen H, et al. *Do we need a biorefinery complexity index? (paper 2AO.7.2). 20th European Biomass Conference. A. Grassi.* Milan (Italy): ETA Florence; 2012.

Kamm B, Gruber PR, Kamm M, editors. *Biorefineries—Industrial Processes and Products; Status Quo and Future Directions.* Weinheim: Wiley-VCH Verlag GmbH & Co. KGaA; 2006.

Kamm B, Kamm M. Principles of biorefineries. Appl Microbiol Biotechnol 2004;64(2):137–145.

Klasson KT, Ackerson MD, Clausen EC, Gaddy JL. Bioreactor design for synthesis gas fermentations. Fuel 1991;70:605–614.

Knoef H. *Handbook Biomass Gasification*. 2nd ed. Enschede (The Netherlands): Biomass Technology Group; 2012.

Liou JS, Balkwill DL, Drake GR, Tanner RS. *Clostridium carboxidivorans* sp. nov., a solvent-producing clostridium isolated from an agricultural settling lagoon, and reclassification of the acetogen *Clostridium scatologenes* strain SL1 as *Clostridium drakei* sp. nov. Int J Syst Evol Microbiol 2005;55:2085–2091.

Lorowitz WH, Bryant MP. Peptostreptococcus productus strain that grows rapidly with CO as the energy source. Appl Environ Microbiol 1984;47:961–964.

Merida W, Maness P-C, Brown RC, Levin DB. Enhanced hydrogen production from indirectly heated, gasified biomass, and removal of carbon gas emissions using a novel biological gas reformer. Int J Hydrogen Energy 2004;29:283–290.

Mohammadi M, Najafpour GD, Younesi H, Lahijani P, Uzir MH, Mohamed AR. Bioconversion of synthesis gas to second generation biofuels: a review. Renew Sustain Energy Rev 2011;15:4255–4273.

Mosier NS, Wyman C, Dale B, Elander R, Lee YY, Holtzapple M, Ladisch M. Features of promising technologies for pretreatment of lignocellulosic biomass. Bioresour Technol 2005;96:673–686.

Moulijn JA, Makkee M, van Diepen A. *Chemical Process Technology*. 1st ed. New York: John Wiley & Sons Ltd; 2001.

Nahar K, Ozores-Hampton M. *Jatropha*: an alternative substitute to fossil fuel. Immokalee (FL, USA), Horticultural Sciences Department, Florida Cooperative Extension Service, Institute of Food and Agricultural Sciences, University of Florida; 2011. Report nr HS1193. p 10.

National Non-Food Crops Centre (NNFCC). *Biorefineries: definitions, examples of current activities and suggestions for UK development*. York (UK): The National Non-Food Crops Centre (NNFCC); 2007. p 30.

Ofori-Boateng C, Teon LK. *Feasibility of Jatropha Oil for Biodiesel: Economic Analysis*. Linköping: World Renewable Energy Congress; 2011. pp 463–470.

Peters MS, Timmerhaus KD. *Plant Design and Economics for Chemical Engineers*. 4th ed. New York: McGraw-Hill, International Editions; 1991.

Sakai S, Nakashimada Y, Yoshimoto H, Watanabe S, Okada H, Nishio N. Ethanol production from H_2 and CO_2 by a newly isolated thermophilic bacterium, Moorella sp. HUC22-1. Biotechnol Lett 2004;26:1607–1612.

Sun Y, Cheng J. Hydrolysis of lignocellulosic materials for ethanol production: a review. Bioresour Technol 2002;83:1–11.

Tanner RS, Miller LM, Yang D. *Clostridium ljungdahlii* sp. nov., an acetogenic species in clostridial rRNA homology group I. Int J Syst Bacteriol 1993;43(2):232–236.

Towler G, Sinnott R. *Chemical Engineering Design—Principles, Practice and Economics of Plant and Process Design*. 2nd ed. Oxford: Butterworth-Heinemann; 2012.

Wright MM, Brown RC. Comparative economics of biorefineries based on the biochemical, and thermochemical platforms. Biofuels Bioprod Biorefin J 2007;1:49–56.

Younesi H, Najafpour GD, Ku Ismail KS, Mohamed AR, Kamaruddin AH. Biohydrogen production in a continuous stirred tank bioreactor from synthesis gas by anaerobic photosynthetic bacterium: Rhodopirillum rubrum. Bioresour Technol 2008;99:2612–2619.

Zinoviev S, Müller-Langer F, Das P, Bertero N, Fornasiero P, Kaltschmitt M, Centi G, Miertus S. Next-generation biofuels: survey of emerging technologies and sustainability issues. ChemSusChem 2010;3:1106–1133.

PART IV

END USES

16

HIGH-EFFICIENCY ENERGY SYSTEMS WITH BIOMASS GASIFIERS AND SOLID OXIDE FUEL CELLS

P.V. Aravind and Ming Liu

Department of Process and Energy, Energy Technology Section, Faculty of Mechanical, Maritime and Materials Engineering, Delft University of Technology, Delft, the Netherlands

ACRONYMS

AFC	alkaline fuel cell
ESP	electrostatic precipitator
FC	fuel cell
GDC	gadolinium-doped ceria
GT	gas turbine
LHV	lower heating value
LSCF	lanthanum strontium cobalt ferrite
LSM	lanthanum strontium manganite
MCFC	molten carbonate fuel cell
PAFC	phosphoric acid fuel cell
PEMFC	polymer electrolyte membrane fuel cell
SOFC	solid oxide fuel cell
TPB	triple-phase boundary
WESP	wet electrostatic precipitator
YSZ	yttria-stabilized zirconia

Biomass as a Sustainable Energy Source for the Future: Fundamentals of Conversion Processes,
First Edition. Edited by Wiebren de Jong and J. Ruud van Ommen.
© 2015 American Institute of Chemical Engineers, Inc. Published 2015 by John Wiley & Sons, Inc.

SYMBOLS

E	reversible (Nernst) voltage	[V]
F	Faraday constant	[C·mol^{-1}]
$\Delta\bar{g}$	molar Gibbs energy change	[J·mol^{-1}]
Δh	molar enthalpy change	[J·mol^{-1}]
I	current	[A]
n	number of electrons	[–]
$\dot{n}_j$	molar flow rate of species j	[mol·s^{-1}]
P	power	[W]
p	pressure	[Pa]
R_u	universal gas constant	[J·mol^{-1}·K^{-1}]
R	electrical resistance	[Ω]
T	temperature	[K]
U_j	utilization factor of species j	[–]
$\dot{U}$	total chemical power	[W]
x	mole fraction	[–]
V	cell voltage	[V]
η_a	anodic polarization losses	[V]
η_c	cathodic polarization losses	[V]
η_{rev}	reversible efficiency	[–]
φ_m	mass flow rate	[kg·s^{-1}]

Subscripts

e	electrical
f	fuel
i	internal
r	reaction

16.1   INTRODUCTION

Fuel cells (FC) are energy conversion devices that directly convert the chemical energy of a fuel into electrical energy. There are five major types of fuel cells, namely, phosphoric acid fuel cells (PAFCs), polymer electrolyte membrane fuel cells (PEMFCs), alkaline fuel cells (AFCs), molten carbonate fuel cells (MCFCs), and solid oxide fuel cells (SOFCs). These FC types operate in different temperature regimes, use different materials, and have different fuel tolerance and performance characteristics (O'Hayre et al., 2009). MCFCs and SOFCs are high-temperature (>600°C) FC that can operate using various fuels, such as H_2, CO, and CH_4. Because syngas produced by biomass gasification contains mainly H_2, CO, and CH_4, biosyngas can be used as a fuel for MCFCs and SOFCs provided that it is sufficiently clean.

High-temperature FC also provide heat, which can be converted into electricity using conventional technologies, such as gas turbines (GT) or steam turbines, resulting in even more efficient energy systems. Moreover, biomass conversion is carbon neutral and offers decentralized energy generation. Although the MCFC is suitable for biosyngas conversion, in this chapter, we will focus on the SOFC as this type of FC has a higher resistance to gas contaminants, which is important when using biosyngas as fuel.

16.2 SOLID OXIDE FUEL CELLS

16.2.1 Principle of an SOFC

SOFCs typically work at temperatures in the range of 600–1000°C. The operation of an SOFC is presented schematically in Figure 16.1. In an SOFC with an oxide ion-conducting electrolyte, the fuel enters the anode chamber and is electrochemically oxidized. Oxygen enters the cathode chamber and is ionized and transported through the solid electrolyte to the anode. The porous anode disperses the fuel gas over its interphase with the electrolyte, allowing the reaction products to diffuse through the interphase to become mixed with the anode flow. The anode catalyzes the chemical and electrochemical reactions and conducts the electrons that are freed. These electrons flow through an external circuit to the cathode, delivering electrical power. The cathode distributes the oxygen at its interphase with the electrolyte and conducts the electrons from the external circuit so that oxygen molecules are reduced into oxide ions. Oxide ions are conducted through the electrolyte to the anode. The electrolyte contains many oxygen vacancies that allow oxygen ions to hop through. The electrolyte mainly prevents the two electrodes from coming into electronic

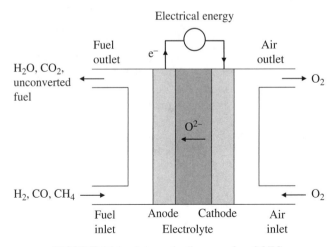

FIGURE 16.1 Schematic diagram of an SOFC.

contact by blocking the electrons and allows the flow of oxide ions from the cathode to the anode to maintain the overall electrical charge balance and to close the current circuit. In addition, the electrolyte determines the operating temperature of the FC.

In a H_2–O_2 SOFC, the electrochemical reactions occurring are

$$H_2 + O^{2-} \rightarrow H_2O + 2e^- \qquad \text{(RX. 16.1)}$$

at the anode and

$$\frac{1}{2}O_2 + 2e^- \rightarrow O^{2-} \qquad \text{(RX. 16.2)}$$

at the cathode. The resulting overall reaction thus is

$$H_2 + \frac{1}{2}O_2 \rightarrow H_2O \qquad \text{(RX. 16.3)}$$

As there are no irreversible combustion processes, energy conversion in FC is not limited to the efficiency of the Carnot cycle; thus, they possess a high reversible efficiency Equation (16.1):

$$\eta_{\text{rev}} = \frac{\Delta \bar{g}_r}{\Delta h_r} \qquad \text{(Eq. 16.1)}$$

In Equation (16.1), $\Delta \bar{g}_r$ is the change of Gibbs energy (J mol^{-1}), and Δh_r is the enthalpy change (J mol^{-1}).

For an FC operating at a constant temperature and pressure, the maximum electrical work that the FC can perform is given by the negative of the Gibbs free energy difference Equation (16.2):

$$\Delta \bar{g}_f = -nFE \qquad \text{(Eq. 16.2)}$$

where n is the number of electrons transferred for each molecule of fuel, F is the Faraday constant (= 96,485 C mol^{-1}), and E is the reversible voltage (V).

The following Nernst equation gives the reversible voltage for the total cell reaction occurring in the FC when hydrogen is used as fuel:

$$E = E^0 + \left(\frac{R_uT}{2F}\right) \times \ln\left(\frac{p_{H_2} \times p_{O_2}^{1/2}}{p_{H_2O}}\right) \qquad \text{(Eq. 16.3)}$$

with E^0 being the standard reversible voltage at the temperature of interest:

$$E^0 = -\frac{\Delta \bar{g}_r^0}{2F} \qquad \text{(Eq. 16.4)}$$

In Equation (16.3), R_u is the universal gas constant $(= 8.3143 \text{ J mol}^{-1} \text{ K}^{-1})$, T is the operating temperature (K), and p_i is the partial pressure of the reactants and products (Pa). In Equation (16.4), $\Delta \bar{g}_r^0$ is the standard-state Gibbs free energy change for the reaction (J mol^{-1}).

However, the useful voltage output (V) under load conditions, i.e., when a current passes through the cell, is given by

$$V = E - I \times R_i - \eta_c - \eta_a \qquad \text{(Eq. 16.5)}$$

where n is the number of electrons transferred for each molecule of fuel, R_i is the internal electrical resistance of the SOFC (Ω), I is the current passing through the cell (A), and η_c and η_a are polarization losses associated with the cathode and anode, respectively. The voltage loss due to internal electrical resistance includes contributions from the electrodes and the electrolyte, with the highest contribution originating from the ionic conduction through the solid electrolyte. For a further explanations of cell voltage losses, readers are referred to O'Hayre et al. (2009) and Singhal and Kendall (2003).

The current flow is always proportional to the number of oxide ions reacting with the fuel molecules and thus to the amount of fuel consumed, as expressed by (16.6)

$$\dot{n}_j = \frac{I}{nF} \qquad \text{(Eq. 16.6)}$$

Example 16.1

For the following reaction in an SOFC working at 800°C:

$$H_2 + 1/2\ O_2 \rightarrow H_2O \text{ with } \Delta \bar{g}_r = -188.6 \text{ kJ mol}^{-1}$$

1. Calculate the reversible voltage.

2. Calculate the hydrogen flow (mol s^{-1}) required to produce 1 ampere current.

Solution

1. According to Equation (16.4), the reversible voltage is

$$E = -\frac{-188.6 \times 1,000}{2 \times 96,485} = 0.977 \text{ V}$$

2. According to Equation (16.6), the hydrogen flow rate is

$$\dot{n}_{H_2} = \frac{1}{2 \times 96,485} = 5.18 \times 10^{-6} \text{ mol·s}^{-1}$$

Example 16.2

An SOFC operating with natural gas produces 2 W electric power (P_e) with an electrical efficiency (η_e) of 52% based on the lower heating value (LHV) of the natural gas.

1. Calculate the heat released by the FC per unit time.

2. If 70% of the available heat can be recovered by heating water from 20 to 70°C, calculate the amount of hot water (g) produced when the FC is operating for 1 h. The specific heat capacity of water, $c_{p,w}$, is 4.186 J g^{-1}·K^{-1}.

Solution

1. The total chemical power of the natural gas ($\dot{U}$) is

$$\dot{U} = \frac{P_e}{\eta_e} = \frac{2}{0.52} = 3.85 \text{ W}$$

The amount of heat (Q) released by the FC is

$$Q = \dot{U}(1-\eta_e) = 3.85 \times (1-0.52) = 1.848 \text{ W}$$

2. 70% heat recovered is 0.7 × 1.848 = 1.30 W.
   The mass flow rate of water φ_m is

$$\varphi_m = \frac{Q}{c_{p,w}\Delta T} = \frac{1.30}{4.186 \times (70-20)} = 6.2 \times 10^{-3} \text{ g·s}^{-1}$$

The amount of hot water produced during 1 h of operation is

$$6.2 \times 10^{-3} \times 3600 = 22.4 \text{ g}$$

16.2.2   Anode

Cermet (ceramic–metallic) anodes are often used for SOFCs to ensure thermal and chemical compatibility in high-temperature SOFC environments. While anodes are necessary for the electrochemical fuel oxidation to take place, they also cause losses in power production. The anodic losses are due to internal electrical resistance, contact resistance, mass transfer limitations, and the limited rate of the electrochemical reactions (Zhu and Deevi, 2003). The internal resistance is the resistance to the transport of electrons and oxide ions within the anode. The contact resistance is caused by poor adherence between the anode and the solid electrolyte and is generally not affected by fuel variations. The mass transport limitation is related to the diffusion of gas-phase species through the porous electrode. The porosity or microstructure of the electrode is an important parameter affecting gas diffusion. Mass transfer losses may become significant at higher current flows and at greater fuel utilization. Anodic activation losses caused by the slow electrochemical reactions are related to the charge transfer processes at the anode and depend on the length of the electrode/electrolyte/gas triple-phase boundary (TPB) and the electrocatalytic activity of the electrode itself. The effective electrochemical reaction zone at the anode of an SOFC is mainly limited to physical TPB for anodes made of nickel/yttria-stabilized zirconia (Y_2O_3-stabilized ZrO_2, YSZ), which is currently the most common anode material for SOFC applications. Nickel serves as an electrocatalyst for the electrochemical oxidation of hydrogen and as a reforming catalyst for carbonaceous fuels. Nickel also provides electronic conductivity for the anode. YSZ is an oxide ion conductor and provides a framework for the dispersion of Ni particles.

Oxides with a perovskite structure are attractive as anode material because of their low sensitivity to sulfur poisoning and carbon deposition (Fergus, 2006). However, due to their lower electronic conductivity, they also require an additional electron collector as is the case for ceria-based anodes. The high electrical conductivity of copper has led to many studies on anodes containing copper due to the expectation that these anodes will present advantages when operated with hydrocarbon fuels. The low catalytic activity of copper reduces carbon deposition. Copper-containing anodes are used along with YSZ or ceria. Although these anodes present advantages such as low carbon deposition, the fact that they sinter at high temperature makes them useful only at a rather low operating temperature of around or below 600°C. Currently, researchers also are developing anodes with new materials, but reliable operation of these anodes has not yet been proven.

State-of-the-art FC are generally supported on electrolytes. Anode-supported cells with thin electrolytes are also under development. Such cells are preferred for low-temperature operation. The application of a thin electrolyte helps to reduce the electrolyte losses that can occur at low temperatures. Selection of the proper anode material and operating parameters such as working temperature are critical issues when SOFCs are used with fuels other than hydrogen such as biosyngas.

16.2.3 Electrolyte

Several ceramic materials have been employed as active SOFC components (Stambouli and Traversa, 2002). The most common electrolyte to date is stabilized zirconia. An example of this is YSZ, which exhibits pure oxide ion conduction with no electronic conduction. The crystalline array of ZrO_2 has two oxide ions for every zirconium ion. However, Y_2O_3 only has 1.5 oxide ions for every yttrium ion, resulting in oxygen vacancies in the crystal structure. Oxide ions from the cathode are able to diffuse through these vacancies until they reach the anode. Other common oxide-based ceramic electrolytes that can be used in SOFCs include samaria-doped ceria and gadolinium-doped ceria. Solid electrolytes are not discussed in detail here as they are not expected to be significantly affected by changes in fuel.

16.2.4 Cathode

Just as the anode, the cathode is also a porous structure that must permit the transport of oxygen molecules (Adler, 2004). Only noble metals or electronic conducting oxides

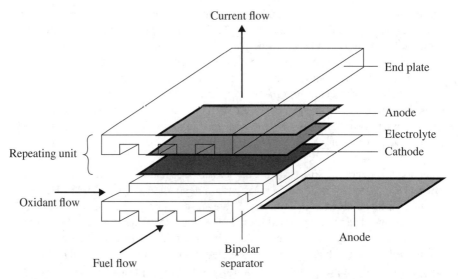

FIGURE 16.2   A typical planar SOFC configuration. (Source: Reproduced with permission from Stambouli and Traversa (2002). © Elsevier Science Ltd.)

can be used as cathode materials for SOFCs because of the high operating temperature. Noble metals are not desirable for practical applications because of their prohibitive cost and insufficient long-term mechanical stability. Perovskite-type lanthanum strontium manganite (LSM) provides excellent thermal expansion matching with zirconia electrolytes and performs well at operating temperatures above 800°C. For operation at lower temperatures, mixed ionic/electronic conducting ceramics, such as the perovskite lanthanum strontium cobalt ferrite (LSCF), are under serious consideration. Currently, several new cathode materials are under development for SOFCs, but a detailed description of such materials is beyond the scope of this book. The reader is referred to the overview works of Skinner (2001) and Jacobson (2010).

16.2.5 Stack Design

To reach sufficient power levels, FC are interconnected to form stacks. There are different stack design concepts based on tubular and flat-plate cells. Figure 16.2 shows a planar SOFC configuration. Interconnects, which are made of ceramic or metallic compounds depending on the operating temperature, connect the cells within the stack.

In comparison to planar designs, tubular designs do not require a specific seal to isolate the oxidant from the fuel, which makes the performance of the tubular cell highly stable over long-term operation. A schematic view is presented in Figure 16.3.

16.3 BIOMASS GASIFIER–SOFC COMBINATION

Considering biosyngas as a fuel for use in SOFCs raises many critical issues. One significant task is to define the tolerance limits of the anode for contaminants such as tar,

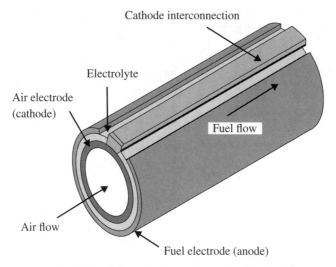

FIGURE 16.3   Schematic view of a tubular SOFC design.

particulates, sulfur compounds, and halides that may be present in the biosyngas. These contaminants may interact with different anode materials in various ways, and therefore, the tolerance levels of different anodes for these contaminants may vary considerably. This, in turn, affects the requirements for gas cleaning systems. Therefore, selecting suitable anodes, gas cleaning systems, and suitable operating parameters is critical for developing feasible and efficient biomass gasifier–SOFC systems.

In the state-of-the-art Ni/YSZ anodes, the electrode performance is most likely affected by poisoning of the anode as a result of chemical reactions and/or deposition of contaminant species on the catalyst surface due to preferential adsorption. A proper understanding of these phenomena enables determination of the cleaning requirements for the gas to be fed to the FC.

Once the influence of contaminants on SOFC performance and the effect of the SOFC operating parameters on these influences are known, the cleaning requirements and the operating parameters of the system components can be determined. Based on this information, it will be possible to select suitable cleaning devices and other system components. Additionally, different types of gasifiers can be compared with respect to their influence on SOFC operation and durability.

16.3.1   SOFC Operation with Biosyngas

A summary of possible reactions inside SOFCs when running with syngas is given in Table 16.1. Hydrogen is one of the main constituents of biosyngas and is the standard fuel for fuel cells. CO and CH_4 also are fuels for SOFCs, but they can contribute to carbon deposition (RX. 16.8, 16.9, and 16.10) under certain circumstances. The carbon deposition tendency is lower when the cell is under an electrical load. This is because the oxygen ions that pass through can oxidize the deposited carbon. Both CO_2 and N_2 (from the gasification air) affect the cell voltage. H_2O assists in the water–

TABLE 16.1   Possible reactions inside SOFCs when running with syngas (enthalpy change of reaction given at 25°C and 1 bar)

	Dry reforming	
$CH_4 + CO_2 \rightleftarrows 2\ CO + 2H_2$	$\Delta_r H^0 = 247\ kJ\ mol^{-1}$	(RX. 16.4)
	Steam reforming	
$CH_4 + H_2O \rightleftarrows CO + 3H_2$	$\Delta_r H^0 = 206\ kJ\ mol^{-1}$	(RX. 16.5)
$CH_4 + 2\ H_2O \rightleftarrows CO_2 + 4H_2$	$\Delta_r H^0 = 165\ kJ\ mol^{-1}$	(RX. 16.6)
	Water–gas shift	
$CO + H_2O \rightleftarrows CO_2 + H_2$	$\Delta_r H^0 = -41\ kJ\ mol^{-1}$	(RX. 16.7)
	Carbon formation	
$CH_4 \rightleftarrows C + 2H_2$	$\Delta_r H^0 = 75\ kJ\ mol^{-1}$	(RX. 16.8)
$2\ CO \rightleftarrows C + CO_2$	$\Delta_r H^0 = -172\ kJ\ mol^{-1}$	(RX. 16.9)
$CO + H_2 \rightleftarrows C + H_2O$	$\Delta_r H^0 = -131\ kJ\ mol^{-1}$	(RX. 16.10)
	Electrochemical reactions	
$H_2 + O^{2-} \rightleftarrows H_2O + 2e^-$	$\Delta_r H^0 = -242\ kJ\ mol^{-1}$	(RX. 16.11)
$CO + O^{2-} \rightleftarrows CO_2 + 2e^-$	$\Delta_r H^0 = -283\ kJ\ mol^{-1}$	(RX. 16.12)
$C + O^{2-} \rightleftarrows CO + 2e^-$	$\Delta_r H^0 = -111\ kJ\ mol^{-1}$	(RX. 16.13)
$C + 2O^{2-} \rightleftarrows CO_2 + 4e^-$	$\Delta_r H^0 = -394\ kJ\ mol^{-1}$	(RX. 16.14)
$CH_4 + O^{2-} \rightleftarrows CO + 2H_2 + 2e^-$	$\Delta_r H^0 = -37\ kJ\ mol^{-1}$	(RX. 16.15)
$CH_4 + O^{2-} \rightleftarrows CO_2 + 2H_2 + 2e^-$	$\Delta_r H^0 = -36\ kJ\ mol^{-1}$	(RX. 16.16)
$\frac{1}{2}O_2 + 2e^- \rightarrow O^{2-}$		(RX. 16.17)

gas shift reaction (RX. 16.7), while CH_4 may be reformed inside the cell into H_2 and CO_2 (RX. 16.4, 16.5, and 16.6).

16.3.2   Impact of Biomass-Derived Contaminants and Gas Cleaning on SOFC Operation

Biosyngas from the biomass gasifier contains various contaminants. These contaminants are mainly tar, particulate matter, alkali metal compounds and other halides, sulfur compounds, and nitrogen compounds. Many of these compounds are likely to poison the FC anode and must be removed from the gas before it enters the FC. Understanding the impact of various gaseous components on the anode performance (Aravind and De Jong, 2012; Aravind et al., 2008; Liu et al., 2013) is essential in the choice of the most suitable system components, such as the type of gasifier and gas cleaning system, for smooth FC operation.

16.3.2.1   Tars   Tar can impact the SOFC in several ways, including catalyst deactivation and FC degradation due to anode-side carbon deposits. Tar can also be reformed and subsequently oxidized, contributing to electricity production. Direct oxidation of certain tar components cannot be ruled out. It is also possible that tar molecules simply pass through the anode without exerting any significant influence. The fate of tar on

SOFC anodes depends on the type of SOFC employed and its operating conditions, such as temperature, biomass-derived product gas moisture content, and cell voltage. It also may depend upon the thermodynamic possibility of carbon deposition, the kinetics of carbon formation, and subsequent reaction steps. Several recent publications describe the influence of tar components on SOFC performance (Aravind et al., 2008; Mermelstein et al., 2010; Singh et al., 2005). Most of these studies have presented promising results regarding the tolerance of SOFCs during comparatively short exposure to tar-loaded gases (Hofmann et al., 2009). However, highly detailed theoretical and experimental studies are required to understand the fate of tar at SOFC anodes.

16.3.2.2 Particulate Matter Solid particulates are always present in the raw biosyngas generated in gasifiers. The size of the particulates present in biosyngas can range from a few micrometers to the submicrometer level (Aravind et al., 2012). Particulates generally include the inorganic material derived from mineral matter in the biomass feedstock, unconverted biomass in the form of char, and material from the gasifier bed if bed materials are employed in the gasification process (Aravind et al., 2012).

Note that these particle sizes (Hindsgaul et al., 1999) match well with the pore sizes of SOFC anodes, which also range from submicrometer level to a few micrometers (Lee et al., 2003). Thus, particulates are likely to block the micropores of the anodes if they are in the solid form at SOFC operating temperatures, thus decreasing the SOFC performance. Hofmann et al. (2008) have reported the effects of particulates from biomass gasifiers on SOFCs with Ni–GDC anodes and found that particulates up to approximately 10 μm in diameter deposited on the surface of the anode, decreasing the performance of the SOFC during the measurement. However, more information is needed on the impact of particulates on SOFC performance. It is likely that particulates should be removed as much as possible, even to the level of a few ppm, to enable smooth long-term SOFC operation with biosyngas.

16.3.2.3 Alkali Compounds Biomass often contains significant amounts of alkali compounds, mainly consisting of potassium and sodium, with the amount of potassium considerably higher than that of sodium. Eutectic potassium and sodium salts in the ash material can vaporize at gasification temperatures above 700°C. Unlike the solid particulates that can be separated by physical means such as barrier filters, the vaporized alkali compounds will remain in the product gas at high temperature. For this reason, simple filtration cannot always remove these compounds. Condensation of the vaporized alkali compounds on particles in the gas stream typically begins at approximately 650°C, with deposition subsequently occurring on cooler surfaces in the system such as heat exchangers or turbine expansion blades.

Of the total alkali content of the biomass, only a minor fraction remains in the gas phase after the gasification process. However, Nurk et al. (2011) reported that an amount of KCl as small as 6 ppm in the SOFC feed can decrease the performance of the SOFC, indicating that the alkali tolerance of SOFCs may be less than a few ppm.

16.3.2.4 Halides The halide gases obtained from gasification mainly contain HCl, with HF and HBr as the other two main constituents. Measurements have indicated that HCl is the prevalent chlorine product (Zevenhoven and Kilpinen, 2001), present up to 200 ppm (Van der Drift, 2001) depending on the type of biomass and the operating conditions of the gasifier. The presence of HCl in the feed to SOFCs can result in the adsorption of chlorine onto the surfaces of nickel particles in the anode causing a loss of nickel particles, thus degrading the FC performance (Xu et al., 2010). However, the impact of halides on SOFC performance has not been studied in detail.

16.3.2.5 Sulfur Compounds In general, biomass fuel contains much less sulfur than coal. The sulfur in biomass is converted to hydrogen sulfide or sulfur oxides during gasification, depending on the gasification system. Sulfur in the biosyngas is likely to be present mainly as H_2S (Aravind et al., 2012), and its content varies from 20 to 200 ppm (Johansson et al., 1999). The use of sorbents such as limestone can reduce the amount of H_2S present in the gas.

H_2S is likely to cause considerable problems in FC operation as it can adsorb on the active sites of the anode, thus inhibiting adsorption of the fuel molecules, which negatively affects the fuel oxidation reaction rate. The effects of sulfur on SOFC performance have been studied extensively, with particular emphasis on short-term poisoning (Cheng et al., 2007), but some uncertainties remain regarding the long-term behavior. For example, the effects of H_2S poisoning may be reversible (Rasmussen and Hagen, 2009). Furthermore, H_2S may also be a fuel for SOFCs under certain circumstances (Aguilar et al., 2004).

16.3.2.6 Nitrogen Compounds The main nitrogen-containing contaminants in biosyngas are NH_3 and HCN. The acceptable levels of ammonia in the outlet of flue gas streams from power plants are typically dictated by local regulations. Thus, ammonia is usually removed to satisfy local environmental regulations. In SOFCs, however, ammonia can be used to produce electrical power. Ammonia dissociates into N_2 and H_2 at the anode, and the H_2 is then electrochemically oxidized (Staniforth and Ormerod, 2003; Wojcik et al., 2003).

HCN is another nitrogen-containing contaminant, which can be present at levels up to a few hundred ppm (Aravind et al., 2012). The impact of HCN on the SOFC anode has not been well studied and should be explored in the future.

16.3.2.7 Other Contaminants A variety of contaminants are present in biosyngas in very small quantities. These contaminants originate from the biomass or from components of the gasifier system or possibly the cooling and the cleaning systems. Some examples are mercury, cadmium, lead, manganese, cobalt, antimony, selenium, beryllium, arsenic, chromium, nickel, and silicon. The presence of these contaminants is often limited to a few ppm or sub-ppm levels (Salo and Mojtahedi, 1998). Their impact on SOFCs has rarely been studied. Future studies are needed to understand the impact of these contaminants on the long-term operation of SOFCs.

NO_x emissions from SOFC systems are generally below 0.5 ppm (tinyurl.com/qhswo7e) and are mainly generated in the afterburner. These low concentrations of NO_x emitted by SOFCs are far below the NO_x emission limits (tinyurl.com/kdfbsbz) for combustion plants in the EU and the United States.

TABLE 16.2   Typical gas cleaning options (Liu et al. 2011)

Contaminants	Low temperature	High temperature
Particulates	Bag filter, cyclone wet scrubber, wet electrostatic precipitator (WESP)	Cyclone, electrostatic precipitator (ESP), bag filter, granular bed filter, rigid barrier filter
Tars	Wet scrubber, WESP, filter	Catalytic cracking (750–900°C) or high-temperature (900–1200°C) thermal cracking
Alkali compounds	Removal as solid particulates	Removal as solid particulates (<600°C), alkali getter (>800°C)
H_2S	Wet scrubber, activated carbon	Sorbents (>300°C, e.g., ZnO)
HCl	Wet scrubber	Sorbents (300–600°C, e.g., Na_2CO_3)

16.3.2.8   Summary of Gas Cleaning   Various systems are available for the removal of the major biosyngas contaminants to certain levels, which may not necessarily meet the requirements for SOFCs. Some of these systems are commercially available, and others are in the research and development stages. These gas cleaning techniques have to be studied in detail to explore the application in gasifier–SOFC systems. Table 16.2 summarizes the general cleaning possibilities considered at both high and low temperatures.

Efforts to design suitable cleaning systems for SOFC operation with biosyngas are still in the early stage. Presently available low-temperature cleaning systems could be the first choice for connecting SOFCs to biomass gasifiers. However, high-temperature gas cleaning systems are preferable, but their development is rather complicated and requires extensive research and development.

16.3.3   Operational Experience with SOFCs Connected to Biomass Gasifiers

A few studies have been reported on tests with FC downstream of a biomass gasifier. The Energy Center of the Netherlands (ECN) tested a Sulzer Hexis 1 kWe SOFC stack downstream of a two-stage gasifier with two fuels, willow and Rofire. Rofire is a mixture of plastics and paper waste. An electrical efficiency of 41% was observed with willow and 36% with Rofire. The longest test was reported to last 48 h, with only a slight degradation in the performance of the stack (Oudhuis et al., 2004). The Paul Scherrer Institute in Switzerland reported the testing of an SOFC stack downstream of an updraft gasifier with a ceramic filter operated at 400°C for gas cleaning. The stack was operated for a period of 100 h with a fuel consisting of 50–60% filtered fuel gas from the gasifier and hydrogen. The performance of this system was reported to remain constant (Sime et al., 2002). Hofmann et al. (2009) investigated the influence of real biosyngas with a tar level of over 10 g mn^{-3} on the performance of SOFCs with Ni–GDC anodes and reported successful SOFC operation for 7 h. A planar SOFC stack fueled by gas obtained by gasifying wood in the Viking gasifier has operated

successfully for 150 h without carbon deposition or significant performance degradation, as reported by Hofmann et al. (2007). Several other successful gasifier–SOFC
experiments have been reported with varying degrees of detail.

In general, gasifier–SOFC experiments have shown that SOFCs can operate with
clean biosyngas and that many of the contaminants can be tolerated to a certain extent.
However, developing the most suitable gas cleaning schemes for a variety of gasifiers
is a highly challenging task.

16.3.4   Biomass Gasifier–SOFC–Gas Turbine Systems

SOFCs running on natural gas have an electrical efficiency of 45–60% (Stambouli and
Traversa, 2002). They operate at high temperature and also produce high-temperature
waste heat. Fuel conversion in the FC is not complete, and the unreacted fuel leaving
the cell can be combusted. If the cell is operated at high pressure, the flue gas can be
passed through a GT to extract mechanical energy, which can then be converted into
electrical energy by a generator as illustrated in Figure 16.4.

The first SOFC–GT hybrid system was installed in the National Fuel Cell Research
Center at the University of California, Irvine (tinyurl.com/o8gl3zq). This installation
succeeded as a proof-of-concept demonstration and determined the operating
windows of the pressurized system running on natural gas. This system has been operated for nearly 2900 h and achieved an electrical efficiency of approximately 53%,
comparable with the value of 57% presented for the Siemens Westinghouse

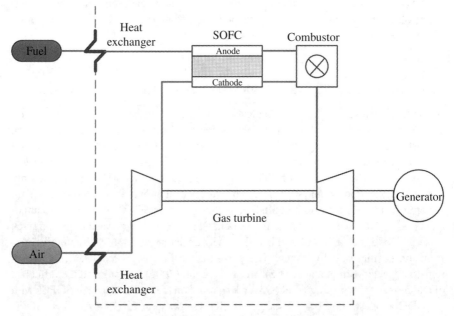

FIGURE 16.4   Schematic view of an SOFC–GT system.

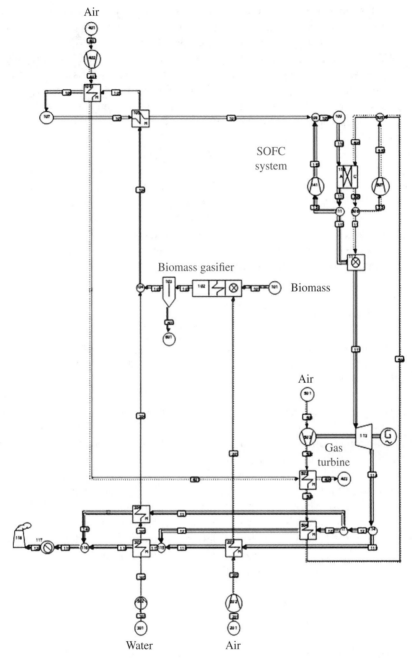

FIGURE 16.5 System layout of a gasifier–SOFC–GT power plant built in Cycle-Tempo. (Source: Reproduced with permission from Aravind et al. (2009). © Elsevier B.V.)

SOFC–GT systems (Roberts and Brouwer, 2006). Different groups have suggested different system designs and characteristics, although the practical application of such systems is still in an early stage of development.

SOFC–GT systems running on biosyngas represent an attractive option for high-efficiency power generation. Designing biomass gasifier–SOFC–GT systems with the highest possible system efficiency requires detailed system studies. Currently, studies on the development of such systems mainly start with thermodynamic evaluations, which have indicated that net electrical efficiencies of 60–70% and total system efficiencies of up to approximately 80% are achievable (Aravind et al., 2009, 2012; Toonssen et al., 2011). An example of a system layout built in the flow sheet computer program Cycle-Tempo (tinyurl.com/mpz8cxv) is shown in Figure 16.5 (Aravind et al., 2009). Studies of such systems clearly indicate the thermodynamic advantages of gasifier–SOFC systems. However, more studies are required to clearly understand how these systems can be optimized. Future studies should compare different system configurations, help select suitable operating parameters for different system components to enable optimal system performance, and reveal the effects of different gas cleaning options on the thermodynamic performance of the system.

16.4 CONCLUDING REMARKS

The SOFC is the most technically demanding part of a biomass gasifier–SOFC system. Thus, the correct choice of an SOFC and its operating conditions is critical to enable smooth functioning of gasifier–SOFC systems. As the fuel reacts at the anode compartment, the critical part of SOFC selection is the anode. The choice of gasifier depends significantly on the scale of application. For decentralized power generation at lower power levels of a few hundred kW, downdraft fixed bed cocurrent gasifiers may represent a good choice, whereas for large-scale applications of a few MW, fluidized bed gasifiers may be a better choice. The choice of gas cleaning technique depends on the tolerance of the anode to the contaminants and the capability of the technique. Studies indicate that systems in which gasifier, SOFCs, and GT are coupled can result in significantly higher system efficiencies when compared with competing systems.

However, the construction of commercially viable power plants based on these concepts requires further research and development of the technologies involved, including the FC and gas cleaning systems, and can only be achieved after a thorough technical and economic evaluation of the entire array of available alternatives.

CHAPTER SUMMARY AND STUDY GUIDE

This chapter deals with the development of highly efficient energy systems based on biomass gasification and SOFCs. In view of developing such energy systems, SOFC components, gas cleaning options, and system integration are addressed.

Achievements obtained thus far and future study topics for developing the energy systems are also discussed.

KEY CONCEPTS

Energy efficiency
Gas cleaning
Electrochemistry
Thermodynamics
Fuel utilization
Current density
Polarization losses
Reversible voltage or Nernst voltage

SHORT-ANSWER QUESTIONS

16.1 Explain the fate of the main gas components in biosyngas when the syngas is allowed to pass through the anode of an SOFC producing current.

16.2 Compare the electrical and total thermal efficiencies of gasifier–SOFC–GT systems with competing power production technologies.

16.3 What is the fate of tars in biosyngas for the following two cases?

a. Biosyngas containing tar is processed in an SOFC.

b. Biosyngas containing tar is processed in a reciprocating engine.

16.4 What are the advantages and disadvantages of high-temperature and near-ambient-temperature gas cleaning systems in the context of gasifier–SOFC systems?

16.5 What are the causes of carbon deposition in SOFCs?

PROBLEMS

16.1 An electrolyte-supported SOFC is operated at atmospheric pressure and 800°C with the following mole fractions of the reactant and product species: x_{H_2} = 0.95 and x_{H_2O} = 0.05 (anode) and x_{O_2} = 0.21 (cathode). At 800°C, the fuel cell has $\Delta \bar{g}_f$ = −188.6 kJ.mol^{-1} and Δh_r = −248.3 kJ·mol^{-1} of H$_2$, and the conductivity of the cell is 5 Ω^{-1}·m^{-1}. The cell active area is 2×10^{-4} m^2, and the electrolyte thickness is 100 μm. If the cell is operated at 0.7 V, then determine the following:

a. The inlet Nernst voltage

b. The rates at which hydrogen and oxygen are consumed

c. The electrical efficiency (fuel to electricity)

16.2 A clean syngas stream contains 20% H$_2$, 16% CO, 1% CH$_4$, 10% CO$_2$, 15% H$_2$O, and 38% N$_2$ (by volume). An SOFC is operated with a fuel with this syngas composition in the anode at atmospheric pressure and 800°C and air ($x_{O_2} = 21\%$, $x_{N_2} = 0.79$) in the cathode. Assume that the cell's reversible voltage is 0.9 V, and all irreversibilities of the electrochemical process are included in the equivalent cell resistance, R_{equiv}, which equals 5.5×10^{-5} Ωm^2 at 800°C. The irreversible cell voltage is 0.8 V. The fuel utilization (the ratio of the fuel used to the total fuel provided to the fuel cell) at the anode $U_f =$ 80% and the oxygen utilization at the cathode $U_{O_2} = 15\%$. The cell's active area is 81 cm^2.

a. Calculate the electricity produced.
b. Determine the amount of biosyngas equivalent required.
c. Calculate the airflow rate.
d. Assume the syngas required is generated from biomass gasification and the total chemical energy input is 34 W. Determine the energy efficiency of the conversion of biomass into electricity.

PROJECTS

P16.1 Visit a local gasifier-based power plant, e.g., processing biomass or coal; note down the syngas composition including contaminants; and design a gas cleaning system to upgrade the syngas for use in SOFCs.

P16.2 Visit a fuel cell lab, record the key characteristics of an SOFC, and measure and compare the performance of the SOFC operating with different fuels, e.g., hydrogen, carbon monoxide, or syngas.

P16.3 Thermodynamically evaluate the effects of biomass gasification technology, e.g., different gasifier types and gasifying agents such as steam and air on gasifier–SOFC system efficiencies.

INTERNET REFERENCES

tinyurl.com/kdfbsbz
http://www2.dmu.dk/atmosphericenvironment/expost/database/docs/elv_combustion.pdf

tinyurl.com/mpz8cxv
www.cycle-tempo.nl

tinyurl.com/o8gl3zq
http://www.nfcrc.uci.edu/3/activities/researchsummaries/Hybrid_FC-GT_Systems/
   Hybrid220kW/HYBRIDfuelCELL_Hybrid_220kwSOFC.pdf

tinyurl.com/qhswo7e
www.hightech.fi/direct.aspx?area=htf&prm1=25&prm2=article

REFERENCES

Adler SB. Factors governing oxygen reduction in solid oxide fuel cell cathodes. Chem Rev 2004;104(10):4791–4843.

Aguilar L, Zha S, Cheng Z, Winnick J, Liu M. A solid oxide fuel cell operating on hydrogen sulfide (H_2S) and sulfur-containing fuels. J Power Sources 2004;135(1–2):17–24.

Aravind PV, De Jong W. Evaluation of high temperature gas cleaning options for biomass gasification product gas for solid oxide fuel cells. Prog Energy Combust Sci 2012;38:737–764.

Aravind PV, Ouweltjes JP, Woudstra N, Rietveld G. Impact of biomass-derived contaminants on SOFCs with Ni/Gadolinia anodes. Electrochem Solid-State Lett 2008;11(2): B24–B28.

Aravind PV, Schilt C, Türker B, Woudstra T. Thermodynamic model of a very high efficiency power plant based on a biomass gasifier, SOFCs, and a gas turbine. IJRED 2012;2(1):51–55.

Aravind PV, Woudstra T, Woudstra N, Spliethoff H. Thermodynamic evaluation of small-scale systems with biomass gasifiers, solid oxide fuel cells with Ni/GDC anodes and gas turbines. J Power Sources 2009;190(2):461–475.

Cheng Z, Zha S, Liu M. Influence of cell voltage and current on sulfur poisoning behavior of solid oxide fuel cells. J Power Sources 2007;172:688–693.

Fergus JW. Oxide anode materials for solid oxide fuel cells. Solid State Ionics 2006;177 (17–18):1529–1541.

Hindsgaul C, Schramm J, Gratz L, Henriksen U, Bentzen JD. Physical and chemical characterization of particles in producer gas from wood chips. Bioresour Technol 1999;73 (2):147–155.

Hofmann P, Panopoulos KD, Aravind PV, Siedlecki M, Schweiger A, Karl J, Ouweltjes JP, Kakaras E. Operation of solid oxide fuel cell on biomass product gas with tar levels >10 g Nm^{-3}. Int J Hydrogen Energy 2009;34:9203–9212.

Hofmann P, Panopoulos KD, Fryda LE, Schweiger A, Ouweltjes JP, Karl J. Integrating biomass gasification with solid oxide fuel cells: effect of real product gas tars, fluctuations and particulates on Ni-GDC anode. Int J Hydrogen Energy 2008;33:2834–2844.

Hofmann P, Schweiger A, Fryda L, Panopoulos KD, Hohenwarter U, Bentzen JD, Ouweltjes JP, Ahrenfeldt J, Henriksen U, Kakaras E. High temperature electrolyte supported Ni-GDC/ YSZ/LSM SOFC operation on two-stage Viking gasifier product gas. J Power Sources 2007;173(1):357–366.

Jacobson AJ. Materials for solid oxide fuel cells. Chem Mater 2010;22(3):660–674.

Johansson EM, Berg M, Kjellström J, Järås SG. Catalytic combustion of gasified biomass: poisoning by sulphur in the feed. Appl Catal B Environ 1999;20:319–332.

Lee JH, Heo JW, Lee DS, Kim J, Kim GH, Lee HW, Song HS, Moon JH. The impact of anode microstructure on the power generating characteristics of SOFC. Solid State Ionics 2003;158(3–4):225–232.

Liu M, Aravind PV, Woudstra T, Cobas VRM, Verkooijen AHM. Development of an integrated gasifier-solid oxide fuel cell test system: a detailed system study. J Power Sources 2011;196(17):7277–7289.

Liu M, Van der Kleij A, Verkooijen AHM, Aravind PV. An experimental study of the interaction between tar and SOFCs with Ni/GDC anodes. Appl Energy 2013;108:149–157.

Mermelstein J, Millan M, Brandon N. The impact of steam and current density on carbon formation from biomass gasification tar on Ni/YSZ, and Ni/CGO solid oxide fuel cell anodes. J Power Sources 2010;195:1657–1666.

Nurk G, Holtappels P, Figi R, Wochele J, Wellinger M, Braun A, Graule T. A versatile salt evaporation reactor system for SOFC operando studies on anode contamination and degradation with impedance spectroscopy. J Power Sources 2011;196(6):3134–3140.

O'Hayre R, Suk-Won C, Colella WG, Prinz FB, editors. *Fuel Cell Fundamentals*. Hoboken (NJ): John Wiley & Sons; 2009.

Oudhuis A, Bos A, Ouweltjes JP, Rietveld G, van der Giesen AB. High efficiency electricity and products from biomass and waste; experimental results and proof of principle of staged gasification and fuel cells. Second World Conference and Technology Exhibition on Biomass for Energy, Industry and Climate Protection. Rome (Italy); May 10–14, 2004. Florence: ETA Florence; 2004.

Rasmussen JFB, Hagen A. The effect of H_2S on the performance of Ni–YSZ anodes in solid oxide fuel cells. J Power Sources 2009;191:534–541.

Roberts RA, Brouwer J. Dynamic simulation of a pressurized 220 kW solid oxide fuel-cell-gas-turbine hybrid system: modeled performance compared to measured results. J Fuel Cell Sci Technol 2006;3(1):18–25.

Salo K, Mojtahedi W. Fate of alkali and trace metals in biomass gasification. Biomass Bioenergy 1998;15(3):263–267.

Sime R, Stucki S, Biollaz S, Wiasmitinow A. Linking wood gasification with SOFC hybrid processes. Proceedings of the 5th European SOFC Forum, Lucerne, Switzerland; July 2002.

Singh D, Hernandez-Pacheco E, Hutton PN, Patel N, Mann MD. Carbon deposition in an SOFC fueled by tar-laden biomass gas: a thermodynamic analysis. J Power Sources 2005;142 (1–2):194–199.

Singhal C, Kendall K. *High Temperature Solid Oxide Fuel Cells*. New York: Elsevier; 2003.

Skinner J. Recent advances in perovskite-type materials for SOFC cathodes. Fuel Cell Bull 2001;4(33):6–12.

Stambouli AB, Traversa E. Solid oxide fuel cells (SOFCs): a review of an environmentally clean and efficient source of energy. Renew Sustain Energy Rev 2002;6(5):433–455.

Staniforth J, Ormerod RM. Clean destruction of waste ammonia with consummate production of electrical power within a solid oxide fuel cell system. Green Chem 2003;5(5):606–609.

Toonssen R, Sollai S, Aravind PV, Woudstra N, Verkooijen AHM. Alternative system designs of biomass gasification SOFC/GT hybrid systems. Int J Hydrogen Energy 2011;36 (16):10414–10425.

Van der Drift A. Ten residual biomass fuels for circulating fluidized-bed gasification. Biomass Bioenergy 2001;20(1):45–56.

Wojcik A, Middleton H, Damopoulos I, Van Herle J. Ammonia as a fuel in solid oxide fuel cells. J Power Sources 2003;118(1–2):342–348.

Xu CC, Gong MY, Zondlo JW, Liu X, Finklea HO. The effect of HCl in syngas on Ni-YSZ anode-supported solid oxide fuel cells. J Power Sources 2010;195(8):2149–2158.

Zevenhoven R, Kilpinen P. *Control of Pollutants in Flue Gases and Fuel Gases*. Espoo (Helsinki): Helsinki University of Technology; 2001.

Zhu WZ, Deevi SC. A review on the status of anode materials for solid oxide fuel cells. Mater Sci Eng A 2003;362(1–2):228–239.

17

SYNTHESIS GAS UTILIZATION FOR TRANSPORTATION FUEL PRODUCTION

J. Ruud van Ommen and Johan Grievink

Department of Chemical Engineering, Product & Process Engineering Group, Faculty of Applied Sciences, Delft University of Technology, Delft, the Netherlands

ACRONYMS

BTL   biomass to liquids
CTL   coal to liquids
DME   dimethyl ether
FT(S)   Fischer–Tropsch (synthesis)
GTL   gas to liquids
LPG   liquefied petroleum gas
MTG   methanol to gasoline
MTO   methanol to olefins
SNG   synthetic natural gas
XTL   "anything" to liquids

SYMBOLS

a reaction rate coefficient in Yates and Satterfield $[mol \cdot s^{-1} \cdot kg_{cat}^{-1} \cdot bar^{-2}]$
expression
b adsorption coefficient in Yates and Satterfield $[bar^{-1}]$
expression

Biomass as a Sustainable Energy Source for the Future: Fundamentals of Conversion Processes,
First Edition. Edited by Wiebren de Jong and J. Ruud van Ommen.
© 2015 American Institute of Chemical Engineers, Inc. Published 2015 by John Wiley & Sons, Inc.

$c_{i,b}$	bulk concentration of component i	$[\text{mol·m}^{-3}]$
d_p	particle diameter	$[\text{m}]$
$D_{i,eff}$	effective diffusion coefficient of component i	$[\text{m}^2\text{·s}^{-1}]$
$E_{A,a}$	activation energy for rate constant a	$[\text{J·mol}^{-1}]$
F	catalyst activity multiplication factor	$[-]$
n	number of carbon atoms in hydrocarbon	$[-]$
p_i	partial pressure of component i	$[\text{Pa or bar}]$
r	reaction rate per unit mass of catalyst	$[\text{mol·s}^{-1}\text{·kg}_{\text{cat}}^{-1}]$
R_u	universal gas constant	$[=8.314\ \text{J·mol}^{-1}\text{·K}^{-1}]$
T	temperature	$[\text{K}]$
Th	Thiele modulus	$[-]$
α	chain growth probability	$[-]$
$\Delta_b H$	adsorption enthalpy in Yates and Satterfield expression	$[\text{J·mol}^{-1}]$
$\Delta_r H$	reaction enthalpy	$[\text{J·mol}^{-1}]$
η	catalyst effectiveness	$[-]$
ρ_{cat}	catalyst density	$[\text{kg·m}^{-3}]$

17.1 INTRODUCTION

A considerable part of the energy we consume is used to fuel vehicles in the form of gasoline, diesel, kerosene, or other liquid fuels derived from crude oil. In the EU, about one third of the energy consumption in 2011 was used for transportation (tinyurl.com/388hn38). Worldwide, various steps are taken to make the transition to more sustainable ways of fueling our cars, trucks, planes, and ships. Some radically different solutions have been proposed, such as using hydrogen or electricity. These have, however, two important drawbacks. First, the energy density is often low, which decreases the range of a vehicle. Second, switching to hydrogen or electricity requires a completely new infrastructure of supplying the fuel. Liquid transportation fuels derived from biomass do not have these disadvantages. Their properties are much closer to those of the currently used fuels, enabling a quick and smooth transition. If the biomass required is grown in a sustainable way (see Chapter 1), these fuels are much more sustainable than crude oil-derived fuels.

Much of the technology that is currently proposed or used for converting biomass into synthetic fuels relies on earlier developments for the production of fuels from natural gas or coal. Often, these routes rely on the production of syngas (a mixture of mainly hydrogen and carbon monoxide) as a convenient intermediate product, as discussed in Chapter 10. In addition to liquid fuels, we will also treat synthetic natural gas (SNG) in this chapter. Although it is not a liquid and not only used for transportation, the production route is similar to that of the other synthetic fuels treated in this chapter. Also, the gaseous compound dimethyl ether (DME) will be briefly treated, as it is strongly related to methanol production. Figure 17.1 gives an overview of the main production routes of these fuels via syngas.

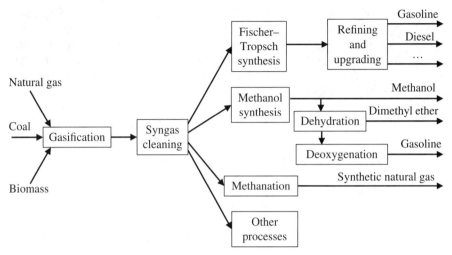

FIGURE 17.1 Schematic overview of the main routes from syngas to synthetic fuels.

The overall scheme for the conversion of various feedstocks into liquid fuels is referred to as "anything" to liquid (XTL). Depending on the feedstock, the names gas to liquids (GTL), coal to liquids (CTL), and biomass to liquids (BTL) are used for the specific variants, with, respectively, gas, coal, and biomass as feedstock. It should be noted here that there are also alternative routes that are not based on syngas, such as pyrolysis discussed in Chapter 11. Using the production of syngas as an intermediate step has the advantage that the process becomes insensitive to the composition of the biomass feed. We will focus on the processing aspects of the various routes. Routes that are still far from commercial implementation, such as the direct production of ethanol from syngas, will not be treated. The most important route is the Fischer–Tropsch synthesis (FTS), which was already used as an example process in Chapter 7 on process design. This process will be discussed in more detail in Section 17.2. Second, we will discuss the production of SNG. Third, we will discuss the synthesis of methanol from syngas and the possible follow-up step to DME. Finally, we will make a comparison of the different routes, with respect to efficiency and economic potential.

17.2 FISCHER–TROPSCH SYNTHESIS

FTS is a series of reactions by which syngas is converted into hydrocarbons. FTS was discovered by Franz Fischer and Hans Tropsch in the 1920s. During the Second World War, it was used in Germany to produce transportation fuels from coal. After the war, interest decreased: a fuel supply system based on crude oil was economically more attractive than one based on coal in combination with FTS. In South Africa, which has large coal reserves, an important reason for starting coal-based FTS was the embargo against the Apartheid policies, which limited its access to crude oil, and the use of FTS has continued there. Since the 1980s, the interest in FTS has been growing again. Sasol and Shell have commissioned large plants for natural gas-based FTS in Qatar in 2006 and 2011, respectively. The capacity of the Sasol plant is 34,000 barrels·day^{-1} (4,600 t·day^{-1}), and that of the Shell plant 140,000 barrels·day^{-1} (19,000 t·day^{-1}). Reasons for the renewed interest in the application of FTS to syngas obtained from coal are that the worldwide coal reserves are much larger than those of oil and gas and that some countries have a lot of coal, but little or no oil or gas reserves; China is a prominent example. Finally, also biomass-based FTS has attracted more interest in recent years. This way of producing biofuel has the advantage that it can rely on the technology developed for fossil feedstocks and thus is more mature than several other ways of producing biomass-based transportation fuels.

In the following sections, FTS technology will be presented by its key constituents in order of an increasing scale: reaction stoichiometry and kinetics, catalyst aspects, and design and operation of different reactor types.

17.2.1 Reaction Stoichiometry and Kinetics

FTS is a way to convert syngas into hydrocarbons at elevated temperature and pressure. It is a complex network of reactions, but the simplified overall reaction scheme is given as

$$(2n+1)H_2 + nCO \rightarrow C_nH_{(2n+2)} + nH_2O \quad n = 1, 2, \ldots, > 100 \qquad \text{(RX. 17.1)}$$

$$\Delta_r H = \sim -170 \text{ kJ } (\text{mol CO})^{-1}$$

The most widely used kinetics for this reaction is the Yates and Satterfield expression (Yates and Satterfield, 1991), which follows the Langmuir–Hinshelwood model. It gives the reaction rate based on CO per catalyst mass as

$$r = \frac{F \, a \, p_{CO} \, p_{H_2}}{(1 + b \, p_{CO})^2} \qquad \text{(Eq. 17.1)}$$

In this equation, F is a catalyst activity multiplication factor that accounts for improvements in catalyst activity since publication of the original parameter values in 1991 (Guettel and Turek, 2009; Vervloet et al., 2012), p_i is the partial pressure of reactant i, a is the reaction rate coefficient (per unit mass of catalyst), and b is the adsorption coefficient of CO Equation (17.2), as reported by Maretto and Krishna (1999), which can be calculated as

$$a = a_0 \exp\left(\frac{E_a}{R_u}\left(\frac{1}{493.15} - \frac{1}{T}\right)\right); \quad b = b_0 \exp\left(\frac{\Delta_b H}{R_u}\left(\frac{1}{493.15} - \frac{1}{T}\right)\right) \qquad \text{(Eq. 17.2)}$$

In this expression, R_u is the universal gas constant and T is the temperature in K. The other required values are given in Table 17.1. Please note that there is a typo in the

TABLE 17.1   Values required to calculate the temperature dependence of the Yates and Satterfield expression for the FT reaction

Symbol	Meaning	Value	Units
a_0	Reaction rate coefficient at T = 493.15 K	$8.88533 \cdot 10^{-3}$	$mol \cdot s^{-1} \cdot kg_{cat}^{-1} \cdot bar^{-2}$
E_a	Activation energy	3.737×10^4	$J \cdot mol^{-1}$
b_0	Adsorption coefficient at T = 493.15 K	2.226	bar^{-1}
$\Delta_b H$	Adsorption enthalpy	-6.837×10^3	$J \cdot mol^{-1}$

TABLE 17.2   Temperature dependence of the Yates and Satterfield expression for the FT reaction rate

T (K)	a $(mol \cdot s^{-1} \cdot kg_{cat}^{-1} \cdot bar^{-2})$	b (bar^{-1})	r $(mmol \ kg^{-1} \cdot s^{-1})$
470	5.65×10^{-3}	5.07	0.324
500	1.00×10^{-2}	1.77	4.54
530	1.67×10^{-2}	0.697	44.8

original publication of Maretto and Krishna (a minus sign is missing in the exponent of a_0); the values in Table 17.1 are correct.

Example 17.1   CO reaction rate as a function of temperature

Plot the reaction rate for CO Equation (17.1) as a function of temperature for a relevant temperature range for FTS (470–530 K). Assume a partial H_2 pressure of 20 bar and a partial CO pressure of 10 bar and a catalyst multiplication factor F of 3.

You will find an increasing reaction rate with temperature. Why is the operating temperature of Fischer–Tropsch (FT) reactors typically limited to about 500 K, while a higher temperature would give a higher reaction rate?

Solution

Substitution of the values given in Table 17.1 in the expressions in Equation (17.2) gives the values for a and b. Subsequently, r can be calculated. In this case, all provided coefficients have the right units for use in the expressions. However, note that that is often not the case: often you will have to convert values before the expression can be used (e.g., the reactant pressures are given in MPa instead of bar). To check your calculation, Table 17.2 gives the intermediate and final values for three temperatures. Figure 17.2 gives the plot for the complete temperature range.

The operating temperature of low-temperature FT reactors is typically limited to around 500 K, since the reaction is strongly exothermal: a further increase of the temperature will lead to a higher reaction rate and even more release of heat, making it very hard to cool and control the reaction. Moreover, conversion is not the only thing that counts: also the selectivity is very important. At higher temperatures, a larger amount of less favorable products—such as methane—will be produced.

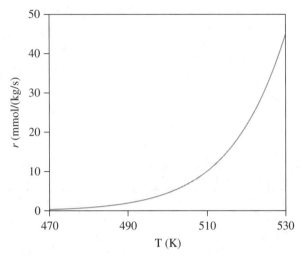

FIGURE 17.2 The FT reaction rate according to the Yates and Satterfield expression as a function of temperature.

The Yates and Satterfield expression just describes the conversion of syngas by FTS; it does not say anything about the structure of the obtained products. The most important products are linear alkanes (*n*-paraffins, hydrocarbons with only single C–C bonds), but also considerable amounts of alkenes (olefins, hydrocarbons with a C=C bond) and branched alkanes can be produced. In addition, small amounts of oxygenated hydrocarbons such as alcohols are produced. FTS can be considered as a polymerization process taking place at the surface of a catalyst. By adding more carbon atoms, the chains become longer and longer (propagation), until the chain growth is terminated (see Figure 17.3a). In every step, the probability for propagation is α, while the probability for termination is $(1-\alpha)$. This means that the parameter α—denoted as the chain growth probability—determines the product distribution (see Figure 17.3b). This distribution is named the Anderson–Schulz–Flory distribution, after the proposers of this mechanism. In most cases, the aim is to produce liquid hydrocarbons, i.e., molecules with a considerable number of carbon atoms. This means that α should be high enough; typical values in practice are around 0.9. The fact that at high α also heavy waxes ($n > 20$) are produced is seen as less of a problem: these can be used in various applications (e.g., as lubricant and for coating) or can be easily cracked to shorter molecules. Ways to increase α are to operate at lower H_2: CO ratio or at lower temperature. The latter has its limitations, since it will also reduce the reaction rate.

17.2.2 Catalyst Aspects

The FTS requires a solid-phase catalyst to enhance the reaction rate. The two main catalyst types applied are cobalt and iron. Most current processes use cobalt: it is more active (can be used at a lower temperature) and less prone to sintering, but it

(a)   (b)

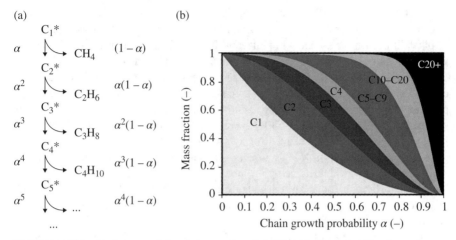

FIGURE 17.3 (a) Scheme of the chain reactions in FTS with the probability for each step, according to the Anderson–Schulz–Flory mechanism. (b) Plot of the resulting product distribution.

is also more expensive than iron and less resistant to poisons. Whereas iron-based catalysts typically have Fe_3O_4 as their main component, cobalt-based catalyst typically consists of alumina support particles with cobalt deposited on it.

Both cobalt and iron are sensitive to high partial pressures of water. To prevent too high partial pressures of water and to increase conversion, water can be removed, e.g., by interstage cooling and condensation or by membranes. In addition, a promoter (aiding compound) can be added to iron (e.g., potassium), so that it catalyzes the water–gas shift reaction in addition to the FT reaction and converts water:

$$CO + H_2O \rightleftharpoons CO_2 + H_2 \quad \Delta_r H = -41 \ kJ \cdot mol^{-1} \qquad (RX. \ 17.2)$$

An important design variable for FTS is the syngas ratio, the ratio of the hydrogen concentration to the carbon monoxide concentration in the feed stream. The water–gas shift reaction (RX. 17.2) makes that an iron-based catalyst can work with a range of syngas ratios. For the more widely used cobalt catalyst, H_2 and CO are only consumed by FTS, in a ratio between 2 (for production of infinitely long hydrocarbon chains) and 3 (for production of methane), depending on α. It can be shown mathematically that the H_2/CO consumption ratio exhibits the remarkably linear result $(3-\alpha)$ (Vervloet et al., 2012). Typically, α-values around 0.9 are desired, giving a syngas ratio around 2.1 based on stoichiometric requirements. However, it may be advantageous to choose a lower syngas ratio, as this will lead to a lower amount of water in the reactor and a larger α-value (Vervloet et al., 2012).

17.2.3 Process Conditions and Reactor Types

FTS is typically carried out at pressures between 10 and 60 bar and temperatures in the range of 480–620 K. In view of the required large scale of FTS reactors, the process is operated in a continuous mode and at steady state. The syngas conversion in a single pass through a reactor that can be achieved in practice is limited for various reasons. Typical limitations are heat removal capacity, avoiding condensation of reaction water, mechanical restrictions to the maximum size of a reactor, and pressure drop considerations. The bulk of the unconverted syngas is recycled with significant cost of recompression. Key considerations in developing FTS reactors are heat removal, pressure drop, and avoidance of diffusion limitations. Often, a distinction is made between high-temperature FTS (580–620 K) and low-temperature FTS (480–530 K). In high-temperature FTS—using iron as a catalyst—the main products are in the gasoline range (C_4–C_{12}); this fraction is gaseous under reaction conditions. Low-temperature FTS—which is receiving more attention these days—mainly produces longer-chain hydrocarbons and is better suited for diesel production. The product mix mostly has a liquid nature at operating conditions; typically, a cobalt catalyst is used for this process.

FTS is a strongly exothermal process ($\Delta_r H \approx -170$ kJ·mol^{-1} CO), which means that proper heat removal is an important aspect of the reactor design. Failing to remove all heat results in increasing temperatures, which increases the reaction rate but decreases the selectivity toward the desired long-chain hydrocarbons. Even worse, it could lead to hot spots and a reactor runaway: an uncontrolled increase in temperature resulting in an explosion. On the other hand, it is crucial to make good use of the heat released to come to an energy-efficient process: the higher the temperature at which heat is released, the more valuable it is.

For high-temperature FTS, different kinds of fluidized bed reactors have been used (see Steynberg et al. (1999) for an overview of these reactors). Because of the larger importance of low-temperature FTS, we will focus on the two reactors mainly used for this: multitubular packed beds and slurry bubble columns (Guettel and Turek, 2009; Hooshyar et al., 2012) (see Figure 17.4).

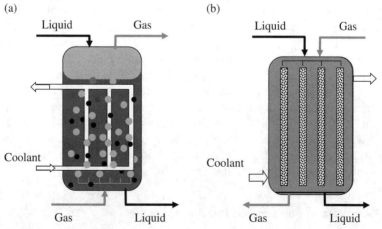

FIGURE 17.4   Schematic of (a) a slurry bubble column reactor and (b) a multitubular packed bed reactor.

A slurry bubble column is a large vessel filled with liquid (the products) in which the catalyst particles are suspended. These particles are small ($\sim$50 μm) such that no internal diffusion limitations occur. The syngas feed enters at the bottom, while the liquid products are removed from the upper part of the reactor. The technical challenge is to effectively separate the small catalyst particles from the product fluid, after which the catalyst can be recycled. Syngas is distributed at the bottom of the column and rises as bubbles through the liquid in upward direction. The reactants from the bubbles (H_2 and CO) dissolve in the liquid and react at the catalyst surface. Slurry bubble columns are typically operated at high gas fractions and high catalyst loadings, both around 30 vol.%. At such high gas fractions, the bubbles typically have a bimodal distribution: small ones of a few millimeters and large ones of a few centimeters in diameter. Especially the large bubbles lead to a vigorous mixing of the slurry, preventing settling of the solids and giving a rather even temperature distribution in the reactor. Because of the large heat production by FTS, the reactor contains a large number of heat exchanging tubes. The removed heat is not lost: it is used to generate steam. The conversion in a slurry bubble column is typically rather high, up to 80–90%.

A packed bed is an amount of particles packed together, in the case of an FT reactor enclosed by a vertical tube. The outer side of the tube is cooled by boiling water. The catalyst particles are porous themselves, but in the packed bed, there is also porosity ($\sim$40 vol.%) between the particles. Typically, a part of the product stream leaving the reactor is recycled to have liquid present right from the entrance of the reactor. This is done to ensure sufficient heat transfer from the particles to the cooled tube wall. To prevent too high a pressure drop, the catalyst particles cannot be too small; typically, particles of a few millimeters in diameter are used. A drawback of this relatively large size is that diffusion limitations inside the particle will occur, so that the central part of the particle is not used as optimal as the outer region due to lower concentrations of the reactants (see Figure 17.5). Another challenge is the removal of heat: the heat transport in the radial direction is poor. Therefore, packed beds for FTS have a maximum diameter of just few centimeters. In order to reach reasonable production rates, ten thousands of tubes containing the catalyst particles are put in parallel and placed in a vessel containing boiling

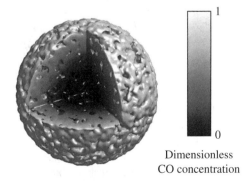

Dimensionless
CO concentration

FIGURE 17.5 Typical dimensionless CO concentration in an FT catalyst particle of 1.5 mm in diameter (based on Vervloet et al. (2012)).

water as a coolant, thus generating steam. This is called a multitubular packed bed reactor. As catalyst development progresses and catalysts become more active, the heat production per unit volume also increases. A possible solution is to replace the randomly packed beds by structured packings that force convective transport in the radial direction, giving better heat transfer (Pangarkar et al., 2008). Such packing can be coated and/or filled with catalysts. The investment costs of such systems—which are not yet used in practice—will be larger, while the benefit is that the better heat transfer enables the use of tubes with larger diameters, so less tubes are needed per vessel. The latter brings down the investment; so there is a trade-off between tube size and degree of structuring of the packing. Currently, commercial multitubular reactors are operated at low conversions per reactor, typically considerably lower than 50%. By recycling and/or putting reactors in series, a much higher overall conversion is reached for the complete process.

Currently, these two reactor types are both operated at large scale (>30,000 barrels·day^{-1}): slurry bubble columns by Sasol and multitubular packed bed by Shell. However, in both cases, the syngas is not produced from biomass but from natural gas. The plants are located very close to natural gas fields, enabling economies of scale. Transportation of liquid fuel is much cheaper than that of natural gas. It is also possible to base relatively large plants on a feedstock of biomass, but this will require a lot of transportation of biomass. According to Hamelinck et al. (2004), the optimum scale for a BTL production plant is about 2000 MW$_{th}$; for comparison, the Sasol plant in Qatar produces about 4000 MW$_{th}$. However, if cheap biomass is locally available, smaller, distributed plants may be an option too, especially since these do not require shipping of raw biomass over long distances. A number of smaller companies are developing small-scale FTS reactor technology, e.g., Rentech, Syntroleum, and Velocys. Some of these companies rely on scale-down versions of the large-scale technology, whereas others propose radically different concepts. For example, Velocys commercializes a millichannel reactor: a reactor with many parallel channels of 1 or a few mm (so an order of magnitude smaller than for a conventional multitubular reactor), filled with particles of around 250 µm diameter (Deshmukh et al., 2011). Because the ratio of the tube diameter to the particle diameter is relatively small, the pressure drop stays limited, even with these small particles: wall effects lead to higher bed porosity. Because of the small channel diameter, a good heat transfer can be ensured, and a highly active catalyst can be applied. Velocys claims to achieve a production of 12 barrels day^{-1} ·kg^{-1} reactor mass, 1.5–4 times higher than the commercialized large-scale technologies (Deshmukh et al., 2011).

Summarizing, the production of liquid transportation fuels from biomass via gasification and FTS has the advantages that it can work with a wide variety of second-generation feedstocks, that it uses mature technology, and that the product is easily integrated in the existing fuel distribution system. It requires, however, large investment costs, and the efficiency is not very high, as will be discussed in Section 17.5. There is an issue with respect to a potential downscaling of FTS technology for regional distributed applications. The purpose of such downscaling is the minimization of the costs of the entire logistic production chain from biomass to energy product(s). One could consider producing the raw hydrocarbon mixture locally in smaller units, while refining could take place in larger refineries.

Example 17.2   Thiele modulus and effectiveness factor

If catalyst particles are relatively large and the reaction rate is fast, the diffusion rate might be too slow to provide the center of the catalyst particle with enough reactant to maximize the reaction rate. This can be expressed by the Thiele modulus (see also Chapter 9):

$$Th = \frac{d_p}{6} \sqrt{\frac{r \, \rho_{cat}}{D_{eff,CO} \, c_{CO,b}}}$$

where d_p is the particle diameter, r is the reaction rate expressed per kg cat, ρ_{cat} is the catalyst density, $D_{eff,CO}$ is the effective CO diffusion coefficient (taking into account catalyst porosity and pore tortuosity), and $c_{CO,b}$ is the bulk concentration of CO. We assume that at 500 K $D_{eff,CO}$ is 5.20×10^{-9} m$^2 \cdot$s^{-1}, ρ_{cat} is 2500 kg$\cdot$m^{-3}, and $c_{CO,b}$ is 300 mol$\cdot$m^{-3}. If the Thiele modulus is small ($\ll 1$), there are no diffusion limitations; if the Thiele modulus becomes larger, diffusion starts to play a role and the effective reaction rate is lower than the intrinsic reaction rate. From the Thiele modulus, we can calculate the catalyst effectiveness factor:

$$\eta \equiv \frac{\text{reaction rate with diffusion limitation}}{\text{reaction rate at bulk conditions}} = \frac{\tanh Th}{Th}$$

Calculate the catalyst effectiveness factor at 500 K with the reaction rate calculated in Example 17.1 ($F = 3$) for catalyst particles of 2 mm, which is a typical size used in a fixed bed. What is your conclusion? What happens if the catalyst activity is increased so that F becomes 10? What is the effectiveness factor for typical slurry catalyst particles with a diameter of 50 μm?

Solution

$$Th = \frac{2.0 \times 10^{-3}}{6} \sqrt{\frac{4.54 \times 10^{-3} \cdot 2500}{5.20 \times 10^{-9} \cdot 300}} = 0.899; \quad \eta = 0.796$$

The effectiveness factor is lower than 1, showing that there are some diffusion limitations. If we increase F, the reaction rate increases with a factor 3.33, leading to an effectiveness of 0.567. Since the reaction becomes faster and the diffusion rate stays the same, the diffusion limitations become worse. For slurry catalyst particles of only 50 μm in diameter, the effectiveness factor is 1.00, so there is no diffusion limitation.

17.3   SYNTHETIC NATURAL GAS SYNTHESIS

An alternative approach for chemical conversion of syngas into fuel is the production of SNG. It is sometimes also called substitute natural gas or green gas, but the latter term can also encompass methane produced by anaerobic digestion. The synthesis of

SNG is closely related to FTS. There is already quite some experience from the production of SNG from coal. This makes it possible to quickly implement this technology to convert biomass into biofuel. Moreover, it has the advantage that the gas produced this way can be easily incorporated in the existing fuel distribution infrastructure: it can be mixed into the existing natural gas networks for applications such as household heating or to fuel cars as is done with compressed natural gas.

SNG synthesis could be seen as FTS with a very low α-value; whereas in FTS an important aim is to minimize the methane production, in SNG synthesis, one wants to maximize it. Thus, the main reaction, referred to as methanation, is

$$3H_2 + CO \rightleftarrows CH_4 + H_2O \quad \Delta_r H = -206 \text{ kJ} \cdot \text{mol}^{-1} \quad \text{(RX. 17.3)}$$

Just like FTS, this reaction is strongly exothermal, making good heat removal a key issue in reactor design. Two main reactor concepts have been proven suitable for the production of SNG, namely, a series of adiabatic fixed bed reactors with intermediate cooling and/or gas recycle and fluidized bed reactors (Kopyscinski et al., 2010).

While in FTS for the production of liquid fuels the temperature must not be too high because of the desire to minimize the production of methane, in methanation, higher reaction temperatures can be used, typically 520–770 K. Pressures can vary widely, but are typically in the range of 20–70 bar. However, operation close to atmospheric pressure is also possible. Methanation reactors are normally not fed with syngas, but with producer gas: a mixture containing mainly H_2, CO, CO_2, H_2O, and CH_4. Nickel is the most widely used catalyst for this reaction, due to its selectivity, activity, and price. However, nickel-based catalysts are very vulnerable to catalyst poisons such as sulfur species (e.g., H_2S, COS, organic sulfur) and chlorine (Kopyscinski et al., 2010). The equilibrium reaction given in (RX. 17.3) is strongly on the right and nearly full conversion can be reached.

The first commercial plant for the production of SNG from coal, the Great Plains Synfuels Plant, is located in North Dakota (United States). It was commissioned in 1984 and converts $18,000 \text{ t} \cdot \text{day}^{-1}$ of lignite coal into 4.8 million m^3 SNG. The methanation takes place in two adiabatic fixed bed reactors in series with internal recycle, the so-called Lurgi process (Kopyscinski et al., 2010). Today, more coal to SNG plants are planned, especially in China.

From a sustainability point of view, it would be attractive to use biomass for the production of SNG. The challenges of using biomass instead of coal arise from the different chemical composition and different kind of impurities in the producer gas such as organic sulfur and from the smaller unit size (Kopyscinski et al., 2010). SNG production from syngas is not yet commercially applied. Worldwide, three research centers are the main players in SNG R&D: the Energy Research Centre of the Netherlands (ECN), the Centre for Solar Energy and Hydrogen Research (ZSW, Germany), and the Paul Scherrer Institute (PSI, Switzerland). Currently, the first plant to produce SNG is being built in Gothenburg, Sweden (tinyurl.com/ pfc2ggu). An experimental 20 MW plant was started up in 2013 and a commercial 80–100 MW plant should be ready in 2016.

> **Example 17.3 Production capacity of an SNG plant**
>
> The SNG plant in Gothenburg has a production capacity of 20 MW biogas. The lower heating value of the biogas is 35 MJ·m^{-3}, and that of gasoline is 33 MJ·L^{-1}.
>
> a. What is the yearly production of biogas in m^3?
> b. The biogas produced will be used as a transportation fuel. How many cars can approximately drive on this gas?
>
> **Solution**
> a. 20 MW of biogas is equivalent to 20 MJ·s^{-1}. Dividing by the lower heating value gives a production of 0.571 m^3·s^{-1}. There are 8760 h in a year, but it is reasonable to assume an uptime for the plant of 8000 h (see also Chapter 7). This means a production of 16 Mm^3 of biogas per year.
> b. Assuming that a car drives on average 18,000 km·year^{-1} and uses 1 L of gasoline per 14 km, this gives a yearly gasoline use of roughly 1300 L, which is equivalent to an energy use of 43 GJ. This corresponds to ca. 1200 m^3 of biogas per year. Thus, the plant can provide approximately 13,000 cars with fuel.

17.4 METHANOL SYNTHESIS

Instead of producing alkanes via FTS, syngas can also be converted into methanol. Methanol is an important chemical. It mostly serves as an intermediate for other chemicals, especially formaldehyde, but it can also serve as a liquid transportation fuel. The reaction equation for the formation of methanol from syngas is

$$CO + 2H_2 \rightleftarrows CH_3OH \quad \Delta_r H = -91 \text{ kJ·mol}^{-1} \qquad \text{(RX. 17.4)}$$

The formation of methanol is an exothermal equilibrium reaction. The equilibrium shifts to the right with decreasing temperature and shifts to the left with increasing temperature. In addition, the equilibrium shifts to the right with increasing pressure. Therefore, it is desirable to carry out the reaction at high pressure and low temperature. However, too low a temperature would lead to a very low reaction rate. Typical conditions used in practice are 490–560 K and 50–100 bar. In the past, operation at higher pressure has been applied as well, but this is less favorable because of high investment and operating costs. In the industrially used temperature and pressure ranges, the conversion in the reactor is rather low (around 50%) (Fiedler et al. 2011), which means that a large portion of the syngas needs to be recycled. The reaction rate can be increased by feeding a mixture of syngas and CO_2 to the reactor. Simultaneous to the methanol formation (RX. 17.4), the water–gas shift reaction (RX. 17.2) will take place. According to Lee (1990), CO_2 concentrations of up to 7 vol.% can improve the conversion to methanol. Higher levels of CO_2 decrease the conversion (Phillips et al., 2011).

Nowadays, mainly, a $Cu/ZnO/Al_2O_3$ catalyst is used. This catalyst enables methanol production with a very high selectivity, greater than 99%. Since the methanol synthesis reaction is quite exothermal, heat removal is an important issue in the design of a suitable reactor. The most widely used reactor type is an adiabatic reactor consisting of a single catalyst bed (Fiedler et al. 2011) to which the syngas is supplied in a distributed way: by injecting cold syngas at several points along the reactor, the reaction is quenched. This leads to a sawtooth-shaped temperature profile. An alternative reactor configuration is the quasi-isothermal reactor, which is very similar to the multitubular reactor employed in FTS. Many parallel tubes filled with catalyst are placed in a column filled with boiling water providing the cooling.

Methanol can be blended with gasoline, but its use is limited because of concerns about corrosion of some metals and its toxicity and solubility in water. An alternative is not to use methanol for combustion engines, but to convert it onboard of a car into hydrogen that can be used in a fuel cell. The conversion of methanol to hydrogen can be efficiently carried out using steam reforming at a relatively low temperature of 500–600 K, making it more efficient than using other liquid transportation fuels to generate hydrogen (Brown, 2001). An alternative is the use of direct methanol fuel cells, but these still have a low efficiency.

Instead of using methanol directly as a fuel, it can also be converted into DME or into gasoline. DME, which is a gaseous compound, can be used in direct DME fuel cells and in diesel engines. This might be attractive, since DME combines high diesel engine performance (cetane number 55) with low emissions (Semelsberger et al., 2006). DME is produced by dehydration of methanol:

$$2CH_3OH \rightleftharpoons CH_3-O-CH_3 + H_2O \quad \Delta_r H = -23.4 \text{ kJ·mol}^{-1} \quad \text{(RX. 17.5)}$$

Various acidic catalysts can be used for this reaction, such as γ-alumina, aluminum silicate, or zinc chloride. The reaction is typically carried out in the gas phase in a fixed bed reactor, at a temperature around 500–600 K and a pressure around 10–20 bar.

Instead of having DME as the final product of the process, it is also possible to produce a mixture of gasoline-range hydrocarbons; this is the so-called methanol-to-gasoline (MTG) process. DME is first converted by deoxygenating into a mixture of light olefins (C_2–C_5), which is finally converted into a mixture of C_{6+} alkanes and olefins and aromatic compounds. The gasoline synthesis, which uses DME as the input, typically takes place in parallel fixed bed reactors containing a ZSM-5 catalyst, operated at 600–700 K and approximately 20 bar (Moulijn et al., 2013). When the prices of olefins (used in, e.g., plastic production) are high compared to that of gasoline, it is more attractive to have olefins as the final product: methanol to olefins (MTO). At the time of writing (summer 2014), MTO is indeed economically more attractive than MTG.

17.5   COMPARISON OF THE DIFFERENT OPTIONS

In the previous sections, we have treated the main routes for synthesizing synthetic fuels from syngas: the FTS, SNG production, methanol production, and briefly

DME production. In this section, we will make a concise comparison of the different routes, based on material available in literature.

Zhang (2010) made a comparison of the different fuels, with an emphasis on the production efficiency (well-to-tank efficiency), the energy content of the synthesized fuel as a fraction of the energy in the biomass feed. He reported that the production efficiency is highest for SNG ($\sim$65%). Gassner and Maréchal (2012) reported an even higher efficiency for SNG ($\sim$70%). They also reported that the overall efficiency can even reach 90% when additional production of heat and electricity is considered (polygeneration). In addition to the high efficiency, SNG has the advantage that it can efficiently be produced in relatively small plants (<100 MW). The drawback is that it is a gaseous fuel. Zhang (2010) reported a production efficiency of approximately 55% for methanol, which has the lowest energy content among the liquid fuels. Regarding the production efficiency, it seems better to convert methanol into DME (production efficiency $\sim$60%). Moreover, DME can be applied in diesel engines, which makes it the fuel with the highest well-to-wheel efficiency (not just taking into account the production efficiency but also the efficiency of the engine it is driving). A disadvantage is that DME is a gaseous fuel that is hard to mix with standard diesel. Alternatives are to fuel diesel engines with 100% DME or to use it in a gasoline engine mixed with liquefied petroleum gas (LPG).

According to Zhang (2010), diesel obtained via FTS has a production efficiency of only approximately 40%. One reason for the low efficiency for the biomass-based production of FT fuel compared to, e.g., methanol is that the reaction is more exothermal. Moreover, it requires more elaborate separation of the crude product than the other routes and produces more water: about 25% of the energy losses are due to condensation of the formed water. Zhang (2010) also states that the production of a wide range of hydrocarbons, including alcohols and aldehydes, contributes to the low efficiency. This is of course true when the only product of interest is diesel, but when FTS is integrated in a biorefinery scheme (see Chapter 15), this is less of a problem. Tock et al. (2010) report much higher production efficiencies ($\sim$60%) for FTS when not just FT diesel but the whole FT crude is considered. In their analysis, the production of FT crude is more efficient and cheaper than that of methanol and DME (both with a production efficiency of $\sim$53%). Moreover, they point out that the choice of the gasification process has a large influence on the overall performance: different synthetic fuels require different gasification technologies to maximize the efficiency. They claim that for liquid fuel synthesis, the best choice is directly heated gasification using a circulating fluidized bed, followed by steam methane reforming. It should be noted that FTS is a relatively complex process that requires large investment costs. This gives a large barrier for developing countries to implement this technology (Zinoviev et al., 2010). For large-scale operation, Haarlemmer et al. (2012) conclude after comparing a large number of studies that use of FTS for biomass conversion will lead to prices between 1.00 and 1.40 € L^{-1} of fuel. It will only become economically viable when crude oil price levels are high or when the environmental benefits of green FTS fuels are valued (Hamelinck et al., 2004). Van Vliet et al. (2009) report that oil prices should stay above $75 per barrel to make biofuel production by FTS profitable.

In conclusion, it can be said that the comparison of the various routes from biomass to synthetic fuels is not straightforward. Many aspects play a role, and it is important to rightly choose the boundary conditions. To determine the efficiency, do we consider the energy content of the produced fuel as the final outcome (well-to-tank efficiency) or the distance that a car can drive on it (well-to-wheel efficiency)? In addition to these efficiencies, there are also other definitions of efficiency, such as the atom efficiency. For example, the carbon efficiency is the amount of carbon atoms in the useful product as a fraction of the carbon atoms in the feedstock. An additional question is how to account for by-products (chemicals, electricity) that are produced. To determine investment costs, do we consider building a plant on an existing site or do we take a "greenfield" approach (see Chapter 7)? Since different authors make different decisions regarding these aspects, one should be very careful in directly comparing publications reporting an economic analysis of the different biofuel routes.

CHAPTER SUMMARY AND STUDY GUIDE

This chapter treats the various routes to convert syngas to transportation fuels. Most developments in this field so far are based on syngas produced from fossil fuels. However, we can benefit from this knowledge and experience when developing processes to produce biofuels from syngas. We pay most attention to the most successful route from syngas to automotive fuels up to now: the FTS, which converts syngas into a range of hydrocarbons. We discuss both the small-scale (reaction and catalyst) and the large-scale aspects (reactor) of the process. Subsequently, we treat the production of SNG from syngas, which has certain similarities with FTS. We also briefly treat the production of methanol and the conversion of methanol into DME and gasoline. Finally, we discuss some literature on comparing which of these routes is the most efficient one. It is not obvious to choose the boundary conditions for such a comparison. As a consequence, there is no clear agreement about the most viable route.

KEY CONCEPTS

Fischer–Tropsch synthesis
Anderson–Schulz–Flory distribution
Diffusion limitations
Slurry bubble column
Multitubular packed bed reactor
Synthetic natural gas (SNG) by methanation
Methanol synthesis, methanol dehydration, methanol to olefins, methanol to gasoline
Well-to-tank efficiency, well-to-wheel efficiency

SHORT-ANSWER QUESTIONS

17.1 What are the advantages of liquid transportation fuels based on biomass over other sustainable energy solutions for transportation purposes?

17.2 Compare the different fuels treated in this chapter: give their advantages and disadvantages.

17.3 What are the hurdles when trying to make a "fair" comparison between the different fuels?

17.4 In multitubular FTS reactors, typically particles of 1 or a few mm are used. Decreasing the particle size would increase the catalyst effectiveness. For what reason(s) is the particle size not reduced?

17.5 The reaction equation for Fischer–Tropsch synthesis (RX. 17.1) shows that for longer hydrocarbons, the Fischer–Tropsch reaction needs H_2/CO in a ratio of about 2:1. However, in practice, often, a lower syngas ratio is used. What are the advantages of using a lower syngas ratio? What are the disadvantages?

17.6 Compare (RX. 17.1) and (RX. 17.3). What agreement and what difference do you observe?

17.7 Why is it currently more attractive to convert methanol to olefins (MTO) than to gasoline (MTG)?

PROBLEMS

17.1 In Example 17.1, we have calculated the reaction rate for Fischer–Tropsch synthesis as a function of the temperature. Assume a fixed bed reactor in which the catalyst, with a density of 2500 $kg \cdot m^{-3}$, fills 60% of the reactor volume (porosity is 0.40). The catalyst effectiveness may be assumed to be 1 (note that is a simplification; typically, it is <1 in fixed beds). The reactor is cooled with cooling water of 465 K. The heat transfer coefficient is 600 $W \cdot m^{-2} \cdot K^{-1}$, and the reactor contains 100 m^2 of heat exchange surface per m^3 of reactor volume.

a. What is the rate of heat generation per m^3 of reactor volume? Plot this as a function of the temperature for the range 470–530 K.

b. What is the rate of heat removal of the heat exchange surface at 470, 500, and 530 K? Plot this for the same temperature range.

c. What do you conclude from comparing the results of a and b?

17.2 Suppose the Fischer–Tropsch reactions take place in identical spherical catalyst particles suspended in a homogeneous fluid phase. The reactants hydrogen (H_2) and carbon monoxide (CO) diffuse from the fluid at the outer surface of a catalyst particle through the pores to the catalytic reactive sites inside the particle, while the reaction products diffuse out of the particle. The main issue of this problem is what would be a suitable molar H_2/CO ratio at the outer surface of the catalyst particles to favor the FT reactions throughout the particles.

From a stoichiometric point of view, this H_2/CO ratio should be close to $3-\alpha$, where α is the chain growth probability (see Section 17.2.2). However, α is sensitive to the amount of hydrogen. The chain growth probability increases with a lower H_2/CO ratio by reducing the termination of chain growth by hydrogen. Another complication is the effect of spatially distributed reaction–diffusion in the catalyst particle. H_2 diffuses faster than CO and is consumed in larger amounts. Thus, the H_2/CO ratio will vary along the radial coordinate of the particle. This is borne out by the radially distributed concentrations of the reactants, which can be obtained by solving the reaction–diffusion equations for a catalyst particle. Solving such equations with realistic kinetics, e.g., Equation (17.1), requires a numerical approach because coupled nonlinear differential equations are involved. Here, we will use a simplified reaction–diffusion equation for which an analytical solution can be obtained to illustrate effects of unequal diffusion. Our gross simplifications (not suitable for real design cases) involve assuming:

- Zero-order kinetics (obtained by ignoring the partial pressure dependencies in Equation (17.1))
- Constant α
- Fickian diffusion, which neglects flux interactions between reactants and products inside the catalyst pores

The analytical solution to the reaction–diffusion equation is

$$c_i(z) = c_i^{(S)} + \frac{s_i}{D_i}\frac{a}{6}\left(r^2 - z^2\right) \qquad \text{for } 0 \leq z \leq r$$

$$c_i(r) = c_i^{(S)} \qquad\qquad\qquad i = CO, H_2$$

$$c_i(0) \geq 0 \quad => \quad \text{physical feasibility}: \quad \frac{|s_i|}{D_i}\frac{a}{6}r^2 \leq c_i^{(S)}$$

a : zero-order reaction rate coefficient $\left(\text{kmol·s}^{-1}\text{·m}_{cat}^{-3}\right)$
$c_i^{(S)}$: concentration of component i at surface $\left(\text{kmol·m}^{-3}\right)$
D_i : Fickian diffusion coefficient of component i $\left(\text{m}^2\text{·s}^{-1}\right)$
r : Radius of spherical catalyst particle (m)
s_i : stoichiometric coefficient of component i $(-)$
Note : $s_i < 0$ for a reactant
z : radial coordinate (from center to surface S) (m)

Solving two equations with four unknowns $\left(c_i^{(S)}, c_i(0), i = H_2, CO\right)$ requires two more specifications:

1. The H_2/CO ratio will increase toward the center of the particle due to faster diffusion of hydrogen. For that reason, the perfect stoichiometric ratio is only imposed as an upper bound in the particle center:

$$z = 0: \quad \frac{c_{H_2}(z)}{c_{CO}(z)} = 3 - \alpha$$

2. At the outer surface, the CO concentration is given, while the H_2 concentration is a degree of freedom to match the previous center condition:

$$z = r: \quad c_{CO}(z) = c_{CO}^{(S)} = 0.2 \quad \left(kmol \cdot m^{-3}\right)$$

Typical numerical values (right order of magnitude) for the model parameters (at 500 K) are:

$$\alpha = 0.9; \quad a = 6.0 \times 10^{-3} \quad \left(kmol \cdot s^{-1} \cdot m_{cat}^{-3}\right)$$
$$D_{CO} = 5.2 \times 10^{-9} (m^2 \cdot s^{-1}); \quad D_{H_2} = 2.7 D_{CO} (m^2 \cdot s^{-1})$$
$$r = 0.5 \times 10^{-3} (m); \quad s_{CO} = -1; \quad s_{H_2} = -(3 - \alpha)$$

a. Given the concentration distribution equations with boundary conditions, derive and report the expressions for the H_2 concentration at the outer surface $\left(c_{H_2}^{(S)}\right)$ and the expressions for the H_2 and CO concentrations in the center.

b. Use these expressions to compute the concentrations in the center and the H_2/CO ratio at the outer surface $(\sigma_{H_2/CO})$. Check by which fraction (μ) this ratio is smaller than the ideal stoichiometric ratio of $3 - \alpha$, where $\mu = \sigma_{H_2/CO}/(3 - \alpha)$.

17.3 A Fischer–Tropsch slurry bubble column reactor is shown in Figure 17.4. Small catalyst particles are suspended in the liquid product (a mix of hydrocarbons and water), while syngas is vigorously bubbled through the liquid. Due to intense mixing, the compositions of the liquid and the gas are uniform within each phase throughout the reactor, i.e., independent of the location in the reactor. The reactants, CO and H_2, are only partially converted in such a reactor. A fraction of the unconverted syngas is recycled and mixed with fresh syngas feed. The main issue is the determination of the composition of fresh syngas meeting a target H_2/CO ratio for the gas phase. This latter target must be a suitable one for the Fischer–Tropsch reactions in the catalyst particles exposed to the gas.

a. Derive the H_2 and CO molar balances over the reactor–recycle–mixer system. Each balance specifies a relation between the concentrations in the fresh syngas and the gas inside the reactor. The balance equations will contain the following parameters:

The degree of conversion of CO in this reactor (X), relative to intake to reactor

The recycle ratio (r) of the unconverted syngas

The conversion of H_2 relative to the CO conversion in reaction stoichiometry $(3-\alpha)$

b. Using these balances, show how the H_2/CO molar ratio $(x_{H_2/CO})$ in the fresh syngas feed can be related to the corresponding ratio in the gas inside the reactor $(\sigma_{H_2/CO})$. The gas in the reactor is well mixed, so the H_2/CO ratio of the syngas exit stream is equal to the ratio inside.

c. Compute the molar H_2/CO ratio $(x_{H_2/CO})$ in the fresh syngas meeting a target H_2/CO ratio in the gas inside the reactor $(\sigma_{H_2/CO})$. This target ratio is expressed as a fraction (μ) of the ideal stoichiometric H_2/CO ratio for the Fischer–Tropsch reactions: $\sigma_{H_2/CO} = \mu(3-\alpha)$.

The following numerical values are applicable:

Chain growth probability: $\alpha = 0.9$

Reduction factor: $\mu = 0.8$

Degree of CO conversion: $X = 0.6$

Recycle fraction for syngas: $r = 0.8$

PROJECTS

P17.1 Develop a process block diagram with molar balances for the conversion of methanol to DME, reaction equation (RX. 17.5), based on thermodynamic considerations.

a. Use Gibbs free energies of the reaction components (see data tables in books on chemical engineering thermodynamics) to derive the Gibbs free energy change of the reaction. Compute the associated reaction equilibrium constant at 500 K. The reaction of methanol to DME occurs in the gas phase at 10 bar(a). When reporting your results, mention the thermodynamic equations used (e.g., from Chapter 5 in this book or other textbooks).

b. Determine the equilibrium conversion of methanol at 500 K, assuming that the reacting species are not involved in any other reactions. Decide if recovery and recycling of unconverted methanol are desirable. If the methanol conversion is sufficiently high (typically >98%), the cost of feed recovery and recycling often is larger than the extra economic benefit derived from a fuller use of the feed.

c. Determine the boiling points of DME, methanol, and water (see data tables in books on chemical engineering thermodynamics) and find out if separation of these three components can be achieved by means of condensations.

d. Suppose DME is formed in a reactor operating in at steady-state continuous flow mode at 500 K and 10 bar. How much reaction heat must be removed from the reactor when $0.5 \, kmol \cdot s^{-1}$ methanol is converted in the reactor? What cooling medium would you prefer: low pressure steam (at 410 K), medium pressure steam (at 480 K), or a cold methanol process stream (at 325 K)?

e. Develop a block diagram for the DME process. Create some alternative configurations and explain which one you prefer.

f. Compute the molar flows in the process block diagram you prefer. The feed rate of pure methanol to the process is $0.5 \, kmol \cdot s^{-1}$. To keep things simple, you can work with the simplifying assumptions that reaction equilibrium fixes the conversion of methanol and that complete physical separation of a component is achieved (100% separation efficiency).

Optional: You can also use a chemical process flow sheeting program to solve the molar balances of the components. In that case, more realistic performance conditions for process units can be set: e.g., methanol conversion is up to 98% of the equilibrium conversion, and separations achieve 99% recovery of a component.

g. Obtain estimates of the prices of feed and products from public sources and determine the economic potential at the input–output level (see Chapter 7 for the definition of economic potential).

h. Write a design report with the results along with the rationales for your design decisions.

INTERNET REFERENCES

tinyurl.com/388hn38
http://epp.eurostat.ec.europa.eu/portal/page/portal/energy/data/main_tables

tinyurl.com/pfc2ggu
http://gobigas.goteborgenergi.se/En/Start

REFERENCES

Brown LF. A comparative study of fuels for on-board hydrogen production for fuel-cell-powered automobiles. Int J Hydrogen Energy 2001;26(4):381–397.

Deshmukh SR, Tonkovich ALY, McDaniel JS, Schrader LD, Burton CD, Jarosch KT, Simpson AM, Kilanowski DR, Leviness S. Enabling cellulosic diesel with microchannel technology. Biofuels 2011;2(3):315–324.

Fiedler EG, Grossmann G, Kersebohm DB, Weiss G, Witte C. Methanol. In: *Ullmann's Encyclopedia of Industrial Chemistry*. Weinheim: Wiley-VCH; 2011.

Gassner M, Maréchal F. Thermo-economic optimisation of the polygeneration of synthetic natural gas (SNG), power and heat from lignocellulosic biomass by gasification and methanation. Energy Environ Sci 2012;5(2):5768–5789.

Guettel R, Turek T. Comparison of different reactor types for low temperature Fischer-Tropsch synthesis: a simulation study. Chem Eng Sci 2009;64(5):955–964.

Haarlemmer G, Boissonnet G, Imbach J, Setier PA, Peduzzi E. Second generation BtL type biofuels—a production cost analysis. Energy Environ Sci 2012;5(9):8445–8456.

Hamelinck CN, Faaij APC, Den Uil H, Boerrigter H. Production of FT transportation fuels from biomass; technical options, process analysis and optimisation, and development potential. Energy 2004;29(11):1743–1771.

Hooshyar N, Vervloet D, Kapteijn F, Hamersma PJ, Mudde RF, van Ommen JR. Intensifying the fischer-tropsch synthesis by reactor structuring—a model study. Chem Eng J 2012; 207–208: 865–870.

Kopyscinski J, Schildhauer TJ, Biollaz SMA. Production of synthetic natural gas (SNG) from coal and dry biomass–a technology review from 1950 to 2009. Fuel 2010;89(8):1763–1783.

Lee S. *Methanol Synthesis Technology*. Boca Raton: CRC Press Inc; 1990.

Maretto C, Krishna R. Modelling of a bubble column slurry reactor for Fischer-Tropsch synthesis. Catal Today 1999;52(2–3): 279–289.

Moulijn JA, Makkee M, van Diepen AE. *Chemical Process Technology*. Chichester: Wiley; 2013.

Pangarkar K, Schildhauer TJ, Ruud van Ommen J, Nijenhuis J, Kapteijn F, Moulijn JA. Structured packings for multiphase catalytic reactors. Ind Eng Chem Res 2008;47(10): 3720–3751.

Phillips SDT, Tarud JK, Biddy MJ, Dutta A. *Gasoline from Wood via Integrated Gasification, Synthesis, and Methanol-to-Gasoline Technologies*. Golden, Colorado: NREL; 2011.

Semelsberger TA, Borup RL, Greene HL. Dimethyl ether (DME) as an alternative fuel. J Power Sourc 2006;156(2): 497–511.

Steynberg AP, Espinoza RL, Jager B, Vosloo AC. High temperature Fischer-Tropsch synthesis in commercial practice. Appl Catal A Gen 1999;186(1–2):41–54.

Tock L, Gassner M, Maréchal F. Thermochemical production of liquid fuels from biomass: thermo-economic modeling, process design and process integration analysis. Biomass Bioenergy 2010;34(12):1838–1854.

van Vliet, OPR, Faaij APC, Turkenburg WC. Fischer-Tropsch diesel production in a well-to-wheel perspective: a carbon, energy flow and cost analysis. Energy Convers Manage 2009;50(4):855–876.

Vervloet D, Kapteijn F, Nijenhuisa J, Ruud van Ommena J. Fischer-Tropsch reaction-diffusion in a cobalt catalyst particle: aspects of activity and selectivity for a variable chain growth probability. Catal Sci Technol 2012;2(6):1221–1233.

Yates IC, Satterfield CN. Intrinsic kinetics of the Fischer-Tropsch synthesis on a cobalt catalyst. Energy Fuels 1991;5(1):168–173.

Zhang W. Automotive fuels from biomass via gasification. Fuel Process Technol 2010;91(8): 866–876.

Zinoviev S, Müller-Langer F, Das P, Bertero N, Fornasiero P, Kaltschmitt M, Centi G, Miertus S. Next-generation biofuels: survey of emerging technologies and sustainability issues. ChemSusChem 2010;3(10):1106–1133.

18

CHEMISTRY OF BIOFUELS AND BIOFUEL ADDITIVES FROM BIOMASS

ISABEL W.C.E. ARENDS

Department of Biotechnology, Biocatalysis Group, Faculty of Applied Sciences, Delft University of Technology, Delft, the Netherlands

ACRONYMS

DMF	dimethylfuran
EFE	ethylfurfuryl ether
EMF	ethyl methoxyfurfural
ETBE	ethyl tertiary butyl ether
EV	ethyl valerate
FAME	fatty acid methyl esters
gVL	γ-valerolactone
HHV	higher heating value
HMF	hydroxymethylfurfural
LA	levulinic acid
MF	methylfuran
MMF	methyl methoxyfurfural
MTBE	methyl tertiary butyl ether
PV	pentyl valerate
VA	valeric acid

Biomass as a Sustainable Energy Source for the Future: Fundamentals of Conversion Processes,
First Edition. Edited by Wiebren de Jong and J. Ruud van Ommen.
© 2015 American Institute of Chemical Engineers, Inc. Published 2015 by John Wiley & Sons, Inc.

18.1 INTRODUCTION

First-generation biodiesel (fatty acid methyl esters (FAME)) and bioethanol have paved the way for the use and public awareness of biofuels in a socio- and technoeconomic context. FAME and bioethanol have achieved commercial status and market acceptance. At present, new generations of biofuels are progressing as alternatives, and enormous efforts are made by knowledge institutes and companies and stimulated by governmental or broader policies and legislations in finding the best biofuel, both in terms of biomass source and the final desired structure of the fuel. The present chapter is meant to provide understanding of the molecular structure of biofuels and of the organic chemistry behind the conversion of biomass to biofuel. With this knowledge in hand, chemical engineers can rank the potential of novel biofuels continuously appearing in the literature by comparing molecules of similar structures and origins. In addition, the potential and greenness of alternative biofuels and their production methods are discussed in view of their atom economy and compliance with the 12 principles of green chemistry. These principles are a possible instrument of predicting how good new biofuels will score from a socioeconomic and technical perspective (Sheldon, 2011).

The emphasis in this chapter is on liquid biofuels with a well-defined organic structure that can replace the current fossil-derived gasoline and diesel components. The concept is to directly transform biomass feedstocks based on, e.g., cellulose and plant oils into organics with a high energy density. The alternative route, gasification of biomass to CO/H_2 and further conversion to liquid fuels, falls outside the scope of this chapter. For an overview of this technology, the reader is referred to Chapters 10 and 15 as well as excellent reviews available in the literature (Demirbas, 2007; Ragauskas et al., 2006). Similarly, the description of thermochemically (i.e., catalytic pyrolysis) produced bio-oils, which have a large potential as, e.g., heavy oil substitutes, can be found in Chapter 11.

In terms of gaseous feedstocks, hydrogen (produced by, e.g., gasification of biomass) and methane (from biogas) have a high energy density (with a HHV of $142 \, MJ \cdot kg^{-1}$ and $56 \, MJ \cdot kg^{-1}$, respectively) and can in principle be considered as biofuels. However, their gaseous nature still necessitates special requirements.

In the global energy market, the main nonfossil options are solar, wind, hydro, biomass, and geothermal energy. These primary energy sources need to be converted into energy carriers such as electricity, hydrogen, and biofuels. Electrical cars are emerging, but efficient storage of large amounts of electrical energy is still a challenge. Similarly, the storage and distribution of hydrogen faces many obstacles. Biofuels offer a nonfossil alternative to the current high energy density liquids used in cars, trucks, and airplanes.

18.2 BIOETHANOL AND BIODIESEL

The massive use of fossil gasoline and diesel is linked to their higher heating value (HHV) of $44 - 48 \, MJ \cdot kg^{-1}$ (tinyurl.com/oms56vw). This heating value can be easily calculated using Dulong's formula (derived originally for coal, and here, it is assumed that no sulfur is present in the fuel):

$$HHV\left(MJ \cdot kg^{-1}\right) = 33.8\left(wt \cdot fraction\,C\right) + 144.2\left[wt \cdot fraction\,H_2\right.$$

$$\left. - \left(wt \cdot fraction\,O_2/8\right)\right] \qquad \text{(Eq. 18.1)}$$

Biofuels are already on the market in the form of ethanol (HHV of 29.7 MJ $\cdot$ kg^{-1}) for gasoline engines and FAME (HHV of 41 MJ $\cdot$ kg^{-1}, as calculated for methyl linoleate) for diesel engines. Usually, these components are used for blending (typically 2 – 10 vol.%) in oil-derived gasoline and diesel. Blending largely circumvents the need for adjustments to existing car engines.

However, the engine performance of fuels is characterized by more than its HHV value. Based on the properties of engines (sparked ignition vs. compressed spontaneous ignition), the organic structures need to meet certain requirements in terms of volatility, stability, and boiling point trajectory. The performance of a fuel in the engine has to be judged along the ten criteria of the ASTM standard (tinyurl.com/pogugy). One of the indicators is the octane number (gasoline, octane number around 95) or cetane number (diesel, cetane ($C_{16}H_{34}$) number around 45). For a chemical perspective on the structure of biofuels, Table 18.1 gives an overview of the elemental composition of traditional fuels versus their biomass counterparts (Petrus and Noordermeer, 2006).

What is immediately noticed is that in classical fuels no oxygen is present. The presence of oxygen in fuels in the form of alcohols and/or ethers lowers their energy density but is reputed to improve the quality of the exhaust. Therefore, gasoline has to be mixed with oxygenates such as ethanol, methyl tertiary butyl ether (MTBE), and ethyl tertiary butyl ether (ETBE), in order to prevent knocking of the engine. Thus, the message is that the complete performance of a biofuel compound has to be taken into account and that simply mimicking the structure of fossil fuels is not necessarily desirable. For example, a number of favorable properties of bioethanol are that it is easily produced through fermentation, has a reasonable HHV of 29.7 MJ $\cdot$ kg^{-1}, is not very toxic, and has a high octane number of 113. This has led to its massive use either in blending or in flex cars. In Figure 18.1, the reaction for fermentative ethanol

TABLE 18.1 Composition of gasoline and diesel compared with FAME, ethanol, and carbohydrates in general

	Gasoline	Diesel	FAME	Ethanol	Carbohydrate
Carbon chain length	5 – 10	12 – 20	19	2	$[5-6]_n$
O/C molar ratio	0	0	0.11	0.5	1
H/C molar ratio	~2	~2	~2	3	2
Polarity	Nonpolar	Nonpolar	Nonpolar	Polar	Polar
Structure	Branched/ aromatic/cyclic/ unsaturated	Linear/ saturated	Linear	C_2H_5OH	Linear/ cyclic polyalcohol

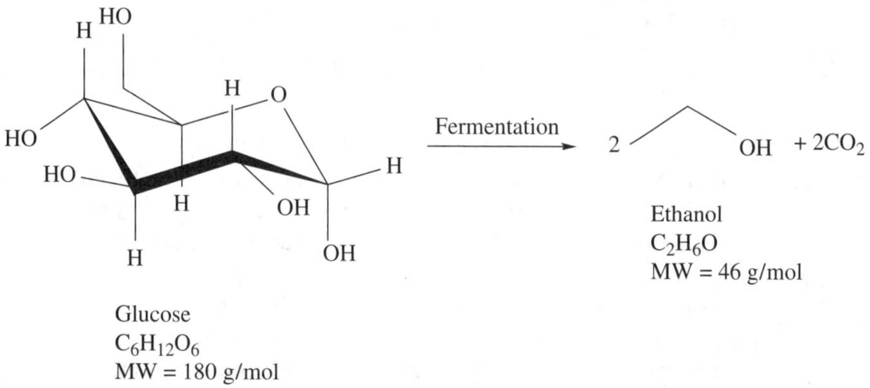

Glucose
$C_6H_{12}O_6$
MW = 180 g/mol

Ethanol
C_2H_6O
MW = 46 g/mol

Atom efficiency: $(2 \times 46/180) \times 100\% = 51\%$

FIGURE 18.1 Atom efficiency of ethanol production.

production is given. The production of every molecule of ethanol is accompanied by the production of one molecule of CO_2. The total atom efficiency of the process (defined as the weight of the atoms in the product divided by weight of the atoms in the feed) is, therefore, only 51%. The atom efficiency is one of the twelve principles of *green chemistry* (see Box 18.1) (Anastas and Warner, 2000). In general, high atom efficiencies are inherent to a sustainable process. Of course, the atom efficiency is only one aspect. A more practical number is the so-called E-factor (see Example 18.1), which takes into account all waste produced by the plant (Sheldon, 2007). The E-factor can range from ~0.1 in oil refineries to >100 in the production of pharmaceuticals.

Biodiesel is the second most abundant renewable liquid fuel. The production capacity in Europe on July 1, 2012 was around 23 million tonnes (tinyurl.com/mjhnewl). As can be seen from Table 18.1, it has a favorable carbon chain length, thus providing an excellent basis for a high-performance biofuel. The starting vegetable oils from which FAME is produced (see Figure 18.2) are too viscous and usually have a too high free fatty acid content to be used directly. The first diesel engine ran on peanut oil (Diesel, 1895). FAME is a product of transesterification. Transesterification lowers the molecular weight and viscosity by converting triglyceryl esters of fatty acids into their methyl esters and glycerol (see Figure 18.2). This technology is rather straightforward: simply boiling a bottle of oil with methanol and some base or acid (as catalyst) can produce biodiesel.

The evaluation of the production of biodiesel with respect to the 12 principles of green chemistry (see Box 18.1) is not so easy. The production processes of biodiesel have to be stringently judged. For instance, in case palm oil-based facilities produce large amounts of residual biomass (waste biomass of lignocellulosic nature), one can argue that the E-factor (see Example 18.1) for these types of plants is unacceptably high (principles 1 and 2). Therefore, this process does not comply with these 12 principles. Furthermore, the use of catalysts is not common practice in biodiesel

---

BOX 18.1: THE TWELVE PRINCIPLES OF GREEN CHEMISTRY

1. Prevention instead of remediation
2. Atom efficiency
3. Less hazardous chemicals
4. Design safer chemical products
5. Safer solvents and auxiliaries
6. Energy efficient by design
7. Renewable raw materials
8. Shorter synthesis
9. Catalytic methodologies
10. Design for degradation
11. Analysis for pollution prevention
12. Inherently safer chemistry

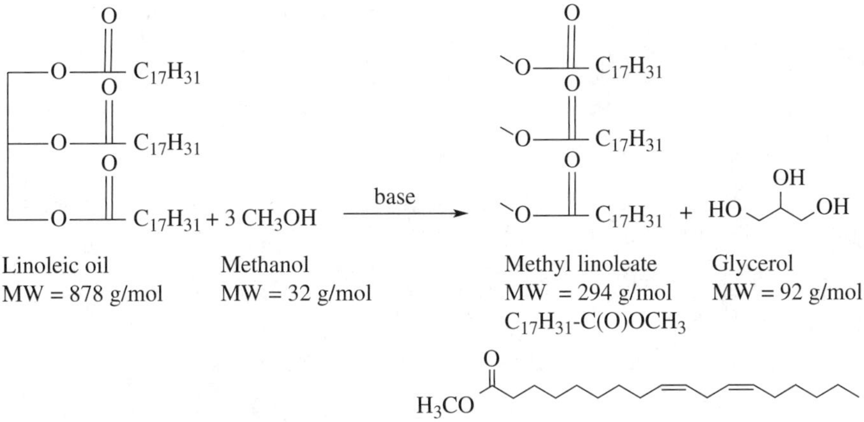

Linoleic oil	Methanol	Methyl linoleate	Glycerol
MW = 878 g/mol	MW = 32 g/mol	MW = 294 g/mol	MW = 92 g/mol
		$C_{17}H_{31}$-C(O)OCH$_3$	

Atom efficiency: $[3 \times 294/(878 + 3 \times 32)] \times 100\% = 91\%$

FIGURE 18.2 Atom efficiency of FAME production. Linoleic acid was used as a model fatty acid compound. It is one of the major components of soybean oil. In practice, biodiesel oils, such as vegetable oils and waste oils, consist of a mixture of saturated (e.g., stearic acid), mono-unsaturated, di-unsaturated, and poly-unsaturated fatty acid esters.

production technology, where often the use of large amounts of bases is the cheapest way to convert oils into FAME. Also, the origin of methanol (the solvent needed for esterification) needs to be judged. In other words, there is no general answer to the question whether biodiesel production complies with the twelve principles, and every biodiesel production facility has to be evaluated separately.

Example 18.1 E-factor calculation for the production of FAME

Beside (the theoretical) atom efficiency, the E-factor is a more practical parameter for determining the greenness of a process. It refers to the kg of waste produced per kg of product. For the process depicted in Figure 18.2 for the production of FAME, make a reasonable assumption of the E-factor, and argue which streams should be denoted as waste. Also, make an estimation of the use of solvents or aqueous streams and additives required during the production (in kg).

Solution

Before we can start making an estimate, we need some details about the process. The cheapest way to make biodiesel is to mix methanol (usually molar excess of about two is sufficient to drive the reaction and obtain good mixing) with KOH as base (in situ formation of $KOCH_3$) and mix this with the heated plant oil. Two layers will be formed, with the FAME as one layer and glycerol/salt/CH_3OH as the other layer. The basic salt solution cannot be recycled and needs to be neutralized with an equimolar amount of acid (HCl) before disposal. The actual amount of waste produced in this step per kg of FAME (3.4 mol) is thus 0.25 kg KCl (3.4 mol). In practice, excess base is used, so probably, a factor of about 5% needs to be added, but as we do not know these details, we will stick to a 1:1 ratio. Washing and drying will be required to clean the biodiesel. The water required for this will go to the wastewater treatment.

Per kg of biodiesel product (in this case 3.4 mol), we can make assumptions about the amounts of waste produced:

1. Glycerol: 1.15 mol (0.10 kg). Officially, this is a (by-)product, not waste, so we can leave it out of the calculation.

2. Excess methanol that cannot be recycled: The extra methanol that is produced as byproduct is $1 \, mol \cdot mol^{-1}$ biodiesel 3.4 mol methanol = 0.1 kg.

3. Salt resulting from neutralization of the "catalyst" used: 3.4 mol KCl (0.25 kg).

4. Wastewater: This will make up most of the waste but is also most difficult to estimate. Moreover, water is usually excluded in determining E-factors.

Thus, in total, a minimum of 0.35 kg waste per kg of biodiesel is produced. The E-factor is, therefore, 0.35. In practice, this number will be much higher due to additional washing steps and downstream handling units. This number is therefore the minimum value.

From the aforementioned example, we learn that for every kg of FAME, at least 0.35 kg of waste is produced. In practice, the real factor will be much higher because the downstream processing of the product will require additional washings and handling. There are many variations on the synthesis, using, e.g., an enzymatic catalyst (Fjerbaek et al., 2009) or ethanol instead of methanol as the alcohol, but the principle stays the same. The main drawback is the need for inexpensive and sustainable feedstocks. Normally, palm, sunflower, rapeseed, and soybean oils are used, but these are also food sources, and as a result, competition between food and fuel markets occurs. One can argue that this competition should be avoided at all times and technologically there are many alternatives to bypass the food/feed discussion. For example, the use of low-quality oils that cannot be used in food applications can offer an alternative, and new catalytic processes can be developed to exploit these less expensive oil streams.

This is a general point in all use of biomass. A drawback of using only the easily processed sugar and triglyceride fractions of a plant is that these fractions form only a part of it. Accordingly, the net energy yield that can be achieved using these fractions is poor, and only specific crops can be used. To improve the energy yield of fuels from biomass, lignocellulosic feedstocks must be considered despite their complexity. This will enable the use of all parts of the current feedstocks (such as waste from palm trees and molasses from sugarcane). In addition, trees, switchgrasses, and other low-ranked biomass on poor lands can be considered as feedstocks. For a detailed discussion of the selection of lignocellulosic streams and pretreatment thereof, we refer to the recent literature reviews (Zinoviev et al., 2010).

As a variation to this traditional transesterification of oils, *Neste* has developed a process to fully reduce oil to a linear alkane component, including the catalytic hydrogenation of the glycerol, which normally ends up as by-product, to propane (tinyurl.com/l5jdxxq).

18.3 CONVERSION OF SUGARS TO HYDROCARBON FUELS

In a worldwide context, the most desirable source of biomass stems from lignocellulose (Dornburg et al., 2010). Lignocellulose can be acquired from biomass waste, as well as from woody or grass samples, and holds the promise of sustainable use. The so-called second-generation biofuels derived from lignocellulosic biomass (see Box 18.2) can thus still use a sugar moiety as molecular basis for a biofuel.

In order to convert sugars into biofuels, two types of organic transformations need to be performed: (1) a substantial reduction of the oxygen content needs to be realized, and (2) C–C bonds need to be created between biomass-derived intermediates to increase the molecular weight for use as biodiesel substitutes (see Table 18.1, C12–C18 range). When looking at a sugar molecule, two main ways can be considered to decrease the O/C ratio and thus increase its energy content, namely, the release of CO_2 and the release of H_2O.

As can be seen from Figure 18.1, the production of two molecules of ethanol from glucose ($C_6H_{12}O_6$) is accompanied by the formation of two CO_2 molecules. Elimination of two CO_2 molecules and one H_2O molecule from $C_6H_{12}O_6$ gives $C_4H_{10}O$. This molecular formula corresponds to 1-butanol, the main product of anaerobic fermentation of sugars by *Clostridia* bacteria. Diethyl ether, which is usually formed by dehydration of two ethanol molecules, is another compound having the same molecular formula. Based on their structure, both these compounds are suitable candidates for biofuels and have been considered as serious alternatives to bioethanol.

BOX 18.2: GENERATIONS OF BIOFUELS

First-generation biofuels are biofuels directly produced from food crops, such as the oils for use in biodiesel or the sugars for producing bioethanol for fermentation. First-generation biofuels have been criticized because they release more CO_2 in their production than they capture CO_2 during their cultivation. In addition, the use of food crops as source for fuel comes with ethical and global dilemmas (Karlsson, 2007).

Second-generation biofuels are produced from nonfood crops, such as lignocellulose (wood), food crop waste, and grasses. They are more cost competitive compared to first-generation biofuels, and life cycle assessments have indicated that their use will result in a decrease of transport-related greenhouse gas emissions (Davis et al., 2009; Pickett et al., 2008). The production of second-generation biofuels must be accompanied by an integral usage of all parts of the biomass, including recycling of biomass parts that have been turned into products and electrical energy generation from residual parts (such as lignin).

Third-generation biofuels are based on improvements in the production of biomass. For the production of these biofuels, advantage is taken of specially engineered energy crops such as algae. In theory, algae can produce more energy per unit area than conventional crops, and they can be grown on land as well as water. Furthermore, they can be engineered to produce not only ethanol but also diesel, gasoline, and kerosene. A major disadvantage at this moment is the cost of production, as well as problems concerning the robustness of the agricultural production.

Fourth-generation biofuels are aimed not only at producing sustainable energy but also at capturing and storing CO_2. Production of these biofuels differs from second- and third-generation production, because at all stages of production, the carbon dioxide must be captured, leading to a reduction in CO_2 emissions.

Elimination of one CO_2 and one H_2O molecule can be realized via fermentation and commonly leads to acetone and acetic acid as products (see Figure 18.3). Alternatively, a molecule of ethyl lactate can be produced from lactic acid and ethanol, which are products of, respectively, anaerobic and aerobic fermentation. Elimination of another molecule of water leads to the formation of γ-valerolactone (gVL). Simply removing three molecules of water form glucose leads to hydroxymethylfurfural (HMF). These glucose-derived molecules are denoted as platform molecules or building blocks. They can play a role as biofuels directly (ethanol, butanol) or can be considered as starting materials for the production of biofuels or biobased additives for fuels, provided cheap technology is available to perform this process.

Lignocellulose feedstocks also contain approximately 25 wt% of hemicellulose, which contains a large percentage of pentose sugars. Xylose (e.g., from straw) can be readily fermented to ethanol. Chemical water elimination from xylose leads to furfural, another platform chemical (see Example 18.2).

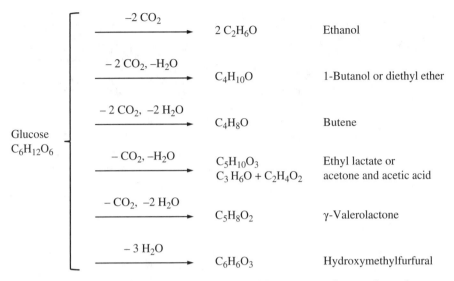

FIGURE 18.3   Possible structures upon release of three or more O atoms from glucose.

Example 18.2   Conversion of xylose

Write down the chemical equation for the fermentation of xylose to ethanol and for the dehydration of xylose to furfural, and determine the atom efficiencies of these reactions. Based on this information, which process is more attractive for biofuel purposes? Which information are you lacking?

Solution
Aerobic fermentation of xylose:

$$C_5H_{10}O_5 + 2O_2 \rightarrow C_2H_5OH + 3CO_2 + 2H_2O$$

The atom efficiency for ethanol production thus is

$$\frac{46}{150 + 2 \cdot 32} \cdot 100\% = 21\%$$

Chemical dehydration of xylose:

$$C_5H_{10}O_5 \rightarrow C_5H_4O_2 + 3H_2O$$

The atom efficiency for furfural production is

$$\frac{96}{150} \cdot 100\% = 64\%$$

Based on this information, furfural production seems most attractive for biofuel production because of its higher atom efficiency.

However, information on the yields of both reactions is also required. For the fermentation reactions, the yield is close to 100%, while for the conversion of xylose to furfural, yields close to 90% have been reported. All together, one could say that based on the atom yield, dehydration of furfural is the better way to make fuels (21% atom yield for ethanol vs. 58% atom yield for furfural).

The additional information we need is the heating value and fuel performance of both fuels. As a first indication, the HHV values for ethanol ($29.7\,MJ \cdot kg^{-1}$) and furfural (calculated according to Dulong's formula as $21\,MJ \cdot kg^{-1}$) can be compared. Putting this information together, the atom yield for furfural production is 2.8 times higher, while the heating value for furfural is 1.4 times lower than for ethanol, pointing toward furfural as a more sustainable fuel source.

Both biotechnological and chemical processes are available to eliminate CO_2 and/or H_2O from biomass constituents. Yeasts and other organisms have specific enzymes that catalyze CO_2 elimination reactions. In chemical transformations however, the elimination of water from carbohydrates is easier than the elimination of CO_2. The elimination of water is commonly carried out using acidic catalysts at elevated temperatures. Full elimination of all oxygen from sugar is only possible through the introduction of extra hydrogen. This involves considerably more process steps and thus significant additional costs. This will be discussed in more detail for selected examples (see Section 18.4).

In general, three strategies can be distinguished for the conversion of sugar streams into biofuels:

1. Direct elimination of CO_2 from sugars or cellulosic streams through fermentation: examples are the production of ethanol and butanol.

2. Release of a combination of CO_2 and H_2O molecules from sugars or cellulosic streams leading to platform chemicals, which can then be converted into biofuels or biobased additives for fuels.

3. Aqueous reforming of glucose streams in the presence of catalysts and at high temperatures; in this process, a combination of products is formed.

In the following, an overview of strategies 2 and 3 will be presented. For more information on the direct fermentation of sugars to biofuels (strategy 1), we refer to Chapter 13.

In 2004, the department of energy in the United States published a top 12 list of platform chemicals that can be derived from biomass. In 2010, Bozell and Petersen published a renewed list of chemical opportunities from the biorefinery (see Figure 18.4) (Bozell and Petersen, 2010).

Here, we highlight those platform chemicals that are most relevant for the production of biofuels, namely, furanics and levulinic acid (LA).

Extremely relevant is the source of the sugar or cellulosic stream. Transport and work-up of these biomass sources are surrounded with challenges, which require

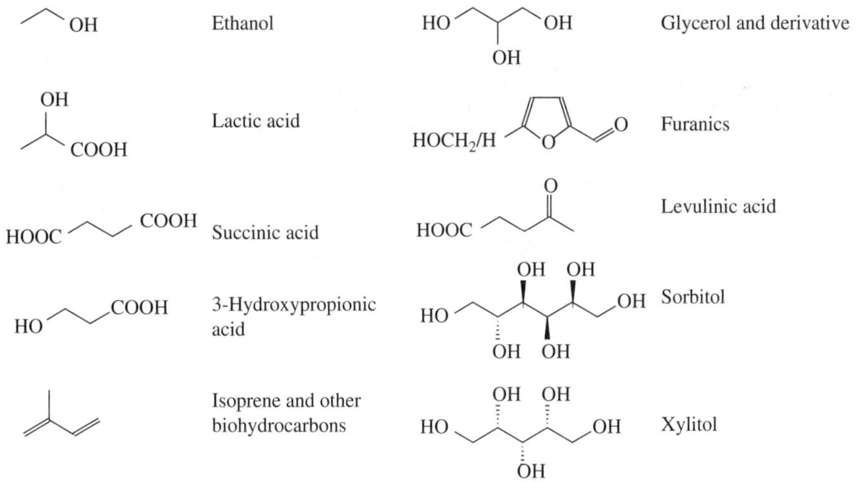

FIGURE 18.4 New top chemical opportunities from a biorefinery. The chemicals on the left are produced through fermentation of sugars, and the chemicals on the right are produced through chemical conversion (mainly dehydration and hydrogenation) of sugars. Based on information from Bozell and Petersen (2010).

alterations to the traditional bulk-scale crude oil and coal-oriented chains. The transition to a biobased economy also requires the development of integrated processes where (ligno)cellulosic streams can be used directly as feed. Current developments aim at the direct fermentation of cellulosic streams to ethanol, thereby circumventing the need to produce glucose as intermediate; there are many reviews available (see for a representative overview, e.g., Blanch, 2012). In addition, chemical catalytic processes are under development, which focus on the processing of cellulosic streams, such as the one-pot bifunctional catalytic conversion of cellulose into sorbitol using a combination of acid hydrolysis and metal-catalyzed hydrogenolysis (Van de Vyver et al., 2011). Another example is the hydrogenolysis of starch and cellulosic feedstocks to LA (Ruppert et al., 2012). Hydrogenolysis has as major disadvantage, being the extra and costly production of hydrogen. Integrated biorefinery examples where, e.g., formic acid as side product is used for the hydrogenolysis are expected to lay out the basis for future developments (Wright and Palkovits, 2012).

18.4 GREENNESS OF THE CONVERSION OF PLATFORM MOLECULES INTO BIOBASED FUEL ADDITIVES

18.4.1 Hydroxymethylfurfural

The most well-studied transformation occurs upon elimination of three water molecules from fructose and/or glucose, leading to the formation of HMF. The conversion of glucose is shown in Figure 18.5.

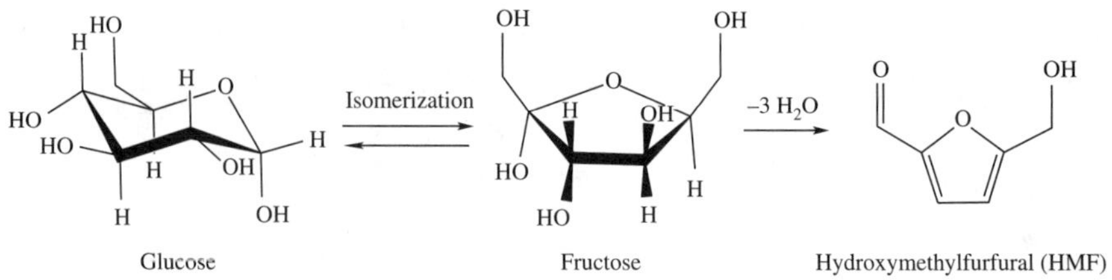

FIGURE 18.5   Glucose to fructose to HMF.

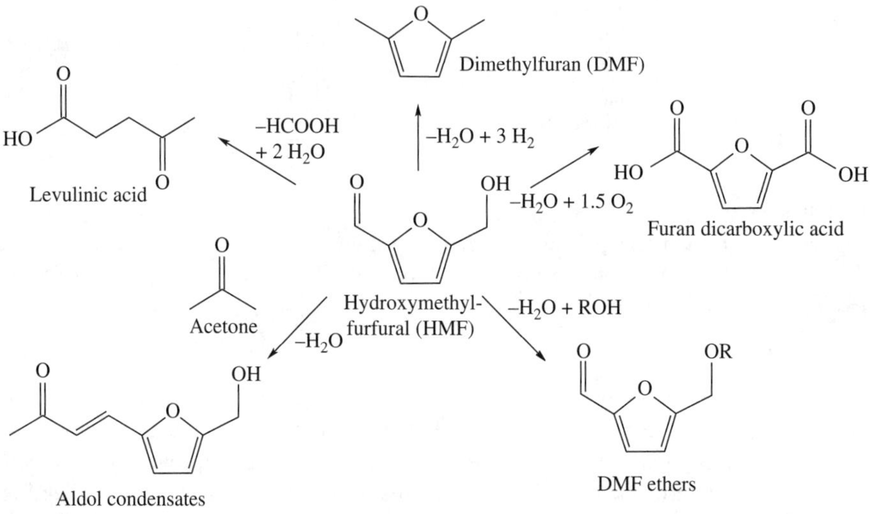

FIGURE 18.6   HMF as platform chemical; regarding the DMF ethers, MMF is the product when R = Me, and EMF is the product when R = Et.

The dehydration principle is a chemical process for which many additives and (acidic) catalysts have been probed to increase the selectivity. Direct conversion of glucose generally results in low conversions and selectivities. Yields for the production of HMF from glucose using mineral or organic acids of up to 45% have been reported (Rosatella et al., 2011). Generally, biphasic systems are applied, which diminish the subsequent hydration of HMF to LA (see Section 18.4.3). The greenness of this process is highly questionable, according to its E-factor: yields of 45%, accompanied by the formation of many by-products, cannot form the basis for a sustainable process. Therefore, significant technological progress is needed for scale-up.

HMF has been proposed as a starting material for the production of fuel additives, notably 2,5-dimethylfuran (DMF), DMF ethers, and aldol condensates. DMF is not soluble in water and can be used as blender in transportation fuels. Aldol condensates can act as precursors for diesel- and kerosene-type fuels. HMF has also been proposed as platform chemical for the production of polymers. In this case, furan dicarboxylic acids are the desired polymer building blocks. An overview of the major outlets of HMF is presented in Figure 18.6. We call this kind of technology platform technology. An alternative is not to produce HMF itself but the corresponding more stable

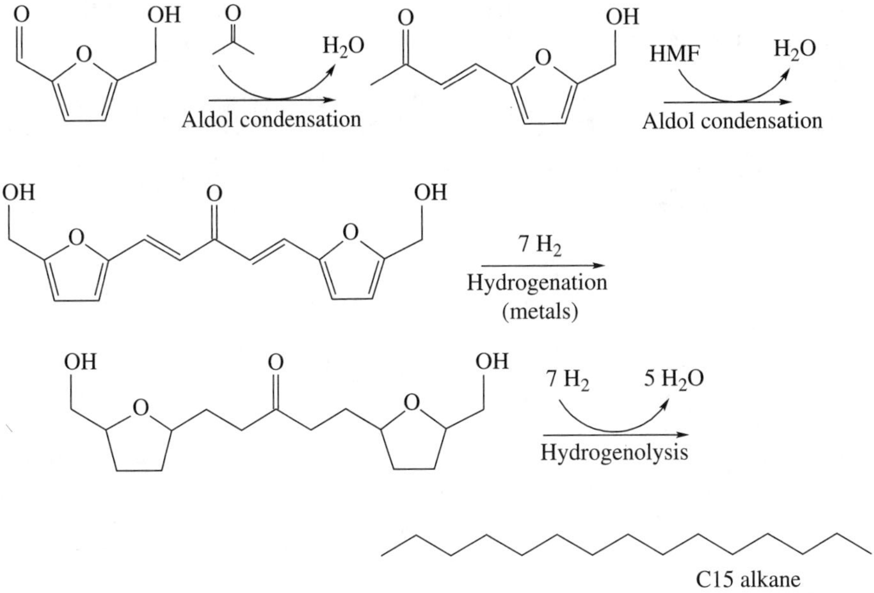

FIGURE 18.7 Conversion of HMF to diesel-range compounds.

ethers, which is done in a process developed by Avantium (Gruter and De Jong, 2009), and which leads to higher yields and selectivities (see Figure 18.6). The production of both methyl methoxyfurfural (MMF) and ethyl methoxyfurfural (EMF) has been reported. Typically, flow processes with low residence times are used. Starting with glucose, dissolved in water/ethanol/10% H_2SO_4, at 195 °C, a selectivity of >90% for HMF plus EMF has been reported (Gruter and De Jong, 2007). Especially, the amounts of by-products in the form of humins seem to be lower, thereby significantly lowering the E-factor for these processes.

The transformation of HMF into DMF or aldol condensates also can be questioned from a green chemistry standpoint. Conversion of HMF into DMF is carried out over heterogeneous catalysts in modest yields of 60 – 79%, and there is an additional need for external hydrogen (Roman-Leshkov et al., 2007).

For the production of diesel-range fuels, HMF needs to undergo chain lengthening (see Figure 18.7). An essential reaction to achieve this is the aldol condensation. A C–C bond reaction of HMF with a ketone, e.g., acetone, over a base catalyst (NaOH) at 25 °C produces a C9 derivative (when acetone is used), which can react with a second molecule of HMF to produce a C15 intermediate. Condensation products can then undergo hydrogenation/dehydration over bifunctional catalysts with metal and acid sites (e.g., Pd/γAl$_2$O$_3$ at 100 – 140 °C and 25 – 52 bar) to produce linear C9 or C15 alkanes that are hydrophobic and separate spontaneously from water, reducing the cost of purification.

A combination of aldol condensation, dehydration, hydrogenation, and hydrogenolysis is thus required to produce the desired end molecules. Especially, aldol condensations are reputed for modest selectivities and massive by-product formation.

In addition, all these routes require huge amounts of hydrogen that has to be produced separately (from biomass or otherwise).

Example 18.3   Greenness of HMF conversion to C15 alkane

How would you assess the greenness of the route for the production of a C15 alkane, shown in Figure 18.7?

Solution
When studying the reactions in Figure 18.7, one learns that this route is in contrast with at least four of the twelve green chemistry principles: (1) a multistep process is required; (2) aldol condensation is a noncatalytic step (requiring a base and therewith lots of salt waste), which is generally hampered by low yields; and (3) because of the additional hydrogen required, a large amount of energy is required for its production. (4) For all these steps, substantial amounts of solvents are needed. Therefore, without addressing the exact E-factor, one can safely say that this is not a green process.

18.4.2   Furfural

Wheat bran and straw available from agricultural residues are rich in xylan-type hemicellulose mainly consisting of linear chains of xylose, combined with arabinose ($\sim$33 wt%). These so-called hemicelluloses are easily depolymerized through hydrolysis in dilute acid. Furfural is one of the molecules that can be obtained by removal of three molecules of water from C5 sugars, as depicted in Figure 18.8. Yields in this case are higher than for HMF (Figure 18.5). In the literature, yields of up to 87% (with 95% selectivity) of furfural from xylose have been reported under optimized laboratory reactor conditions (Marcotullio and De Jong, 2011). The advantage compared to HMF is that furfural contains less oxygen. A disadvantage is that it stems from C5 sugars, rather than C6 sugars, which are less abundant.

In a recent publication by researchers from Shell (Lange et al., 2012), the use of furfural as platform molecule for biofuels has been evaluated. From a range of candidates, methylfuran (MF) and ethylfurfuryl ether (EFE) apparently had the most favorable properties. Figure 18.9 depicts the conversion of furfural to MF and EFE. In both cases, hydrogen is required to upgrade furfural. MF was positively evaluated in road tests in 10 vol.% blends with gasoline.

18.4.3   Levulinic Acid/gVL/Valeric Acid as C5 Platform

Cellulosic streams can be directly converted into LA/formaldehyde mixtures through acid hydrolysis, as shown in Figure 18.10 (Alonso et al., 2011; Wright and Palkovits, 2012). This is another C5 platform, next to furfural, which leads to favorable C/O and C/H ratios and which has the advantage of having a linear structure. In Figure 18.10, the possible reactions for LA are shown. Hydrogenation leads to gVL and further hydrogenation to valeric acid (VA).

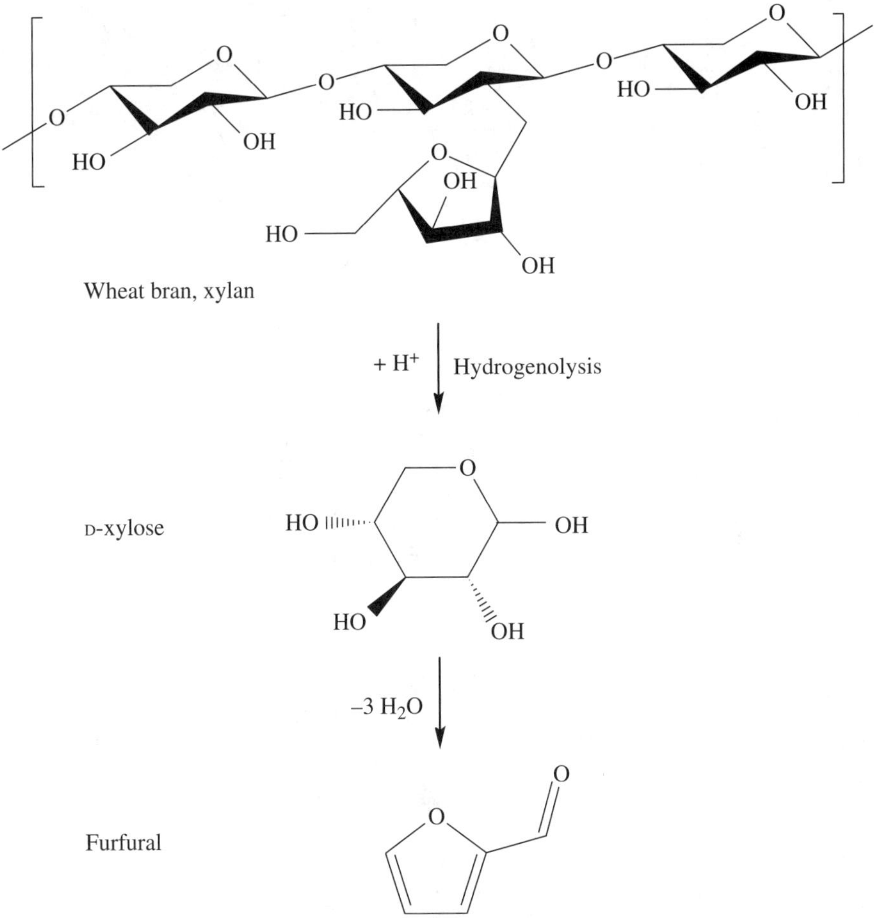

FIGURE 18.8   Conversion of xylan into furfural via xylose.

The direct processing of cellulose is hampered by solubility issues; cellulose is not soluble in aqueous mixtures, and efficient separation of LA and gVL from these mixtures requires special solvents. Interestingly, the combination of hydrolysis and hydrogenation seems to overcome some of these challenges. Integrated biorefinery approaches using the formed formic acid as hydrogen source for gVL production, combined with recycling of H_2SO_4 streams, have been proposed, and these seem to overcome some of the major disadvantages of cellulose processing, namely, low yields and the production of sulfuric acid (needed as catalyst) as waste. These integrated solutions are definitely needed in order to make the production of these C5 platform molecules acceptable both from a green chemistry and an economic standpoint.

Several LA derivatives, such as gVL and its hydrogenated analogue MeTHF, have been considered as fuel candidates. However, Shell researchers reported that the fuel blending properties of these compounds are poor and instead proposed VA derivatives

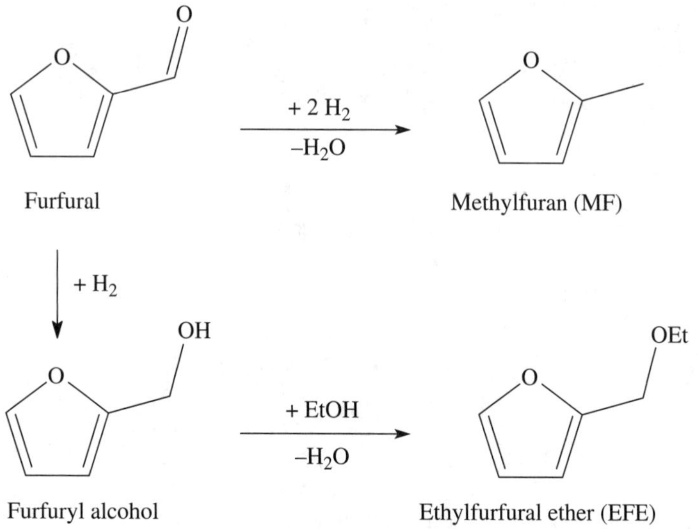

FIGURE 18.9 Furfural conversion to MF and EFE.

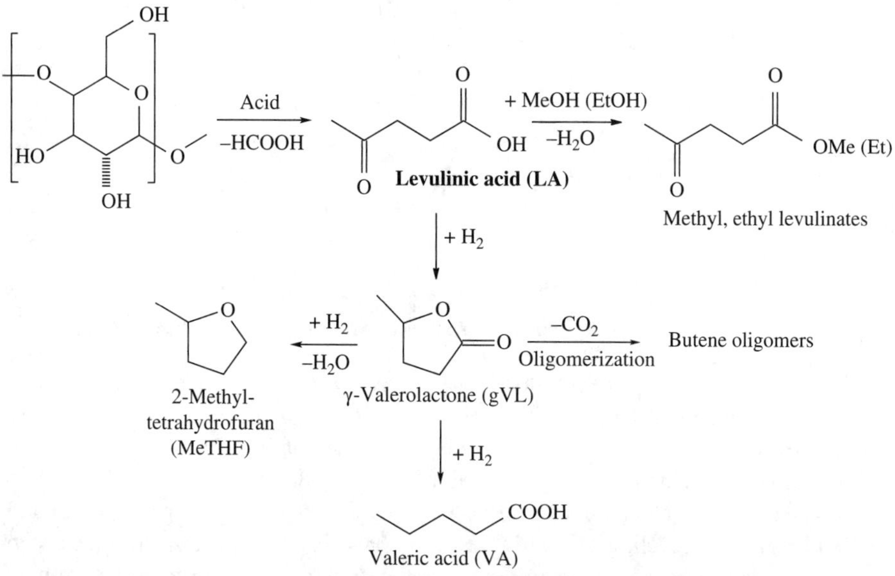

FIGURE 18.10 LA/gVL/VA as C5 platform from lignocellulose.

as fuel blenders (Lange et al., 2010). Notably, gVL does seem to be a suitable platform chemical for applications such as solvents and chemicals. In addition, gVL can be converted to kerosene- and diesel-range hydrocarbons through decarboxylation to butenes and subsequent butene oligomerization (Bond et al., 2010). In view of the many reaction steps required in the latter case, it can be questioned whether this will lead to a sustainable process.

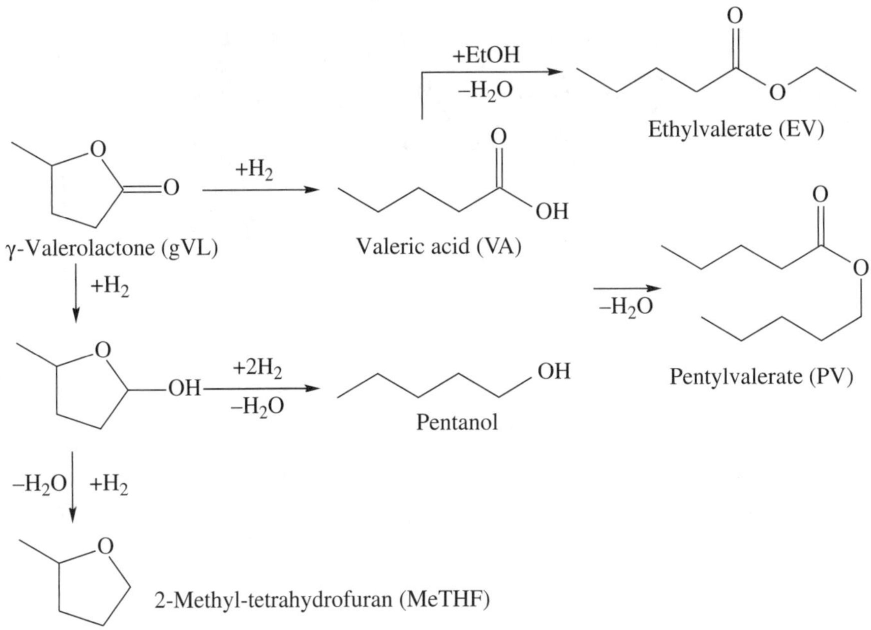

FIGURE 18.11 Conversion of gVL to the VA biofuels EV and PV.

Valeric acid derivatives, such as ethyl valerate, are reported to have favorable fuel properties (Lange et al., 2010). Noteworthy is the direct formation of pentyl valerate (PV) from gVL over a bifunctional acidic and hydrogenation catalyst (e.g., Pt/ZSM-5), as depicted in Figure 18.11. VA biofuels have acceptable energy densities and a more appropriate (lower) polarity than, e.g., ethanol and butanol. Depending on the chain length, they can be either employed for gasoline applications (ethyl valerate (EV) and propyl valerate have boiling points around 150 °C) or diesel-range applications (PV, boiling point above 200 °C). PV has a better volatility and cold-flow property match with diesel than FAME, albeit with a lower energy density. EV was evaluated positively in a road test, using a blend of 15 vol.% EV in regular gasoline.

Example 18.4 Production of gVL

In the scheme depicted in Figure 18.12, a possible process integration concept for gVL production is given (see Alonso et al., 2011).
  Describe the advantages and disadvantages of this concept.

Answer
In this concept, the formic acid formed during cellulose hydrolysis to form LA is used in the third step for the hydrogenation of LA to gVL. In this way, less additional hydrogen is needed. Secondly, streams of sulfuric acid are recycled, thus making this, hopefully, a salt-free process.

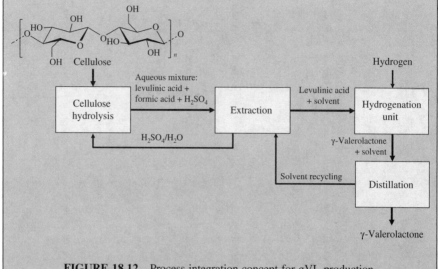

FIGURE 18.12 Process integration concept for gVL production.

Disadvantages of this process are the need for a solvent; solvent recycling always leads to solvent losses, which are generally in the range of 20%. The handling of sulfuric acid streams is accompanied by environmental, corrosion, and safety issues.

18.5 DIRECT AQUEOUS REFORMING OF SUGARS LEADING TO A RANGE OF ALKANES

Aqueous-phase reforming (APR) of sugars and polyols was initially developed as a technology to produce renewable hydrogen ($+CO_x$). However, nowadays, by an appropriate choice of catalyst, substantial amounts of light alkanes and monofunctional compounds can be produced as well (Alonso et al., 2010).

To illustrate the principle of this technology, which is promoted by the company Virent (tinyurl.com/q89rwxe), the equation is given for the conversion of glucose into its fully saturated analogue hexane:

$$19\,C_6H_{14}O_6 \rightarrow 13\,C_6H_{14} + 36\,CO_2 + 42\,H_2O \qquad \text{(RX. 18.1)}$$

The fact that no external hydrogen is needed makes this a relatively cheap technology. However, from this equation, it follows that carbon atom efficiency can never exceed the value of 68%. Another disadvantage is that selectivities for saturated C4–C6 alkanes are low, because generally lower alkanes (<C4) are formed.

Typically, supported Pt catalysts are used. The nature of both the metal and the support significantly influences the APR reactions of aqueous solutions of sugars with C/O ratios of 1. As the acidity of the catalytic system increases, e.g., with a solid acid catalyst support such as SiO_2 and Al_2O_3, the selectivity to alkanes increases due to the

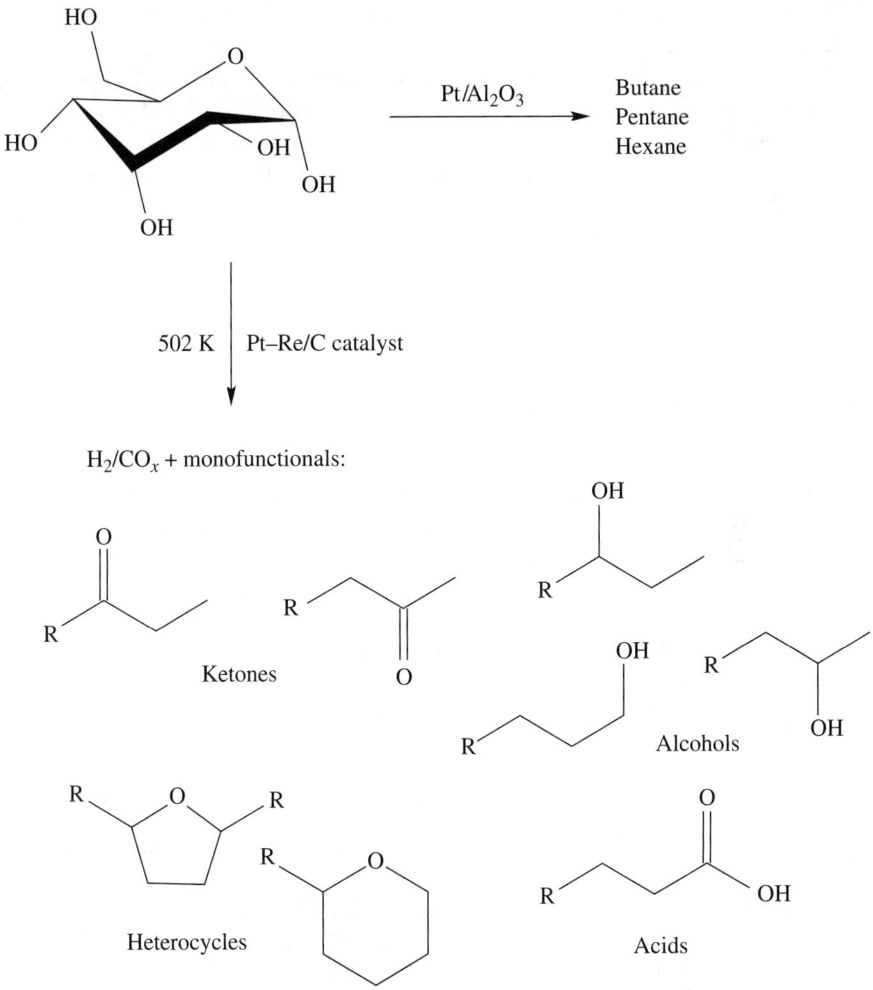

FIGURE 18.13 Aqueous reforming of glucose. Treatment of aqueous streams of glucose over catalysts at elevated temperatures leads to a plethora of either nonfunctionalized or monofunctionalized molecules. The oxygen ends up as either CO_2 or water.

increased rates of dehydration and hydrogenation compared to hydrogenolysis and reforming. This can be exploited to selectively produce light alkanes (see Figure 18.13). By operating under conditions that favor C–O bond cleavage (i.e., high oxygenate feed concentration, elevated system pressure, low temperature), a Pt–Re/C catalyst is able to partially deoxygenate polyols and produce monofunctional intermediates, which are predominantly 2-ketones, secondary alcohols, heterocycles, and carboxylic acids. In a next step, these monofunctionals can be used as feedstock for a variety of upgrading strategies: (i) C – C coupling through aldol condensation of ketones over base catalysts, (ii) C – C coupling of alcohols using bifunctional metal/base catalysts in the presence of hydrogen, and (iii) C – C coupling through ketonization of carboxylic acids in the presence of base catalysts (Corma et al., 2008). As noted previously, these upgrading processes are all multistep processes with modest yields and thus are not in compliance with the 12 principles of green chemistry.

A variant on this process is the catalytic depolymerization process, developed by Alphakat (tinyurl.com/kbqsza3). This process is principally based on direct liquefaction of biomass. In this process, organic biomass is cracked into light oil with the aid of ion-exchange catalysts at a temperature of 370 °C under atmospheric pressure. The biomass produces a mixture of oil (>40 wt%), water, coal, and gas. In this way, biodiesel has been produced with an estimated cost price of 0.23–0.40 € · l^{-1}. Although this process is cost-effective, its greenness can be questioned, because less than half of the carbon in the biomass is effectively used.

18.6 FUTURE GENERATIONS OF BIOFUEL

Currently, biotechnology is progressing rapidly to implement technologies that allow the direct fermentation of lignocellulosic streams. In this way, fermentation processes can overcome their dependence on classical (food-competing) sugar steams. Biotechnological production of biofuels as promoted by, e.g., Amyris (tinyurl.com/mzf3sux) leans on the synthetic biology approach to make microbes synthesize a large variety of organic molecules, including diesel-type biofuels. In theory, all synthetic pathways can be incorporated. For terpenes, which have a favorable C/H ratio of 0.8, promising results have been obtained. These so-called biohydrocarbons have recently been added to the top 10 list of platform chemicals to be produced from sugars (see Figure 18.4).

Apart from lignocellulosic streams, algae form another source of biomass. Variations on classical biodiesel synthesized from soybean oil, palm oil, etc. can be delivered by, e.g., algae oil. Apart from using the oil from the algae, photosynthetic algae or cyanobacteria can also be regarded as microbial factories, which take their energy from sunlight. Biotechnological processes involving cyanobacteria reactors producing ethanol directly from seawater, sugars, and light have been promoted (tinyurl. com/pg8mk9g). Synthetic Genomics (tinyurl.com/lajphg8) is at the forefront of promoting these photogenerated biofuels. They claim that microbes can be engineered to perform practically any chemical reaction. Ideally, genetically engineered photosynthetic cells, when exposed to sunlight and carbon dioxide, could produce and secrete energy-rich fats, which could then be refined directly into biodiesel fuel. To circumvent contamination of microbial strains, algae have to be grown in photobioreactors. In addition, in order to make this technology viable, microbes have to be reused multiple times, and fat produced by the algae should be released instead of stored. The production and harvesting of algae are surrounded by many agricultural and societal dilemmas, and the future will learn which technologies will be viable.

CHAPTER SUMMARY AND STUDY GUIDE

The technology to upgrade sugars to biofuels is rapidly progressing. Many fermentation, thermochemical, and catalytic methodologies are under development to upgrade lignocellulosic biomass and its constituents, C6 and C5 sugars, to biofuels. Besides bioethanol and FAME, a variety of biofuels have come into sight that have favorable boiling points, polarities, energy densities, and octane and cetane numbers, such as HMF and DMF condensates, furfural and derivatives, levulinates, and valerates. These alternative biofuels can be produced via various routes, but the cost of these upgrading technologies is still significant. In addition, atom efficiencies are low (<80%), additional hydrogen is often needed, and many steps are required. Apart from that, an even larger hurdle to be taken is the choice and pretreatment of biomass. Suitable sugar and lignocellulose streams need to be generated that can be converted to biofuels at reasonable costs.

The viability of this so-called platform technology will very much depend on the concept of biorefineries, in which a combination of biofuels and chemicals is produced. Notably, integration of reaction steps in which cellulose pretreatment and catalytic conversion are conducted in one process is required in order to bring commercial application of biofuels, other than ethanol and biodiesel, closer to the market.

After reading this chapter, the reader should be able to judge whether new biofuels that are proposed in patents or literature can form a valuable and sustainable lead compound. A convenient way to do this is from a green chemistry perspective. Is there a real atom-efficient route and a low E-factor? Are all the parts of the biomass really used, and are not too many steps involved? Is this biofuel a gasoline or a diesel substitute or perhaps even a kerosene substitute? Furthermore, it is important to understand the difference between a fermentation route and a catalyzed chemical reaction.

KEY CONCEPTS

Carbohydrates/sugars
Biofuels
Atom efficiency
E-factor
Platform chemical
Biofuel generations

SHORT-ANSWER QUESTIONS

18.1 Make a top 5 list of the most promising biofuels mentioned in this chapter.

18.2 What is the atom efficiency of HMF production?

18.3 What is the E-factor for HMF production when the yield is 60% (at 100% conversion) and supposing a 1/5 water/organic mixture is used for the reaction. Assume that maximally 80% of the solvent can be recycled and that 1 kg glucose is dissolved in 100 l water.

18.4 Which principles of green chemistry are in accordance with the production of biofuels?

18.5 For the production of which biofuels (e.g., from the top 5 list) is extra hydrogen needed? Where would the hydrogen come from?

18.6   What are the advantages and disadvantages of aqueous reforming?

18.7   Would you consider the lignin from lignocellulose processing as waste? What would you do with it?

PROBLEMS

18.1   Write out the mechanism for aldol condensation of HMF and acetone. What are possible by-products?

18.2   Look up the conditions and yields for the conversion of HMF to C18 molecules. What is the most difficult step?

18.3   Also for the production of kerosene, biomass alternatives are considered. One option is to use isobutene as starting material. Which technology do you propose to convert isobutene into kerosene? Is this an atom-efficient technology?

18.4   Look up three alternative ways to produce kerosene from biomass.

18.5   Look up details of production of DMF ethers by Avantium as listed in their patents. Which catalyst is used? What is the yield? Estimate the E-factor.

18.6   Go one step deeper into calculating an E-factor by making a rough process design using the flowsheeting package Aspen Plus^{TM} including energy and mass balances, e.g., for the example in Problem 18.5.

PROJECTS

P18.1   Make biodiesel yourself, and measure the E-factor by weighing the mass of the materials going in and the mass of the products coming out. An alternative approach is to measure the weight of the by-products (including waste) and compare this with the weight of the product.

P18.2   Try to calculate the amount of energy needed to produce a biofuel, and compare this with the heating value of the biofuel.

INTERNET REFERENCES

tinyurl.com/kbqsza3
www.alphakat.de (The production of synthetic diesel from biomass)

tinyurl.com/lajphg8
www.syntheticgenomics.com

tinyurl.com/l5jdxxq
www.nesteoil.fi (NExBTL whitepaper)

tinyurl.com/mjhnewl
www.ebb-eu.org

tinyurl.com/mzf3sux
www.amyris.com

tinyurl.com/oms56vw
http://webbook.nist.gov/chemistry

tinyurl.com/pg8mk9g
www.algenolbiofuels.com

tinyurl.com/pogugy
www.astm.org/Standards/petroleum-standards.html

tinyurl.com/qadwaj6
www.acs.org/greenchemistry

tinyurl.com/q89rwxe
www.virent.com

REFERENCES

Alonso DM, Bond JQ, Dumesic JA. Catalytic conversion of biomass to biofuels. Green Chem 2010;12(9):1493–1513.

Alonso DM, Wettstein SG, Bond JQ, Root TW, Dumesic JA. Production of biofuels from cellulose and corn stover using alkylphenol solvents. ChemSusChem 2011;4(8):1078–1081.

Anastas PT, Warner JC. *Green Chemistry: Theory and Practice*. New York: Oxford University Press; 2000.

Blanch HW. Bioprocessing for biofuels. Curr Opin Biotechnol 2012;23:390–395.

Bond JQ, Alonso DM, Wang D, West RM, Dumesic JA. Integrated catalytic conversion of γ-valerolactone to liquid alkenes for transportation fuels. Science 2010;327(5669): 1110–1114.

Bozell JJ, Petersen GR. Technology development for the production of biobased products from biorefinery carbohydrates—the US department of energy's "top 10" revisited. Green Chem 2010;12(4):539–554.

Corma A, Renz M, Schaverien C. Coupling fatty acids by ketonic decarboxylation using solid catalysts for the direct production of diesel, lubricants, and chemicals. ChemSusChem 2008;1(8–9):739–741.

Davis SC, Anderson-Teixeira KJ, Delucia EH. Life-cycle analysis and the ecology of biofuels. Trends Plant Sci 2009;14(3):140–146.

Demirbas A. Progress and recent trends in biofuels. Prog Energy Combust Sci 2007;33:1–18.

Diesel R. Method and apparatus for converting heat into work. US patent 542,846. 1895.

Dornburg V, van Vuuren D, van de Ven G, Langeveld H, Meeusen M, Banse M, van Oorschot M, Ros J, van den Born GJ, Aiking H, Londo M, Mozaffarian H, Verweij P, Lysen E,

Faaij A. Bioenergy revisited: key factors in global potentials of bioenergy. Energy Environ Sci 2010;3:258–267.

Fjerbaek L, Christensen KV, Norddahl B. A review of the current state of biodiesel production using enzymatic transesterification. Biotechnol Bioengineer 2009;102(5):1298–1315.

Gruter GJM, Dautzenberg F. Method the synthesis of 5-alkoxymethylfurfural ethers and their use. EP 1834950 A1. 2007.

Gruter GJM, De Jong E. Furanics: novel fuel options from carbohydrates. Biofuels Technol 2009;1:11–17.

Karlsson M. Sustainable bioenergy: a framework for decision makers. UN-Energy; 2007. Available at www.fao.org/docrep/010/a1094e/a1094e00.htm. Accessed May. Accessed May 30, 2014.

Lange J-P, Price R, Ayoub PM, Louis J, Petrus L, Clarke L, Gosselink H. Valeric biofuels: a platform of cellulosic transportation fuels. Angew Chem Int Ed Engl 2010;49:4479–4483.

Lange J-P, van der Heide E, van Buijtenen J, Price R. Furfural: a promising platform for lignocellulosic biofuels. ChemSusChem 2012;5:150–166.

Marcotullio G, De Jong W. Furfural formation from D-xylose: the use of different halides in dilute aqueous acidic solutions allows for exceptionally high yields. Carbohydr Res 2011;346:1291–1293.

Petrus L, Noordermeer MA. Biomass to biofuels, a chemical perspective. Green Chem 206;8:861–867.

Pickett J, Anderson D, Bowles D, Bridgwater T, Jarvis P, Mortimer N. *Sustainable biofuels: prospects and challenges.* The Royal Society. London; 2008. Available at http://royalsociety.org. Accessed May 21, 2014.

Ragauskas AJ, Williams CK, Davison BH, Britovsek G, Cairney J, Eckert CA, Frederick WJ Jr, Hallett JP, Leak DJ, Liotta CL, Mielenz JR, Murphy R, Templer R, Tschaplinski T. The path forward for biofuels and biomaterials. Science 2006;311:484–489.

Roman-Leshkov Y, Barrett CJ, Liu ZY, Dumesic JA. Production of dimethylfuran for liquid fuels from biomass-derived carbohydrates. Nature 2007;447(7147):982–985.

Rosatella AA, Simeonov SP, Frade RFM, Afonso CAM. 5-Hydroxymethylfurfural (HMF) as a building block platform: biological properties, synthesis and synthetic applications. Green Chem 2011;13:754–793.

Ruppert AM, Weinberg K, Palkovits R. Hydrogenolysis goes bio: from carbohydrates and sugar alcohols to platform chemicals. Angew Chem Int Ed Engl 2012;51(11):2564–2601.

Sheldon RA. The E factor: fifteen years on. Green Chem 2007;9(12):1273–1283.

Sheldon RA. Utilisation of biomass for sustainable fuels and chemicals: molecules, methods and metrics. Catal Today 2011;167(1):3–13.

Van de Vyver S, Geboers J, Jacobs PA, Sels BF. Recent advances in the catalytic conversion of cellulose. ChemCatChem 2011;3(1):82–94.

Wright WRH, Palkovits R. Development of heterogeneous catalysts for the conversion of levulinic acid to γ-valerolactone. ChemSusChem 2012;5(9):1657–1667.

Zinoviev S, Müller-Langer F, Das P, Bertero N, Fornasiero P, Kaltschmitt M, Centi G, Miertus S. Next-generation biofuels: survey of emerging technologies and sustainability issues. ChemSusChem 2010;3:1106–1133.

INDEX

Note: Page numbers in *italics* refer to Figures; those in **bold** to Tables

Biomass as a Sustainable Energy Source for the Future: Fundamentals of Conversion Processes,
First Edition. Edited by Wiebren de Jong and J. Ruud van Ommen.
© 2015 American Institute of Chemical Engineers, Inc. Published 2015 by John Wiley & Sons, Inc.